Lecture Notes in Artificial Intelligence 8726

Subseries of Lecture Notes in Computer Science

LNAI Series Editors

Randy Goebel
University of Alberta, Edmonton, Canada
Yuzuru Tanaka
Hokkaido University, Sapporo, Japan
Wolfgang Wahlster
DFKI and Saarland University, Saarbrücken, Germany

LNAI Founding Series Editor

Joerg Siekmann
DFKI and Saarland University, Saarbrücken, Germany

T0214972

Toon Calders   Floriana Esposito
Eyke Hüllermeier   Rosa Meo (Eds.)

Machine Learning and Knowledge Discovery in Databases

European Conference, ECML PKDD 2014
Nancy, France, September 15-19, 2014
Proceedings, Part III

 Springer

Volume Editors

Toon Calders
Université Libre de Bruxelles, Faculty of Applied Sciences
Department of Computer and Decision Engineering
Av. F. Roosevelt, CP 165/15, 1050 Brussels, Belgium
E-mail: toon.calders@ulb.ac.be

Floriana Esposito
Università degli Studi "Aldo Moro", Dipartimento di Informatica
via Orabona 4, 70125 Bari, Italy
E-mail: floriana.esposito@uniba.it

Eyke Hüllermeier
Universität Paderborn, Department of Computer Science
Warburger Str. 100, 33098 Paderborn, Germany
E-mail: eyke@upb.de

Rosa Meo
Università degli Studi di Torino, Dipartimento di Informatica
Corso Svizzera 185, 10149 Torino, Italy
E-mail: meo@di.unito.it

ISSN 0302-9743   e-ISSN 1611-3349
ISBN 978-3-662-44844-1   e-ISBN 978-3-662-44845-8
DOI 10.1007/978-3-662-44845-8
Springer Heidelberg New York Dordrecht London

Library of Congress Control Number: 2014948041

LNCS Sublibrary: SL 7 – Artificial Intelligence

© Springer-Verlag Berlin Heidelberg 2014

Typesetting: Camera-ready by author, data conversion by Scientific Publishing Services, Chennai, India

Printed on acid-free paper

Springer is part of Springer Science+Business Media (www.springer.com)

Preface

The European Conferences on Machine Learning (ECML) and on Principles and Practice of Knowledge Discovery in Data Bases (PKDD) have been organized jointly since 2001, after some years of mutual independence. Going one step further, the two conferences were merged into a single one in 2008, and these are the proceedings of the 2014 edition of ECML/PKDD. Today, this conference is a world-wide leading scientific event. It aims at further exploiting the synergies between the two scientific fields, focusing on the development and employment of methods and tools capable of solving real-life problems.

ECML PKDD 2014 was held in Nancy, France, during September 15–19, co-located with ILP 2014, the premier international forum on logic-based and relational learning. The two conferences were organized by Inria Nancy Grand Est with support from LORIA, a joint research unit of CNRS, Inria, and Université de Lorraine.

Continuing the tradition, ECML/PKDD 2014 combined an extensive technical program with a demo track and an industrial track. Recently, the so-called Nectar track was added, focusing on the latest high-quality interdisciplinary research results in all areas related to machine learning and knowledge discovery in databases. Moreover, the conference program included a discovery challenge, a variety of workshops, and many tutorials.

The main technical program included five plenary talks by invited speakers, namely, Charu Aggarwal, Francis Bach, Lise Getoor, Tie-Yan Liu, and Raymond Ng, while four invited speakers contributed to the industrial track: George Hébrail (EDF Lab), Alexandre Cotarmanac'h (Twenga), Arthur Von Eschen (Activision Publishing Inc.) and Mike Bodkin (Evotec Ltd.).

The discovery challenge focused on "Neural Connectomics and on Predictive Web Analytics" this year. Fifteen workshops were held, providing an opportunity to discuss current topics in a small and interactive atmosphere: Dynamic Networks and Knowledge Discovery, Interactions Between Data Mining and Natural Language Processing, Mining Ubiquitous and Social Environments, Statistically Sound Data Mining, Machine Learning for Urban Sensor Data, Multi-Target Prediction, Representation Learning, Neural Connectomics: From Imaging to Connectivity, Data Analytics for Renewable Energy Integration, Linked Data for Knowledge Discovery, New Frontiers in Mining Complex Patterns, Experimental Economics and Machine Learning, Learning with Multiple Views: Applications to Computer Vision and Multimedia Mining, Generalization and Reuse of Machine Learning Models over Multiple Contexts, and Predictive Web Analytics.

Nine tutorials were included in the conference program, providing a comprehensive introduction to core techniques and areas of interest for the scientific community: Medical Mining for Clinical Knowledge Discovery, Patterns in Noisy and Multidimensional Relations and Graphs, The Pervasiveness of

Machine Learning in Omics Science, Conformal Predictions for Reliable Machine Learning, The Lunch Is Never Free: How Information Theory, MDL, and Statistics are Connected, Information Theoretic Methods in Data Mining, Machine Learning with Analogical Proportions, Preference Learning Problems, and Deep Learning.

The main track received 481 paper submissions, of which 115 were accepted. Such a high volume of scientific work required a tremendous effort by the area chairs, Program Committee members, and many additional reviewers. We managed to collect three highly qualified independent reviews per paper and one additional overall input from one of the area chairs. Papers were evaluated on the basis of their relevance to the conference, their scientific contribution, rigor and correctness, the quality of presentation and reproducibility of experiments. As a separate organization, the demo track received 24 and the Nectar track 23 paper submissions.

For the second time, the conference used a double submission model: next to the regular conference track, papers submitted to the Springer journals *Machine Learning* (MACH) and *Data Mining and Knowledge Discovery* (DAMI) were considered for presentation in the conference. These papers were submitted to the ECML/PKDD 2014 special issue of the respective journals, and underwent the normal editorial process of these journals. Those papers accepted for the of these journals were assigned a presentation slot at the ECML/PKDD 2014 conference. A total of 107 original manuscripts were submitted to the journal track, 15 were accepted in DAMI or MACH and were scheduled for presentation at the conference. Overall, this resulted in a number of 588 submissions, of which 130 were selected for presentation at the conference, making an overall acceptance rate of about 22%.

These proceedings of the ECML/PKDD 2014 conference contain the full papers of the contributions presented in the main technical track, abstracts of the invited talks and short papers describing the demonstrations, and the Nectar papers. First of all, we would like to express our gratitude to the general chairs of the conference, Amedeo Napoli and Chedy Raïssi, as well as to all members of the Organizing Committee, for managing this event in a very competent and professional way. In particular, we thank the demo, workshop, industrial, and Nectar track chairs. Special thanks go to the proceedings chairs, Élisa Fromont, Stefano Ferilli and Pascal Poncelet, for the hard work of putting these proceedings together. We thank the tutorial chairs, the Discovery Challenge organizers and all the people involved in the conference, who worked hard for its success. Last but not least, we would like to sincerely thank the authors for submitting their work to the conference and the reviewers and area chairs for their tremendous effort in guaranteeing the quality of the reviewing process, thereby improving the quality of these proceedings.

July 2014   Toon Calders
                                                                     Floriana Esposito
                                                                     Eyke Hüllermeier
                                                                           Rosa Meo

Organization

ECML/PKDD 2014 Organization

Conference Co-chairs

Amedeo Napoli   Inria Nancy Grand Est/LORIA, France
Chedy Raïssi   Inria Nancy Grand Est/LORIA, France

Program Co-chairs

Toon Calders   Université Libre de Bruxelles, Belgium
Floriana Esposito   University of Bari, Italy
Eyke Hüllermeier   University of Paderborn, Germany
Rosa Meo   University of Turin, Italy

Local Organization Co-chairs

Anne-Lise Charbonnier   Inria Nancy Grand Est, France
Louisa Touioui   Inria Nancy Grand Est, France

Awards Committee Chairs

Johannes Fürnkranz   Technical University of Darmstadt, Germany
Katharina Morik   University of Dortmund, Germany

Workshop Chairs

Bettina Berendt   KU Leuven, Belgium
Patrick Gallinari   LIP6 Paris, France

Tutorial Chairs

Céline Rouveirol   University of Paris-Nord, France
Céline Robardet   University of Lyon, France

Demonstration Chairs

Ricard Gavaldà   UPC Barcelona, Spain
Myra Spiliopoulou   University of Magdeburg, Germany

Publicity Chairs

Stefano Ferilli   University of Bari, Italy
Pauli Miettinen   Max-Planck-Institut, Germany

Panel Chairs

Jose Balcazar   UPC Barcelona, Spain
Sergei O. Kuznetsov   HSE Moscow, Russia

Industrial Chairs

Michael Berthold   University of Konstanz, Germany
Marc Boullé   Orange Labs, France

PhD Chairs

Bruno Crémilleux   University of Caen, France
Radim Belohlavek   University of Olomouc, Czech Republic

Nectar Track Chairs

Evimaria Terzi   Boston University, USA
Pierre Geurts   University of Liège, Belgium

Sponsorship Chairs

Francesco Bonchi   Yahoo ! Research Barcelona, Spain
Jilles Vreeken   Saarland University/Max-Planck-Institut,
                                Germany

Proceedings Chairs

Pascal Poncelet   University of Montpellier, France
Élisa Fromont   University of Saint Etienne, France
Stefano Ferilli   University of Bari, Italy

EMCL PKDD Steering Committee

Fosca Giannotti   University of Pisa, Italy
Michèle Sebag   Université Paris Sud, France
Francesco Bonchi   Yahoo! Research Barcelona, Spain
Hendrik Blockeel   KU Leuven, Belgium and Leiden University,
                                The Netherlands

Katharina Morik	University of Dortmund, Germany
Tobias Scheffer	University of Potsdam, Germany
Arno Siebes	Utrecht University, The Netherlands
Dimitrios Gunopulos	University of Athens, Greece
Michalis Vazirgiannis	École Polytechnique, France
Donato Malerba	University of Bari, Italy
Peter Flach	University of Bristol, UK
Tijl De Bie	University of Bristol, UK
Nello Cristianini	University of Bristol, UK
Filip Železný	Czech Technical University in Prague, Czech Republic
Siegfried Nijssen	LIACS, Leiden University, The Netherlands
Kristian Kersting	Technical University of Dortmund, Germany

Area Chairs

Hendrik Blockeel	KU Leuven, Belgium
Henrik Boström	Stockholm University, Sweden
Ian Davidson	University of California, Davis, USA
Luc De Raedt	KU Leuven, Belgium
Janez Demšar	University of Ljubljana, Slovenia
Alan Fern	Oregon State University, USA
Peter Flach	University of Bristol, UK
Johannes Fürnkranz	TU Darmstadt, Germany
Thomas Gärtner	University of Bonn and Fraunhofer IAIS, Germany
João Gama	University of Porto, Portugal
Aristides Gionis	Aalto University, Finland
Bart Goethals	University of Antwerp, Belgium
Andreas Hotho	University of Würzburg, Germany
Manfred Jaeger	Aalborg University, Denmark
Thorsten Joachims	Cornell University, USA
Kristian Kersting	Technical University of Dortmund, Germany
Stefan Kramer	University of Mainz, Germany
Donato Malerba	University of Bari, Italy
Stan Matwin	Dalhousie University, Canada
Pauli Miettinen	Max-Planck-Institut, Germany
Dunja Mladenić	Jozef Stefan Institute, Slovenia
Marie-Francine Moens	KU Leuven, Belgium
Bernhard Pfahringer	University of Waikato, New Zealand
Thomas Seidl	RWTH Aachen University, Germany
Arno Siebes	Utrecht University, The Netherlands
Myra Spiliopoulou	Magdeburg University, Germany
Jean-Philippe Vert	Mines ParisTech, France
Jilles Vreeken	Max-Planck-Institut and Saarland University, Germany

Marco Wiering   University of Groningen, The Netherlands
Stefan Wrobel   University of Bonn & Fraunhofer IAIS,
                                 Germany

Program Committee

Foto Afrati	Wray Buntine	Wei Ding
Leman Akoglu	Robert Busa-Fekete	Ying Ding
Mehmet Sabih Aksoy	Toon Calders	Stephan Doerfel
Mohammad Al Hasan	Rui Camacho	Janardhan Rao Doppa
Omar Alonso	Longbing Cao	Chris Drummond
Aijun An	Andre Carvalho	Devdatt Dubhashi
Aris Anagnostopoulos	Francisco Casacuberta	Ines Dutra
Annalisa Appice	Michelangelo Ceci	Sašo Džeroski
Marta Arias	Loic Cerf	Tapio Elomaa
Hiroki Arimura	Tania Cerquitelli	Roberto Esposito
Ira Assent	Sharma Chakravarthy	Ines Faerber
Martin Atzmüller	Keith Chan	Hadi Fanaee-Tork
Chloe-Agathe Azencott	Duen Horng Chau	Nicola Fanizzi
Antonio Bahamonde	Sanjay Chawla	Elaine Faria
James Bailey	Keke Chen	Fabio Fassetti
Elena Baralis	Ling Chen	Hakan
Daniel Barbara'	Weiwei Cheng	Ferhatosmanoglou
Christian Bauckhage	Silvia Chiusano	Stefano Ferilli
Roberto Bayardo	Vassilis Christophides	Carlos Ferreira
Aurelien Bellet	Frans Coenen	Cèsar Ferri
Radim Belohlavek	Fabrizio Costa	Jose Fonollosa
Andras Benczur	Bruno Cremilleux	Eibe Frank
Klaus Berberich	Tom Croonenborghs	Antonino Freno
Bettina Berendt	Boris Cule	Élisa Fromont
Michele Berlingerio	Tomaz Curk	Fabio Fumarola
Indrajit Bhattacharya	James Cussens	Patrick Gallinari
Marenglen Biba	Maria Damiani	Jing Gao
Albert Bifet	Jesse Davis	Byron Gao
Enrico Blanzieri	Martine De Cock	Roman Garnett
Konstantinos Blekas	Jeroen De Knijf	Paolo Garza
Francesco Bonchi	Colin de la Higuera	Eric Gaussier
Gianluca Bontempi	Gerard de Melo	Floris Geerts
Christian Borgelt	Juan del Coz	Pierre Geurts
Marco Botta	Krzysztof Dembczyński	Rayid Ghani
Jean-François Boulicaut	François Denis	Fosca Giannotti
Marc Boullé	Anne Denton	Aris Gkoulalas-Divanis
Kendrick Boyd	Mohamed Dermouche	Vibhav Gogate
Pavel Brazdil	Christian Desrosiers	Marco Gori
Ulf Brefeld	Luigi Di Caro	Michael Granitzer
Björn Bringmann	Jana Diesner	Oded Green

Tias Guns
Maria Halkidi
Jiawei Han
Daniel Hernandez
    Lobato
José Hernández-Orallo
Frank Hoeppner
Jaakko Hollmén
Geoff Holmes
Arjen Hommersom
Vasant Honavar
Xiaohua Hu
Minlie Huang
Eyke Hüllermeier
Dino Ienco
Robert Jäschke
Frederik Janssen
Nathalie Japkowicz
Szymon Jaroszewicz
Ulf Johansson
Alipio Jorge
Kshitij Judah
Tobias Jung
Hachem Kadri
Theodore Kalamboukis
Alexandros Kalousis
Pallika Kanani
U Kang
Panagiotis Karras
Andreas Karwath
Hisashi Kashima
Ioannis Katakis
John Keane
Latifur Khan
Levente Kocsis
Yun Sing Koh
Alek Kolcz
Igor Kononenko
Irena Koprinska
Nitish Korula
Petr Kosina
Walter Kosters
Georg Krempl
Konstantin Kutzkov
Sergei Kuznetsov

Nicolas Lachiche
Pedro Larranaga
Silvio Lattanzi
Niklas Lavesson
Nada Lavrač
Gregor Leban
Sangkyun Lee
Wang Lee
Carson Leung
Jiuyong Li
Lei Li
Tao Li
Rui Li
Ping Li
Juanzi Li
Lei Li
Edo Liberty
Jefrey Lijffijt
shou-de Lin
Jessica Lin
Hsuan-Tien Lin
Francesca Lisi
Yan Liu
Huan Liu
Corrado Loglisci
Eneldo Loza Mencia
Chang-Tien Lu
Panagis Magdalinos
Giuseppe Manco
Yannis Manolopoulos
Enrique Martinez
Dimitrios Mavroeidis
Mike Mayo
Wannes Meert
Gabor Melli
Ernestina Menasalvas
Roser Morante
João Moreira
Emmanuel Müller
Mohamed Nadif
Mirco Nanni
Alex Nanopoulos
Balakrishnan
    Narayanaswamy
Sriraam Natarajan

Benjamin Nguyen
Thomas Niebler
Thomas Nielsen
Siegfried Nijssen
Xia Ning
Richard Nock
Niklas Noren
Kjetil Nørvåg
Eirini Ntoutsi
Andreas Nürnberger
Salvatore Orlando
Gerhard Paass
George Paliouras
Spiros Papadimitriou
Apostolos Papadopoulos
Panagiotis Papapetrou
Stelios Paparizos
Ioannis Partalas
Andrea Passerini
Vladimir Pavlovic
Mykola Pechenizkiy
Dino Pedreschi
Nikos Pelekis
Jing Peng
Ruggero Pensa
Fabio Pinelli
Marc Plantevit
Pascal Poncelet
George Potamias
Aditya Prakash
Doina Precup
Kai Puolamaki
Buyue Qian
Chedy Raïssi
Liva Ralaivola
Karthik Raman
Jan Ramon
Huzefa Rangwala
Zbigniew Raś
Chotirat
    Ratanamahatana
Jan Rauch
Soumya Ray
Steffen Rendle
Achim Rettinger

Fabrizio Riguzzi

Céline Robardet

Marko Robnik Sikonja

Pedro Rodrigues

Juan Rodriguez

Irene Rodriguez-Lujan

Fabrice Rossi

Juho Rousu

Céline Rouveirol

Stefan Rüping

Salvatore Ruggieri

Yvan Saeys

Alan Said

Lorenza Saitta

Ansaf Salleb-Aouissi

Scott Sanner

Vítor Santos Costa

Raul Santos-Rodriguez

Sam Sarjant

Claudio Sartori

Taisuke Sato

Lars Schmidt-Thieme

Christoph Schommer

Matthias Schubert

Giovanni Semeraro

Junming Shao

Junming Shao

Pannaga Shivaswamy

Andrzej Skowron

Kevin Small

Padhraic Smyth

Carlos Soares

Yangqiu Song

Mauro Sozio

Alessandro Sperduti

Eirini Spyropoulou

Jerzy Stefanowski

Jean Steyaert

Daniela Stojanova

Markus Strohmaier

Mahito Sugiyama

Johan Suykens

Einoshin Suzuki

Panagiotis Symeonidis

Sandor Szedmak

Andrea Tagarelli

Domenico Talia

Pang Tan

Letizia Tanca

Dacheng Tao

Nikolaj Tatti

Maguelonne Teisseire

Evimaria Terzi

Martin Theobald

Jilei Tian

Ljupco Todorovski

Luis Torgo

Vicenç Torra

Ivor Tsang

Panagiotis Tsaparas

Vincent Tseng

Grigorios Tsoumakas

Theodoros Tzouramanis

Antti Ukkonen

Takeaki Uno

Athina Vakali

Giorgio Valentini

Guy Van den Broeck

Peter van der Putten

Matthijs van Leeuwen

Maarten van Someren

Joaquin Vanschoren

Iraklis Varlamis

Michalis Vazirgiannis

Julien Velcin

Shankar Vembu

Sicco Verwer

Vassilios Verykios

Herna Viktor

Christel Vrain

Willem Waegeman

Byron Wallace

Fei Wang

Jianyong Wang

Xiang Wang

Yang Wang

Takashi Washio

Geoff Webb

Jörg Wicker

Hui Xiong

Jieping Ye

Jeffrey Yu

Philip Yu

Chun-Nam Yu

Jure Zabkar

Bianca Zadrozny

Gerson Zaverucha

Demetris Zeinalipour

Filip Železný

Bernard Zenko

Min-Ling Zhang

Nan Zhang

Zhongfei Zhang

Junping Zhang

Lei Zhang

Changshui Zhang

Kai Zhang

Kun Zhang

Shichao Zhang

Ying Zhao

Elena Zheleva

Zhi-Hua Zhou

Bin Zhou

Xingquan Zhu

Xiaofeng Zhu

Kenny Zhu

Djamel Zighed

Arthur Zimek

Albrecht Zimmermann

Indre Zliobaite

Blaz Zupan

Demo Track Program Committee

Martin Atzmueller
Bettina Berendt
Albert Bifet
Antoine Bordes
Christian Borgelt
Ulf Brefeld
Blaz Fortuna

Jaakko Hollmén
Andreas Hotho
Mark Last
Vincent Lemaire
Ernestina Menasalvas
Kjetil Nørvåg
Themis Palpanas

Mykola Pechenizkiy
Bernhard Pfahringer
Pedro Rodrigues
Jerzy Stefanowski
Grigorios Tsoumakas
Alice Zheng

Nectar Track Program Committee

Donato Malerba
Dora Erdos
Yiannis Koutis

George Karypis
Louis Wehenkel
Leman Akoglu

Rosa Meo
Myra Spiliopoulou
Toon Calders

Additional Reviewers

Argimiro Arratia
Rossella Cancelliere
Antonio Corral
Joana Côrte-Real
Giso Dal
Giacomo Domeniconi
Roberto Esposito
Pedro Ferreira
Asmelash Teka Hadgu
Isaac Jones
Dimitris Kalles
Yoshitaka Kameya
Eamonn Keogh
Kristian Kersting
Rohan Khade
Shamanth Kumar
Hongfei Li

Elad Liebman
Babak Loni
Emmanouil Magkos
Adolfo Martínez-Usó
Dimitrios Mavroeidis
Steffen Michels
Pasquale Minervini
Fatemeh Mirrashed
Fred Morstatter
Tsuyoshi Murata
Jinseok Nam
Rasaq Otunba
Roberto Pasolini
Tommaso Pirini
Maria-Jose
    Ramirez-Quintana
Irma Ravkic

Kiumars Soltani
Ricardo Sousa
Eleftherios
    Spyromitros-Xioufis
Jiliang Tang
Eleftherios Tiakas
Andrei Tolstikov email
Tiago Vinhoza
Xing Wang
Lorenz Weizsäcker
Sean Wilner
Christian Wirth
Lin Wu
Jinfeng Yi
Cangzhou Yuan
Jing Zhang

Sponsors

Gold Sponsor
Winton   http://www.wintoncapital.com

Silver Sponsors

Deloitte	http://www.deloitte.com
Xerox Research Centre Europe	http://www.xrce.xerox.com

Bronze Sponsors

EDF	http://www.edf.com
Orange	http://www.orange.com
Technicolor	http://www.technicolor.com
Yahoo! Labs	http://labs.yahoo.com

Additional Supporters

Harmonic Pharma	http://www.harmonicpharma.com
Deloitte	http://www.deloitte.com

Lanyard

Knime	http://www.knime.org

Prize

Deloitte	http://www.deloitte.com
Data Mining and Knowledge Discovery	http://link.springer.com/journal/10618
Machine Learning	http://link.springer.com/journal/10994

Organizing Institutions

Inria	http://www.inria.fr
CNRS	http://www.cnrs.fr
LORIA	http://www.loria.fr

Invited Talks Abstracts

Scalable Collective Reasoning Using Probabilistic Soft Logic

Lise Getoor

University of California, Santa Cruz
Santa Cruz, CA, USA
getoor@cs.umd.edu

Abstract. One of the challenges in big data analytics is to efficiently learn and reason collectively about extremely large, heterogeneous, incomplete, noisy interlinked data. Collective reasoning requires the ability to exploit both the logical and relational structure in the data and the probabilistic dependencies. In this talk I will overview our recent work on probabilistic soft logic (PSL), a framework for collective, probabilistic reasoning in relational domains. PSL is able to reason holistically about both entity attributes and relationships among the entities. The underlying mathematical framework, which we refer to as a hinge-loss Markov random field, supports extremely efficient, exact inference. This family of graphical models captures logic-like dependencies with convex hinge-loss potentials. I will survey applications of PSL to diverse problems ranging from information extraction to computational social science. Our recent results show that by building on state-of-the-art optimization methods in a distributed implementation, we can solve large-scale problems with millions of random variables orders of magnitude faster than existing approaches.

Bio. In 1995, Lise Getoor decided to return to school to get her PhD in Computer Science at Stanford University. She received a National Physical Sciences Consortium fellowship, which in addition to supporting her for six years, supported a summer internship at Xerox PARC, where she worked with Markus Fromherz and his group. Daphne Koller was her PhD advisor; in addition, she worked closely with Nir Friedman, and many other members of the DAGS group, including Avi Pfeffer, Mehran Sahami, Ben Taskar, Carlos Guestrin, Uri Lerner, Ron Parr, Eran Segal, Simon Tong.

In 2001, Lise Getoor joined the Computer Science Department at the University of Maryland, College Park.

Network Analysis in the Big Data Age: Mining Graph and Social Streams

Charu Aggarwal

IBM T.J. Watson Research Center, New York
Yorktown, NY, USA
charu@us.ibm.com

Abstract. The advent of large interaction-based communication and social networks has led to challenging streaming scenarios in graph and social stream analysis. The graphs that result from such interactions are large, transient, and very often cannot even be stored on disk. In such cases, even simple frequency-based aggregation operations become challenging, whereas traditional mining operations are far more complex. When the graph cannot be explicitly stored on disk, mining algorithms must work with a limited knowledge of the network structure. Social streams add yet another layer of complexity, wherein the streaming content associated with the nodes and edges needs to be incorporated into the mining process. A significant gap exists between the problems that need to be solved, and the techniques that are available for streaming graph analysis. In spite of these challenges, recent years have seen some advances in which carefully chosen synopses of the graph and social streams are leveraged for approximate analysis. This talk will focus on several recent advances in this direction.

Bio. Charu Aggarwal is a Research Scientist at the IBM T. J. Watson Research Center in Yorktown Heights, New York. He completed his B.S. from IIT Kanpur in 1993 and his Ph.D. from Massachusetts Institute of Technology in 1996. His research interest during his Ph.D. years was in combinatorial optimization (network flow algorithms), and his thesis advisor was Professor James B. Orlin. He has since worked in the field of data mining, with particular interests in data streams, privacy, uncertain data and social network analysis. He has published over 200 papers in refereed venues, and has applied for or been granted over 80 patents. Because of the commercial value of the above-mentioned patents, he has received several invention achievement awards and has thrice been designated a Master Inventor at IBM. He is a recipient of an IBM Corporate Award (2003) for his work on bio-terrorist threat detection in data streams, a recipient of the IBM Outstanding Innovation Award (2008) for his scientific contributions to privacy technology, and a recipient of an IBM Research Division Award (2008) for his scientific contributions to data stream research. He has served on the program committees of most major database/data mining conferences, and served as program vice-chairs of the SIAM Conference on Data Mining, 2007, the IEEE ICDM Conference, 2007, the WWW Conference 2009, and the IEEE ICDM Conference, 2009. He served as an associate editor of the IEEE Transactions on Knowledge and Data Engineering Journal from 2004 to 2008. He is an associate editor of the ACM TKDD Journal, an action editor of the Data Mining and Knowledge Discovery Journal, an associate editor of the ACM SIGKDD Explorations, and an associate editor of the Knowledge and Information Systems Journal. He is a fellow of the ACM (2013) and the IEEE (2010) for contributions to knowledge discovery and data mining techniques.

Big Data for Personalized Medicine: A Case Study of Biomarker Discovery

Raymond Ng

University of British Columbia
Vancouver, B.C., Canada
mg@cs.ubc.ca

Abstract. Personalized medicine has been hailed as one of the main frontiers for medical research in this century. In the first half of the talk, we will give an overview on our projects that use gene expression, proteomics, DNA and clinical features for biomarker discovery. In the second half of the talk, we will describe some of the challenges involved in biomarker discovery. One of the challenges is the lack of quality assessment tools for data generated by ever-evolving genomics platforms. We will conclude the talk by giving an overview of some of the techniques we have developed on data cleansing and pre-processing.

Bio. Dr. Raymond Ng is a professor in Computer Science at the University of British Columbia. His main research area for the past two decades is on data mining, with a specific focus on health informatics and text mining. He has published over 180 peer-reviewed publications on data clustering, outlier detection, OLAP processing, health informatics and text mining. He is the recipient of two best paper awards from 2001 ACM SIGKDD conference, which is the premier data mining conference worldwide, and the 2005 ACM SIGMOD conference, which is one of the top database conferences worldwide. He was one of the program co-chairs of the 2009 International conference on Data Engineering, and one of the program co-chairs of the 2002 ACM SIGKDD conference. He was also one of the general co-chairs of the 2008 ACM SIGMOD conference. For the past decade, Dr. Ng has co-led several large scale genomic projects, funded by Genome Canada, Genome BC and industrial collaborators. The total amount of funding of those projects well exceeded $40 million Canadian dollars. He now holds the Chief Informatics Officer position of the PROOF Centre of Excellence, which focuses on biomarker development for end-stage organ failures.

Machine Learning for Search Ranking and Ad Auction

Tie-Yan Liu

Microsoft Research Asia
Beijing, P.R. China
tyliu@microsoft.com

Abstract. In the era of information explosion, search has become an important tool for people to retrieve useful information. Every day, billions of search queries are submitted to commercial search engines. In response to a query, search engines return a list of relevant documents according to a ranking model. In addition, they also return some ads to users, and extract revenue by running an auction among advertisers if users click on these ads. This "search + ads" paradigm has become a key business model in today's Internet industry, and has incubated a few hundred-billion-dollar companies. Recently, machine learning has been widely adopted in search and advertising, mainly due to the availability of huge amount of interaction data between users, advertisers, and search engines. In this talk, we discuss how to use machine learning to build effective ranking models (which we call learning to rank) and to optimize auction mechanisms. (i) The difficulty of learning to rank lies in the interdependency between documents in the ranked list. To tackle it, we propose the so-called listwise ranking algorithms, whose loss functions are defined on the permutations of documents, instead of individual documents or document pairs. We prove the effectiveness of these algorithms by analyzing their generalization ability and statistical consistency, based on the assumption of a two-layer probabilistic sampling procedure for queries and documents, and the characterization of the relationship between their loss functions and the evaluation measures used by search engines (e.g., NDCG and MAP). (ii) The difficulty of learning the optimal auction mechanism lies in that advertisers' behavior data are strategically generated in response to the auction mechanism, but not randomly sampled in an i.i.d. manner. To tackle this challenge, we propose a game-theoretic learning method, which first models the strategic behaviors of advertisers, and then optimizes the auction mechanism by assuming the advertisers to respond to new auction mechanisms according to the learned behavior model. We prove the effectiveness of the proposed method by analyzing the generalization bounds for both behavior learning and auction mechanism learning based on a novel Markov framework.

Bio. Tie-Yan Liu is a senior researcher and research manager at Microsoft Research. His research interests include machine learning (learning to rank, online learning, statistical learning theory, and deep learning), algorithmic game theory, and computational economics. He is well known for his work on learning to rank for information retrieval. He has authored the first book in this area, and published tens of highly-cited papers on both algorithms and theorems of learning to rank. He has also published extensively on other related topics. In particular, his paper won the best student paper award of SIGIR (2008), and the most cited paper award of the Journal of Visual Communication and Image Representation (2004-2006); his group won the research break-through award of Microsoft Research Asia (2012). Tie-Yan is very active in serving the research community. He is a program committee co-chair of ACML (2015), WINE (2014), AIRS (2013), and RIAO (2010), a local co-chair of ICML 2014, a tutorial co-chair of WWW 2014, a demo/exhibit co-chair of KDD (2012), and an area/track chair of many conferences including ACML (2014), SIGIR (2008-2011), AIRS (2009-2011), and WWW (2011). He is an associate editor of ACM Transactions on Information System (TOIS), an editorial board member of Information Retrieval Journal and Foundations and Trends in Information Retrieval. He has given keynote speeches at CCML (2013), CCIR (2011), and PCM (2010), and tutorials at SIGIR (2008, 2010, 2012), WWW (2008, 2009, 2011), and KDD (2012). He is a senior member of the IEEE and the ACM.

Beyond Stochastic Gradient Descent for Large-Scale Machine Learning

Francis Bach

INRIA, Paris
Laboratoire d'Informatique de l'Ecole Normale Superieure
Paris, France
francis.bach@inria.fr

Abstract. Many machine learning and signal processing problems are traditionally cast as convex optimization problems. A common difficulty in solving these problems is the size of the data, where there are many observations ("large n") and each of these is large ("large p"). In this setting, online algorithms such as stochastic gradient descent which pass over the data only once, are usually preferred over batch algorithms, which require multiple passes over the data. In this talk, I will show how the smoothness of loss functions may be used to design novel algorithms with improved behavior, both in theory and practice: in the ideal infinite-data setting, an efficient novel Newton-based stochastic approximation algorithm leads to a convergence rate of $O(1/n)$ without strong convexity assumptions, while in the practical finite-data setting, an appropriate combination of batch and online algorithms leads to unexpected behaviors, such as a linear convergence rate for strongly convex problems, with an iteration cost similar to stochastic gradient descent.
(joint work with Nicolas Le Roux, Eric Moulines and Mark Schmidt)

Bio. Francis Bach is a researcher at INRIA, leading since 2011 the SIERRA project-team, which is part of the Computer Science Laboratory at Ecole Normale Superieure. He completed his Ph.D. in Computer Science at U.C. Berkeley, working with Professor Michael Jordan, and spent two years in the Mathematical Morphology group at Ecole des Mines de Paris, then he joined the WILLOW project-team at INRIA/Ecole Normale Superieure from 2007 to 2010. Francis Bach is interested in statistical machine learning, and especially in graphical models, sparse methods, kernel-based learning, convex optimization vision and signal processing.

Industrial Invited Talks Abstracts

Making Smart Metering Smarter by Applying Data Analytics

Georges Hébrail

EDF Lab
CLAMART, France
georges.hebrail@edf.fr

Abstract. New data is being collected from electric smart meters which are deployed in many countries. Electric power meters measure and transmit to a central information system electric power consumption from every individual household or enterprise. The sampling rate may vary from 10 minutes to 24 hours and the latency to reach the central information system may vary from a few minutes to 24h. This generates a large amount of - possibly streaming - data if we consider customers from an entire country (ex. 35 millions in France). This data is collected firstly for billing purposes but can be processed with data analytics tools with several other goals. The first part of the talk will recall the structure of electric power smart metering data and review the different applications which are considered today for applying data analytics to such data. In a second part of the talk, we will focus on a specific problem: spatio-temporal estimation of aggregated electric power consumption from incomplete metering data.

Bio. Georges Hébrail is a senior researcher at EDF Lab, the research centre of Electricité de France, one of the world's leading electric utility. His background is in Business Intelligence covering many aspects from data storage and querying to data analytics. From 2002 to 2010, he was a professor of computer science at Telecom ParisTech, teaching and doing research in the field of information systems and business intelligence, with a focus on time series management, stream processing and mining. His current research interest is on distributed and privacy-preserving data mining on electric power related data.

Ads That Matter

Alexandre Cotarmanac'h

VP Platform & Distribution
Twenga
alexandre.cotarmanach@twenga.com

Abstract. The advent of realtime bidding and online ad-exchanges has created a new and fast-growing competitive marketplace. In this new setting, media-buyers can make fine-grained decisions for each of the impressions being auctioned taking into account information from the context, the user and his/her past behavior. This new landscape is particularly interesting for online e-commerce players where user actions can also be measured online and thus allow for a complete measure of return on ad-spend.

Despite those benefits, new challenges need to be addressed such as:

- the design of a real-time bidding architecture handling high volumes of queries at low latencies,
- the exploration of a sparse and volatile high-dimensional space,
- as well as several statistical modeling problems (e.g. pricing, offer and creative selection).

In this talk, I will present an approach to realtime media buying for online e-commerce from our experience working in the field. I will review the aforementioned challenges and discuss open problems for serving ads that matter.

Bio. Alexandre Cotarmanac'h is Vice-President Distribution & Platform for Twenga.

Twenga is a services and solutions provider generating high value-added leads to online merchants that was founded in 2006.

Originally hired to help launch Twenga's second generation search engine and to manage the optimization of revenue, he launched in 2011 the affinitAD line of business and Twenga's publisher network. Thanks to the advanced contextual analysis which allows for targeting the right audience according to their desire to buy e-commerce goods whilst keeping in line with the content offered, affinitAD brings Twenga's e-commerce expertise to web publishers. Alexandre also oversees Twenga's merchant programme and strives to offer Twenga's merchants new services and solutions to improve their acquisition of customers.

With over 14 years of experience, Alexandre has held a succession of increasingly responsible positions focusing on advertising and web development. Prior to joining Twenga, he was responsible for the development of Search and Advertising at Orange. Alexandre graduated from Ecole polytechnique.

Machine Learning and Data Mining in Call of Duty

Arthur Von Eschen

Activision Publishing Inc.
Santa Monica, CA, USA
Arthur.VonEschen@activision.com

Abstract. Data science is relatively new to the video game industry, but it has quickly emerged as one of the main resources for ensuring game quality. At Activision, we leverage data science to analyze the behavior of our games and our players to improve in-game algorithms and the player experience. We use machine learning and data mining techniques to influence creative decisions and help inform the game design process. We also build analytic services that support the game in real-time; one example is a cheating detection system which is very similar to fraud detection systems used for credit cards and insurance. This talk will focus on our data science work for Call of Duty, one of the bestselling video games in the world.

Bio. Arthur Von Eschen is Senior Director of Game Analytics at Activision. He and his team are responsible for analytics work that supports video game design on franchises such as Call of Duty and Skylanders. In addition to holding a PhD in Operations Research, Arthur has over 15 years of experience in analytics consulting and R&D with the U.S. Fortune 500. His work has spanned across industries such as banking, financial services, insurance, retail, CPG and now interactive entertainment (video games). Prior to Activision he worked at Fair Isaac Corporation (FICO). Before FICO he ran his own analytics consulting firm for six years.

Algorithms, Evolution and Network-Based Approaches in Molecular Discovery

Mike Bodkin

Evotec Ltd.
Oxfordshire, UK
Mike.Bodkin@evotec.com

Abstract. Drug research generates huge quantities of data around targets, compounds and their effects. Network modelling can be used to describe such relationships with the aim to couple our understanding of disease networks with the changes in small molecule properties. This talk will build off of the data that is routinely captured in drug discovery and describe the methods and tools that we have developed for compound design using predictive modelling, evolutionary algorithms and network-based mining.

Bio. Mike did his PhD in protein de-novo design for Nobel laureate sir James Black before taking up a fellowship in computational drug design at Cambridge University. He moved to AstraZeneca as a computational chemist before joining Eli Lilly in 2000. As head of the computational drug discovery group at Lilly since 2003 he recently jumped ship to Evotec to work as the VP for computational chemistry and cheminformatics. His research aims are to continue to develop new algorithms and software in the fields of drug discovery and systems informatics and to deliver and apply current and novel methods as tools for use in drug research.

Table of Contents – Part III

Main Track Contributions

Demo Track Contributions

Nectar Track Contributions

FLIP: Active Learning for Relational Network Classification

Tanwistha Saha, Huzefa Rangwala, and Carlotta Domeniconi

Department of Computer Science
George Mason University
Fairfax, Virginia, USA
tsaha@gmu.edu, {rangwala,carlotta}@cs.gmu.edu

Abstract. Active learning in relational networks has gained popularity in recent years, especially for scenarios when the costs of obtaining training samples are very high. We investigate the problem of active learning for both single- and multi-labeled relational network classification in the absence of node features during training. The problem becomes harder when the number of labeled nodes available for training a model is limited due to budget constraints. The inability to use a traditional learning setup for classification of relational data, has motivated researchers to propose *Collective Classification* algorithms that jointly classifies *all* the test nodes in a network by exploiting the underlying correlation between the labels of a node and its neighbors. In this paper, we propose active learning algorithms based on different query strategies using a collective classification model where each node in a network can belong to either one class (single-labeled network) or multiple classes (multi-labeled network). We have evaluated our method on both single-labeled and multi-labeled networks, and our results are promising in both the cases for several real world datasets.

1 Introduction

In recent years, *relational learning* has gained popularity because of the ability to represent many real world datasets as a graph representing the interaction pattern between the instances in that datasets. Social networks of individuals, protein-protein interaction networks in biological domain and citation networks of scientific articles are only a few examples of this representation. The objective of relational learning is to efficiently and accurately classify nodes in a network by using the latent relational information.

The first step towards building a classification model is to acquire a representative training set. However, acquiring training samples can be expensive due to the cost of querying labels through interactions with a human or *oracle*. Active learning aims to learn a model with minimal querying cost and can also prioritize the acquisition of labeled samples under budget constraints. Previously, different active learning strategies have been developed for selecting the most informative sample(s) to improve the generalization performance of a classifier. Even

T. Calders et al. (Eds.): ECML PKDD 2014, Part III, LNCS 8726, pp. 1–18, 2014.

though active learning approaches for single-labeled datasets have been extensively studied, algorithms for multi-labeled datasets has not yet been explored. The task becomes even more challenging for multi-labeled networks because traditional active learning strategies do not take into account the explicit relational information.

Collective Classification methods jointly classify *all* the test nodes in a network by leveraging the complex and implicit correlations between multiple entities and their labels. These methods are applicable towards networks which have topological features [10,16], but may or may not have node features [10]. In either case, optimizing the cost of acquiring training data involves the use of active learning algorithms. Most of the work on active learning for relational data using collective classification, assume that node features are available during the learning process [2,17].

In this paper we have developed a pool-based active learning strategy for single- and multi-labeled networks based on the intuition that, unlabeled instances which are harder to classify undergo multiple changes in their predicted labels during consecutive iterations of collective inference. We refer to this changing of labels of an instance as *flipping*, and our method as FLIP. We also investigate the situation when *only a subset of the labels* of a multi-labeled instance can be queried during each round of active learning, thereby, creating a challenge regarding which subset of labels to choose. We propose a method called FLIP-per-label for pool-based active learning to address this real-world situation. Our contributions in this paper are summarized as follows:

1. Active learning strategy for single labeled and multi-labeled networks (FLIP)
2. Active learning strategy for querying a subset of labels of an instance for multi-labeled network (FLIP-per-label)

These methods were developed assuming no node features are available to us during learning. We experimented with six real world single-labeled and three multi-labeled networks, and our results show statistically significant improvements over random sampling and other baselines for most of the datasets.

2   Related Work

Previous active learning algorithms [1,2,4,12] determine informative examples to query based on one or more of the following properties: (i) maximum entropy based on classifier's prediction of its label; (ii) least confidence of the classifier on its label; (iii) maximum disagreement between multiple classifiers (e.g., ensemble) predicting its label. Zhu *et al.* [21] have deviated from this approach by combining active learning with semi-supervised learning using Gaussian field and harmonic functions, and employing Empirical Risk Minimization (ERM) framework. Macskassy proposed a method [11] that uses graph-based metrics (e.g., clustering co-efficient, betweenness centrality and degree of a node) to identify a set of informative examples and then apply ERM on that same set of examples to identify the *single most informative* instance from the pool. Most of these methods [11,21] had the benefit of using the features of the instances during training phase. This is often a challenge for relational networks because the features may not be available for mining due to privacy concerns. In this work, we focus on developing methods that learn from the structural properties of networks.

Our method is inspired by Zhu et al. and Macskassy [11,21], but unlike their approaches which focus on matrix based methods (that rely heavily on rich instance features), our method explores the use of the collective inference procedure within the active learning strategy. Bilgic et al. [2] were the first to propose an active learning approach that uses the disagreement score between a content-only classifier (i.e., trained using node features) and a collective classifier. Shi et al. [17] proposed a batch algorithm that combines node features and link information for active sample selection strategy. The algorithm by Ji et al. [6] selects instances that minimize the total variance of the distribution of the unlabeled samples and the total prediction error. Kuwadekar and Neville [8] use a probabilistic relational model [14] to select informative instances. All these methods rely on the use of both the node features and the structure of the network.

In case of multi-label learning when the labels are correlated and there is explicit link structure between the instances of the dataset, it becomes challenging to leverage these multiple correlations during training [20]. Recently proposed work of Ghamrawi and McCallum [5], Kong et al. [7], Saha et al. [15] and Wang and Sukthankar [19], have developed multi-labeled collective classification algorithms, most of which [5,7,15] require nodes features to be available during learning. Wang and Sukthankar [19] proposed a collective classifier that derives social context features from the network structure during learning, whereas, our work is focused on building an intelligent active learning model using one-vs-rest multi-label version of a state-of-the-art collective classifier [10].

3 Methods

3.1 Definition and Notations

Given a network $\mathcal{G}=(\mathcal{V}, \mathcal{E})$ where $\mathcal{V}$ is the set of nodes and $\mathcal{E}$ is the set of edges, each node $v_i \in \mathcal{V}$ ($|\mathcal{V}| = n$ is the total number of nodes in the network) can have either one label (for single-labeled network) or multiple labels (for multi-labeled networks). The label(s) of any node v_i in the graph is represented as $\mathbf{y}_i = (y_{i1}, \cdots, y_{iK})$ where $y_{ik} = 1$ if node v_i belongs to the class k ($k \in \{1, \cdots, K\}$; K is the total number of classes). For single-labeled network, a node can belong to only one class. Hence, only one element, say y_{ik}, of the label vector $\mathbf{y}_i$ can have value 1 (if v_i belongs to the class k) and the rest of the values are 0. For multi-labeled network, $y_{ik} = 1 \ \forall k \in M$, where M is the set of classes to which this node belongs ($|M| \leq K$). Given the set of labeled nodes L and the set of unlabeled nodes U, the objective of *collective classification* is to predict the labels of *all* the unlabeled nodes [16]. $\mathcal{Y} = \{\mathbf{y}_1, \cdots, \mathbf{y}_n\}$ represents the set of label vectors of all the n nodes in network. l and u represent the indices of a labeled node and an unlabeled node respectively, such that $l \in L$ and $u \in U$. $\mathbf{Y}_L$ is the label matrix for training data, where each row is the label vector $\mathbf{y}_l$, $\forall l \in L$.

$\mathbf{Y}_U$ is the predicted output label matrix of the unlabeled data, where each row is the predicted label vector $\mathbf{y}_u$, $\forall u \in U$. For active learning setup, we denote $\mathbf{P}_U$ as the class probability matrix of all the unlabeled nodes such that each row of $\mathbf{P}_U$ is a vector $\mathbf{p}_u = (p_{u1}, \cdots, p_{uK})$ and denote $\mathbf{P}_L$ as the class probability matrix of all the labeled nodes in training set L, such that $\mathbf{P}_L = \mathbf{Y}_L$. p_{uk} is equal to $P(y_{ik} = 1|N_i)$, where $i = u$ and N_i is the set of neighboring nodes of node v_i. PO represents the set of indices of all the nodes in unlabeled pool. The class conditional probability $\mathbf{P}$ of all the labeled and unlabeled nodes is given by, $\mathbf{P} = [\mathbf{P}_L; \mathbf{P}_U]$, where $\mathbf{P}_U$ is predicted by active learner.

3.1.1   Weighted Vote Relational Neighbor

Macskassy and Provost [10] proposed the weighted vote relational neighbor (wvRN) algorithm to classify nodes in a relational network using only the network structure. Given $v_i \in U$, the wvRN classifier estimates the probability of node v_i having class label k (i.e., $y_{ik} = 1$) as a weighted average of the labels of it's neighboring nodes.

$$P(y_{ik} = 1|N_i) = \frac{1}{Z} \sum_{v_j \in N_i} w_{ij} \cdot P(y_{jk} = 1|N_j). \tag{1}$$

N_i is the set of neighbors of v_i in $\mathcal{G}$, Z is the normalization constant ($Z = \sum_{k=1}^{K} P(y_{ik}|N_i)$) and w_{ij} is the weight on the edge, e_{ij}. In the collective classification algorithm, the *bootstrap* phase assigns a class probability to all the test nodes $v_i \in U$ by estimating the class prior probability as:

$$P(y_{ik} = 1) = \frac{1}{|L|} \sum_{j \in L} \mathbb{I}(y_{jk} = 1), \tag{2}$$

where $\mathbb{I}(\cdot)$ is an indicator function with value 1 if the arguement is true.

Relaxation labeling is a collective inference method based on the approach by Chakrabarti *et al.* [3]. During each iteration of collective inference, relaxation labeling [10,3] is used to update the prediction probability estimates from previous iterations. At step $t+1$ the predicted labels of *all* the test nodes are updated based on their estimation at step t. The update rule is given by [10,19]:

$$\mathbf{P}_i^{(t+1)} = \beta^{(t+1)} \cdot \mathcal{M}_\mathcal{R}(v_i^{(t)}) + (1 - \beta^{(t+1)}) \cdot \mathbf{P}_i^{(t)} \tag{3}$$

A simulated annealing based technique [10] is used to reduce the influence of neighbors by giving more weight to a node's current estimate. In Equation (3), $\mathcal{M}_\mathcal{R}(\cdot)$ is the relational model (here, wvRN), $\beta^1 = \gamma$ and $\beta^{(t+1)} = \beta^t \cdot \alpha$ where γ and α are constants in the range $(0, 1]$, both γ and α values are chosen closer to 1, $\mathbf{P}_i^{(t)}$ is a vector with class probability values for node v_i at step t.

3.1.2   Multi-label Weighted Vote Relational Neighbor

The wvRN algorithm was designed for classifying single-labeled networks. We used wvRN's approach and developed a one-vs-rest algorithm to implement a multi-label weighted vote relational neighbor (ML-wvRN) classifier. We use relaxation labeling for the collective inference, and estimate the probability for a test node to belong to K different classes. We select only those classes that have higher probabilities (the number of selected classes is equal to the total number of classes of the test node and is always $\leq K$) and assign the corresponding labels to the test node. A similar approach was taken by Tang *et al.* [18] in multi-label classification. This classifier is referred to as ML-wvRN-RL.

Algorithm 1. FLIP

Input: $\mathcal{G} = (\mathcal{V}, \mathcal{E})$: the network, CC: collective classifier, *maxiter*: number of iterations for the approximate inference of CC, b: batchsize, B: budget, PO: pool, T: Initial training set with $l\%$ of $|\mathcal{V}|$ as labeled samples
Output: L: updated training set
1: Initialize $L \leftarrow T$
2: **while** $|L| < B$ **do**
3:    Run CC with L as labeled data and PO as unlabeled data to predict labels for instances in the pool PO
4:    $S \leftarrow 0$
5:    **for** each node v_i s.t. $i \in PO$ **do**
6:       $S[i] = \sum_{t=1}^{maxiter} \sum_{k=1}^{K} |y_{ik}^t - y_{ik}^{t-1}|$ where y_{ik}^t is the predicted label of node v_i in the t-th iteration of inference through CC, such that $y_{ik}^t \in \{0,1\} \ \forall k \in \{1, 2, \cdots, K\}$
7:    **end for**
8:    Sort S in descending order of values
9:    Pick b nodes from pool PO having top b values in S
10:   Add the indices of these b nodes to set L, and remove these indices from pool PO
11:   $|L| \leftarrow |L| + b$
12: **end while**

3.2 Active Learning Using Iterative Classification Algorithm

For collective classification using iterative inference [9,13] in relational networks, the label information is propagated from the training node to the test nodes through multiple iterations. This causes the predicted label of a test instance to undergo multiple changes before it finally converges to a particular label. In our approach we monitor the frequency of label change for all the unlabeled instances in the pool set to identify the most informative ones. During the inference steps, certain nodes that are harder to classify, change their label(s) more frequently than others. We run multiple iterations of inference on the pool set and aggregate the total changes in labels for each node in the pool set. This aggregation or frequency of label changes is defined as *FLIP score* and an instance with a high *FLIP score* is considered to be a likely candidate for selection by our active learning algorithm (due to it's uncertainty in converging to a fixed label(s)). We refer to this approach of picking a batch of instances based on their *FLIP scores*, as FLIP and describe it in Algorithm 1. For single-labeled networks, wvRN with relaxation labeling (wvRN-RL) is used as the collective classifier (CC) in Algorithm 1. For multi-labeled networks, ML-wvRN-RL is used.

3.2.1 Case Study
We performed a case study on a derived co-authorship network from DBLP[1].

---

[1] http://www.informatik.uni-trier.de/~ley/db/

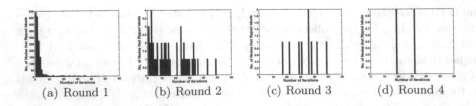

(a) Round 1	(b) Round 2	(c) Round 3	(d) Round 4

Fig. 1. Distribution of number of Instances which changed labels across number of Iterations in wvRN algorithm's collective inference step during Active Learning

Details of this single-labeled 2-class network and the experimental setup is mentioned in Section 4.1. We randomly sampled 2% of the nodes as training and 30% as testing sets, and left the rest in the pool. We used wvRN-RL [10] as collective classifer with training and pool sets; and performed collective inference on nodes in the pool. In Figure 3.2, we show the distribution of instances in the pool that changed labels across four rounds of active learning. For each round we chose 100 instances from pool with the highest *FLIP scores* across 50 iterations of collective inference and added them to the training set. For each round, the total number of instances that flipped labels, gradually decreased over consecutive iterations of the inference phase. These plots motivated us to use the inference steps of collective classification in order to identify informative instances from pool.

3.3   Active Learning for Multi-labeled Networks

We use the FLIP algorithm for multi-labeled networks with ML-wvRN-RL as the collective classifier. We aggregate each label's *FLIP scores* and select those instances which have highest total *FLIP score*. Rest of the steps are same as shown in Algorithm 1. This method is also referred by FLIP in our experiments with multi-labeled networks.

3.3.1   Active Learning with Per-Label Cost

We proposed a method to optimize the cost associated with querying *each label* within the multi-labeled relational networks. Specifically, in each iteration either a single label or a subset of labels of selected samples are queried, while the labels of the remaining instances are inferred in the subsequent rounds. Querying a *subset* of labels of a multi-labeled instance makes more efficient utilization of budget compared to querying all the labels. We refer to this method as FLIP-per-label (Algorithm 2). Intuitively, this method iteratively chooses a (node,label) pair for which the label has flipped maximum number of times.

The number of (node,label) pairs chosen in each round of active learning depends on a parameter *batchsize*, i.e., the total number of *labels* to be queried. Unlike FLIP, in this case array S contains the *FLIP scores* for all possible (node,label) pairs. Additionally, a list called $PAIR$ (of (node,label) pairs) is maintained. We initialize $\mathbf{P}_U = 0.5 \times \mathbf{1}_{|U| \times K}$ assuming all unobserved or *missing* labels are equally likely with probability score $= 0.5$. We set $|U| = |PO|$, i.e., number of unlabeled nodes in the pool. In line 3 of Algorithm 2, the learner predicts the probability scores of all the classes (and thereby, labels) for each of the instances in the pool. It queries labels based on the probability values in $\mathbf{P}_U$ (line 17), thereby, updating matrix $\mathbf{P}_L$ (line 19) depending on which instances have their specific label(s) queried. If the predicted class conditional probability of an instance lies between lower and upper probability thresholds (l_{tr} and u_{tr} respectively), then the corresponding class label is assumed to be non-informative, and hence, not queried. We set $l_{tr} = 0.3$ and $u_{tr} = 0.7$, assuming that any label k of node v_i having probability score $P(y_{ik} = 1|N_i) \le 0.3$ can not be the true class of v_i. Similarly, any label k of node v_i having probability score $P(y_{ik} = 1|N_i) \ge 0.7$ can be considered as the true label of v_i. In either of the two cases, the label k is queried by the learner. The algorithm is referred as FLIP-PL in the plots of Section 4.

Algorithm 2. FLIP-per-label

Input: $\mathcal{G} = (\mathcal{V}, \mathcal{E})$: the network, K: number of classes, CC: collective classifier, *maxiter*: number of iterations for the approximate inference of CC, b: batchsize, B: budget, PO: pool, T: Initial training set with $l\%$ of $|\mathcal{V}|$ as labeled samples, $\mathbf{Y}_L$: Label matrix of nodes in training set T, l_{tr}: lower threshold of probability score, u_{tr}: upper threshold of probability score
Output: L: updated training set
1: Initialize $L \leftarrow T; \mathbf{P}_L = \mathbf{Y}_L; \mathbf{P}_U = 0.5 \times \mathbf{1}_{|PO| \times K}; total \leftarrow \emptyset$
2: **while** $total < B$ **do**
3:   Run CC with L as labeled data and $U = PO$ as unlabeled data to predict labels for unlabeled instances in the pool PO, update $\mathbf{P}_U$ from Equations (1)-(3)
4:   $S \leftarrow 0, index \leftarrow 1, PAIR \leftarrow 0$
5:   **for** each node v_i s.t. $i \in PO$ **do**
6:     **for** each label $k \in \{1, 2, \cdots, K\}$ **do**
7:       $S[index] = \sum_{t=1}^{maxiter} |y_{ik}^t - y_{ik}^{t-1}|$ where y_{ik}^t is the predicted label of node v_i in the t-th iteration of inference through CC, such that $y_{ik}^t \in \{0, 1\}$
8:       $PAIR[index] \leftarrow (i, k)$ /* store the (node,label) pair */
9:       $index \leftarrow index + 1$
10:     **end for**
11:   **end for**
12:   $SS \leftarrow Sort(S, descend)$
13:   $pair \leftarrow (i, k)$ pairs selected from $PAIR$ s.t. $pair[1] = (i, k)$ has the *FLIP* score $= SS[1]$ and $pair[index] = (i', k')$ has *FLIP* score $= SS[index]$
14:   $count \leftarrow 0$
15:   **for** each $(i, k) \in pair$ **do**
16:     Get $P(y_{ik} = 1|N_i)$ from updated $\mathbf{P}_U$ computed in line 3 using Equations (1)-(3)
17:     **if** $P(y_{ik} = 1|N_i) \ge u_{tr}$ or $P(y_{ik} = 1|N_i) \le l_{tr}$ **then**
18:       Query label k of node v_i in the pool PO
19:       Add node v_i to L with v_i's k-th label disclosed to CC, remove i from pool PO, Update $\mathbf{P}_L$ with the queried label information of v_i
20:       $count \leftarrow count + 1$
21:       **if** $count == b$ **then**
22:         **break**
23:       **end if**
24:     **end if**
25:   **end for**
26:   $total \leftarrow total + b$
27: **end while**

3.4   Variants of FLIP for Single- and Multi-labeled Networks

The active learning algorithm relies on computing the *FLIP scores* for each of the instances in the pool during the inference phase. However, several instances may end up having the same *FLIP scores*. As such, we developed different tie-breaking strategies and variations of the FLIP algorithm.

3.4.1   Betweenness Centrality

Macskassy [11] proposed several graph-based metrics to select informative instances for single labeled datasets, out of which betweenness centrality metric was found to be most useful. For a node v_i, the shortest path betweenness $c_B(v_i)$, is defined as follows [11]:

$$c_B(v_i) = \sum_{v_a, v_b \in \mathcal{V}} \frac{\sigma(v_a, v_b | v_i)}{\sigma(v_a, v_b)} \tag{4}$$

where $\sigma(v_a, v_b | v_i)$ represents the number of shortest paths that pass through node v_i and $\sigma(v_a, v_b)$ represents number of shortest paths between any pair of nodes v_a, v_b in the graph. Nodes with high betweenness centrality scores can be considered as *information hubs* in the network (making them important for collective inference). Upon encountering a tie on *FLIP scores*, nodes with high betweenness centrality are finally chosen. This is referred by FLIP-BC.

3.4.2   Hops from Training Nodes

For our active learning algorithm, the objective is to select instances that can disperse the labeled information over the entire network. To this end, upon encountering a tie in *FLIP score* we choose only those nodes that are at a greater distance (hops) away from *any* of the training nodes. We refer to this as FLIP-H.

3.4.3   Absolute difference in Probability Score

For each node in the pool, our active learner predicts the class conditional probabilities with values in $[0, 1]$ (see Equation (1)). However, when the learner is most uncertain about the class of an instance, it is more likely to predict a score ≈ 0.5 for each of the K classes. Instead of computing *FLIP score* for v_i as the count/frequency of *flips*, we compute the deviation of class conditional probabilities from 0.5 for each of the labels at the end of the maximum allowed iterations ($t = maxiter$) and assign the score to $S[i]$.

$$S[i] = \sum_{k=1}^{K} |P^{maxiter}(y_{ik} = 1 | N_i) - 0.5| \tag{5}$$

where $P^t(y_{ik} = 1 | N_i)$ at $t = 0$, is initialized according to Equation (2). We query those instances, which have lowest scores in S. If $P^{maxiter}(y_{ik} = 1 | N_i) \approx 0.5$ then learner is uncertain about its true label and $S[i] \approx 0$ and v_i can be considered as an informative sample. Intuitively, this approach is very similar to FLIP, because $P^{maxiter}(y_{ik} = 1|N_i) \approx 0.5$ will most likely cause node v_i to flip its labels several times. We refer to this method as FLIP-A.

3.4.4 Entropy

Entropy is a measure of uncertainty in any system. Since, we are predicting the labels of the instances in the pool, we assume that instances which have the highest entropy due to their predicted labels, are most difficult to classify and are expected to be the most informative instances for learning a model. For a node v_i in the pool, the entropy $c_E(v_i)$ due to collective inference is:

$$c_E(v_i) = -\sum_{k=1}^{K} P^{maxiter}(y_{ik} = 1|N_i) \cdot \log(P^{maxiter}(y_{ik} = 1|N_i)) \qquad (6)$$

This entropy is used as utility score in identifying informative nodes from the pool. We refer this baseline as Entropy.

3.5 Variants of FLIP-per-label for Multi-labeled Networks

The subtle difference between FLIP and FLIP-per-label required us to use a different set of variants compared to that defined in Section 3.4.

3.5.1 Cumulative FLIP score

This is a two-phase selection procedure. In first phase, node v_i is selected based on its overall FLIP score computed as $S[i] = \sum_{t=1}^{maxiter} \sum_{k=1}^{K} |y_{ik}^{t} - y_{ik}^{t-1}|$ (line 6, Algorithm 1). In second phase, any label k of v_i that has probability score $P^{maxiter}(y_{ik} = 1|N_i) \leq l_{tr}$ or $P^{maxiter}(y_{ik} = 1|N_i) \geq u_{tr}$ was selected for annotation ($P^{maxiter}(y_{ik} = 1|N_i)$ is the probability of class k after maxiter iterations during collective classification). This method is referred as FLIP-PL-ALL.

3.5.2 Betweenness Centrality

The nodes with high betweenness centrality scores (computed using Equation (4)) in the pool are selected first. For each such node v_i, the labels which have probability scores $P^{maxiter}(y_{ik} = 1|N_i) \leq l_{tr}$ or $P^{maxiter}(y_{ik} = 1|N_i) \geq u_{tr}$ are selected for annotation. This is referred by BC-FLIP-PL. We also propose a baseline method that uses betweenness centrality score to identify nodes from pool, and then for each such node v_i, randomly selects $\lceil 0.5 \times K \rceil$ labels which have high FLIP scores. This method is referred as BC-RAND.

3.5.3 Entropy

For a node v_i in the pool, the entropy $c_E(v_i)$ is measured according to Equation (6) (Section 3.4.4). This is used as a utility score in identifying informative nodes from the pool. For each such multi-labeled node v_i, $\lceil 0.5 \times K \rceil$ labels which have high FLIP scores are queried. This method is referred by Ent-FLIP-PL.

Table 1. Description of datasets (single-labeled and multi-labeled networks)

Type	Name	$|\mathcal{V}|$	$|\mathcal{E}|$	K	ADN†	ACC†	ALN †	(+)/(-)†
				Single-label				
Binary	DBLP(B)-binary	5329	21880	2	7.2117	0.8127	1	2935/2394
	IMDB - prod	1176	37174	2	63.2211	0.3963	1	564/612
Multi-class	Cora	2708	5278	7	3.8981	0.2407	1	NA
	DBLP(B)-multiclass	5329	21880	6	7.2117	0.8127	1	NA
	Industry - pr	2189	11666	12	10.6587	0.5425	1	NA
	Flickr	7971	478980	7	120.1807	0.2955	1	NA
				Multi-label				
NA	DBLP(A)	10314	47200	6	8.1526	0.9999	1.6191	NA
	DBLP(B)	5329	21880	6	7.2117	0.8127	1.2211	NA
	IMDB - actor	2411	12255	22	10.1697	0.4720	3.6838	NA

† ADN = Average degree per node, ACC = Average clustering co-efficient, ALN = Average number of labels per node, (+)/(-) = Numbers of instances in positive/negative class

4   Experimental Protocol

4.1   Datasets

We evaluate the performance of our algorithms on six single-labeled datasets and three multi-labeled datasets. All datasets used in this paper can be downloaded from the website[2]. Characteristics of these datasets are provided in Table 1. We validate our algorithms using the framework described by Bilgic et al. [2]. First we select 30% of the nodes randomly from each network as test samples, and keep these nodes as well as all the edges connected to these nodes separate from the network during active learning. The remaining nodes are split into pool and training set. We found that instead of choosing samples for training randomly, if we choose samples that have high betweenness centrality score then the performance of the classifier improves. The performance of the learner is measured w.r.t. the test set after putting the test nodes and edges back in the network. We choose only 2% of the total number of instances having high betweenness centrality scores as the initial training set.

We have extracted four different co-authorship networks (two single-labeled networks named as *DBLP(B)-binary*, *DBLP(B)-multiclass* and two multi-labeled networks named as *DBLP(B)* and *DBLP(A)*) of computer science researchers from the DBLP[3] bibliographic database as done by Kong et al. [7]. *DBLP(B)* is a dataset consisting of authors publishing in different computer science areas whereas *DBLP(A)* consists of authors from specific disciplines of computer science (Data Mining/AI). *DBLP(B)-binary* dataset is derived from *DBLP(B)* by considering "Networking" area as the positive class and rest of the classes as negative class. *DBLP(B)-multiclass* is derived by labeling each author (node) in *DBLP(B)* with the research area (label) in which the author has published most of his/her papers. We created an undirected version of the *Cora* citation network of papers (without any node features) belonging to one

---

[2] http://www.cs.gmu.edu/~tsaha/Projects/
[3] http://www.informatik.uni-trier.de/~ley/db/

of the seven AI related research areas from the original dataset used by Lu and Getoor [9]. *Industry-pr* network comprises of 2189 companies that co-occurred with at least one other company in the PR Newswire release dataset[4]. The labels of the companies are based on Yahoo!'s 12 industry sectors. *IMDB-prod* network contains movies (a link exists if two movies shared a production company) released in the United States between 1996 and 2001, with class labels identifying whether the opening weekend box-office receipts will exceed 2 million dollars or not [10]. Another network of actors (referred as *IMDB-actor*) who acted in movies released between 1990 − 2012, was created by us. There is an edge between two actors if they have acted together in a movie. The labels of an actor are the multiple genres of the movies in which that actor acted. Since this network is multi-labeled, each actor has one or more genre(s) as his/her label. *Flickr* network was created by sampling the seven most populated classes from the original network by Tang *et al.* [18]. This is a single-labeled connected network of 7971 individuals belonging to 7 specific interest groups (classes).

4.2 Comparative Methods

Both FLIP and FLIP-per-label algorithms and their variants are compared w.r.t. several baseline methods listed in Table 2. BC is the baseline method that chooses instances with high betweenness centrality score as informative samples. A tie is resolved by random selection. For single-labeled networks, Random baseline model randomly selects a batch of nodes as informative samples, and queries any label for each of those nodes. For multi-labeled networks, $\lceil 0.5 \times K \rceil$ labels were chosen for querying in Random. We also experimented our active learning paradigm with a link based classifier [9] without using any node features, but the results were considerably poor, because such classifiers use both node and topological features to learn a model (results not reported here due to space).

Table 2. Comparing methods for FLIP (single-labeled and multi-labeled networks) and FLIP-per-label (multi-labeled networks)

Type of Method	Algorithm	Type of Classification	Publication
Variants of FLIP	FLIP	Single and multi-label	This paper (Algorithm 1)
	FLIP-BC	Single and multi-label	This paper (Section 3.4.1)
	FLIP-H	Single and multi-label	This paper (Section 3.4.2)
	FLIP-A	Single and multi-label	This paper (Section 3.4.3)
	BC	Single and multi-label	Baseline ([11])
	Entropy	Single and multi-label	Baseline (Section 3.4.4)
	Random	Single and multi-label	Baseline ([2,11])
Variants of FLIP-per-label	FLIP-PL	Multi-label	This paper (Algorithm 2)
	FLIP-PL-ALL	Multi-label	This paper (Section 3.5.1)
	BC-FLIP-PL	Multi-label	This paper (Section 3.5.2)
	Ent-FLIP-PL	Multi-label	This paper (Section 3.5.3)
	BC-RAND	Multi-label	Baseline (Section 3.5.2)
	Random	Multi-label	Baseline (Section 4.2)

---

[4] http://netkit-srl.sourceforge.net/data.html

4.3   Validation Protocol

For single-labeled networks, we report the 0/1 loss (error) on the test set for all the comparing methods. The performances reported for all single-labeled and multi-labeled networks are an average of 10 independent runs. For each round of active learning in single-labeled networks, we choose batch size $b = 5$ for IMDB-prod and Industry-pr networks, $b = 10$ for Cora network, and $b = 20$ for DBLP(B) and Flickr networks. The value of batch size, b was determined depending on the total number of nodes in the network. For multi-labeled networks we use hamming loss (lower the better) and micro-F1 score (higher the better) as evaluation metrics [7,15]. For each round of active learning using FLIP, we choose the batchsize $b = 5$ for IMDB-actor, $b = 20$ for DBLP(B) and $b = 30$ for DBLP(A) networks, respectively. For FLIP-per-label, $b = 110$ for IMDB-actor, $b = 60$ for DBLP(B) and $b = 90$ for DBLP(A) are used. We conducted 30 and 50 rounds of active learning for single-labeled and multi-labeled networks, respectively, in order to observe the convergence of all the comparing methods.

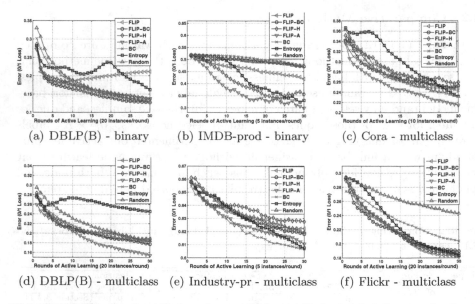

(a) DBLP(B) - binary   (b) IMDB-prod - binary   (c) Cora - multiclass

(d) DBLP(B) - multiclass   (e) Industry-pr - multiclass   (f) Flickr - multiclass

Fig. 2. Performance of active learning methods on all single-labeled networks (best viewed in color print)

5   Results and Discussion

5.1   Active Learning for Single-Labeled Networks

Figure 4.3 shows the performance of all the active learning methods on six single-labeled networks. The X-axis represents the rounds of active learning that have been carried out and Y axis reports the classification error (or 0/1 loss) on

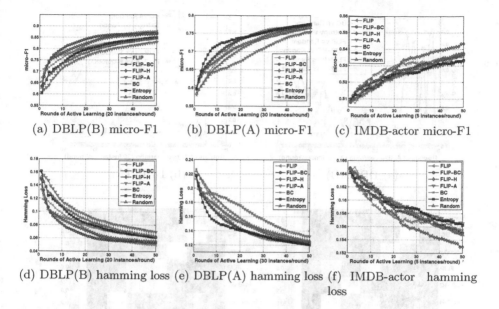

(a) DBLP(B) micro-F1   (b) DBLP(A) micro-F1   (c) IMDB-actor micro-F1

(d) DBLP(B) hamming loss (e) DBLP(A) hamming loss (f) IMDB-actor hamming loss

Fig. 3. Performance of FLIP on multi-labeled networks w.r.t. micro-F1 score (↑) and Hamming Loss (↓) (best viewed in color print)

the test nodes. We can see that most of the active learning strategies perform well compared to random sampling and other two baseline methods (BC and Entropy). For the IMDB-prod, DBLP(B)-multiclass and Flickr networks, all the active learning algorithms outperform random sampling. These three networks consist of instances that belong to uncorrelated classes, but still have a fully connected structure that can propagate label information over the entire network. So, when *FLIP score* is the only criterion for identifying informative samples in this network, then nodes in the pool undergo multiple changes in their labels through consecutive iterations which causes FLIP-based active learning methods to perform well. The high error values for Industry-pr network is because it has 12 classes and the number of samples in the dataset is only 2189, resulting on an average ≈ 182 instances per class. Hence, the classification task is harder for this dataset. For brevity, we do not include statistical significance tests for these results in this paper. Details can be found in the submitted supplementary file.

5.2 Active Learning for Multi-labeled Networks

5.2.1 Performance of FLIP

Figures 5.1 shows the performance of different methods w.r.t. micro-F1 scores and hamming loss for DBLP and IMDB-actor networks, respectively. For DBLP(B) network, the performance of all the active learning methods are better in comparison to the baselines. The good performance of all FLIP-based methods (except FLIP-A) on this dataset is due to the lack of correlation between

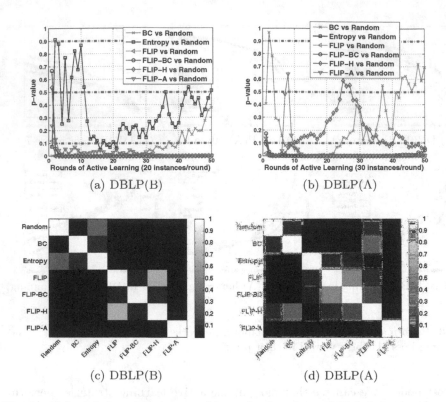

(a) DBLP(B)   (b) DBLP(A)

(c) DBLP(B)   (d) DBLP(A)

Fig. 4. p-value plots and heatmaps of hamming loss for multi-labeled networks with FLIP: $p <= 0.1$ denotes random sampling is significantly worse, $p >= 0.9$ denotes random sampling is not significantly worse compared to other methods at 10% significance level (best viewed in color print)

the classes to which nodes of this network belong. Hence, when we are using only *FLIP* score to query *all* the labels of an instance, the learner is abruptly fed with lots of information that help improve its performance. FLIP-A selects instances which have class conditional probability values closer to 0.5 for *all* the classes. For single-labeled network, this is a good metric to identify informative instances, however, for multi-labeled networks this is misleading because it ends up choosing those instances for which *all* the classes have conditional probability values ≈ 0.5. For DBLP(A) network, since all the labels of this network are highly correlated, we see a less promising performance from the FLIP-based active learning methods in the first few rounds. For the IMDB-actor network, FLIP-H is the best performer. The sparsity of this network and fewer instances per class, poses a hard task for the classifier. When the hops from a training node are considered as tie-breaking criterion for the *FLIP score*, it enables nodes from farther apart zones in the network to get added to the training set, thereby improving the overall diversity of the training set.

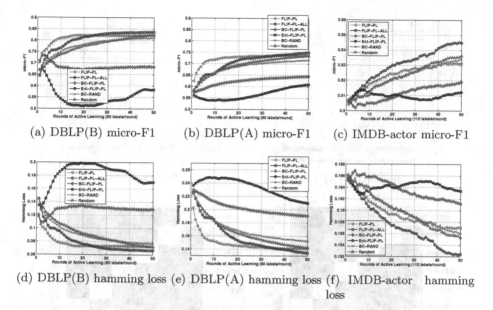

(a) DBLP(B) micro-F1   (b) DBLP(A) micro-F1   (c) IMDB-actor micro-F1

(d) DBLP(B) hamming loss (e) DBLP(A) hamming loss (f) IMDB-actor hamming
loss

Fig. 5. Performance of FLIP-per-label on multi-labeled networks w.r.t. micro-F1 ($\uparrow$)
and Hamming Loss ($\downarrow$) (best viewed in color print)

To assess the statistical significance of the results and to compare between
the different methods, we performed paired t-tests following the work of Bilgic
$et\ al.$ [2]. Figures 4(a)-(b) show the p-value plots for DBLP(B) and DBLP(A)
networks. The X axis corresponds to the rounds of active learning, and the
Y axis corresponds to the p-value for the paired t-test of the hamming loss
resulting from 10 independent runs of the corresponding pair of methods at
each round of active learning. If the p-value for a A vs B plot lies below 0.1
then model A wins over model B at 10% significance level.

In order to observe how each of the active learning approaches are performing
individually, we show a heatmap of p-values comparing the algorithms in a pairwise
fashion. For example, consider Figure 4(c) where each block represents a p-value
obtained from the corresponding pair of algorithms (along the rows and columns of
the figure). For each algorithm, we aggregate the performance measures obtained
across all the active learning rounds (50 rounds for multi-labeled networks) for each
of the 10 independent runs. Darker intensity colored boxes indicate that p-value
lies below 0.1 and the corresponding pair of methods show significantly different
results from one another. Lighter intensity colored boxes suggest otherwise.

We have performed pairwise t-tests on each pair of models for the hamming
loss (Figures 4(a)-(b) for comparison with Random model and Figures 4(c)-(d)
for comparison with every other method). For brevity, we include results for
IMDB-actor network in the submitted supplementary file. Figure 4(c)-(d) shows
that, for both the DBLP(B) and DBLP(A) networks, all the active learning

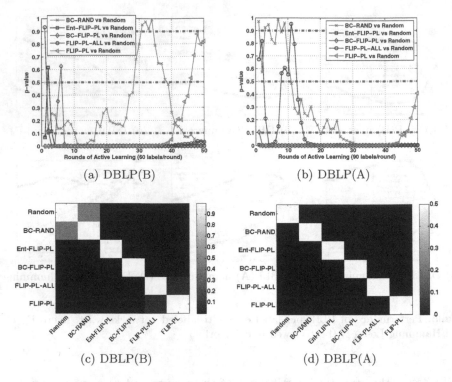

(a) DBLP(B)   (b) DBLP(A)

(c) DBLP(B)   (d) DBLP(A)

Fig. 6. p-value plots and heatmaps of hamming loss for DBLP(B) and DBLP(A) networks with FLIP-per-label method: $p <= 0.1$ denotes random sampling is significantly worse, $p >= 0.9$ denotes random sampling is not significantly worse compared to other methods at 10% significance level (best viewed in color print)

methods are significantly better than random sampling (`Random` model) and also statistically significant in comparison to one another.

5.2.2   Performance of FLIP-Per-Label

At each round of active learning, instead of querying *all* the labels of an instance, we queried labels based on (node,label) pairs that were identified as top candidates by the `FLIP-per-label` algorithms. Figure 5.2.1 shows the performance of `FLIP-PL` and `FLIP-PL-ALL` algorithms for three multi-labeled networks w.r.t. other methods. The poor performance of `BC-FLIP-PL` and `Ent-FLIP-PL` is due to the fact that we initialized the class conditional probability of all pooled nodes with 0.5 in the beginning of active learning. This accounts for high entropy values but low overall *FLIP scores* in the first few rounds, thereby choosing samples that are not quite informative for both `BC-FLIP-PL` and `Ent-FLIP-PL` methods. Figure 5.2.1 supports these through the corresponding p-value plots and respective heatmaps that compare the performance of different models against random sampling, and also against one another for DBLP datasets (we include results for IMDB-actor dataset in the supplementary file).

6 Conclusion

Active learning for relational network classification is an emerging field of research. In this paper, we developed several active learning algorithms based on a state-of-the-art relational classifier. Our contribution can be summarized as follows: (*i*) our methods rely on network structure for collective classification of single-labeled and multi-labeled networks; (*ii*) we propose a different scenario in active learning on networks for spending the budget (FLIP-per-label) when only a subset of all possible labels can be queried for instances in multi-labeled networks. To the best of our knowledge, we are the first to propose these two categories of active learning algorithms in multi-labeled relational network datasets. FLIP shows good results on single-labeled as well as multi-labeled networks, in comparison to multiple baselines. FLIP-per-label tackles more restricted budget situations in active learning, and the results are promising for several real world multi-labeled networks.

Acknowledgements. Huzefa Rangwala is supported by NSF Career Award (IIS 1252318). These experiments were run on ARGO, a research computing cluster provided by the Office of Research Computing at George Mason University, Virginia, USA. (URL: http://orc.gmu.edu)

References

1. Bilgic, M., Getoor, L.: Effective label acquisition for collective classification. In: Proceeding of the 14th ACM SIGKDD, pp. 43–51. ACM (2008)
2. Bilgic, M., Mihalkova, L., Getoor, L.: Active learning for networked data. In: Proceedings of the 27th International Conference on Machine Learning (2010)
3. Chakrabarti, S., Dom, B., Indyk, P.: Enhanced hypertext categorization using hyperlinks. ACM SIGMOD Record 27, 307–318 (1998)
4. Dagan, I., Engelson, S.P.: Committee-based sampling for training probabilistic classifiers. In: ICML, vol. 95, pp. 150–157 (1995)
5. Ghamrawi, N., McCallum, A.: Collective multi-label classification. In: Proceedings of the 14th ACM CIKM, pp. 195–200 (2005)
6. Ji, M., Han, J.: A variance minimization criterion to active learning on graphs. In: AISTATS (2012)
7. Kong, X., Shi, X., Philip, S.: Multi-label collective classification. In: Proceedings of SIAM International Conference on Data Mining (SDM), pp. 618–629 (2011)
8. Kuwadekar, A., Neville, J.: Relational active learning for joint collective classification models. In: Proceedings of the 28th ICML 2011, vol. 11, pp. 385–392 (2011)
9. Lu, Q., Getoor, L.: Link-based classification. In: Workshop at International Conference on Machine Learning (ICML), vol. 20, p. 496 (2003)
10. Macskassy, S., Provost, F.: Classification in networked data: A toolkit and a univariate case study. The Journal of Machine Learning Research 8, 935–983 (2007)

11. Macskassy, S.A.: Using graph-based metrics with empirical risk minimization to speed up active learning on networked data. In: Proceedings of 15th ACM SIGKDD, pp. 597–606. ACM (2009)
12. McCallum, A., Nigam, K., et al.: Employing em in pool-based active learning for text classification. In: Proceedings of ICML 1998, pp. 350–358 (1998)
13. Neville, J., Jensen, D.: Iterative classification in relational data. In: Proc. AAAI-2000 Workshop, pp. 13–20 (2000)
14. Neville, J., Jensen, D.: Relational dependency networks. The Journal of Machine Learning Research 8, 653–692 (2007)
15. Saha, T., Rangwala, H., Domeniconi, C.: Multi-label collective classification using adaptive neighborhoods. In: ICMLA, vol. 1, pp. 427–432. IEEE (2012)
16. Sen, P., Namata, G., Bilgic, M., Getoor, L., Galligher, B., Eliassi-Rad, T.: Collective classification in network data. AI Magazine 29(3), 93 (2008)
17. Shi, L., Zhao, Y., Tang, J.: Batch mode active learning for networked data. ACM Transactions on Intelligent Systems and Technology (TIST) 3(2), 33 (2012)
18. Tang, L., Liu, H.: Relational learning via latent social dimensions. In: Proceedings of the 15th ACM SIGKDD, pp. 817–826. ACM (2009)
19. Wang, X., Sukthankar, G.: Multi-label relational neighbor classification using social context features. In: Proceedings of the 19th ACM SIGKDD, pp. 464–472 (2013)
20. Zhang, M., Zhou, Z.: A review on multi-label learning algorithms (2013)
21. Zhu, X., Lafferty, J., Ghahramani, Z.: Combining active learning and semi-supervised learning using gaussian fields and harmonic functions. In: ICML 2003 Workshop, pp. 58–65 (2003)

Clustering via Mode Seeking by Direct Estimation of the Gradient of a Log-Density

Hiroaki Sasaki[1], Aapo Hyvärinen[2,3], and Masashi Sugiyama[1]

[1] Graduate School of Information Science and Engineering,
Tokyo Institute of Technology, Tokyo, Japan
[2] Department of Computer Science and HIIT,
University of Helsinki, Helsinki, Finland
[3] Cognitive Mechanisms Laboratories, ATR, Kyoto, Japan
sasaki@sg.cs.titech.ac.jp, sugi@cs.titech.ac.jp,
aapo.hyvarinen@helsinki.fi

Abstract. *Mean shift clustering* finds the *modes* of the data probability density by identifying the zero points of the density gradient. Since it does not require to fix the number of clusters in advance, the mean shift has been a popular clustering algorithm in various application fields. A typical implementation of the mean shift is to first estimate the density by kernel density estimation and then compute its gradient. However, since a good density estimation does not necessarily imply an accurate estimation of the density gradient, such an indirect two-step approach is not reliable. In this paper, we propose a method to *directly* estimate the gradient of the log-density without going through density estimation. The proposed method gives the global solution analytically and thus is computationally efficient. We then develop a mean-shift-like fixed-point algorithm to find the modes of the density for clustering. As in the mean shift, one does not need to set the number of clusters in advance. We experimentally show that the proposed clustering method significantly outperforms the mean shift especially for high-dimensional data.

Keywords: Log-Density Gradient Estimation, Mean Shift, Clustering, High-Dimensional Data.

1 Introduction

Seeking the *modes* of a probability density has led to a powerful clustering algorithm called the *mean shift* [6,8,11]. In the mean shift algorithm, all input samples are initially regarded as candidates of the modes of the density and they are iteratively updated and merged. Finally, clustering is performed by associating the input samples with the obtained modes. An advantage of the mean shift is that the number of clusters does not need to be specified in advance. Thanks to this extremely useful property, the mean shift has been successfully employed in various applications such as image segmentation [8,24,26] and object tracking [7,9].

T. Calders et al. (Eds.): ECML PKDD 2014, Part III, LNCS 8726, pp. 19–34, 2014.
© Springer-Verlag Berlin Heidelberg 2014

In mode seeking, a central technical challenge is accurate estimation of the gradient of a density. The mean shift takes a two-step approach: kernel density estimation (KDE) is first used to approximate the density and then its gradient is computed. However, such a two-step approach performs poorly because a good estimator of the density does not necessarily mean a good estimator of the density gradient. In particular, KDE tends to produce a smooth density estimate and therefore the modes in a multi-modal density could be collapsed. Furthermore, KDE itself tends to perform poorly in high-dimensional problems [8].

To overcome this problem, we propose a method called the *least-squares log-density gradient* (LSLDG), which *directly* estimates the gradient of a log-density by least-squares without going through density estimation. The proposed method can be regarded as a non-parametric extension of *score matching* [14,21], which has originally been developed for least-squares parametric density estimation with intractable partition functions. We then derive a fixed-point algorithm to find the modes of the density, which is our proposed clustering algorithm called *LSLDG clustering*.

All tuning parameters included in LSLDG such as the Gaussian kernel width and the regularization parameter can be objectively optimized by cross-validation in terms of the squared error. Furthermore, since LSLDG clustering inherits the same algorithmic structure as the original mean shift, it does not require the number of clusters to be fixed in advance. Thus, LSLDG clustering does not involve *any* tuning parameters to be manually determined, which is a significant advantage over standard clustering algorithms such as *spectral clustering* [19], because clustering is an unsupervised learning problem and appropriately controlling tuning parameters is generally very hard. A recent study based on *information-maximization clustering* [22] provided an information-theoretic mean to determine tuning parameters objectively, but it still requires the user to fix the number of clusters in advance.

The remainder of this paper is structured as follows. We derive a method to directly estimate the gradient of a log-density in Section 2, and then use it for finding clusters in the data in Section 3. Various possibilities for extension are discussed in Section 4. The usefulness of the proposed method is experimentally investigated in Section 5. Finally this paper is concluded in Section 6.

2   Direct Estimation of the Gradient of a Log-Density

In this section, we propose a method to estimate the log-density gradient.

2.1   Problem Formulation

Let us consider a probability distribution on $\mathbb{R}^d$ with density $p^*(\boldsymbol{x})$, which is unknown but n i.i.d. samples $\mathcal{X} = \{\boldsymbol{x}_i\}_{i=1}^n$ are available:

$$\mathcal{X} = \{\boldsymbol{x}_i\}_{i=1}^n \overset{\text{i.i.d}}{\sim} p^*(\boldsymbol{x}).$$

Our goal is to estimate the gradient of the logarithm of the density $p^*(x)$ with respect to x from $\mathcal{X}$:

$$g^*(x) = (g_1^*(x), \ldots, g_d^*(x))^\top = \nabla \log p^*(x) = \frac{\nabla p^*(x)}{p^*(x)}.$$

A naive approach to estimate $g^*(x)$ is to first obtain a density estimate $\widehat{p}(x)$ and then compute its log-gradient $\nabla \log \widehat{p}(x)$. However, this two-step approach does not work well because a good density estimate $\widehat{p}(x)$ does not necessarily provide an accurate estimate of its log-density gradient $\nabla \log \widehat{p}(x)$. For example, in Figures 1(a) and (b), density estimation is performed very well by KDE, but its log-density gradient produces oscillated errors. These errors become more prominent especially in higher-dimensional data (Figure 1(c)).

Below, we describe a method to directly estimate the log-density gradient $\nabla \log p^*(x)$ without going through density estimation. Our proposed method is based on the mathematics of score matching [14]; the difference is that our goal is to approximate the gradient of the log-density instead of model parameter estimation.

2.2   Least-Squares Log-Density Gradient

Our basic idea is to directly fit a model $g(x) = (g_1(x), \ldots, g_d(x))^\top$ to the true log-density gradient $g^*(x)$ under the squared loss:

$$
\begin{aligned}
J_j(g_j) &= \int \left(g_j(x) - g_j^*(x) \right)^2 p^*(x)\mathrm{d}x - \int g_j^*(x)^2 p^*(x)\mathrm{d}x \\
&= \int g_j(x)^2 p^*(x)\mathrm{d}x - 2 \int g_j(x) g_j^*(x) p^*(x)\mathrm{d}x \\
&= \int g_j(x)^2 p^*(x)\mathrm{d}x - 2 \int g_j(x) \partial_j p^*(x)\mathrm{d}x \\
&= \int g_j(x)^2 p^*(x)\mathrm{d}x + 2 \int \partial_j g_j(x) p^*(x)\mathrm{d}x,
\end{aligned}
$$

where ∂_j denotes the partial derivative with respect to the j-th variable of x and the last equality follows from *integration by parts* under some conditions [14]. Then the empirical approximation of J_j is given as

$$\widehat{J}_j(g_j) = \frac{1}{n} \sum_{i=1}^{n} g_j(x_i)^2 + \frac{2}{n} \sum_{i=1}^{n} \partial_j g_j(x_i). \tag{1}$$

As the model $g_j(x)$, we use the following linear-in-parameter model, which is related to using an exponential family for density modeling:

$$g_j(x) = \sum_{i=1}^{n} \theta_{i,j} \psi_{i,j}(x) = \boldsymbol{\theta}_j^\top \boldsymbol{\psi}_j(x),$$

where $\boldsymbol{\theta}_j$ denotes the parameter vector and $\psi_{i,j}(\boldsymbol{x})$ is a basis function. The derivative of this model is given by

$$\partial_j g_j(\boldsymbol{x}) = \sum_{i=1}^{n} \theta_{i,j} \partial_j \psi_{i,j}(\boldsymbol{x}) = \boldsymbol{\theta}_j^\top \boldsymbol{\varphi}_j(\boldsymbol{x}),$$

where $\boldsymbol{\varphi}_j(\boldsymbol{x}) = (\partial_j \psi_{1,j}(\boldsymbol{x}), \dots, \partial_j \psi_{n,j}(\boldsymbol{x}))$.

Adding an ℓ_2-regularizer to (1), we can compactly express the optimization problem as

$$\widehat{\boldsymbol{\theta}}_j = \arg\min_{\boldsymbol{\theta}_j} \left[\boldsymbol{\theta}_j^\top \boldsymbol{G}^{(j)} \boldsymbol{\theta}_j + 2\boldsymbol{\theta}_j^\top \boldsymbol{h}_j + \lambda \boldsymbol{\theta}_j^\top \boldsymbol{\theta}_j \right], \tag{2}$$

where $\lambda \geq 0$ is the regularization parameter, and $\boldsymbol{G}^{(j)}$ and $\boldsymbol{h}_j$ are defined by

$$\boldsymbol{G}^{(j)} = \frac{1}{n} \sum_{i=1}^{n} \boldsymbol{\psi}_j(\boldsymbol{x}_i) \boldsymbol{\psi}_j(\boldsymbol{x}_i)^\top, \quad \boldsymbol{h}_j = \frac{1}{n} \sum_{i=1}^{n} \boldsymbol{\varphi}_j(\boldsymbol{x}_i).$$

As in score matching for an exponential family [15], the optimization problem (2) can be solved analytically as

$$\widehat{\boldsymbol{\theta}}_j = -(\boldsymbol{G}^{(j)} + \lambda \boldsymbol{I})^{-1} \boldsymbol{h}_j,$$

where $\boldsymbol{I}$ denotes the identity matrix. Finally, we obtain the estimator $\widehat{g}_j$ as

$$\widehat{g}_j(\boldsymbol{x}) = \sum_{i=1}^{n} \widehat{\theta}_{i,j} \psi_{i,j}(\boldsymbol{x}) = \widehat{\boldsymbol{\theta}}_j^\top \boldsymbol{\psi}_j(\boldsymbol{x}).$$

We call this method the *least-squares log-density gradient* (LSLDG).

2.3 Model Selection by Cross-Validation

The performance of LSLDG depends on the choice of the regularization parameter λ and parameters included in the basis function $\boldsymbol{\psi}_j$. They can be objectively chosen via cross-validation as follows:

1. Divide the samples $\mathcal{X} = \{\boldsymbol{x}_i\}_{i=1}^{n}$ into N disjoint subsets $\{\mathcal{X}_i\}_{i=1}^{N}$.
2. For $i = 1, \dots, N$
   (a) Compute the LSLDG estimator $\widehat{g}_j^{(i)}$ from $\mathcal{X} \backslash \mathcal{X}_i$ (i.e., all samples except $\mathcal{X}_i$).
   (b) Compute its hold-out error for $\mathcal{X}_i$:

$$\mathrm{CV}^{(i)} = \frac{1}{|\mathcal{X}_i|} \sum_{\boldsymbol{x} \in \mathcal{X}_i} \sum_{j=1}^{d} \left[\widehat{g}_j^{(i)}(\boldsymbol{x})^2 + 2\partial_j \widehat{g}_j^{(i)}(\boldsymbol{x}) \right],$$

where $|\mathcal{X}_i|$ denotes the cardinality of $\mathcal{X}_i$.

3. Compute the average hold-out error as

$$\text{CV} = \frac{1}{N} \sum_{i=1}^{N} \text{CV}^{(i)}. \tag{3}$$

4. Choose the model that minimizes (3) with respect to λ and parameters in ψ_j, and compute the final LSLDG estimator $\widehat{g}_j$ with the chosen model using all samples $\mathcal{X}$.

3   Clustering via Mode Seeking

In this section, we derive a clustering algorithm based on LSLDG. Our basic idea follows the same line as the *mean shift* algorithm [6,8,11], i.e., to assign each data sample to a nearby *mode* of the density.

3.1   Gradient-Based Approaches

A naive implementation of this idea is to use *gradient ascent* for each data sample to let it converge to one of the modes of the density in the vicinity:

$$\boldsymbol{x}_i \longleftarrow \boldsymbol{x}_i + \varepsilon \widehat{\boldsymbol{g}}(\boldsymbol{x}_i),$$

where $\varepsilon > 0$ is the step size.
  Since

$$\boldsymbol{g}(\boldsymbol{x}) = \nabla \log p(\boldsymbol{x}) = \frac{\nabla p(\boldsymbol{x})}{p(\boldsymbol{x})} \propto \nabla p(\boldsymbol{x}),$$

the gradient of the log-density $\log p(\boldsymbol{x})$ keeps the same direction as the gradient of the original density $p(\boldsymbol{x})$. However, due to $p(\boldsymbol{x})$ in the denominator, the log-gradient vector gets longer when $p(\boldsymbol{x}) < 1$ and shorter when $p(\boldsymbol{x}) > 1$. This is practically suitable adjustment because $p(\boldsymbol{x}) < 1$ ($p(\boldsymbol{x}) > 1$) often means that the current point $\boldsymbol{x}$ is far from (close to) a mode. Indeed, the faster convergence of gradient ascent with the log-density was asserted in the same way [11].
  To further increase the speed of convergence, using a *quasi-Newton* method is also promising:

$$\boldsymbol{x}_i \longleftarrow \boldsymbol{x}_i + \varepsilon \widehat{\boldsymbol{Q}} \widehat{\boldsymbol{g}}(\boldsymbol{x}_i),$$

where $\widehat{\boldsymbol{Q}}$ is an estimate of the inverse Hessian matrix.

3.2   Fixed-Point Approach

In the gradient-based approaches, choosing the step size parameter ε is a crucial problem. To avoid this problem, we develop a *fixed-point method*, in analogy to the original mean-shift method.

To easily derive a fixed-point equation, we focus on the basis function of the following form:

$$\psi_{i,j}(\boldsymbol{x}) = \frac{1}{\sigma^2}[\boldsymbol{c}_i - \boldsymbol{x}]_j \phi_i(\boldsymbol{x}),$$

where σ^2 is a constant, $\boldsymbol{c}_i$ is a d-dimensional constant vector, $\phi_i(\boldsymbol{x})$ is a "mother" basis function, and $[\cdot]_j$ denotes the j-th element of a vector. A typical choice of the mother basis function $\phi_i(\boldsymbol{x})$ is the Gaussian function:

$$\phi_i(\boldsymbol{x}) = \exp\left(-\frac{\|\boldsymbol{x} - \boldsymbol{c}_i\|^2}{2\sigma^2}\right), \tag{4}$$

where the Gaussian center $\boldsymbol{c}_i$ may be fixed at sample $\boldsymbol{x}_i$. In experiments, we only use 100 Gaussian centers chosen randomly from $\mathcal{X}$. This reduction of Gaussian centers significantly decreases the computational costs without sacrificing the performance, as shown in Section 5.2.

For this model, the LSLDG solution can be expressed as

$$\widehat{g}_j(\boldsymbol{x}) = \sum_{i=1}^{n} \widehat{\theta}_{i,j} \psi_{i,j}(\boldsymbol{x}) = \frac{1}{\sigma^2} \sum_{i=1}^{n} \widehat{\theta}_{i,j}[\boldsymbol{c}_i - \boldsymbol{x}]_j \phi_i(\boldsymbol{x})$$

$$= \frac{1}{\sigma^2} \sum_{i=1}^{n} \widehat{\theta}_{i,j} \phi_i(\boldsymbol{x})[\boldsymbol{c}_i]_j - \frac{[\boldsymbol{x}]_j}{\sigma^2} \sum_{i=1}^{n} \widehat{\theta}_{i,j} \phi_i(\boldsymbol{x}).$$

If $\sum_{i=1}^{n} \widehat{\theta}_{i,j} \phi_i(\boldsymbol{x}) \neq 0$, setting $\widehat{g}_j(\boldsymbol{x})$ to zero yields

$$[\boldsymbol{x}]_j = \frac{\sum_{i=1}^{n} \widehat{\theta}_{i,j} \phi_i(\boldsymbol{x})[\boldsymbol{c}_i]_j}{\sum_{i=1}^{n} \widehat{\theta}_{i,j} \phi_i(\boldsymbol{x})}. \tag{5}$$

We propose to use this equation as a fixed-point update formula by iteratively substituting the right-hand side to the left-hand side. In the vector-matrix form, the update formula is compactly expressed as

$$\boldsymbol{x}_i \longleftarrow \boldsymbol{B}\phi(\boldsymbol{x}_i)./(\widehat{\boldsymbol{\Theta}}^\top \phi(\boldsymbol{x}_i)),$$

where $B_{j,i} = \widehat{\theta}_{i,j}[\boldsymbol{c}_i]_j$, $\widehat{\Theta}_{i,j} = \widehat{\theta}_{i,j}$, $\phi(\boldsymbol{x}) = (\phi_1(\boldsymbol{x}), \ldots, \phi_n(\boldsymbol{x}))^\top$, and "./" denotes the element-wise division.

This update formula is similar to the one used in the original mean shift algorithm [8, Eq.(20)], which corresponds to $\widehat{\theta}_{i,j} = 1/n$:

$$\boldsymbol{x} \longleftarrow \frac{\sum_{i=1}^{n} \phi_i(\boldsymbol{x})\boldsymbol{c}_i}{\sum_{i=1}^{n} \phi_i(\boldsymbol{x})},$$

where ϕ_i is typically chosen as the Gaussian function (4). Thus, the proposed method can be regarded as a weighted variant of the mean shift algorithm, where the weights $\widehat{\theta}_{i,j}$ are learned by LSLDG. A similar weighted mean shift method has already been studied in [6], but the weights were determined heuristically.

The mean shift update was proven to be equivalent to gradient ascent with an adaptive step size [6]. LSLDG-based clustering also inherits this property. Indeed, if $[\boldsymbol{x}]_j \sum_{i=1}^{n} \widehat{\theta}_{i,j} \phi_i(\boldsymbol{x})$ is subtracted from and added to the numerator of Eq.(5) (thus the equation remains the same), we obtain

$$[\boldsymbol{x}]_j = [\boldsymbol{x}]_j + \varepsilon_j(\boldsymbol{x}) \widehat{g}_j(\boldsymbol{x}),$$

where

$$\varepsilon_j(\boldsymbol{x}) = \frac{\sigma^2}{\sum_{i=1}^{n} \widehat{\theta}_{i,j} \phi_i(\boldsymbol{x})}.$$

This shows that our fixed-point update rule can be regarded as gradient ascent with an adaptive step size $\varepsilon_j(\boldsymbol{x})$.

If $\phi_i(\boldsymbol{x})$ is set to be the Gaussian function (4), $\sum_{i=1}^{n} \widehat{\theta}_{i,j} \phi_i(\boldsymbol{x})$ can actually be regarded as an estimate of the original log-density $\log p^*(\boldsymbol{x})$. More specifically, we can easily see that the partial derivative of $\phi_i(\boldsymbol{x})$ with respect to the j-th variable of $\boldsymbol{x}$ is $\psi_{i,j}(\boldsymbol{x})$:

$$\partial_j \phi_i(\boldsymbol{x}) = \psi_{i,j}(\boldsymbol{x}).$$

Then we have

$$\partial_j \log p^*(\boldsymbol{x}) = g_j^*(\boldsymbol{x}) \approx \widehat{g}_j(\boldsymbol{x}) = \sum_{i=1}^{n} \widehat{\theta}_{i,j} \psi_{i,j}(\boldsymbol{x})$$

$$= \sum_{i=1}^{n} \widehat{\theta}_{i,j} \partial_j \phi_i(\boldsymbol{x}) = \partial_j \sum_{i=1}^{n} \widehat{\theta}_{i,j} \phi_i(\boldsymbol{x}).$$

This implies that $\sum_{i=1}^{n} \widehat{\theta}_{i,j} \phi_i(\boldsymbol{x})$ is an estimate of $\log p^*(\boldsymbol{x})$ up to a constant. Therefore, when $\log p^*(\boldsymbol{x})$ is small (large), the proposed fixed-point algorithm adaptively increases (decreases) the step size $\varepsilon_j(\boldsymbol{x})$ to more aggressively (conservatively) ascend the gradient. This step-size adaptation would be reasonable because small (large) $\log p^*(\boldsymbol{x})$ often means that the current solution is far from (close to) a mode.

4   Extensions

In the previous section, we focused on the simplest setting to clearly convey the essence of the proposed idea. However, we can easily extend the proposed method to various directions. In this section, we discuss such possibilities.

4.1   Common Basis Functions

When the basis function is common to all dimensions, i.e., $\psi_j(\boldsymbol{x}) = \psi(\boldsymbol{x})$ for $j = 1, \ldots, d$, the matrix $\boldsymbol{G}^{(j)}$ becomes independent of j as $\boldsymbol{G} = \frac{1}{n} \sum_{i=1}^{n} \psi(\boldsymbol{x}_i) \psi(\boldsymbol{x}_i)^{\top}$. Then, matrix inverse has to be computed only once for all dimensions:

$$(\widehat{\boldsymbol{\theta}}_1, \ldots, \widehat{\boldsymbol{\theta}}_d) = -(\boldsymbol{G} + \lambda \boldsymbol{I})^{-1} (\boldsymbol{h}_1, \ldots, \boldsymbol{h}_d).$$

This significantly speeds up the computation particularly when the dimensionality d is high.

4.2 Multi-task Learning

The above common-basis setup allows us to employ the *regularized multi-task* method [10], by regarding the estimation problem of $g_j^*(\boldsymbol{x})$ as the j-th task. The basic idea of regularized multi-task learning is that, if $g_j^*(\boldsymbol{x})$ and $g_{j'}^*(\boldsymbol{x})$ are similar to each other, the corresponding parameters $\boldsymbol{\theta}_j$ and $\boldsymbol{\theta}_{j'}$ are imposed to be close to each other. This idea can be implemented in the regularization framework as

$$\min_{\boldsymbol{\theta}_1,\ldots,\boldsymbol{\theta}_d} \left[\sum_{j=1}^d \left(\boldsymbol{\theta}_j^\top \boldsymbol{G}^{(j)} \boldsymbol{\theta}_j + 2\boldsymbol{\theta}_j^\top \boldsymbol{h}_j + \lambda_j \boldsymbol{\theta}_j^\top \boldsymbol{\theta}_j \right) + \gamma \sum_{j,j'=1}^d \gamma_{j,j'} \|\boldsymbol{\theta}_j - \boldsymbol{\theta}_{j'}\|^2 \right],$$

where $\lambda_j > 0$ is the ordinary regularization parameter for the j-th task, $0 \leq \gamma_{j,j'} \leq 1$ is the similarity between the j-th task and the j'-th task, and $\gamma > 0$ controls the strength of this multi-task regularizer. A notable advantage of this regularization approach is that the solution can be obtained analytically. When the task similarity $\gamma_{j,j'}$ is unknown, task similarity and solutions may be iteratively learned. More specifically, starting from $\gamma_{j,j'} = 1$ for all $j, j' = 1, \ldots, d$, the solutions $\boldsymbol{\theta}_1, \ldots, \boldsymbol{\theta}_d$ are computed. Then, task similarity is updated, e.g., by $\gamma_{j,j'} = \exp(-\|\boldsymbol{\theta}_j - \boldsymbol{\theta}_{j'}\|^2)$ for $j, j' = 1, \ldots, d$, and the solutions $\boldsymbol{\theta}_1, \ldots, \boldsymbol{\theta}_d$ are computed again.

4.3 Sparse Estimation

Instead of the ℓ_2-regularizer $\lambda\|\boldsymbol{\theta}_j\|^2$, the ℓ_1-regularizer $\lambda\|\boldsymbol{\theta}_j\|_1$ may be used to obtain a sparse solution [25]. The entire regularization path (i.e., the solutions for all $\lambda \geq 0$) can also be computed efficiently, based on the piece-wise linearity of the solution path with respect to λ [12].

4.4 Bregman Loss

The squared loss can be generalized to the *Bregman loss* [3]. More specifically, for f being a differentiable and strictly convex function and $C_j^{(f)} = \int f(g_j^*(\boldsymbol{x}))p^*(\boldsymbol{x})\mathrm{d}\boldsymbol{x}$,

$$
\begin{aligned}
J_j^{(f)}(g_j) &= \int \left(f(g_j^*(\boldsymbol{x})) - f(g_j(\boldsymbol{x})) - f'(g_j(\boldsymbol{x}))(g_j^*(\boldsymbol{x}) - g_j(\boldsymbol{x})) \right) p^*(\boldsymbol{x})\mathrm{d}\boldsymbol{x} - C_j^{(f)} \\
&= \int \left(-f(g_j(\boldsymbol{x})) + f'(g_j(\boldsymbol{x}))g_j(\boldsymbol{x}) \right) p^*(\boldsymbol{x})\mathrm{d}\boldsymbol{x} - \int f'(g_j(\boldsymbol{x}))\partial_j p^*(\boldsymbol{x})\mathrm{d}\boldsymbol{x} \\
&= \int \left(-f(g_j(\boldsymbol{x})) + f'(g_j(\boldsymbol{x}))g_j(\boldsymbol{x}) + \partial_j f'(g_j(\boldsymbol{x})) \right) p^*(\boldsymbol{x})\mathrm{d}\boldsymbol{x},
\end{aligned}
$$

where $f'(t)$ is the derivative of $f(t)$ with respect to t and the last equality follows again from integration by parts. The empirical approximation of $J_j^{(f)}$ is given as

$$\widehat{J}_j^{(f)}(g_j) = \frac{1}{n} \sum_{i=1}^n \left(-f(g_j(\boldsymbol{x}_i)) + f'(g_j(\boldsymbol{x}_i))g_j(\boldsymbol{x}_i) + \partial_j f'(g_j(\boldsymbol{x}_i)) \right).$$

When $f(t) = t^2$, the Bregman loss is reduced to the squared loss and we can recover the LSLDG criterion (1). On the other hand, $f(t) = -\log t$ gives the *Kullback-Leibler loss* [17], $f(t) = t \log t - (1 + t) \log(1 + t)$ gives the *logistic loss* [23], and $f(t) = (t^{1+\alpha} - t)/\alpha$ for $\alpha > 0$ gives the *power loss* [2]. Although each choice has its own specialty, e.g., the power loss possesses high robustness against outliers, the squared loss was shown to be endowed with the highest numerical stability in terms of the *condition number* [16].

4.5 Blurring Mean Shift

Fukunaga and Hostetler originally proposed a mean shift algorithm for updating not only the data points but also the density estimation at each iteration [11]. Later, this algorithm was named the *blurring mean shift* [4,6]. Combined with the idea of the blurring mean shift, another possible algorithm for LSLDG clustering is to re-estimate the log-density gradient at each iteration for new data points. This algorithm hopefully works well as the blurring mean shift does [4].

5 Experiments

In this section, we demonstrate the usefulness of the proposed LSLDG method.

A MATLAB implementation of LSLDG and its clustering algorithm based on the fixed-point approach is available from

http://sugiyama-www.cs.titech.ac.jp/~sugi/software/LSLDG/index.html

5.1 Illustration of Log-Density Gradient Estimation

We first illustrate how LSLDG estimates log-density gradients using $n = 1,000$ samples drawn from $p(x)$, where either

- $p(x)$ is the standard normal density, or
- $p(x)$ is a mixture of two Gaussians with means 2 and -2, variances 1 and 1, and mixing coefficients 0.5 and 0.5.

As described in Section 2.3, the Gaussian width σ and the regularization parameter λ are chosen by 5-fold cross-validation from the following candidate set:

$$\{10^{-2}, 10^{-1.5}, 10^{-1}, 10^{-0.5}, 10^0, 10^{0.5}, 10^1\}. \tag{6}$$

We compare the performance of the proposed method with Gaussian KDE, where the Gaussian width is chosen by likelihood cross-validation from the same candidate set in (6).

The results for the Gaussian data are presented in the upper row of Figure 1. Figure 1(a) shows that LSLDG gives a nice smooth estimate, while the estimate by KDE is rather oscillating. Note that KDE still works well as a density estimator as illustrated in Figure 1(b). This clearly illustrates that a good density

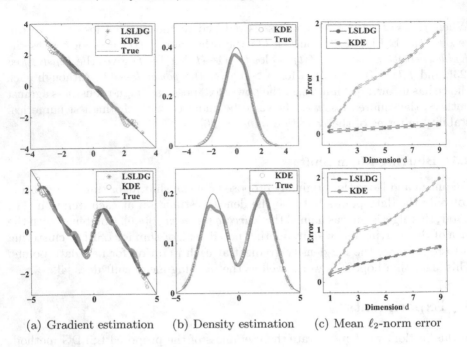

(a) Gradient estimation   (b) Density estimation   (c) Mean ℓ_2-norm error

Fig. 1. LSLDG vs. KDE for (upper row) Gaussian data and (lower row) data sampled from a mixture of two Gaussians. (a) Profiles of the true log-density gradient and its estimates obtained by LSLDG and KDE. (b) True and estimated densities by KDE. (c) Averages and standard deviations of mean ℓ_2-norm errors to the true log-density gradient as functions of input dimensionality over 100 runs.

estimate (obtained by KDE) does not necessarily yield a good estimate of the log-density gradient. We repeated this experiment 100 times and the mean ℓ_2-norm error to the true log-density gradient, $\frac{1}{n}\sum_{i=1}^{n}\|g(x_i) - g^*(x_i)\|$, is plotted in Figure 1(c) as a function of the input dimensionality. This shows that while the error of KDE increases sharply as a function of dimensionality, that of LSLDG increases only mildly. This implies that the advantage of directly estimating the log-density gradient is more prominent in high-dimensional cases. Similar tendencies can be observed also for the Gaussian mixture data in the lower row of Figure 1, where the added dimensions in the lower plot of Figure 1(c) simply follow the standard normal distribution.

5.2   Illustration of Clustering

Next, we illustrate the behavior of LSLDG clustering on $1,000$ samples gathered from the mixture of three Gaussians whose means are $(0, 2)$, $(-2, -2)$, and $(2, -2)$, and covariance matrices are the identity matrix. The mixing coefficients are 0.4, 0.3, and 0.3. Figure 2 illustrates the transition of data samples over update iterations, showing that all points converge to the nearest modes within 47 iterations.

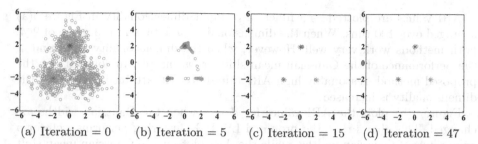

(a) Iteration = 0   (b) Iteration = 5   (c) Iteration = 15   (d) Iteration = 47

Fig. 2. Transition of data points toward the modes. The blue, red, and green symbols represent the three centers of the Gaussian mixture model.

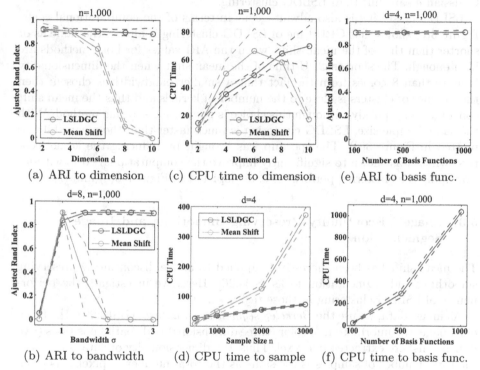

(a) ARI to dimension   (c) CPU time to dimension   (e) ARI to basis func.

(b) ARI to bandwidth   (d) CPU time to sample   (f) CPU time to basis func.

Fig. 3. Means and standard deviations of clustering performance over 100 runs measured by ARI as functions of (a) dimensionality of data and (b) the Gaussian width (when dimensionality is 8). CPU time is also compared with respect to (c) dimensionality and (d) sample size. (e) ARI and (f) CPU time for LSLDG clustering are plotted as functions of the number of basis functions.

We compare the performance of the proposed method with the Gaussian mean shift [5,6]. To investigate the effect of high dimensionality, further dimensions following the standard normal distribution are added to data points. We measure the clustering performance by the *adjusted Rand index* (ARI) [13], which takes the maximum value 1 when clustering is perfect.

ARI values are plotted as a function of input dimensionality in Figure 3(a) averaged over 100 runs. When the dimensionality of data is in the range of 2–4, both methods work very well. However, when the dimensionality is beyond 4, the performance of the Gaussian mean shift drops sharply. In contrast, for the proposed method, reasonably high ARI values are still attained even when the dimensionality is increased.

Figure 3(b) plots the ARI values for $d = 8$ when the Gaussian widths are changed. This shows that the proposed LSLDG clustering performs well for a wide range of Gaussian widths, while the ARI plot for the Gaussian mean shift is peaky. This implies that selection of Gaussian widths is much harder for the Gaussian mean shift than LSLDG clustering.

LSLDG clustering is also advantageous in terms of the computational costs. Figure 3(c) shows that CPU time of LSLDG clustering is almost the same as or shorter than that of the mean shift, when the ARI values for both methods are high enough. The shorter CPU time of the mean shift when the dimensionality is more than 8 comes from the fact that a smaller bandwidth is chosen; then the number of clusters is close to the number of kernels and thus the mean shift converges very quickly, although this choice is poor as a clustering method. With the same sample size, LSLDG clustering is much faster than the mean shift, as plotted in Figure 3(d). The speedup was brought by reducing the kernel centers, which was shown to significantly improve the computational costs without worsening the clustering performance, as depicted in Figures 3(e) and (f).

5.3   Image Discontinuity Preserving Smoothing and Image Segmentation

The mean shift has been successively applied to image discontinuity preserving smoothing and segmentation tasks [8,24,26]. Here, we investigate the performance of LSLDG clustering in those tasks.

As image data, we use the *Berkeley segmentation dataset* (BSD500) [1].[1] From one image, the information of color (three dimensions) and spatial positions (two dimensions) are extracted per pixel. Thus, the dimensionality of data is five, and the total number of samples is the same as the total number of pixels. As often assumed in the mean shift [8], for image data, we use the following mother basis function:

$$\phi_i(x) = \exp\left(-\frac{\|x^c - c_i^c\|^2}{2\sigma_c^2}\right) \exp\left(-\frac{\|x^s - c_i^s\|^2}{2\sigma_s^2}\right), \tag{7}$$

where x^c and x^s denote the elements for colors and spatial positions in a data vector x, respectively. c_i^c and c_i^s are the Gaussian centers. For the two Gaussian widths σ_c and σ_s, cross-validation is performed as in Section 2.3. In this experiment, we use a reduced image (11 by 16 or 16 by 11 pixels) as the Gaussian

---

[1] http://www.eecs.berkeley.edu/Research/Projects/
CS/vision/grouping/resources.html

Fig. 4. Examples of images after LSLDG clustering. The left-hand figure in each pair is the input image, and the right-hand one is the image after LSLDG clustering.

Table 1. Mean ARI values for 200 images. The numbers in the parentheses are standard deviations. The difference between the methods is statistically significant at level 1% by the t-test.

Mean Shift	LSLDGC
0.10(0.06)	**0.13(0.06)**

centers in (7). For the Gaussian mean shift, (7) is employed as a Gaussian kernel, and the two Gaussian widths are cross-validated based on the likelihood.

Six examples of color images after LSLDG clustering are shown in Figure 4. In the results, some of the segments, such as grass, are cleanly smoothed out, while the edges outlining the objects are preserved. These properties are similar to the results for the mean shift [8].

Next, to clarify the difference from the mean shift, we compare the performance measured by ARI. In this experiment, the input images are reduced to 81 by 121 (or 121 by 81) pixels. Since this benchmark dataset contains several ground truths per image, we simply computed the mean ARI value to all the ground truths.

The ARI values are summarized in Table 1, showing that LSLDG clustering outperforms the original mean shift on image segmentation.

5.4   Performance Comparison to Existing Clustering Methods

Finally, we compare LSLDG clustering to existing clustering methods using accelerometric sensor and speech data.

Table 2. Mean ARI for various methods over 100 runs. The standard deviations are indicated in the parentheses. The best method in terms of the average ARI and methods judged to be comparable to the best one by the t-test at the significance level 1% are described in boldface.

Accelerometry ($d = 5$, $n = 300$, and $c = 3$)

KM	SC	Mean Shift	LSLDGC
0.50(0.03)	0.20(0.26)	0.51(0.05)	**0.61(0.13)**

Speech ($d = 50$, $n = 400$, and $c = 2$)

KM	SC	Mean Shift	LSLDGC
0.00(0.00)	0.00(0.00)	0.00(0.00)	**0.13(0.02)**

For comparison, we employ K-means (KM) [18], spectral clustering (SC) [20,19] with the Gaussian similarity, and Gaussian mean shift. Since the user has to set the number of clusters in advance for KM and SC, we set it at the true number of clusters in each dataset. For the Gaussian mean shift, the Gaussian width is chosen by likelihood cross-validation. For LSLDG, in this experiment, we modify the linear-in-parameter model as

$$g_j(\boldsymbol{x}) = \sum_{i=1}^{n} \theta_i \psi_{i,j}(\boldsymbol{x}) = \boldsymbol{\theta}^\top \boldsymbol{\psi}_j(\boldsymbol{x}).$$

The main difference from the model introduced in Section 2.2 is that the coefficients θ_i do not depend on j, namely, the dimensionality of data. This modification considerably decreases the computational costs to higher dimensional data.

In this experiment, we used the following two datasets, where d denotes the dimensionality of data, n denotes the number of samples, and c denotes the number of true clusters:

1. *Accelerometry* ($d = 5, n = 300$, and $c = 3$). The *ALKAN* dataset[2], which contains 3-axis (i.e., x-, y-, and z-axes) accelerometric data.
2. *Speech* ($d = 50, n = 400$, and $c = 2$). An in-house speech dataset, which contains short utterance samples recorded from 2 male subjects speaking in French with sampling rate 44.1kHz.

The details of the two datasets can be seen in [22]. For each dataset, as preprocessing, the variance was normalized after centering in the element-wise manner.

The experimental results are described in Table 2. For the accelerometry dataset, LSLDG clustering shows the best performance among all the methods in the table. In addition to the superior performance, another advantage is that LSLDG clustering does not include any parameters which have to be manually tuned. On the other hand, KM and SC require the users to fix the number of clusters beforehand, which largely influences the clustering performance. Thus,

---

[2] http://alkan.mns.kyutech.ac.jp/web/data.html

LSLDG clustering would be easier to use in practice. For the speech dataset, LSLDG outperforms the existing clustering methods again (Table 2). Since the dimensionality of the dataset, $d = 50$, is much higher than the accelerometry dataset ($d = 5$), LSLDG seems to perform well on high-dimensional data, while the mean shift does not work well on high-dimensional data, as already indicated in Section 5.2.

6 Conclusions

In this paper, we developed a method to directly estimate the log-density gradient, and constructed a clustering algorithm on it. The proposed log-density gradient estimator can be regarded as a non-parametric extension of score matching [14,21], and the proposed clustering algorithm can be regarded as an extension of the mean shift algorithm [6,8,11]. The key advantage compared to the original mean shift is that the proposed clustering method works well on high-dimensional data for which the mean shift works poorly. Furthermore, we showed experimentally that the proposed method outperforms existing clustering methods.

Acknowledgments. H.S. was supported by KAKENHI 23120004. A.H. was supported by the Academy of Finland CoE program, and the Japanese MIC project "Novel and innovative R&D making use of brain structures". M.S. was supported by KAKENHI 25700022 and AOARD.

References

1. Arbelaez, P., Maire, M., Fowlkes, C., Malik, J.: Contour detection and hierarchical image segmentation. IEEE Transactions on Pattern Analysis and Machine Intelligence 33(5), 898–916 (2011)
2. Basu, A., Harris, I.R., Hjort, N.L., Jones, M.C.: Robust and efficient estimation by minimising a density power divergence. Biometrika 85(3), 549–559 (1998)
3. Bregman, L.M.: The relaxation method of finding the common point of convex sets and its application to the solution of problems in convex programming. USSR Computational Mathematics and Mathematical Physics 7(3), 200–217 (1967)
4. Carreira-Perpiñán, M.Á.: Fast nonparametric clustering with gaussian blurring mean-shift. In: ICML 2006, pp. 153–160. ACM (2006)
5. Carreira-Perpiñán, M.Á.: Gaussian mean-shift is an EM algorithm. IEEE Transactions on Pattern Analysis and Machine Intelligence 29(5), 767–776 (2007)
6. Cheng, Y.: Mean shift, mode seeking, and clustering. IEEE Transactions on Pattern Analysis and Machine Intelligence 17(8), 790–799 (1995)
7. Collins, R.T.: Mean-shift blob tracking through scale space. In: CVPR 2003, vol. 2, pp. 234–240. IEEE (2003)
8. Comaniciu, D., Meer, P.: Mean shift: A robust approach toward feature space analysis. IEEE Transactions on Pattern Analysis and Machine Intelligence 24(5), 603–619 (2002)
9. Comaniciu, D., Ramesh, V., Meer, P.: Real-time tracking of non-rigid objects using mean shift. In: CVPR 2000, vol. 2, pp. 142–149. IEEE (2000)
10. Evgeniou, T., Pontil, M.: Regularized multi-task learning. In: Proceedings of the Tenth ACM SIGKDD International Conference on Knowledge Discovery and Data Mining (KDD 2004), pp. 109–117. ACM (2004)

11. Fukunaga, K., Hostetler, L.: The estimation of the gradient of a density function, with applications in pattern recognition. IEEE Transactions on Information Theory 21(1), 32–40 (1975)
12. Hastie, T., Rosset, S., Tibshirani, R., Zhu, J.: The entire regularization path for the support vector machine. Journal of Machine Learning Research 5, 1391–1415 (2004)
13. Hubert, L., Arabie, P.: Comparing partitions. Journal of Classification 2(1), 193–218 (1985)
14. Hyvärinen, A.: Estimation of non-normalized statistical models by score matching. Journal of Machine Learning Research 6, 695–709 (2005)
15. Hyvärinen, A.: Some extensions of score matching. Computational Statistics & Data Analysis 51(5), 2499–2512 (2007)
16. Kanamori, T., Suzuki, T., Sugiyama, M.: Computational complexity of kernel-based density-ratio estimation: A condition number analysis. Machine Learning 90(3), 431–460 (2013)
17. Kullback, S., Leibler, R.A.: On information and sufficiency. The Annals of Mathematical Statistics 22, 79–86 (1951)
18. MacQueen, J.B.: Some methods for classification and analysis of multivariate observations. In: Proceedings of the 5th Berkeley Symposium on Mathematical Statistics and Probability, vol. 1, pp. 281–297. University of California Press, Berkeley (1967)
19. Ng, A.Y., Jordan, M.I., Weiss, Y.: On spectral clustering: Analysis and an algorithm. In: Dietterich, T.G., Becker, S., Ghahramani, Z. (eds.) NIPS, pp. 849–856. MIT Press, Cambridge (2002)
20. Shi, J., Malik, J.: Normalized cuts and image segmentation. IEEE Transactions on Pattern Analysis and Machine Intelligence 22(8), 888–905 (2000)
21. Sriperumbudur, B., Fukumizu, K., Gretton, A., Hyvärinen, A.: Density estimation in infinite dimensional exponential families. arXiv preprint arXiv:1312.3516 (2013)
22. Sugiyama, M., Niu, G., Yamada, M., Kimura, M., Hachiya, H.: Information-maximization clustering based on squared-loss mutual information. Neural Computation 26(1), 84–131 (2014)
23. Sugiyama, M., Suzuki, T., Kanamori, T.: Density ratio matching under the Bregman divergence: A unified framework of density ratio estimation. Annals of the Institute of Statistical Mathematics 64(5), 1009–1044 (2012)
24. Tao, W., Jin, H., Zhang, Y.: Color image segmentation based on mean shift and normalized cuts. IEEE Transactions on Systems, Man, and Cybernetics, Part B: Cybernetics 37(5), 1382–1389 (2007)
25. Tibshirani, R.: Regression shrinkage and subset selection with the lasso. Journal of the Royal Statistical Society, Series B 58(1), 267–288 (1996)
26. Wang, J., Thiesson, B., Xu, Y., Cohen, M.: Image and video segmentation by anisotropic kernel mean shift. In: Pajdla, T., Matas, J(G.) (eds.) ECCV 2004. LNCS, vol. 3022, pp. 238–249. Springer, Heidelberg (2004)

Local Policy Search in a Convex Space and Conservative Policy Iteration as Boosted Policy Search

Bruno Scherrer[1] and Matthieu Geist[2]

[1] Inria, Villers-lès-Nancy, F-54600, France,
Université de Lorraine, LORIA, UMR 7503, Vandœuvre-lès-Nancy, F-54506, France
[2] Supélec – IMS-MaLIS Research Group & UMI 2958 (GeorgiaTech-CNRS), Metz, France

Abstract. Local Policy Search is a popular reinforcement learning approach for handling large state spaces. Formally, it searches locally in a parameterized policy space in order to maximize the associated value function averaged over some pre-defined distribution. The best one can hope in general from such an approach is to get a local optimum of this criterion. The first contribution of this article is the following surprising result: if the policy space is convex, *any* (approximate) *local optimum* enjoys a *global performance guarantee*. Unfortunately, the *convexity* assumption is strong: it is not satisfied by commonly used parameterizations and designing a parameterization that induces this property seems hard. A natural solution to alleviate this issue consists in deriving an algorithm that solves the local policy search problem using a boosting approach (constrained to the convex hull of the policy space). The resulting algorithm turns out to be a slight generalization of conservative policy iteration; thus, our second contribution is to highlight an original connection between local policy search and approximate dynamic programming.

1 Introduction

We consider the reinforcement learning problem [24] formalized through Markov Decision Processes (MDP) [21], in the situation where the state space is large and approximation is required. On the one hand, Approximate Dynamic Programming (ADP) is a standard approach for handling large state spaces. It consists in mimicking in an approximate form the standard algorithms that were designed to optimize globally the policy (maximizing the associated value function for each state). On the other hand, Local Policy Search (LPS) consists in parameterizing the policy (often called an "actor") and locally maximizing the associated expected value function. This can be done for example using a (natural) gradient ascent [3,10]—possibly with a critic [25,20], expectation-maximization (EM) [12], or even directly using some black-box optimization algorithm [9]. LPS methods work particularly well in practice: the just cited papers describe applications to standard benchmarks and applications such as robotics, that are competitive with the ADP approach. Surprisingly, gradient-based and EM approaches, that are usually prone to be stuck in local optima, do not seem to be penalized in applications to Reinforcement Learning. Even more surprisingly, it was shown [10] that a natural gradient ascent in the policy space can outperform ADP on the Tetris game.

Following the seminal works by [4], it has been shown that ADP algorithms enjoy global performance guarantees, bounding the loss of using the computed policy instead

T. Calders et al. (Eds.): ECML PKDD 2014, Part III, LNCS 8726, pp. 35–50, 2014.

of using the optimal one as a function of the approximation errors involved along the iterations: see [18] for approximate policy iteration (API), [19] for approximate value iteration (AVI), or more generally [22] for approximate modified policy iteration. To the best of our knowledge, similar general guarantees do not exist in the literature for LPS algorithms. In general though, the best one can hope for LPS is to get a local optimum of the optimized fitness (that is, a local maximum of the averaged value function), and the important question of the loss with respect to the optimal policy remains open. As for instance mentioned as the main "future work" in [6], where the convergence of a family of natural actor-critic algorithms is proven, "*[i]t is important to characterize the quality of converged solutions.*" The motivation of this paper is to deepen the understanding on the LPS approach.

Our main contribution (Theorem 3, Section 3) is to show that if the policy space on which one performs LPS is a convex subset of the full space of stochastic policies— equivalently this means that if two policies are taken in the space, then their stochastic mixture also belongs to the space—, then *any (approximate) local optimum of the expected value function enjoys a global performance guarantee*, similar to—actually slightly better than (see Section 5)—the one provided for ADP algorithms. After explaining that designing parameterizations that imply the convexity assumption seems particularly difficult, we will propose in Section 4 an algorithmic solution based on a boosting approach (seen as a functional gradient ascent) that can do LPS in the convex hull of a space of deterministic policies. The algorithm we will then obtain happens to be a slight generalization of the Conservative Policy Iteration algorithm [11] that was originally introduced from an ADP viewpoint. Thus, another contribution of our work amounts to draw an original connexion between ADP and LPS. Section 5 will discuss our analysis; notably, a comparison to similar bounds for ADP is proposed and the practical consequences of our result are discussed. Section 6 opens some perspectives. The next section provides the necessary background and states formally what we mean by local policy search.

2   Background and Notations

Write Δ_X the set of probability distributions over a countable set X and Y^X the applications from X to the set Y. By convention, all vectors are column vectors, except distributions which are row vectors (for left multiplication). We consider a discounted MDP $\mathcal{M} = \{\mathcal{S}, \mathcal{A}, P, r, \gamma\}$ [21,5], with $\mathcal{S}$ the countable state space[1], $\mathcal{A}$ the countable action space, $P \in (\Delta_{\mathcal{S}})^{\mathcal{S} \times \mathcal{A}}$ the Markovian dynamics ($P(s'|s, a)$ denotes the probability of transiting to s' from the (s, a) couple), $r \in \mathbb{R}^{\mathcal{S} \times \mathcal{A}}$ the bounded reward function and $\gamma \in [0, 1)$ the discount factor.

A stochastic policy $\pi \in (\Delta_{\mathcal{A}})^{\mathcal{S}}$ associates to each state s a probability distribution $\pi(.|s)$ over the action space $\mathcal{A}$. We say that a policy space Π that is a subset of $(\Delta_{\mathcal{A}})^{\mathcal{S}}$ is convex (or equivalently stable by stochastic mixture) if it satisfies:

$$\forall \pi, \pi' \in \Pi, \quad \forall \alpha \in (0, 1), \quad (1 - \alpha)\pi + \alpha\pi' \in \Pi.$$

---

[1] These results can easily be extended to the case of non-countable state space and compact action space. We chose the countable space setting for the ease and clarity of exposition.

For a given policy π, we define $r_\pi \in \mathbb{R}^S$ as

$$r_\pi(s) = \sum_{a \in \mathcal{A}} \pi(a|s)r(s,a) = \mathbb{E}_{a \sim \pi(.|s)}[r(s,a)]$$

and $P_\pi \in (\Delta_S)^S$ as

$$P_\pi(s'|s) = \sum_{a \in \mathcal{A}} \pi(a|s)P(s'|s,a) = \mathbb{E}_{a \sim \pi(.|s)}[P(s'|s,a)].$$

The value function v_π quantifies the quality of a policy π for each state s by measuring the expected cumulative reward received for starting in this state and then following the policy:

$$v_\pi(s) = \mathbb{E}\left[\sum_{t \geq 0} \gamma^t r_\pi(s_t)|s_0 = s, s_{t+1} \sim P_\pi(.|s_t)\right].$$

The Bellman operator T_π of policy π associates to each function $v \in \mathbb{R}^S$ the function defined as

$$[T_\pi v](s) = \mathbb{E}\left[r_\pi(s) + \gamma v(s')|s' \sim P_\pi(.|s)\right],$$

or more compactly $T_\pi v = r_\pi + \gamma P_\pi v$. The value function v_π is known to be the unique fixed point of T_π.

It is also well-known that there exists a policy π_* that is optimal in the sense that it satisfies $v_{\pi_*}(s) \geq v_\pi(s)$ for all states s and policies π. The value function v_* is the unique fixed point of the following nonlinear Bellman equation:

$$v_* = Tv_* \text{ with } Tv = \max_{\pi \in \mathcal{A}^S} T_\pi v$$

where the max is taken componentwise. Given any function $v \in \mathbb{R}^S$, we say that a policy π' is greedy with respect to v if $T_{\pi'}v = Tv$, and we write $\mathcal{G}(\pi)$ for the set of policies that are greedy with respect to the value v_π of some policy π. The notions of optimal value function and greedy policies are fundamental to optimal control because of the following property: any policy π_* that is greedy with respect to the optimal value is an optimal policy and its value v_{π_*} is equal to v_*. Therefore, an equivalent characterization of the optimality of some policy π is that it is greedy with respect to its own value:

$$\pi \in \mathcal{G}(\pi). \tag{1}$$

For any distribution μ, we define the γ-weighted occupancy measure[2] induced by the policy π when the initial state is sampled from μ as $d_{\mu,\pi} = (1-\gamma)\mu(I - \gamma P_\pi)^{-1}$ (we recall μ to be a row vector by convention) with $(I - \gamma P_\pi)^{-1} = \sum_{t \geq 0}(\gamma P_\pi)^t$. It can easily be seen that $\mu v_\pi = \frac{1}{1-\gamma}d_{\mu,\pi}r_\pi$. For any two distributions μ and ν, we write $\left\|\frac{\mu}{\nu}\right\|_\infty$ for the smallest constant C satisfying $\mu(s) \leq C\nu(s)$, for any $s \in S$ (this constant is actually the supremum norm of the componentwise ratio, thus the notation).

---

[2] When it exists, this measure tends to the stationary distribution of P_π when the discount factor tends to 1.

From an algorithmic point of view, Dynamic Programming methods compute the optimal value policy pair (v_*, π_*) in an iterative way. When the problem is large and cannot be solved exactly, Approximate Dynamic Programming (ADP) refers to noisy implementations of these exact methods, where the noise is due to approximations at each iteration. For instance, Approximate Value and Policy Iteration respectively correspond to the following schemes:

$$v_{k+1} = Tv_k + \epsilon_k \quad \text{and} \quad \begin{cases} v_k = v_{\pi_k} + \epsilon_k \\ \pi_{k+1} \in \mathcal{G}(v_k) \end{cases}.$$

In the Local Policy Search (LPS) context on which we focus in this paper, we write Π the space where we perform the search. For a predefined distribution ν of interest, the problem addressed by LPS can be cast as follows:

find $\pi \in \Pi$ s.t. π is a local maximum of $J_\nu(\pi) = \mathbb{E}_{s \sim \nu}[v_\pi(s)]$.

Assume that we are able to (approximately) find such a locally optimal policy π. A natural question is: can we say something about the distance between the value of this policy v_π and that of the optimal policy $v_* = v_{\pi_*}$? Quite surprisingly, and in contrast with most optimization problems, we are going to provide a condition on the policy space Π that allows to give a nontrivial performance guarantee; this is the aim of the next section.

3 Main Result

In order to state our main result, we need to define a relaxation of the set of policies that are greedy with respect to some given policy.

Definition 1 (μ-**weighted** ϵ-**greedy policies**). *We write $\mathcal{G}_\Pi(\pi, \mu, \epsilon)$ for the set of policies which are ϵ-greedy respectively to π (in μ-expectation), formally defined as*

$$\mathcal{G}_\Pi(\pi, \mu, \epsilon) = \{\pi' \in \Pi \text{ such that } \forall \pi'' \in \Pi, \ \mu T_{\pi'} v_\pi + \epsilon \geq \mu T_{\pi''} v_\pi\}.$$

This is indeed a relaxation of $\mathcal{G}$, as it can be observed that for all policies π and π',

$$\pi' \in \mathcal{G}(\pi) \quad \Leftrightarrow \quad \forall \mu \in \Delta_S, \ \pi' \in \mathcal{G}_\Pi(\pi, \mu, 0)$$
$$\Leftrightarrow \quad \exists \mu \in \Delta_S, \ \mu > 0, \ \pi' \in \mathcal{G}_\Pi(\pi, \mu, 0).$$

We are now ready to state our first important result.

Theorem 1. *Let π be some policy in Π. The following two properties are equivalent:*

$$\forall \pi' \in \Pi, \quad \lim_{\alpha \to 0} \frac{\nu v_{(1-\alpha)\pi + \alpha \pi'} - \nu v_\pi}{\alpha} \leq \epsilon. \tag{2}$$

$$\pi \in \mathcal{G}_\Pi(\pi, d_{\nu,\pi}, (1-\gamma)\epsilon). \tag{3}$$

Equation (3) says that the policy π is approximately greedy with respect to itself, and can be thus seen as a relaxed version of the optimality Equation (1); as we will show below, this will allow us to provide a global performance guarantee for the policy π. Equation (2) says that π is an approximate local optimum of $\pi \mapsto J_\nu(\pi)$ if π is allowed to move in the convex hull of the policy space Π: indeed, whatever the direction we look at in this space, the slope of the improvement—locally around π—is bounded by ϵ. Theorem 1 thus has the following corollary.

Corollary 1. *Assume that the space Π is convex. Then any policy π that is an ϵ-local optimum of $\pi \mapsto J_\nu(\pi)$ (in the sense of Equation (2)) satisfies the relaxed Bellman Equation (3).*

We now turn to the proof of Theorem 1. The following technical (but simple) lemma will be useful for the proof.

Lemma 1. *For any policies π and π', we have*

$$v_{\pi'} - v_\pi = (I - \gamma P_{\pi'})^{-1}(T_{\pi'} v_\pi - v_\pi).$$

Proof. The proof uses the fact that the linear Bellman Equation $v_\pi = r_\pi + \gamma P_\pi v_\pi$ implies $v_\pi = (I - \gamma P_\pi)^{-1} r_\pi$. Then,

$$
\begin{aligned}
v_{\pi'} - v_\pi &= (I - \gamma P_{\pi'})^{-1} r_{\pi'} - v_\pi \\
&= (I - \gamma P_{\pi'})^{-1}(r_{\pi'} + \gamma P_{\pi'} v_\pi - v_\pi) \\
&= (I - \gamma P_{\pi'})^{-1}(T_{\pi'} v_\pi - v_\pi).
\end{aligned}
$$
$\square$

Proof (Proof of Theorem 1). For any α and any $\pi' \in \Pi$, write $\pi_\alpha = (1 - \alpha)\pi + \alpha\pi'$. Using Lemma 1, we have:

$$\nu(v_{\pi_\alpha} - v_\pi) = \nu(I - \gamma P_{\pi_\alpha})^{-1}(T_{\pi_\alpha} v_\pi - v_\pi).$$

By observing that $r_{\pi_\alpha} = (1 - \alpha)r_\pi + \alpha r_{\pi'}$ and $P_{\pi_\alpha} = (1 - \alpha)P_\pi + \alpha P_{\pi'}$, it can be seen that $T_{\pi_\alpha} v_\pi = (1 - \alpha)T_\pi v_\pi + \alpha T_{\pi'} v_\pi$. Thus, using the fact that $v_\pi = T_\pi v_\pi$, we get:

$$
\begin{aligned}
T_{\pi_\alpha} v_\pi - v_\pi &= (1 - \alpha)T_\pi v_\pi + \alpha T_{\pi'} v_\pi - v_\pi \\
&= \alpha(T_{\pi'} v_\pi - v_\pi).
\end{aligned}
$$

In parallel, we have

$$
\begin{aligned}
(I - \gamma P_{\pi_\alpha})^{-1} &= (I - \gamma P_\pi + \alpha\gamma(P_\pi - P_{\pi'}))^{-1} \\
&= (I - \gamma P_\pi)^{-1}(I + \alpha M),
\end{aligned}
$$

where M is bounded (the exact form of the matrix M does not matter). Put together, we obtain

$$\nu(v_{\pi_\alpha} - v_\pi) = \alpha\nu(I - \gamma P_\pi)^{-1}(T_{\pi'} v_\pi - v_\pi) + O(\alpha^2).$$

Taking the limit, we obtain

$$\lim_{\alpha \to 0} \frac{\nu(v_{\pi_\alpha} - v_\pi)}{\alpha} = \nu(I - \gamma P_\pi)^{-1}(T_{\pi'}v_\pi - v_\pi)$$

$$= \frac{1}{1-\gamma}d_{\nu,\pi}(T_{\pi'}v_\pi - v_\pi),$$

and the result follows.

A second important step in our analysis consists in showing that a relaxed optimality characterization as the one of Equation (3) implies a global performance guarantee. To state this result, we first need to define the "ν-greedy-complexity" of our policy space, which measures how good Π was designed so as to approximate the greedy operator, for a starting distribution ν.

Definition 2 (ν-**greedy-complexity**). *We define $\mathcal{E}_\nu(\Pi)$ the ν-greedy-complexity of the policy space Π as*

$$\mathcal{E}_\nu(\Pi) = \max_{\pi \in \Pi} \min_{\pi' \in \Pi} (d_{\nu,\pi}(Tv_\pi - T_{\pi'}v_\pi)).$$

Since $Tv_\pi - T_\pi v_\pi = Tv_\pi - v_\pi \geq 0$, we have $\mathcal{E}_\nu(\Pi) \geq 0$ for any policy space Π. In the limit case where Π contains all (deterministic) policies, we have $\mathcal{E}_\nu(\Pi) = 0$.

Given this definition, we are ready to state our second important result.

Theorem 2. *If $\pi \in \mathcal{G}_\Pi(\pi, d_{\nu,\pi}, \epsilon)$, then for any policy π' and for any distribution μ over $\mathcal{S}$, we have*

$$\mu v_{\pi'} \leq \mu v_\pi + \frac{1}{(1-\gamma)^2}\left\|\frac{d_{\mu,\pi'}}{\nu}\right\|_\infty (\mathcal{E}_\nu(\Pi) + \epsilon).$$

Notice that this theorem is actually a slight[3] generalization of Theorem 6.2 of [11]. We provide the proof for the sake of completeness.

Proof. Using again Lemma 1 and the fact that $Tv_\pi \geq T_{\pi'}v_\pi$, we have

$$\mu(v_{\pi'} - v_\pi) = \mu(I - \gamma P_{\pi'})^{-1}(T_{\pi'}v_\pi - v_\pi)$$

$$= \frac{1}{1-\gamma}d_{\mu,\pi'}(T_{\pi'}v_\pi - v_\pi) \leq \frac{1}{1-\gamma}d_{\mu,\pi'}(Tv_\pi - v_\pi).$$

Since $Tv_\pi - v_\pi \geq 0$ and $d_{\nu,\pi} \geq (1-\gamma)\nu$, we get

$$\mu(v_{\pi'} - v_\pi) \leq \frac{1}{1-\gamma}\left\|\frac{d_{\mu,\pi'}}{\nu}\right\|_\infty \nu(Tv_\pi - v_\pi)$$

$$\leq \frac{1}{(1-\gamma)^2}\left\|\frac{d_{\mu,\pi'}}{\nu}\right\|_\infty d_{\nu,\pi}(Tv_\pi - v_\pi).$$

---

[3] Theorem 2 holds for any policy π', not only for the optimal one, and the error term is split up (which is necessary to provide a more general result).

Using $d_{\nu,\pi}(Tv_\pi - v_\pi) = (d_{\nu,\pi}Tv_\pi - d_{\nu,\pi}v_\pi)$, we get

$$\mu(v_{\pi'} - v_\pi) \leq \frac{1}{(1-\gamma)^2} \left\| \frac{d_{\mu,\pi'}}{\nu} \right\|_\infty \times$$

$$\left(d_{\nu,\pi}Tv_\pi - \max_{\pi' \in \Pi} d_{\nu,\pi}T_{\pi'}v_\pi + \max_{\pi' \in \Pi} d_{\nu,\pi}T_{\pi'}v_\pi - d_{\nu,\pi}v_\pi \right)$$

$$\leq \frac{1}{(1-\gamma)^2} \left\| \frac{d_{\mu,\pi'}}{\nu} \right\|_\infty (\mathcal{E}_\nu(\Pi) + \epsilon). \qquad \square$$

The first main result of the paper is a straightforward combination of Corollary 1 and Theorem 2.

Theorem 3. *Assume that the space Π is convex. Then any policy π that is an ϵ-local optimum of $\pi \mapsto J_\nu(\pi)$ (in the sense of Equation (2)) enjoys the following global performance guarantee:*

$$\mathbb{E}_{s \sim \mu}[v_*(s) - v_\pi(s)] \leq \frac{1}{1-\gamma} \left\| \frac{d_{\mu,\pi_*}}{\nu} \right\|_\infty \left(\frac{\mathcal{E}_\nu(\Pi)}{1-\gamma} + \epsilon \right).$$

4 About the Convex Policy Space Assumption

The remarkable result of the previous section—a connection between *local* optimality and *global* guarantee—relies on the assumption that the policy space Π is convex. Though this assumption may look mild at first sight, we are going to argue that it is in fact strong. We will then propose a natural algorithmic approach for performing Local Policy Search on the convex hull of some (not necessarily convex) policy space Π.

4.1 A Strong Assumption

A common approach (for continuous actions mainly) is to parameterize a mapping from state to actions and to put it as the mean of a Gaussian distribution, that is

$$\pi_\theta(a|s) \propto \exp\left(-\frac{1}{2} \|a - u_\theta(s)\|_{\Sigma^{-1}}^2 \right),$$

with here u_θ the parameterized state to action mapping and Σ a predefined covariance matrix. Obviously, the space of such policies is not convex, since a mixture of Gaussian distributions is in general not a Gaussian distribution. Another common approach (for discrete actions) is to adopt a parameterized Gibbs distribution, that is

$$\pi_\theta(a|s) \propto \exp\left(\theta^\top \psi(s, a) \right),$$

where $\theta^\top \psi(s, a)$ can be seen as a parameterized state-action or score function. Here again, the resulting policy space is not convex in general.

In fact, we consider that it is an open problem to design a non-trivial parameterization that defines a convex policy space (by non-trivial, we mean a space that is neither simply a convex combination of a *small* number of policies nor the *full* convex hull of $\mathcal{A}^S$). Even in a one-state situation, the answer does not seem obvious: this requires to find distributions that are stable by mixture and we did not manage to find any satisfying solution. An alternative approach, that we develop next, is to consider for Π the convex hull of a set of parameterized policies.

4.2 Boosting

Let $\mathcal{P}$ be a space of policies and $\Pi = \mathrm{co}(\mathcal{P})$ denote its convex hull. We propose to use boosting for finding a local maximum of $J_\nu(\pi)$ on Π. More precisely, we propose to apply the AnyBoost.L1 algorithm [17]: it sees boosting as a gradient ascent in function space and constrains the search in the convex hull of the base policy space. Let $\nabla J_\nu(\pi)$ be the functional gradient (according to π) of the LPS objective function. Applied to our problem, AnyBoost.L1 works as follows. At iteration k, we have a policy π_{k-1}, and perform the following steps:

1. compute $h_k \in \mathrm{argmax}_{h \in \mathcal{P}} \langle \nabla J_\nu(\pi_{k-1}), h \rangle$,
2. update the policy: $\pi_k = (1 - \alpha_k)\pi_{k-1} + \alpha_k h_k$, with $\alpha_k \in (0, 1)$ the learning rate.

The basic idea is to perform a functional gradient ascent on $J_\nu(\pi)$. However, the gradient $\nabla J_\nu(\pi_{k-1})$ does not generally belong to $\mathcal{P}$, so we search for a policy h with greatest inner product with $\nabla J_\nu(\pi_{k-1})$. This corresponds to the first step. The second step updates the policy as a mixture of the old one and of the computed h_k, the mixture weight α_k being the learning rate of the gradient ascent. In order to obtain a more practical algorithm, one has to rephrase the optimization problem of the first step.

Proposition 1. *We have that*

$$\mathop{\mathrm{argmax}}_{h \in \mathcal{P}} \langle \nabla J(\pi), h \rangle = \mathop{\mathrm{argmin}}_{h \in \mathcal{P}} d_{\nu,\pi}(Tv_\pi - T_h v_\pi).$$

In particular, assume that $\mathcal{P}$ is a space of deterministic policies and define $q_\pi = T_a v_\pi$ the state-action value function of a policy π (writing with a slight abuse of notation T_a the Bellman operator for the policy associating action a to any state), then

$$\mathop{\mathrm{argmax}}_{h \in \mathcal{P}} \langle \nabla J(\pi), h \rangle = \mathop{\mathrm{argmin}}_{h \in \mathcal{P}} \sum_{s \in S} d_{\nu,\pi}(s) \left(\max_{a \in A} q_\pi(s, a) - q_\pi(s, h(s)) \right).$$

This process can be seen as an approximate version of the greedy step of the Policy Iteration algorithm and may be implemented through a weighted classification problem, or through an ℓ_p-regression of the q_π function.

Proof (Proof of Proposition 1). The functional gradient of J_ν is

$$\nabla J_\nu(\pi) = \frac{1}{1-\gamma} \sum_{s \in S} d_{\nu,\pi}(s) \sum_{a \in A} \nabla \pi(a|s) q_\pi(s, a).$$

This is a rather direct extension of the classic policy gradient theorem [25]. Then, we need to compute its inner product with a function h of $\mathcal{P}$:

$$\langle \nabla J(\pi), h \rangle = \frac{1}{1-\gamma} \langle \sum_s d_{\nu,\pi}(s) \sum_a \nabla \pi(a|s) q_\pi(s, a), h \rangle$$

$$= \frac{1}{1-\gamma} \sum_s d_{\nu,\pi}(s) \sum_a \langle \nabla \pi(a|s), h \rangle q_\pi(s, a)$$

$$= \frac{1}{1-\gamma} \sum_s d_{\nu,\pi}(s) \sum_a h(a|s) q_\pi(s, a)$$

$$= \frac{1}{1-\gamma} d_{\nu,\pi}(T_h v_\pi).$$

Eventually, this allows concluding:

$$\underset{h \in \mathcal{H}}{\operatorname{argmax}} \langle \nabla J, h \rangle = \underset{h \in \mathcal{H}}{\operatorname{argmax}} \, d_{\nu,\pi}(T_h v_\pi)$$

$$= \underset{h \in \mathcal{H}}{\operatorname{argmin}} \, d_{\nu,\pi}(T v_\pi - T_h v_\pi). \qquad \Box$$

4.3 Connection to CPI

Thus, the boosting approach to LPS consists in computing a mixture of policies, each new component of the mixture being the solution of an approximation of the greedy policy respectively to the preceding estimated mixture. It turns out that Conservative Policy Iteration (CPI) [11] is a specific case of this general algorithm, the only difference being that CPI chooses specific values for the learning rate (such as guaranteeing improvements).

If the algorithm resulting from this boosting approach is not really new, it provides some clarifications about LPS, API and CPI. First, this shows that CPI can be derived as an LPS approach, whereas it was originally derived from an API viewpoint, with the desire to fix the potential policy degradation problem of API [11]. This draws a connection between API and LPS that has not yet been documented in the literature, and highlights the fact that CPI is at the frontier of these two approaches. Second, it provides some leads of improvement for CPI (which has strong guarantees but is in general slow).   One could also choose the learning rates according to the boosting optimization theory, or use related heuristics or even some line search. Last but not least, AnyBoost.L1 is perhaps the more natural way to search for a local maximum of J_ν on a convex policy space. Looking for alternative algorithms performing LPS in convex policy spaces is an interesting research direction.

5 Discussion

In this section, we discuss the relations of our analyses with previous works, we compare this guarantee with the standard ones of approximate dynamic programming (focusing particularly on the API algorithm) and we discuss some practical and theoretical consequences of our analysis.

5.1 Closely Related Analysis

A performance guarantee very similar to the one we provide in Theorem 2 was first derived for CPI by [11]. This result of the literature was certainly considered specific to the CPI algorithm, that has unfortunately not been used widely in practice probably because of its somewhat complex implementation. In contrast, we show in this paper that such a performance guarantee is valid for any method that finds a policy that satisfies a relaxed Bellman identity like that given Equation (3), among which CPI naturally arises, as shown in section 4.

Though the main result of our paper is Theorem 3, and since Theorem 2 appears in a very close form in [11], our main technical contribution is Theorem 1 that highlights a deep connection between local optimality and a relaxed Bellman optimality characterization. A result, that is similar in flavor, is derived by [10] for the Natural Policy Gradient algorithm: Theorem 3 there shows that natural gradient updates are moving the policy towards the solution of a (DP) update. The author writes: *"The natural gradient could be efficient far from the maximum, in that it is pushing the policy toward choosing greedy optimal actions"*. Though there is an obvious connection with our work, the result there is limited since—similarly to the work we have just mentioned on CPI—*(i)* it seems to be specific to the natural gradient approach (though our result is general), and *(ii)* it is not exploited so as to connect with a global performance guarantee.

5.2   Relations to Bounds of Approximate Dynamic Programming

The performance guarantee of any approximate dynamic programming algorithm implies *(i)* a (quadratic) dependency on the average horizon $\frac{1}{1-\gamma}$, *(ii)* a concentration coefficient (which quantifies the divergence between the worst discounted average future state distribution when starting from the measure of interest, and the distribution used to control the estimation errors), and *(iii)* an error term linked to the estimation error encountered at each iteration (which can be due to the approximation of value functions and/or policies). Depending on what quantity is estimated, a comparison of these estimation errors may be hard. To ease the comparison, the following discussion focuses on the API algorithm. Note however that several aspects of our comparison holds for other ADP algorithms.

API generates a sequence of policies: at each iteration, a new policy is one that is approximately greedy with respect to the value of the previous policy. This can be achieved through an ℓ_p-regression of the state-action value function [5,18,13] or through a weighted classification problem [14,7,16]. Whatever the approach, the sequence of policies belongs (implicitly for ℓ_p-regression or explicitly for classification) to some space $\mathcal{P}$ that is typically a set of *deterministic policies*. For an initial policy π_0 and a given distribution ν, the API algorithm iterates as follows:

$$\text{pick } \pi_{k+1} \in \mathcal{P}$$
$$\text{such as (approximately) minimizing } \nu(Tv_{\pi_k} - T_{\pi_{k+1}}v_{\pi_k}).$$

This is similar to CPI/boosted LPS, up to the fact that *(i)* it uses ν instead of $d_{\nu,\pi}$ to approximate the greedy policy and *(ii)* it is optimistic (in the sense that $\alpha_k = 1$). To provide the API bound, we need an alternative concentration coefficient as well as some new error characterizing the quality of the space $\mathcal{P}$. Let $C_{\mu,\nu}$ be the concentration coefficient defined as

$$C_{\mu,\nu} = (1-\gamma)^2 \sum_{i=0}^{\infty} \sum_{j=0}^{\infty} \gamma^{i+j} \sup_{\pi \in \mathcal{A}^S} \left\| \frac{\mu(P_{\pi_*})^i (P_\pi)^j}{\nu} \right\|_\infty.$$

Consider the measure of the complexity of the policy space $\mathcal{P}$, similar to $\mathcal{E}_\nu$:

$$\mathcal{E}'_\nu(\mathcal{P}) = \max_{\pi \in \mathcal{P}} \min_{\pi' \in \mathcal{P}} (\nu(Tv_\pi - T_{\pi'}v_\pi)).$$

Let also e be an estimation error term that tends to zero as the number of samples tends to infinity (at a rate depending on the chosen approximator). The performance guarantee of API [18,1,15,16,8] can be expressed as follows:

$$\limsup_{k \to \infty} \mu(v^* - v_{\pi_k}) \le \frac{C_{\mu,\nu}}{(1-\gamma)^2}(\mathcal{E}'_\nu(\mathcal{P}) + e).$$

This bound is to be compared with the result of Theorem 3, regarding the three terms involved: the average horizon, the concentration coefficient and the greedy error term. Each term is discussed now, a brief summary being provided in Table 1. As said in Section 5.1, the LPS bound is really similar to the CPI one, and the bounds of CPI and a specific instance of API have been compared by [8]. Our discussion can be seen as complementary: we consider API more generally, we provide some new elements of comparison, and we illustrate the methods empirically.

Table 1. Comparison of the performance guarantees for LPS and API

	bounded term	horizon term	concentration term	error term
LPS	$\mu(v_* - v_\pi)$	$\frac{1}{(1-\gamma)^2}$	$\left\|\frac{d_{\mu,\pi_*}}{\nu}\right\|_\infty$	$\mathcal{E}_\nu(\Pi) + \epsilon(1-\gamma)$
API	$\limsup_{k \to \infty} \mu(v^* - v_{\pi_k})$	$\frac{1}{(1-\gamma)^2}$	$C_{\mu,\nu}$	$\mathcal{E}'_\nu(\mathcal{P}) + e$

Horizon Term. Both bounds have a quadratic dependency on the average horizon $\frac{1}{1-\gamma}$. For approximate dynamic programming, this bound can be shown to be tight [23], the only known solution to improve this being to introduce non-stationary policies [23]. The tightness of this bound for policy search is an open question. However, we suggest later in Section 5.3 a possible way to improve on this.

Concentration Coefficients. Both bounds involve a concentration coefficient. They can be compared as follows.

Theorem 4. *We always have that:* $\left\|\frac{d_{\mu,\pi_*}}{\nu}\right\|_\infty \le \frac{1}{1-\gamma}C_{\mu,\nu}$. *Also, if there always exists a ν such that* $\left\|\frac{d_{\mu,\pi_*}}{\nu}\right\|_\infty < \infty$ *(by choosing $\nu = d_{\mu,\pi_*}$), there might not exist a ν such that $C_{\mu,\nu} < \infty$.*

Proof. Consider the inequality of the first part. By using the definition of d_{μ,π_*} and eventually the fact that $d_{\mu,\pi_*} \ge (1-\gamma)\nu$, we have

$$C_{\mu,\nu} = (1-\gamma)^2 \sum_{i=0}^{\infty} \sum_{j=0}^{\infty} \gamma^{i+j} \sup_{\pi \in (\Delta_A)^S} \left\|\frac{\mu(P_{\pi_*})^i(P_\pi)^j}{\nu}\right\|_\infty$$

$$\ge (1-\gamma)^2 \left\|\sum_{i,j=0}^{\infty} \gamma^{i+j} \frac{\mu(P_{\pi_*})^{i+j}}{\nu}\right\|_\infty = (1-\gamma) \left\|\sum_{i=0}^{\infty} \gamma^i \frac{d_{\mu,\pi_*}(P_{\pi_*})^i}{\nu}\right\|_\infty$$

$$\ge (1-\gamma)^2 \left\|\sum_{i=0}^{\infty} \gamma^i \frac{\mu(P_{\pi_*})^i}{\nu}\right\|_\infty = (1-\gamma) \left\|\frac{d_{\mu,\pi_*}}{\nu}\right\|_\infty.$$

Let us concentrate on the second part. Consider an MDP with N states and N actions, with $\mu = \delta_1$ being a dirac on the first state, and such that from here action $a \in [1; N]$ leads in state a deterministically. Write $c = \sup_{\pi \in \mathcal{A}^S} \left\| \frac{\mu P_\pi}{\nu} \right\|_\infty$ the first term defining $C_{\mu,\nu}$. For any π, we have $\mu P_\pi \leq c\nu$. Thus, for any action a we have $\delta_a \leq c\nu \Rightarrow 1 \leq c\nu(a)$. Consequently, $1 = \sum_{i=1}^N \nu(i) \geq \frac{1}{c} \sum_{i=1}^N 1 \Leftrightarrow c \geq N$. This being true for arbitrary $N \in \mathbb{N}$, we get $c = \infty$ and thus $C_{\mu,\nu} = \infty$. $\qquad\square$

The second part of this result tells that we may have $\left\| \frac{d_{\mu,\pi_*}}{\nu} \right\|_\infty \ll C_{\mu,\nu}$, which is clearly in favor of LPS (and CPI, which involves the same concentration as LPS).

Error Terms. Both bounds involve an error term. The terms ϵ (LPS) and e (API) can be made arbitrarily small by increasing the computational effort (the time devoted to run the algorithm and the amount of samples used), though nothing more can be said in general without studying a specific algorithmic instance (*e.g.*, type of local search for LPS or type of regressor/classifier for API). The terms defining the "greedy complexity" of policy spaces can be partially compared. Because they use different distributions that can be compared ($d_{\nu,\pi} \geq (1-\gamma)\nu$), we have for all policy spaces Π [8],

$$\mathcal{E}'_\nu(\Pi) \leq \frac{\mathcal{E}_\nu(\Pi)}{1-\gamma}.$$

However, this result does not take into account the fact that LPS (or CPI for the discussion of [8]) works with *stochastic policies* while API works with *deterministic policies*. This make these terms not comparable in general.

Experiments. To get a more precise picture of the relative practical performance of API and LPS, we ran both algorithms on many randomly generated MDPs. In order to assess their quality, we consider finite problems where the exact value function can be computed. More precisely, we consider Garnet problems first introduced by [2], which are a class of randomly constructed finite MDPs. They do not correspond to any specific application, but are totally abstract while remaining representative of the kind of MDP that might be encountered in practice. In our experiments, a Garnet is parameterized by 4 parameters and is written $G(n_S, n_A, b, p)$: n_S is the number of states, n_A is the number of actions, b is a branching factor specifying how many possible next states are possible for each state-action pair (b states are chosen uniformly at random and transition probabilities are set by sampling uniform random $b - 1$ cut points between 0 and 1) and p is the number of features (for linear value function approximation). The reward is state-dependent: for a given randomly generated Garnet problem, the reward for each state is uniformly sampled between 0 and 1. Features are chosen randomly: Φ is a $n_S \times p$ feature matrix of which each component is randomly and uniformly sampled between 0 and 1. The discount factor γ is set to 0.99 in all experiments.

The algorithms API and LPS need to repeatedly compute $\mathcal{G}_\Pi$. In other words, they must be able to make calls to an approximate greedy operator applied to the value v_π of some policy π for some distribution ν or $d_{\nu,\pi}$. To implement this operator, we compute a noisy estimate of the value v_π with a uniform white noise $u(\iota)$ of amplitude ι, then project this estimate onto $\mathcal{H}$, the space spanned by the p chosen features, with respect to the μ-quadratic norm (projection that we write $\Pi_{\mathcal{H},\mu}$), and then applies the (exact) greedy operator on this projected estimate. In a nutshell, one call to the approximate greedy operator $\mathcal{G}_\Pi(\pi, \mu, \epsilon)$ amounts to compute $\mathcal{G}(\Pi_{\mathcal{H},\mu}(v_\pi + u(\iota)))$, with $\mu = \nu$ (API) or $\mu = d_{\nu,\pi}$ (LPS).

In our experiments, we consider Garnet problems with $n_s \in \{50, 100, 200\}$ states, with $n_a \in \{2, 5\}$ actions, and branching factors in $b \in \{1, 2, 10\}$. For each of the 2×3^2 resulting possible combinations, we generated 30 i.i.d. random MDPs $(M_i)_{1 \le i \le 30}$. For each such MDP M_i, we make 30 i.i.d. runs of *(i)* API and *(ii)* LPS with a gradient step-size of 0.1. For each run and algorithm, we compute the distance between the value of the output policy and that of the optimal policy $(\Delta_j)_{1 \le j \le 30}$. Figure 1 displays learning curves with statistics on these random variables. On this large set of problems, LPS

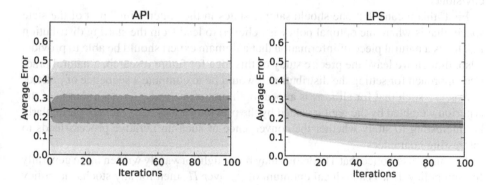

Fig. 1. Learning curves for API and LPS. MDPs are i.i.d. with the distribution of M_1. Conditioned on an MDP M_i, the error measures are i.i.d. with the distribution of Δ_1. The central line is an estimate of the overall average error $E[\Delta_1]$. The three grey regions (from dark to light) are estimates of the variability (across MDPs) of the average error $Std[E[\Delta_1|M_1]]$, the average (across MDPs) of the standard deviation of the error $E[Std[\Delta_1|M_1]]$, and the variability (across MDPs) of the standard deviation of the error $Std[Std[\Delta_1|M_1]]$.

significantly outperforms API, both on average and in terms of variability (across runs and problems). This confirms the importance of the better concentration coefficient of LPS, since it is in theory the main advantage of LPS over API.

5.3 Practical and Theoretical Consequences of Our Analysis

Finally, this section provides a few important consequences of our analysis and of Theorem 3 in particular.

Rich Policy and Equivalence between Local and Global Optimality. If the policy space is very rich, one can easily show that any local optimum is actually global (this result being a direct corollary of Theorem 3).

Theorem 5. *Let $\nu > 0$ be a distribution. Assume that the policy space is rich in the sense that $\mathcal{E}_\nu(\Pi) = 0$, and that π is an (exact) local optimum of J_ν ($\epsilon = 0$). Then, we have $v_\pi = v_*$.*

If this result is well-known in the case of tabular policies, it is to our knowledge new in such a general case (acknowledging that $\mathcal{E}_\nu(\Pi) = 0$ is a rather strong assumption).

Choice of the Sampling Distribution. Provided the result of Theorem 3, and as also mentioned about CPI by [11] since it satisfies a similar bound, if one wants to optimize the policy according to a distribution μ (that is, such that $\mu(v_* - v_\pi)$ is small), then one should optimize the fitness J_ν with the distribution $\nu \simeq d_{\mu,\pi_*}$ (so as to minimize the coefficient $\left\| \frac{d_{\mu,\pi_*}}{\nu} \right\|_\infty$). Ideally, one should sample states based on trajectories following the optimal policy π_* starting from states drawn according to μ. This is in general not realistic since we do not know the optimal policy π_*, but practical solutions may be envisioned.

First, this means that one should sample states in the "interesting" part of the state space, that is where the optimal policy is believed to lead from the starting distribution μ. This is a natural piece of information that a domain expert should be able to provide. Also, though we leave the precise study of this idea for future research, a natural practical approach for setting the distribution ν would be to compute a sequence of policies $\pi_1, \pi_2, \ldots$ such that for all i, π_i is a local optimum of $\pi \mapsto J_{d_{\nu,\pi_{i-1}}}(\pi)$, that is of the criterion weighted by the region visited by the previous policy π_{i-1}. It may particularly be interesting to study whether the convergence of such an iterative process leads to interesting guarantees.

One may also notice that Theorem 3 may be straightforwardly written more generally for any policy. If π is an ϵ-local optimum of J_ν over Π, then for any stochastic policy π' we have

$$\mu v_{\pi'} \leq \mu v_\pi + \frac{1}{1-\gamma} \left\| \frac{d_{\mu,\pi'}}{\nu} \right\|_\infty \left(\frac{\mathcal{E}_\nu(\Pi)}{1-\gamma} + \epsilon \right).$$

Therefore, one can sample trajectories according to an acceptable (and known) controller π' so as to get state samples to optimize $J_{d_{\nu,\pi'}}$. More generally, if we know where a good policy π' leads the system to from some initial distribution μ, we can learn a policy π that is guaranteed to be approximately as good (and potentially better).

A Better Learning Problem? With the result of Theorem 3, we have a squared dependency of the bound on the effective average horizon $\frac{1}{1-\gamma}$. For approximate dynamic programming, it is known that this dependency is tight [5,23]. At the current time, this is an open question for policy search. However, we can improve the bound. We have shown that the ϵ-local optimality of a policy π implies that it satisfies a relaxed Bellman global optimality characterization, $\pi \in \mathcal{G}_\Pi(\pi, d_{\nu,\pi}, \epsilon)$, which in turns implies Theorem 3. The following result, involving a slightly simpler relaxed Bellman equation, can be proved similarly to Theorem 2:

If $\pi \in \mathcal{G}_\Pi(\pi, \nu, \epsilon)$ then $\mu v_{\pi'} \leq \mu v_\pi + \frac{1}{1-\gamma} \left\| \frac{d_{\mu,\pi'}}{\nu} \right\|_\infty (\mathcal{E}_\nu(\Pi) + \epsilon).$

A policy satisfying this relaxed Bellman equation would have an improved dependency on the horizon ($\frac{1}{1-\gamma}$ instead of $\frac{1}{(1-\gamma)^2}$). At the current time, we do not know whether there exists an efficient algorithm for computing a policy satisfying $\pi \in \mathcal{G}_\Pi(\pi, \nu, \epsilon)$. The above guarantee suggests that solving such a problem may improve over traditional policy search and approximate dynamic programming approaches.

6 Conclusion

In the past years, local policy search algorithms have been shown to be practical viable alternatives to the more traditional approximate dynamic programming field. The derivation of global performance guarantees for such approaches, probably considered as a desperate case, was to our knowledge never considered in the literature. In this article, we have shown a surprising result: *any Local Policy Search algorithm*, as long as it is able to *provide an approximate local optimum* of $J_\nu(\pi)$, *enjoys a global performance guarantee* similar to the ones of approximate dynamic programming algorithms. However, this relies on a strong convex policy space assumption, not satisfied by most standard local policy search algorithms. Weakening this hypothesis is an interesting research direction (yet difficult, as convexity is at the core of our analysis).

In order to handle this issue, we proposed to apply AnyBoost.L1 to local policy search. If it is a slight generalization of conservative policy iteration and is thus not a new algorithm, our work provides an original connexion between local policy search, boosting and approximate dynamic programming. Moreover, this suggests some open problems. First, AnyBoost.L1 (and thus CPI) is a rather natural approach to handle convex policy spaces. An interesting alternative would be to study the question of the parameterization of a convex space. If we were able to come up with a non-trivial parameterization, we could use many of the LPS algorithms of the literature (for instance actor-critic algorithms). Our analysis also suggests that it may be better to design algorithms that looks for a policy π satisfying $\pi \in \mathcal{G}_\Pi(\pi, \nu, \epsilon)$ instead of searching for a local maximum of J_ν, as it leads to a better bound (linear dependency on the average horizon). Working in that direction constitutes interesting future research. Last but not least, our experiments on Garnet problems showed that LPS outperforms API. Deepening the comparison of these approaches in larger problems constitutes natural future work.

References

1. Antos, A., Szepesvari, C., Munos, R.: Learning near-optimal policies with Bellman-residual minimization based fitted policy iteration and a single sample path. Machine Learning Journal 71, 89–129 (2008)
2. Archibald, T., McKinnon, K., Thomas, L.: On the Generation of Markov Decision Processes. Journal of the Operational Research Society 46, 354–361 (1995)
3. Baxter, J., Bartlett, P.L.: Infinite-horizon gradient-based policy search. Journal of Artificial Intelligence Research (JAIR) 15, 319–350 (2001)
4. Bertsekas, D., Tsitsiklis, J.: Neuro-Dynamic Programming. Athena Scientific (1996)
5. Bertsekas, D.P.: Dynamic Programming and Optimal Control. Athena Scientific (1995)
6. Bhatnagar, S., Sutton, R.S., Ghavamzadeh, M., Lee, M.: Incremental natural actor-critic algorithms. In: Advances in Neural Information Processing Systems, NIPS (2007)
7. Fern, A., Yoon, S., Givan, R.: Approximate Policy Iteration with a Policy Language Bias: Solving Relational Markov Decision Processes. Journal of Artificial Intelligence Research (JAIR) 25, 75–118 (2006)
8. Ghavamzadeh, M., Lazaric, A.: Conservative and Greedy Approaches to Classification-based Policy Iteration. In: Conference on Artificial Intelligence, AAAI (2012)

9. Heidrich-Meisner, V., Igel, C.: Evolution strategies for direct policy search. In: Rudolph, G., Jansen, T., Lucas, S., Poloni, C., Beume, N. (eds.) PPSN 2008. LNCS, vol. 5199, pp. 428–437. Springer, Heidelberg (2008)
10. Kakade, S.: A Natural Policy Gradient. In: Advances in Neural Information Processing Systems, NIPS (2001)
11. Kakade, S., Langford, J.: Approximately optimal approximate reinforcement learning. In: International Conference on Machine Learning, ICML (2002)
12. Kober, J., Peters, J.: Policy Search for Motor Primitives in Robotics. Machine Learning pp. 171–203 (2011)
13. Lagoudakis, M., Parr, R.: Least-squares policy iteration. Journal of Machine Learning Research (JMLR) 4, 1107–1149 (2003)
14. Lagoudakis, M., Parr, R.: Reinforcement learning as classification: Leveraging modern classifiers. In: International Conference on Machine Learning, ICML (2003)
15. Lazaric, A., Ghavamzadeh, M., Munos, R.: Finite-sample analysis of least-squares policy iteration. Journal of Machine Learning Research 13, 3041–3074 (2011)
16. Lazaric, A., Ghavamzadeh, M., Munos, R.: Analysis of a classification-based policy iteration algorithm. In: International Conference on Machine Learning, ICML (2010)
17. Mason, L., Baxter, J., Bartlett, P., Frean, M.: Boosting algorithms as gradient descent in function space. Tech. rep., Australian National University (1999)
18. Munos, R.: Error bounds for approximate policy iteration. In: International Conference on Machine Learning, ICML (2003)
19. Munos, R.: Performance bounds in Lp norm for approximate value iteration. SIAM Journal on Control and Optimization (2007)
20. Peters, J., Schaal, S.: Natural Actor-Critic. Neurocomputing 71, 1180–1190 (2008)
21. Puterman, M.L.: Markov Decision Processes: Discrete Stochastic Dynamic Programming. Wiley-Interscience (1994)
22. Scherrer, B., Gabillon, V., Ghavamzadeh, M., Geist, M.: Approximate Modified Policy Iteration. In: International Conference on Machine Learning, ICML (2012)
23. Scherrer, B., Lesner, B.: On the Use of Non-Stationary Policies for Stationary Infinite-Horizon Markov Decision Processes. In: Advances in Neural Information Processing Systems, NIPS (2012)
24. Sutton, R., Barto, A.: Reinforcement Learning, An introduction. The MIT Press (1998)
25. Sutton, R.S., McAllester, D.A., Singh, S.P., Mansour, Y.: Policy Gradient Methods for Reinforcement Learning with Function Approximation. In: Advances in Neural Information Processing Systems, NIPS (1999)

Code You Are Happy to Paste: An Algorithmic Dictionary of Exponential Families

Olivier Schwander

Département Signal et Systèmes Électroniques (SSE),
Laboratoire des Signaux et Systèmes (L2S),
CNRS-SUPELEC-PARIS SUD, France
olivier.schwander@supelec.fr

Abstract. We describe a library and a companion website designed to ease the usage of exponential families in various programming languages. Implementation of mathematical formulas in computer programs is often error-prone, difficult to debug and difficult to read afterwards. Moreover, this implementation is heavily dependent of the programming language used and often needs an important knowledge of the idioms of the language. In our system, formulas are described in a high-level language and mechanically exported to the chosen target language and a LaTeX export allows to quickly review correctness of formulas. Although our system is not limited by design to exponential families, we focus on this kind of formulas since they are of great interest for machine learning and statistical modeling applications. Besides, exponential families are a good usecase of our dictionary: among other usages, they may be used with generic algorithms for mixture models such as Bregman Soft Clustering, in which case lots of formulas from the canonical decomposition of the family need to be implemented. We thus illustrate our library by generating code which can be plugged into generic Expectation-Maximization schemes written in multiple languages.

1 Introduction

Except rare theoretical breakthroughs, machine learning research often needs to be validated with experiments and implementations in some programming language (common languages for this use are typically Matlab, Python, R, C, C++ or even Fortran). This implementation step goes through the translation of the mathematical formulas appearing in the new method into computer code. Although one may expect this translation to be straightforward, it usually needs some non trivial knowledge about the used language: the syntax for creating matrices and vectors and the syntax to access elements; the mathematical operators and common functions (like sqrt, exp, sin); the name of mathematical constants; the availability of special mathematical functions (like Γ, erf, etc); the various headers needed to enable access to the previous features (#include, import, etc); and finally the options to give to the compiler or the interpreter (to locate libraries and files).

T. Calders et al. (Eds.): ECML PKDD 2014, Part III, LNCS 8726, pp. 51–65, 2014.
© Springer-Verlag Berlin Heidelberg 2014

Although the general structure of any implementation seems to be similar, a lot of small differences appear between languages. We can study these subtle differences by looking at the implementation of the probability density function of the Gaussian distribution in three different languages: Python (Fig. 1), Matlab (Fig. 2) and C (Fig. 3). Since these three versions come from the reference libraries for numerical computation, they are supposed to be idiomatic and respectful of the usages of each language. We notice the square is made with three different syntax: ** in Python, ^ in Matlab and u*u (where u is a temporary variable holding the quantity to square) in C. We can also remark the various needs regarding the headers: various import for sqrt, exp and pi in Python, nothing in Matlab and one include in C for sqrt, exp, M_PI and fabs (the use of an absolute value around the standard deviation is rather surprising here, but this is beyond the scope of our study).

```
   7 import math
  24 from numpy import exp
  28 from numpy import pi
2112 _norm_pdf_C = math.sqrt(2*pi)
2116 def _norm_pdf(x):
2117     return exp(-x**2/2.0) / _norm_pdf_C
```

Fig. 1. Gaussian in Python (extract from the file scipy/stats/distributions.py from the library Scipy 0.13.3)

```
30 function pdf = stdnormal_pdf (x)
40   pdf = (2 * pi)^(- 1/2) * exp (- x .^ 2 / 2);
42 endfunction
```

Fig. 2. Gaussian in Matlab (extract from the file scripts/statistics/ distributions/normpdf.m from Octave 3.8.1 since the Matlab sources are not available; some lines which check parameters are removed)

This language-specific knowledge is often not problematic at first glance since one tends to use a well-known language for the first experiments of a new method but may become a problem if some parts need to be rewritten in another language for performance reasons or to collaborate with other people using other languages. It also renders more difficult the path between a first research prototype and a real-scale application. Finally, and perhaps more importantly, the source code implementing the formula is often difficult to read: for the original programmer, bugs and mistakes are harder to find and for a newcomer wanting to study the implementation, the code is barely understandable and nearly impossible to use in another application without a lot of work.

```
 22 #include <math.h>
118 double
119 gsl_ran_gaussian_pdf (const double x, const double sigma)
120 {
121   double u = x / fabs (sigma);
122   double p = (1 / (sqrt (2 * M_PI) * fabs (sigma))) * exp (-u * u / 2);
123   return p;
124 }
```

Fig. 3. Gaussian in C (extract from the file `randist/gauss.c` from the Gnu Scientific Library (GSL) 1.16)

The library introduced in this paper allows to describe mathematical formulas in a programming language-agnostic way: the work of translating formulas into a computer-understandable implementation only needs to be made once, facilitating the choice of the most well-suited programming language. A first prototype may be exported to Python and then to C in order to work on a large real life dataset. Another researcher may generate Matlab code and then plug the formula into its own code base. And an engineer in a company may just take the C export and use it for an industrial application or the company may also design its own exporter backend to generate code suited to proprietary internal tools. Since the description language is not really more readable than a programming language, a LaTeX export is provided, allowing to easily proofread the formula. This library is aimed at any people who may want to use mathematical formulas inside computer code and can be used at hand with copy-pasting or in a more clean way by integrating it in a build process.

Beside the library itself, we also present *Code-Formula*, a web application demonstrating our library, which is designed both for educational purpose and to offer an encyclopedia of mathematical functions which can be picked-up when one needs an out-of-the-box implementation of a mathematical formula. This website is inspired by other online dictionaries of mathematical objects, like the Online Encyclopedia of Integer Sequences [8], the Digital Library of Mathematical Functions [5] or the Dynamic Dictionary of Mathematical Functions [2], but to the best of our knowledge, it is the first one focusing on the algorithmic side instead of mathematical properties.

Although the previous remarks can apply to a wide variety of formulas from mathematical science, computer science, physics or engineering fields, we chose to limit ourselves to a dictionary of exponential families for some reasons: first, it is better at first sight to limit the goals of the project to a reasonable set of objects; second, exponential families are widely used in a large variety of fields, including, but not limited to, machine learning; last, and perhaps most importantly, in the recent years, a lot of work has been devoted to the design of generic algorithms for mixtures of exponential families, where the precise family is a parameter of the algorithm, and a few implementations have been worked-on, in Java (jMEF [4]), in Python (pyMEF [7]), in C (`libmef` [6]) or even in R.

Each of these implementations has been confronted to the same kind of work: translating formulas into code. We hope that using our library and website, the implementation of such libraries may be done in a semi-automatic way.

This article in organized as follows: after this introduction detailing motivation and goals of the project, the architecture of the library is described. Then, a few examples of exponential families described using our library are given along the utilization of exported code to plug into a generic Expectation-Maximization method for mixture of exponential families. Finally, the website containing the encyclopedia itself is described.

2   Architecture

2.1   General view

The general architecture of the system is described in Fig. 4: mathematical formulas are described in a high-level frontend, then processed by the code of the library and finally passed to the backends which are in charge of generating programming code. Currently, the language used in the fronted is the same as for the library itself, that is OCaml (but very little knowledge of this language is needed to effectively write formulas). This choice has been made for facility reasons, avoiding the need of writing a parser for a domain specific language and because OCaml, although not well-known in the machine learning field, is well suited for this kind of task. Although a LaTeX frontend may look appealing, this is not feasible for two reasons: first a LaTeX parser is nearly impossible to write, even for the subset of the language expressing the mathematical formulas; second LaTeX formulas carry very few semantic, since the language is designed for display, not for computation.

The core part of the library provides a set of tools to manipulate the formulas, like changing the names of variables inside a formula but in the future other frontends may be added. So far, four backends are available: Python, Matlab, C and LaTeX, the later allowing to easily proofread formulas and to use them directly in publications or documentation.

The source code of this library, called `Formula`, can be browsed online on `http://hub.darcs.net/oschwand/formula` and downloaded on the web-page related to this article: `http://www.lix.polytechnique.fr/~schwander/ecml2014/`.

2.2   Frontend

The frontend is responsible for the translation between a human-understandable description of the formula into a data structure representing the formula which can be passed to export backends. If we stick to the example given in the introduction, that is, the probability density function of the Gaussian distribution

$$f(\sigma, \mu, x) = \frac{1}{\sqrt{2\pi\sigma^2}} \exp\left(-\frac{(x-\mu)^2}{\sigma^2}\right) \tag{1}$$

Frontends   Backends

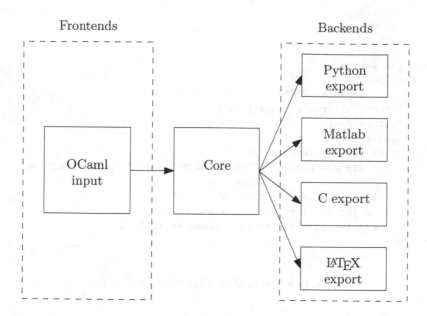

Fig. 4. General architecture of Code you are Happy to Paste

the steps will be the following: first describe the formula in OCaml (Fig. 5) and then compile the description which will be represented as an abstract syntax tree (Fig. 6; for brevity, it is simply a centered and normalized Gaussian). In addition to the formula itself, each description embeds its own documentation, with a description of the formula and with names and properties for the variables used inside.

In the description in Fig. 5, we build a function (line 2) called f, described as *Gaussian PDF* (line 3), taking three arguments (σ, μ and x, line 4) and returning a real number (line 5). After this header, we define three variables (lines 7, 8 and 9), each of them bearing a name and if necessary a documentation and a mathematical property (which is used only for documentation purposes). Finally, the formula itself is described, using a straightforward syntax similar to the one used in many languages (the `Syntax` keyword means the mathematical operators work on nodes of the syntax tree instead of numbers and the ! are used to convert numbers into nodes).

2.3 Backends

Four backends are available so far, trying to cover various use-cases of scientific computing. In each case, the goal is to produce idiomatic code with as less differences as possible as with handmade code.

Latex The LaTeX output does not need more comments since most of the formulas in this document have been generated using our library. A particular attention has been paid to generate nice looking formula, especially by minimizing the number of parentheses.

```
1  let gaussian =
2    Func.def "f"
3      ~doc:"Gaussian PDF"
4      ~args:["\\sigma"; "\\mu"; "x"]
5      ~return:Real
6      Syntax.(
7        let x     = real "x" in
8        let sigma = real ~doc:"standard deviation" ~prop:"positive"
9                         "\\sigma" in
10       let mu    = real ~doc:"mean" "\\mu" in
11       !1 / (sqrt (!2 * pi * sigma ** !2)) *
12         exp (- ((x - mu) ** !2 / sigma ** !2))
13     )
```

Fig. 5. Description of the Gaussian distribution

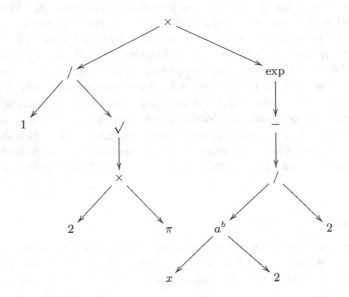

Fig. 6. Tree representing a centered and normalized Gaussian PDF ($\mu = 0, \sigma = 1$)

Python The Python backend outputs code relying on the library numpy which is the standard for scientific computing in Python. This library provides basic mathematical functions along with powerful vectors and matrices operations.

Python.def ~doc:true gaussian

```python
def f(sigma, mu, x):
  """ Gaussian PDF

  x:  (real)
  \sigma: standard deviation (real, positive)
  \mu: mean (real)
  """

  return 1 / numpy.sqrt(2 * numpy.pi * sigma**2) * \
    numpy.exp(- (x - mu)**2 / sigma**2)
```

Matlab The Matlab backend generates code using only built-in functions of Matlab.

Matlab.def ~doc:true gaussian

```matlab
% Gaussian PDF
%
% x:  (real)
% \sigma: standard deviation (real, positive)
% \mu: mean (real)

function f(sigma, mu, x)
  1 / sqrt(2 * pi * sigma^2) * exp(- (x - mu)^2 / sigma^2)
end
```

C The C backend is a little more subtle. First, the C language requires explicit typing indications in the code: thus we need to do the translation between real numbers (on the description side) into double (on the C side), and the same for integers and int.

C.def ~doc:true gaussian

```c
/* Gaussian PDF

  x:  (real)
  \sigma: standard deviation (real, positive)
  \mu: mean (real)
*/
double f(double sigma, double mu, double x) {
  return 1 / sqrt(2 * M_PI * pow(sigma, 2)) * \
    exp(- pow(x - mu, 2) / pow(sigma, 2));
}
```

The Gaussian PDF is too simple to highlight the others subtleties of this backend, we thus add another example, estimating the mean of a set of values. For a straightforward mean function

$$\text{mean}(X) = \frac{1}{|X|} \left(\sum_{i=1}^{|X|} X_i \right) \tag{2}$$

described by

```
let mean =
  Func.def "mean"
    ~args:["X"]
    ~return:Real
    Syntax.(
      let x = var "X" (Vector Real) in
      !1 / (length x) * sum x
    )
```

we get the following C code:

```
double mean(gsl_vector* X) {
  double tmp1 = 0;
  for(unsigned int i=0; i<X->size; i++)
    tmp1 += gsl_vector_get(X, i);

  return 1 / (X->size) * tmp1;
}
```

First, we made the choice to rely on the Gnu Scientific Library (GSL) for all vectors and matrices operations: in addition to the data structures themselves we get also common mathematical operations on vectors and matrices, simplifying the generated code. Second, since GSL does not provide any function to sum the elements, we need to rewrite the formula to replace the $\sum$ operation by a temporary variable which is populated using a `for` loop.

3 Exponential Families

In order to present the content of our encyclopedia, we give a quick recall on exponential families before showing examples of formula descriptions and exported code.

3.1 Definition

Exponential families are an ubiquitous class of distributions and many widely used distributions belong to this class.

An exponential family is a set of distributions whose probability mass or probability density functions admit the following canonical decomposition:

$$p(x; \theta) = \exp(\langle t(x), \theta \rangle - F(\theta) + k(x)) \tag{3}$$

with

- $t(x)$ the sufficient statistic,
- θ the natural parameters,
- $\langle \cdot, \cdot \rangle$ the inner product,
- F the log-normalizer, which is strictly convex and differentiable,
- $k(x)$ the carrier measure.

Since this log-normalizer F is a strictly convex and differentiable function, it admits a dual representation, the convex conjugate F^*, by the Legendre-Fenchel transform:

$$F^*(\eta) = \sup_{\theta} \{ \langle \theta, \eta \rangle - F(\theta) \} \tag{4}$$

We get the maximum for $\theta = (\nabla F)^{-1}(\eta)$ and F^* can be computed with:

$$F^*(\eta) = \langle \eta, (\nabla F)^{-1}(\eta) \rangle - F((\nabla F)^{-1}(\eta)) \tag{5}$$

Many generic information-geometric algorithms (like Bregman Hard Clustering or Bregman Soft Clustering [1]) rely on the knowledge of this decomposition and thus the implementation of these algorithms require to translate these formulas into computer code. Translating these formulas from a language-agnostic description allows to factorize the effort and is less error-prone than ad-hoc manual work.

3.2   Examples

We describe here the full canonical decomposition of two exponential families, the Gaussian distribution and the Laplace law. For brevity, we only give the description of each formula and the LATEX export. The reader can find all the source code related to this article on the webpage http://www.lix.polytechnique.fr/~schwander/ecml2014/. The same content can also be retrieved in the website described in Section 4.

Gaussian distribution This is the opportunity to introduce new syntactic features: functions can take real vectors as arguments and return them using the type **Vector Real**. Inside the formula, elements of the vector can be accessed using the @ operator (like **theta@0**).

```
let f =
  Func.def "F"
    ~doc:"Log-normalizer"
    ~args:["\\theta"]
    ~return:Real
    Syntax.(
      let theta = var ~doc:"natural parameter"
        ~prop:"dimension 2" "\\theta" (Vector Real)
      in
      - (!1 / !4 * ((theta@0) ** !2 / (theta@1))) +
        !1 / !2 * log(- (pi / (theta@1)))
    )
```

$$F(\theta) = -\frac{1}{4}\frac{\theta_0^2}{\theta_1} + \frac{1}{2}\log\left(-\frac{\pi}{\theta_1}\right) \qquad (6)$$

```
let gradF =
  Func.def "\\nabla F"
    ~doc:"Gradient log normalizer"
    ~args:["\\theta"]
    ~return:(Vector Real)
    Syntax.(
      let theta = var ~doc:"natural parameter"
        ~prop:"dimension 2" "\\theta" (Vector Real)
      in
      vector [
        - (theta@0) / (!2 * (theta@1));
        - !1 / (!2 * (theta@1)) +
          (theta@0) ** !2 / (!4 * (theta@1) ** !2);
      ]
    )
```

$$\nabla F(\theta) = \left(\frac{-\theta_0}{2\theta_1}, \frac{-1}{2\theta_1} + \frac{\theta_0^2}{4\theta_1^2}\right) \qquad (7)$$

```
let g =
  Func.def "F^\\star"
    ~doc:"Dual log-normalizer"
    ~args:["\\eta"]
    ~return:Real
    Syntax.(
      let eta = var ~doc:"expectation parameter"
        ~prop:"dimension 2" "\\eta" (Vector Real)
      in
```

```
  - (!1 / !2) * log((eta@0) ** !2 - (eta@1))
)
```

$$F^*(\eta) = -\frac{1}{2}\log\left(\eta_0^2 - \eta_1\right) \tag{8}$$

```
let t =
  Func.def "t"
    ~doc:"Sufficient statistic"
    ~args:["x"]
    ~return:(Vector Real)
    Syntax.(
      let x = real ~doc:"observation" "x" in
      vector [x; x ** !2]
    )
```

$$t(x) = \left(x, x^2\right) \tag{9}$$

Laplace distribution Since the Laplace distribution is of order 1 (with only one scalar parameter), the descriptions are much simpler since we do not need to deal with vectors.

```
let pdf =
  Func.def "f"
    ~doc:"Centered Laplace PDF"
    ~args:["\\sigma"; "x"]
    ~return:Mathset.Real
    Syntax.(
      let x     = real "x" in
      let sigma = real ~doc:"standard deviation" ~prop:"positive"
                       "\\sigma" in
      !1 / (!2 * sigma) *
        exp (- (abs x) / sigma)
    )
```

$$f(\sigma, x) = \frac{1}{2\sigma}\exp\left(\frac{-|x|}{\sigma}\right) \tag{10}$$

```
let f =
  Func.def "F"
    ~doc:"Centered Laplace log-normalizer"
    ~args:["\\theta"]
    ~return:Mathset.Real
```

```
Syntax.(
  let theta = real ~doc:"natural parameter" "\\theta" in
  log (- !2 / theta)
)
```

$$F(\theta) = \log \frac{-2}{\theta} \tag{11}$$

```
let grad_f =
  Func.def "\\nabla F"
    ~doc:"Centered Laplace gradient log-normalizer"
    ~args:["\\theta"]
    ~return:Mathset.Real
    Syntax.(
      let theta = real ~doc:"natural parameter" "\\theta" in
      - !1 / theta
    )
```

$$\nabla F(\theta) = \frac{-1}{\theta} \tag{12}$$

```
let g =
  Func.def "\\nabla F^\\star"
    ~doc:"Centered Laplace dual log-normalizer"
    ~args:["\\eta"]
    ~return:Mathset.Real
    Syntax.(
      let eta = real ~doc:"expectation parameter" "\\eta" in
      - log eta
    )
```

$$\nabla F^{\star}(\eta) = -\log\eta \tag{13}$$

```
let grad_g =
  Func.def "\\nabla F^\\star"
    ~doc:"Centered Laplace dual log-normalizer"
    ~args:["\\eta"]
    ~return:Mathset.Real
    Syntax.(
      let eta = real ~doc:"expectation parameter" "\\eta" in
      - !1 / eta
    )
```

$$t(x) = |x| \tag{14}$$

3.3   Mixture Models

In order to learn mixtures of exponential families, we use an Expectation-Maximization (EM) instance [3] called Bregman Soft Clustering [1], allowing to pass the family as an argument of the algorithm. As usual, this is an iterative algorithm where two steps are repeated until convergence of the log-likelihood of the mixture: expectation step and maximization step. See [4] for more details about the exponential family version of these two steps.

Expectation step

$$p(i|x_t, \eta) = \frac{\omega_i \exp\left(F^\star(\eta_i) + \langle t(x_t) - \eta_i, \nabla F^\star(\eta_i)\rangle\right)}{\sum_{j=1}^k \omega_j \exp\left(F^\star(\eta_j) + \langle t(x_t) - \eta_j, \nabla F^\star(\eta_j)\rangle\right)} \tag{15}$$

Maximization step

$$\omega_i = \frac{1}{N} \sum_{t=1}^N p(i|x_t, \eta) \tag{16}$$

$$\eta_i = \sum_{t=1}^N \frac{p(i|x_t, \eta)}{\sum_{t=1}^N p(i|x_t, \eta)} t(x_t) \tag{17}$$

Currently, theses two steps need to be implemented by hand in each target language since our description language is not expressive enough to manipulate functions inside the formula (we would need to pass $F^\star$, $\nabla F^\star$, t as arguments to the function, or let them as free variable). Nonetheless, as soon as these steps are implemented, with a `while` loop around to iterate, it can be plugged after the automatically generated formulas, forming a full EM iterative scheme.

4   Website

The *Code-Formula* website (accessible through `http://www.lix.polytechnique.fr/~schwander/codeformula`) is designed to spread knowledge about the exponential families. Following the ideas introduced by precursor online dictionaries of mathematical objects, we think the online format is way more suitable for this kind of content than static documents.

Each page on the site (see the screenshot Fig. 7) shows a card about an exponential family, with a list of formulas related to the family. Each formula is presented first with a rendered version of the the latex output followed by exports in the supported languages.

The goal is to become the reference about decomposition of exponential families, serving to diffuse knowledge, demonstrating our description library but also as a direct source for picking-up pre-made implementations of formulas of interest, for researchers and companies.

64   O. Schwander

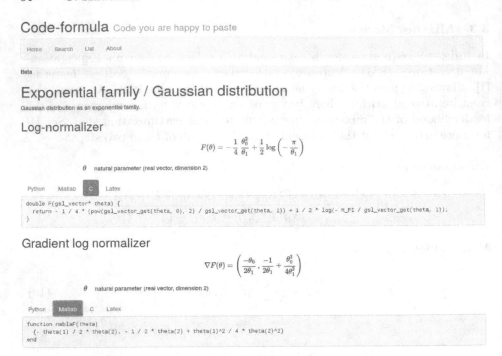

Fig. 7. Gaussian page on Code-Formula, the online encyclopedia of exponential families

5 Conclusion

We presented *Formula*, a library to describe mathematical formulas and to automatically generate code implementing these formulas, and *Code-Formula*, a
website showing an online dictionary of exponential families. Both are aimed
at reducing the time between chalk board work to real implementation of an
algorithm. This is obviously useful for research purposes, easing the first implementation of a new method and also easing a re-implementation of a work
by other researchers, but this may also be useful for students or for companies
seeking to build a real-world implementation of a method.

There are a lot of perspectives which are under work: on the website side,
enlarge the content (contributions are obviously welcomed); on the library side,
it should be interesting to be able to generate the headers needed to execute
the generated code along with necessary compilation flags; it may also be interesting to render the description language expressive enough to directly describe
formulas using other functions (to be able to write the update rule of the EM
algorithm for example). On the short term, new export backends are under work,
like R and Julia.

Acknowledgments. The author would like to thank Frank Nielsen for the insightful discussions about dictionaries of information geometric objects, in par-

ticular for distances and distributions, and James Regis for providing hosting at the LIX laboratory.

References

1. Banerjee, A., Merugu, S., Dhillon, I.S., Ghosh, J.: Clustering with Bregman divergences. The Journal of Machine Learning Research 6, 1705–1749 (2005)
2. Benoit, A., Chyzak, F., Darrasse, A., Gerhold, S., Mezzarobba, M., Salvy, B.: The Dynamic Dictionary of Mathematical Functions (DDMF). In: Fukuda, K., van der Hoeven, J., Joswig, M., Takayama, N. (eds.) ICMS 2010. LNCS, vol. 6327, pp. 35–41. Springer, Heidelberg (2010)
3. Dempster, A.P., Laird, N.M., Rubin, D.B.: Maximum likelihood from incomplete data via the EM algorithm. Journal of the Royal Statistical Society. Series B (Methodological), 1–38 (1977)
4. Nielsen, F., Garcia, V.: Statistical exponential families: A digest with flash cards. CoRR 09114863 (2009)
5. Olver, F.W.J., Lozier, D.W., Boisvert, R.F., Clark, C.W. (eds.): NIST Digital Library of Mathematical Functions, http://dlmf.nist.gov/
6. Schwander, O., Nielsen, F.: Fast learning of gamma mixture models with k-MLE. In: Hancock, E., Pelillo, M. (eds.) SIMBAD 2013. LNCS, vol. 7953, pp. 235–249. Springer, Heidelberg (2013)
7. Schwander, O., Nielsen, F.: PyMEF – A framework for exponential families in Python. In: 2011 IEEE Statistical Signal Processing Workshop (SSP), pp. 669–672 (2011)
8. Neil Sloane. The On-Line Encyclopedia of Integer Sequences, http://oeis.org/

Statistical Hypothesis Testing in Positive Unlabelled Data

Konstantinos Sechidis[1], Borja Calvo[2], and Gavin Brown[1]

[1] School of Computer Science, University of Manchester, Manchester M13 9PL, UK
{sechidik,gavin.brown}@cs.manchester.ac.uk
[2] Department of Computer Science and Artificial Intelligence,
University of the Basque Country, Spain
borja.calvo@ehu.es

Abstract. We propose a set of novel methodologies which enable valid statistical hypothesis testing when we have only *positive and unlabelled* (PU) examples. This type of problem, a special case of semi-supervised data, is common in text mining, bioinformatics, and computer vision. Focusing on a generalised likelihood ratio test, we have 3 key contributions: (1) a proof that assuming all unlabelled examples are negative cases is sufficient for *independence* testing, but not for power analysis activities; (2) a new methodology that compensates this and enables power analysis, allowing sample size determination for observing an effect with a desired power; and finally, (3) a new capability, *supervision determination*, which can determine *a-priori* the number of labelled examples the user must collect before being able to observe a desired statistical effect. Beyond general hypothesis testing, we suggest the tools will additionally be useful for information theoretic feature selection, and Bayesian Network structure learning.

1 Introduction

Learning from Positive-Unlabelled (PU) data is a special case of semi-supervised learning, where we have a small number of examples from the positive class, and a large number of unlabelled examples which could be positive *or* negative. The objective in this situation is to perform standard machine learning activities despite this data restriction. The problem has been referred to in the literature under several names, including *partially supervised classification* [15], *positive example based learning* [19] and *positive unlabelled* learning [9]. A typical application has been text classification — given a number of query documents belonging to a particular class (e.g. academic articles about machine learning), plus a corpus of unlabelled documents, the task is to classify new documents as relevant to the query or not.

Most work in the PU area is concerned with classification, rather than theory. Denis [8] is an interesting exception, which generalised Valiant's PAC learning framework to PU data, and concluded that learning from positive and unlabelled examples is possible, but that we must have some additional prior knowledge

T. Calders et al. (Eds.): ECML PKDD 2014, Part III, LNCS 8726, pp. 66–81, 2014.

about the underlying distribution of examples. We make use of this observation, exploring how *statistical hypothesis testing* manifests in PU data, and how such prior knowledge can be incorporated.

In this context, we focus on the G-test [18], a generalised likelihood ratio test used for testing independence of categorical variables, which is closely related to the mutual information (Sec. 3). The G-test and the mutual information are used extensively, for example in life sciences to test whether two observed natural processes are independent [16]. In machine learning and in data mining they also have a large number of applications, for instance in structure learning of a Bayesian Network [2] or in feature selection [4]. The main contributions[1] of our work are the following

- A proof that, the common assumption of all unlabelled examples being negative is sufficient for *testing independence* (Sec. 4.1), but *insufficient* for more advanced activities such as power analysis (Sec. 4.2).
- A methodology for a-priori power analysis in the PU scenario, enabling *sample size determination* for observing an effect with a desired power (Sec. 5).
- A novel capability: *supervision determination*, which can determine the minimum number of labelled data to achieve a desired power (Sec. 5).

In a general hypothesis testing scenario, our results make clear the implications of using the G-test under the PU constraint, and leads to more cost-effective experimental design. In wider machine learning activities, there are several applications. For example, constraint-based learning of Bayesian Network structures: the decision on whether to include an arc between two nodes is often made with a hypothesis test such as χ^2, or the mutual information, both of which are core to our work. Another example is information theoretic feature selection, Guyon et al. [13, pg 68] discuss how the statistical viewpoint on feature selection allows decision-making on the relevance/redundancy of a feature to be made in a principled manner. Our methods permit these activities under the PU data constraint.

2   Background on the Positive Unlabelled Problem

Positive-Unlabelled data refers to situations where we have a small number of examples labelled as the positive class, and a large number of entirely unlabelled examples, which could be either positive *or* negative. Whilst classification is well explored in the PU scenario, an area in need of attention is *statistical hypothesis testing*: including independence tests, and more complex activities such as power analysis. We now introduce the formal framework of Elkan & Noto [10] for reasoning over PU data, which we build upon in our work.

2.1   Positive Unlabelled Framework

Assume that a dataset $\mathcal{D}$ is drawn i.i.d. from the joint distribution $p(X, Y, S)$, where the features X are categorical, the class Y is binary, and S is a further

---

[1] Matlab code for all methods and the supplementary material with all the proofs and extra results are available in www.cs.man.ac.uk/~gbrown/posunlabelled/.

random variable with possible values 's^+' and 's^-', indicating if the example is labelled (s^+) or not (s^-). Thus $p(x|s^+)$ is the probability of X takes the value x from its alphabet $\mathcal{X}$ conditioned on the labelled set. The same shorthand notation is used for Y, where the positive class is indicated by 'y^+', and the negative class by 'y^-'.

In this context, Elkan & Noto formalise the *selected completely at random* assumption, saying that the examples for the labelled set are selected completely at random from all the positive examples:

$$p(s^+|x, y^+) = p(s^+|y^+) \quad \forall\ x \in \mathcal{X}.$$

Thus, the probability of a positive example being labelled is *independent* of the input x. Perhaps most interestingly, Elkan & Noto proceed to show that this assumption has been followed either explicitly or *implicitly* in most research on PU data.

2.2 A Naive Approach–Assuming Unlabelled Examples are Negative

One approach to learn from this data is to simply assume that any unlabelled examples are negative. This approach, while seemingly naive, has proven to be useful for classification. Elkan & Noto [10] show that a probabilistic classifier trained on such data predicts posterior probabilities that differ from the true values by a constant factor; they suggest a number of ways to estimate this factor using a validation set. In a different context, Blanchard et al. [3] use the same assumption and prove that semi-supervised novelty detection can be reduced to Neyman-Pearson binary classification using the nominal and unlabelled samples as the two classes, in their terminology.

2.3 Incorporating Prior Knowledge

Another general approach follows the theoretical work of Denis [8], incorporating prior knowledge of the class distributions to augment the learning. For example, Calvo et al. [5] build PU naive bayes classifiers, and propose a Bayesian solution to deal with uncertainty in the distribution of the positive class.

At first glance, in the PU learning environment estimating $p(x|y^-)$ seems impossible without negative data. However, with a neat rearrangement of the marginal $p(x)$ and some extra information, it turns out to be possible. The marginal is $p(x) = p(x, y^-) + p(x, y^+)$, which can be rearranged:

$$p(x|y^-) = \frac{p(x) - p(x|y^+)p(y^+)}{1 - p(y^+)}. \tag{1}$$

Denis et al. [9] exploited this to construct a PU Naive Bayes classifier, estimating $p(x|y^+)$ by maximum likelihood on just the labelled set, i.e. assuming $\hat{p}(x|y^+) \approx \hat{p}(x|s^+)$, and estimating $p(x|y^-)$ using equation (1). The prior $p(y^+)$

was provided as a user-specified parameter, $\tilde{p}$. Thus, the missing conditional probability $p(x|y^-)$ is estimated as,

$$\hat{p}(x|y^-) \approx \frac{\hat{p}(x) - \hat{p}(x|s^+)\tilde{p}}{1 - \tilde{p}},$$

where $\hat{p}(x)$, $\hat{p}(x|s^+)$ denote the maximum likelihood estimates of the respective probabilities. Although this seems a heuristic approach, we will now show with the following Lemma that under the *selected completely at random assumption* it is indeed valid.

Lemma 1. *Assuming data is selected completely at random, the conditional distribution of x given y = 1 is equal to the conditional distribution of x given that it is labelled.*

$$p(x|y^+) = p(x|s^+) \quad \forall \ x \in \mathcal{X}.$$

The proof of this Lemma and all the proofs of this work are available in the supplementary material.

2.4   Summary

In our work we will explore both approaches in the context of statistical hypothesis testing. By using the naive but common assumption that all the unlabelled examples are negative we can perform a test of independence between X against either the true labels (Y) or the assumed ones (S). In Section 4 we prove that these two cases have precisely the same false positive rate but different true positive rates (i.e. statistical power). While in Section 5, by using prior-knowledge, we derive a correction factor for the test that brings these into parity – identical true positive and false positive rates. As a consequence we can also perform positive unlabelled sample size determination, and determine the number of labeled examples needed to observe a desired statistical effect with a specified power. Before that, in Section 3 we review the likelihood ratio test that this work builds upon.

3   Hypothesis Testing

3.1   The G-test of Independence

In fully observed categorical data, the G-test can be used to determine statistical independence between categorical variables [18]. It is a generalised likelihood ratio test, where the test statistic can be calculated from sample data counts arranged in a contingency table. Denote by $O_{x,y}$ the observed count of the number of times the random variable X takes on the value x from its alphabet $\mathcal{X}$, while Y takes on $y \in \mathcal{Y}$; and by $O_{x,\cdot}$ and $O_{\cdot,y}$ the marginal counts. The estimated expected frequency of (x, y), assuming X, Y are independent, is given by

$E_{x,y} = \hat{p}(x)\hat{p}(y)N = \frac{O_{x,\cdot}O_{\cdot,y}}{N}$. The *G-statistic* can now be defined as

$$G = 2\sum_{x\in\mathcal{X}}\sum_{y\in\mathcal{Y}} O_{x,y}\ln\frac{O_{x,y}}{E_{x,y}} = 2N\sum_{x\in\mathcal{X}}\sum_{y\in\mathcal{Y}} \hat{p}(x,y)\ln\frac{\hat{p}(x,y)}{\hat{p}(x)\hat{p}(y)} = 2N\hat{I}(X;Y) \quad (2)$$

where $\hat{I}(X;Y)$ is the maximum likelihood estimator of the mutual information between X and Y [17]. Under the null hypothesis that X and Y are statistically independent, G is known to be asymptotically χ^2-distributed, with $\nu = (|\mathcal{X}| - 1)(|\mathcal{Y}| - 1)$ degrees of freedom [18]. For a given dataset, we calculate (2) and check to see whether it exceeds the critical value defined by a significance level α read from a standard statistical table giving the CDF of the χ^2-distribution — if the critical value is not exceeded, the variables are judged to be independent.

3.2 Power Analysis

With such a test, it is common to perform a *power analysis* [6]. The *power* of a test is the probability that the test will reject the null hypothesis when the alternative hypothesis is true. This is also known as the *true positive rate*, or the probability of *not* committing a Type-II error. An a-priori power analysis would take a given sample size N, a required significance level α, an effect size ω, and would then compute the power of the statistical test. However, to do this we need a test statistic with a known distribution under the alternative hypothesis.

It is known that the G-statistic (2) has a large-sample *non-central χ^2* distribution under the alternative hypothesis (i.e. when X and Y are dependent) as presented by Agresti [1, Section 16.3.5]. Agresti shows that the χ^2 non-centrality parameter (λ) has the same form as the G-statistic, but with sample values replaced by population values. In other words, the non-centrality parameter under the alternative hypothesis is given by $\lambda = 2NI(X;Y)$. Thus λ is a parameter, and G is a random variable following a distribution defined by λ.

One important usage of a-priori power analysis is *sample size determination*. In this prospective procedure we specify the significance level of the test (e.g. $\alpha = 0.05$), the desired power (e.g. a false negative rate of 0.01) and the desired effect size – from this we can determine the minimum number of examples required to detect that effect.

It turns out that the effect size of the G-test can be naturally expressed as a function of the *mutual information*. More specifically, the effect size (ω) is the square root of the non-centrality parameter divided by the sample size [6], thus we have $\omega = \sqrt{2I(X;Y)}$. Therefore, to understand hypothesis testing in PU data, we must understand the properties of mutual information in such data.

4   Hypothesis Testing in Positive Unlabelled Data

In this section we will focus in PU data by adopting the very common assumption that all unlabelled examples are negative, and exploring the consequences for hypothesis testing.

4.1 Testing for Independence in Positive Unlabelled Data

In positive unlabelled data, it is not immediately obvious how to apply the G-test described in the previous section, since the variable Y is only partially observed. The 'naive' approach would be to assume all unlabelled examples as negatives, and test for independence in the usual manner. This is in effect testing independence between X and S, the variable describing whether an example is labelled. While this is arguably a rather significant assumption, it is in fact sufficient to answer the question of whether X, Y are *independent*. This is proved formally with the following simple theorem.

Theorem 1. *In the positive unlabelled scenario, under the selected completely at random assumption, a variable X is independent of the class label Y if and only if X is independent of S, so it holds $X \perp\!\!\!\perp Y \Leftrightarrow X \perp\!\!\!\perp S$.*

Intuitively, we can describe variable S as a noisy copy of Y, with no false positives but potentially a large number of false negatives. So instead of checking the independence with the actual variable Y we can check with the noise version S. The proof of the Theorem 1 is available in the supplementary material, though the theorem can be also experimentally verified with a simple 'sanity check' experiment. We generated data as so: X, Y are independent Bernoulli variables each with $p = 0.5$ and take $N = 1000$ observations. For the PU case, all negative examples have their labels removed, and we randomly remove a fraction $(1 - c)$ of the positive labels, where c is the fraction of all positive examples that are labelled, also written as $c = p(s^+)/p(y^+)$. To test independence we apply the G-test with a significance level of $\alpha = 0.01$ to test the assertion that $X \perp\!\!\!\perp Y$ in the supervised case, and $X \perp\!\!\!\perp S$ in the PU case. Since the null hypothesis $(X \perp\!\!\!\perp Y)$ is true, we expect a false positive rate of 1% in both cases — this is verified below in Figure 1a (over *100,000* repeats) holding for all supervision levels $p(s^+)$ and in Figure 1b holding for different sample sizes when we fix the supervision level to be $p(s^+) = 0.1$. The slight fluctuation comes from the limited

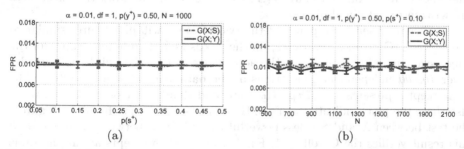

(a)   (b)

Fig. 1. Figure for Type-I error. (a) Type-I error changing as a function of the probability of an example being labeled $p(s^+)$, for fixed $\alpha = 0.01$, $N = 1000$. (b) Type-I error changing as a function of N, for fixed $\alpha = 0.01$, $p(s^+) = 0.10$.

sample size. While Theorem 1 tells us that the two possible tests, $G(X; Y)$ and $G(X; S)$, are equivalent for observing independencies, it says nothing about how well the naive $G(X;S)$ test will perform when the null hypothesis is *false*. In this case we must compare the tests in terms of their *power* to detect a given effect.

4.2   Comparing the Power of the Tests

In order to compare the power of the two tests we must examine their *non-centrality parameters*. Section 3.2 presents that the non-centrality parameter for the G-test is $\lambda = 2NI(X;Y)$. Therefore, the power of the tests depends on the population values of the mutual informations $I(X;Y)$ and $I(X;S)$. With the following theorem we prove an inequality between these two quantities

Theorem 2. *In the positive unlabelled scenario, under the selected completely at random assumption, when X and Y are dependent random variables ($X \not\perp Y$) we have $I(X;Y) > I(X;S)$.*

A direct consequence of the theorem is the following corollary.

Corollary 1. *The derived test under the the naive assumption, $G(X;S)$, is less powerful than $G(X;Y)$. In other words using the noisy copy S of Y, will result in a test $G(X;S)$ which will have a higher false negative rate than $G(X;Y)$.*

The proof of the Theorem 2 is available in the supplementary material, here we will give an experimental verification. As a sanity check we should explore how the two tests perform when we have an actual effect to observe. To create pairs of X and Y with a specific effect we generate data as follows. Firstly, generate a random sample $\mathbf{x} = \{x_1, .., x_N\}$, where each $x_i \in \{0, 1\}$ and $p(x = 1) = 0.2$. Then create an identical copy of this sample as $\mathbf{y}_{i=1}^N$. This creates a dataset where $\mathbf{x}, \mathbf{y}$ are by definition completely dependent. We then corrupt this dependency by picking a random fraction of the examples, and setting a new value for each selected x_i by drawing a binary random variable with parameter $p = 0.5$. It is clear that by varying the number of examples which are corrupted by noise, we generate random variables with different mutual informations. For example when we corrupt 60% of the examples with noise, we can calculate analytically that the resulting variables have $I(X;Y) = 0.053$ (in this work the effect sizes are written to 3 decimal places).

In order to observe the power of the two tests we will plot figures similar to the figures in Gretton & Györfi [12]. In the x-axis we have different values for the effect size, while in the y-axis is the acceptance rate of the null hypothesis $\mathcal{H}_0$ (over *10,000* independent generations of the data, each of size $N = 200$). The y-intercept represents $1 - (Type\ I\ error)$, and should be close to $1 - \alpha$, while elsewhere the plots indicate the Type II error. As we observe from the Figure 2 the test between X and S is less powerful than the test between X and Y, and this result verifies the Corollary 1. Furthermore the intercepts are at the same value (close to the design parameter $1 - \alpha$), which again verifies that the tests have the same Type-I error, but as can be seen different Type-II error.

Given this corollary, it is interesting to ask how we might modify our practice with $G(X;S)$ to achieve a desired power, in spite of the partially observed variable. In the next section we will show how we can incorporate prior knowledge to address this question.

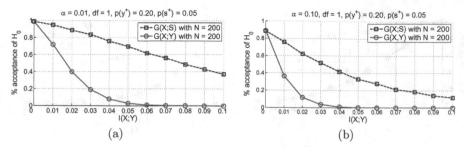

Fig. 2. Figure for comparing the Type-II error of the two tests using (a) $\alpha = 0.01$ and (b) $\alpha = 0.10$

5 Incorporating Prior Knowledge for Power Analysis

In order to use $G(X;S)$ for a-priori power analysis activities, we should quantify the amount of power that we lose by adopting the naive assumption that all the unlabelled examples are negative. In this section we show how to incorporate prior knowledge to calculate this quantity.

Cohen [6] proposed (with appropriate caution) several conventional effect sizes, facilitating cross-experiment comparison. In the case of χ^2 tests, a "medium" effect is $\omega = 0.3$. Since $\omega = \sqrt{2I(X;Y)}$, this translates to $I(X;Y) = 0.045$. With PU data, the key problem emerges here in that the standard effect size is naturally expressed in terms of $I(X;Y)$, whereas our test $G(X;S)$ in terms of $I(X;S)$. In order to deal with this problem, we will incorporate a user's prior knowledge of $p(y^+)$, and correct the non-centrality parameter of the test in such a way that we can use it for a-priori power analysis.

Theorem 3. *The non-centrality parameter of the G-test between X and S takes the form:*

$$\lambda_{G(X;S)} = \kappa \lambda_{G(X;Y)} = \kappa 2NI(X;Y),$$

where $\kappa = \dfrac{1-p(y^+)}{p(y^+)} \dfrac{p(s^+)}{1-p(s^+)} = \dfrac{1-p(y^+)}{p(y^+)} \dfrac{N_{s+}}{N-N_{s+}}.$

Again the proof is in the supplementary material, and here we will give an empirical verification following the same experimental setup as the one described in Section 4.2. Thus as a sanity check in Figure 3 we observe that if we increase the sample size of the test between X and S by a factor κ, the two tests have the same power, and this result verifies Theorem 3. No matter what the sample size is, the intercepts are always at the same value (close to the design parameter $1 - \alpha$), which again verifies that the tests have the same Type-I error.

We see that the non-centrality parameter $\lambda_{G(X;S)}$ is a function of: the sample size, the desired effect size and additionally a *correction factor*, κ, which depends on the number of labelled examples that we have (N_{S+}). When we have full supervision, in other words when $p(s^+) = p(y^+)$, the κ takes the maximum value 1. In any other PU case, where $p(s^+) < p(y^+)$, the value is $\kappa < 1$.

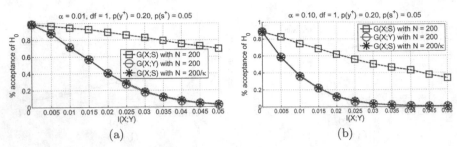

Fig. 3. Figure for comparing the Type-II error of the two tests and the G-test between X and S with corrected sample size , using (a) $\alpha = 0.01$, (b) $\alpha = 0.10$.

In PU data, the prior probability $p(y^+)$ is in general unknown. Elkan & Noto [10] suggest an estimator for this parameter, which could potentially be used before an a-priori power analysis. A different way is to introduce a prior belief over that parameter; we will represent our prior belief as $\tilde{p}$, and the correction factor is re-written as

$$\kappa = \frac{1 - \tilde{p}}{\tilde{p}} \frac{p(s^+)}{1 - p(s^+)} = \frac{1 - \tilde{p}}{\tilde{p}} \frac{N_{S+}}{N - N_{S+}}.$$

This correction factor enables us to use the $G(X; S)$ test in place of the $G(X; Y)$ for power analysis activities, such as sample size determination. Taking advantage of the extra degree of freedom in $p(s^+)$, we can also determine the *required level of supervision* (i.e. number of labelled examples), following the same procedure as in sample size determination. These capabilities will be empirically evaluated in the next section.

6   Experiments for a-priori Power Analysis

In this section we will show the capabilities of the G-test between X and S when the non-centrality parameter is corrected with the κ presented in Theorem 3, including sample size determination under the PU constraint, and a novel capability — determining the minimum number of labelled examples necessary to achieve statistical significance. We separate these experiments in two parts, the first one where we have perfect prior knowledge and the second where we use uncertain prior knowledge.

6.1   Perfect Prior Knowledge

In this section firstly we will provide some theoretical predictions for sample size and supervision determination, and then we will verify them empirically.

Theoretical Predictions for Sample Size Determination
Figure 4a shows how classical power analysis changes under the PU constraint. The illustration is for significance level $\alpha = 0.01$, a required power of 0.99,

$p(y^+) = 0.2$, and binary features (degrees of freedom $\nu = 1$). For the reader's interest, all the figures and tables of this work are reproduced in the supplementary material with $\nu = 9$, meaning $|\mathcal{X}| = 10$.

In Figure 4a we see the dashed line, which shows classical sample size determination – this is a standard result. The solid line shows the PU case, when we can obtain labels only for 5% of the examples (i.e. $p(s^+) = 0.05$).

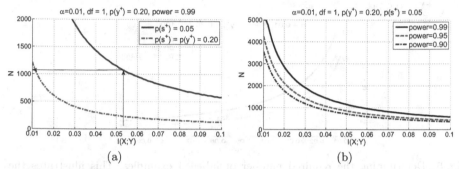

(a)   (b)

Fig. 4. Figures for sample size determination. (a) Contrasting classical power analysis ($p(s^+) = p(y^+)$) with PU power analysis to determine the minimum sample size. Arrows show that with 5% supervision ($p(s^+) = 0.05$), we need $N \geq 1077$ examples to achieve the desired power in order to observe a supervised effect $I(X;Y) = 0.053$. (b) Sample size determination under the PU constraint. Given a required statistical power, this illustrates the minimum total number of examples (N) needed, assuming we can only label 5% of the instances. For example, if we wish to detect a mutual information as low as 0.04, we need $N \geq 1430$ to have a power of 99%.

The figure can be interpreted as follows: if we wish to detect a dependency with mutual information as low as $I(X;Y) = 0.053$, with power 99%, in the fully supervised case (dashed line) we require $N \geq 227$. However in the PU scenario (solid line) with $p(s^+) = 0.05$, this a-priori power analysis indicates we need $N \geq 1077$. Note that the required increase is not a simple multiple of the supervision level: with only 1/4 of the positive examples being labelled one might assume we need a sample 4× larger, which would be 908, however this is insufficient for the required power as shown by the figure. In this case, $\kappa = 0.2105$, and the required increase is a multiple of that factor: $227 \times (1/\kappa) \approx 1078$. The above results are expanded upon in Figure 4b, showing the required N to obtain different power levels.

Theoretical Predictions for Supervision Determination

For power analysis in the PU constraint, we are able to use the same methodologies as in sample size determination to *determine the necessary level of supervision*, i.e. the number of labelled examples. This may have implications in active learning [7], where we can request the labels of particular examples — this methodology allows us to predict when we have sufficient labels to have statistically significant results.

Figure 5 presents the a-priori PU power analysis, allowing us to determine the minimum level of supervision to achieve a certain statistical power. The y-axis is N_{S+}, the number positive examples that have labels. This shows just one scenario, with $\alpha = 0.01$, $N = 1000$, when the true prior is $p(y^+) = 0.2$. As an illustration, the solid line predicts that to detect a dependency as low as $I(X;Y) = 0.053$, with power greater than 99%, we will need to label at least 54 examples or in other words the probability of an example being labeled should be $p(s^+) \geq 0.054$.

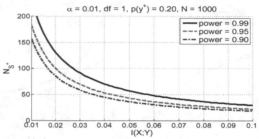

Fig. 5. Determining the required number of labelled examples. This illustrates the required number of labelled examples (N_{S+}), assuming $N = 1000$. For example, to detect a mutual information dependency as low as 0.02, in order to have a power of 95%, we need labels for 101 examples, which means that we need to label at least half of the positive examples.

Verifying the Theoretical Predictions

To verify the theoretical predictions of required sample size and supervision level, we generate binary variables with a very small dependency (i.e. very small effect size) and observe the ability of a test to reject the null hypothesis — or in other words the False Negative Rate (Type-II error). Since the power is given by $1 - FNR$, any prediction of required N to achieve a particular power will translate directly to a corresponding FNR.

As a sanity check, we first verify the classical sample size determination for the G-test. Figure 4a (dashed line) predicts that we will need $N \geq 227$ to detect an underlying effect size of $I(X;Y) = 0.053$, with $\alpha = 0.01$ and power 99%. Figure 6a shows the FNR over *10,000* repeats. Note that the FNR crosses below the 1% rate when $N \approx 225$. The next experiment verifies the PU sample size prediction. As before, the negative examples all have their labels removed, and we randomly remove a fraction $(1-c)$ of the positive labels. Figure 4a (solid line) predicted that to detect an effect as small as $I(X;Y) = 0.053$, with $\alpha = 0.01$ we would require $N \geq 1077$ to achieve an FNR below 1%. The FNR again over *10,000* repeats is shown in Figure 6b, supporting the theory as the FNR crosses 1% when $N \approx 1080$.

Finally we verify the predictions from Figure 5. We generate PU data as before, introducing noise such that the true underlying variables have $I(X;Y) = 0.053$. Figure 7 shows the FNR, verifying that when we provide labels to the example with probability less than 0.054, or in other words when we label less than 54 examples the Type-II error is greater than 1%, and agrees with Figure 5.

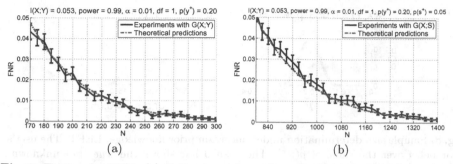

Fig. 6. Figures for FNR. (a) Full supervision, when the true mutual information is $I(X;Y) = 0.053$. This verifies the theoretical prediction from Fig. 4a. (b) Supervision level $p(s^+) = 0.05$, supporting the predictions of Fig. 4a.

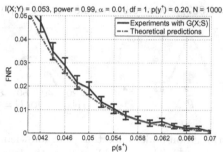

Fig. 7. FNR for varying levels of supervision in the PU constraint, with required power 99%, verifying Figure 5 (solid line), which predicted we would need $p(s^+) \geq 0.054 \Leftrightarrow N_{s+} \geq 54$ to get $FNR < 0.01$.

6.2 Uncertain Prior Knowledge

The previous section assumed we somehow knew the exact value of $p(y^+)$. In a more realistic scenario prior knowledge may be provided as a *distribution* over possible values. We model $\tilde{p}$ as a generalised Beta distribution, between a minimum and a maximum value [14], and use Monte-Carlo simulation to explore the resultant uncertainty in the required sample/supervision sizes.

Figure 8 presents sample size determination when we have uncertain prior knowledge. The dashed vertical line indicates the perfect prior knowledge situation from the previous section. If we use a sample size less than this, we have an increased false negative rate. On the other hand, choosing a larger size will achieve at least the desired power, but at the cost of collecting more data.

Figure 9 presents how this uncertainty would translate to the required number of labeled examples. The same principle of choosing a value over/under the dashed line applies: here if we select $N_{s+} > 54$ we are unnecessarily increasing our cost of label collection. In Figure 10 we observe the behavior when we underestimate (first row) or overestimate (second row) the $p(y^+)$. A general conclusion is that the uncertainty in the prior translates quite directly to an uncertainty of

$I(X;Y) = 0.053$, power $= 0.99$, $\alpha = 0.01$, df $= 1$, $p(y^+) = 0.20$

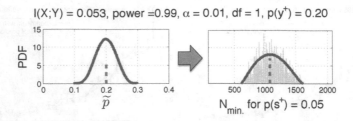

Fig. 8. Sample size determination under uncertain prior knowledge. LEFT: The user's prior belief over the value of $p(y^+)$. The dashed line shows the *true* (but unknown) value in the data. RIGHT: The resultant uncertainty in the required sample size when we have only 5% of the examples being labeled, we plot both the histogram and the generalized beta distribution best fits to the data.

$I(X;Y) = 0.053$, power $= 0.99$, $\alpha = 0.01$, df $= 1$, $p(y^+) = 0.20$

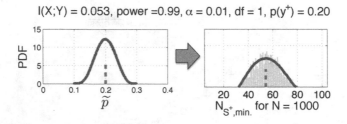

Fig. 9. Supervision determination under uncertain prior knowledge. LEFT: The user's prior belief over the value of $p(y^+)$. The dashed line shows the *true* (but unknown) value in the data. RIGHT: The resultant uncertainty in the minimum number of required labeled examples when we have only $N = 1000$. The dashed line indicates the the true value with no uncertainty in $p(y^+)$.

a similar form over the minimum number of samples and a minimum amount of supervision.

7   Guidance for Practitioners

Guidance for practitioners depends on the conditions in a given application. To ensure an effect would not be missed when indeed present, one should *overestimate* the value of $p(y^+)$, hence leading to a larger number of examples/labels being collected. Conversely, if collection of examples/labels is a costly matter, one can take a more risky, but informed, decision, using less examples/labels. Furthermore, to achieve a desired statistical power, choosing to fix the amount of supervision or the sample size is application-dependent.

Under our framework we can generate tables for sample size and supervision determination under the PU constraint similar with that used in the literature, e.g. Table 1(a). For a given effect size (ω), degrees of freedom (df), significance level (α), level of desired power, prevalence ($p(y^+)$) and fixed supervision level ($p(s^+)$), in Table 1(b) we can observe the minimum sample size is needed in

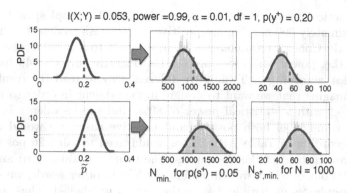

Fig. 10. A-priori power analysis under uncertain prior knowledge, when we underestimate (first row) and overestimate (second row) the prior

order to observe the effect with the given power. For the effect we followed the three levels of Cohen [6] small ($\omega = 0.10 \Leftrightarrow I(X;Y) = 0.005$), medium ($\omega = 0.30 \Leftrightarrow I(X;Y) = 0.045$) and large ($\omega = 0.50 \Leftrightarrow I(X;Y) = 0.125$).

Table 1. Sample size required for $df = 1$ and $\alpha = 0.01$

(a) Traditional

Power	Small	Medium	Large
0.70	962	107	39
0.80	1168	130	47
0.90	1488	166	60
0.95	1782	198	72
0.99	2404	268	97

(b) PU with $p(y^+) = 0.20$, $p(s^+) = 0.05$

Power	Small	Medium	Large
0.70	4566	508	183
0.80	5548	617	222
0.90	7068	786	283
0.95	8462	941	339
0.99	11415	1269	457

A new type of table can be generated when we fix the sample size and we want to determine the minimum amount of supervision (or in other words, the minimum number of labelled examples) that we need in order to observe a specific effect with a desired statistical power. Table 2 presents the minimum number of labelled positive examples that we need when we have similar conditions as before but now we fix the sample size to be $N = 3000$.

Table 2. Labelled positive examples required for a PU test with $p(y^+) = 0.20$, $N = 3000$, $df = 1$ and $\alpha = 0.01$

Power	Small	Medium	Large
0.70	223	27	10
0.80	267	33	12
0.90	331	41	15
0.95	388	49	18
0.99	501	66	24

So in practical terms: if we assume we had 3000 examples, and we know that approximately 600 of them are positive – if we wish to detect a "medium" sized effect (in Cohen's terminology), then, in order to achieve a false negative rate of 5% (i.e. power 0.95), we only need to identify correctly 49 from those 600 examples, according to Table 2. A different way to read the results is the following: imagine that we want to design an experiment in order to observe a medium effect with a statistical power of 80%, and the prevalence is $p(y^+) = 0.20$. If we could label both positive and negative cases, we would need 130 examples according to Table 1(a). So we would need to label 26 positives and 104 negatives. Instead of this we can use the results of Table 1(b) and collect 617 examples out of which we will label only 5%; in other words, we will label only 31 examples as positive and keep the rest as unlabelled. Thus, instead of labelling 104 negative examples, we can label 5 more positive examples and keep 586 as unlabelled. This approach can be useful when it is expensive or difficult to label examples, while it is cheaper to collect unlabelled. Since in the PU context labelling samples is expensive this methodology can be used to save resources.

Our results can be used in any research involving hypothesis testing in PU data. Our framework has been described in terms of the G-test, and the mutual information as an effect size. We can use the same framework to derive similar expressions for the χ^2-test, and the ϕ-coefficient as an effect size. Since both G and χ^2 are used extensively in behavioral sciences and biology, our work may have strong relevance in experimental design for partially supervised data [6,11]. The proposed methods can be used in several machine learning applications. Structure learning of a Bayesian network or Markov Blanket discovery in PU data, would use our corrected G-test to decide whether we add an arc or not, since the same correction factor κ can be derived for the conditional independence test. Furthermore our power analysis methodology would provide guidance in controlling the FNR, preventing potential underfitting of the model; a recent work for the fully supervised case is Bacciu et al. [2] – our framework generalises this to PU data. Another potential application area is information theoretic feature selection. We can apply a wide variety of feature selection criteria in PU data; a recent work for fully supervised data is Brown et al. [4].

8   Conclusions and Future Work

In this work we developed a set of novel methodologies, enabling statistical hypothesis testing activities in PU data. We proved that a very common assumption, of all unlabelled examples being negative, is sufficient for detecting *independence*. However, a G-test using this assumption is less powerful than the fully supervised version, indicating the assumption is invalid for more complex power analysis activities. We solve this problem by deriving a *correction factor* for the test, incorporating prior knowledge from the user. Using this, we can perform sample size determination, and have a novel capability: determining *the required number of labelled examples*. Experimental evidence supports all theoretical predictions. As a future work we will investigate how our framework can be extended to fully semi-supervised data and how the principles can apply to other types of hypothesis test.

Acknowledgments. The research leading to these results has received funding from EPSRC Anyscale project EP/L000725/1 and the European Union's Seventh Framework Programme (FP7/2007-2013) under grant agreement n° 318633. This work was supported by EPSRC grant [EP/I028099/1]. Sechidis gratefully acknowledges the support of the Propondis Foundation.

References

1. Agresti, A.: Categorical Data Analysis, 3rd edn. Wiley Series in Probability and Statistics. Wiley-Interscience (2013)
2. Bacciu, D., Etchells, T., Lisboa, P., Whittaker, J.: Efficient identification of independence networks using mutual information. Computational Statistics 28(2), 621–646 (2013)
3. Blanchard, G., Lee, G., Scott, C.: Semi-Supervised Novelty Detection. Jour. of Mach. Learn. Res. 11 (March 2010)
4. Brown, G., Pocock, A., Zhao, M., Lujan, M.: Conditional likelihood maximisation: A unifying framework for information theoretic feature selection. Jour. of Mach. Learn. Res. 13, 27–66 (2012)
5. Calvo, B., Larrañaga, P., Lozano, J.: Learning Bayesian classifiers from positive and unlabeled examples. Patt. Rec. Letters 28, 2375–2384 (2007)
6. Cohen, J.: Statistical Power Analysis for the Behavioral Sciences, 2nd edn. Routledge Academic (1988)
7. Cohn, D., Atlas, L., Ladner, R.: Improving generalization with active learning. Machine Learning 15(2), 201–221 (1994)
8. Denis, F.: PAC learning from positive statistical queries. In: Richter, M.M., Smith, C.H., Wiehagen, R., Zeugmann, T. (eds.) ALT 1998. LNCS (LNAI), vol. 1501, pp. 112–126. Springer, Heidelberg (1998)
9. Denis, F., Laurent, A., Gilleron, R., Tommasi, M.: Text classification and co-training from positive and unlabeled examples. In: International Conf. on Machine Learning, Workshop: The Continuum from Labeled to Unlabeled Data (2003)
10. Elkan, C., Noto, K.: Learning classifiers from only positive and unlabeled data. In: SIGKDD Int. Conf. on Knowledge Discovery and Data Mining (2008)
11. Ellis, P.: The Essential Guide to Effect Sizes: Statistical Power, Meta-Analysis, and the Interpretation of Research Results. Camb. Univ. Press (2010)
12. Gretton, A., Györfi, L.: Consistent nonparametric tests of independence. The Journal of Machine Learning Research 99, 1391–1423 (2010)
13. Guyon, I., Gunn, S., Nikravesh, M., Zadeh, L.: Feature Extraction: Foundations and Applications. Springer-Verlag New York, Inc., Secaucus (2006)
14. Hahn, G., Shapiro, S.: Statistical Models in Engineering. Wiley Series on Systems Engineering and Analysis Series. John Wiley & Sons (1967)
15. Liu, B., Lee, W., Yu, P., Li, X.: Partially supervised classification of text documents. In: International Conf. on Machine Learning, pp. 387–394 (2002)
16. Nielsen, F.G., Kooyman, M., Kensche, P., Marks, H., Stunnenberg, H., Huynen, M., et al.: The pinkthing for analysing chip profiling data in their genomic context. BMC Research Notes 6(1), 133 (2013)
17. Paninski, L.: Estimation of entropy and mutual information. Neural Computation 15(6), 1191–1253 (2003)
18. Sokal, R., Rohlf, F.: Biometry: The principles and practice of Statistics in Biological data, 3rd edn. W. H. Freeman & Co (1995)
19. Yu, H., Han, J., Chang, K.: PEBL: positive example based learning for web page classification using svm. In: SIGKDD Int. Conf. on Knowledge Discovery and Data Mining (2002)

Students, Teachers, Exams and MOOCs: Predicting and Optimizing Attainment in Web-Based Education Using a Probabilistic Graphical Model

Bar Shalem[1], Yoram Bachrach[2], John Guiver[2], and Christopher M. Bishop[2]

[1] Bar-Ilan University, Ramat Gan, Israel
[2] Microsoft Research, Cambridge, UK

Abstract. We propose a probabilistic graphical model for predicting student attainment in web-based education. We empirically evaluate our model on a crowdsourced dataset with students and teachers; Teachers prepared lessons on various topics. Students read lessons by various teachers and then solved a multiple choice exam. Our model gets input data regarding past interactions between students and teachers and past student attainment. It then estimates abilities of students, competence of teachers and difficulty of questions, and predicts future student outcomes. We show that our model's predictions are more accurate than heuristic approaches. We also show how demographic profiles and personality traits correlate with student performance in this task. Finally, given a limited pool of teachers, we propose an approach for using information from our model to maximize the number of students passing an exam of a given difficulty, by optimally assigning teachers to students. We evaluate the potential impact of our optimization approach using a simulation based on our dataset, showing an improvement in the overall performance.

1 Introduction

Recent years have marked an enormous leap in the use of the Internet and web-based technology. This technology had a huge impact on education, where web-based and online training are emerging as a new paradigm in learning [26]. Distant learning technology makes it easier to access educational resources, reduces costs and allows extending participation in education [28,2,40]. Intelligent online educational technologies enable a deep analysis of student solutions and allows automatic tailoring of content or the difficulty of exercises to the specific student [11]. One innovation that could affect higher education is massive open online courses (MOOCs), online training geared to allow large-scale participation by providing open access to resources [36,16]. MOOC providers offer a wide selection of courses, some already attracting many students. [1]

---

[1] See, for example the report on Peter Norvig and Sebastian Thrun's online artificial intelligence course, with its "100,000 student classroom", in http://www.ted.com/talks/peter_norvig_the_100_000_student_classroom.html.

T. Calders et al. (Eds.): ECML PKDD 2014, Part III, LNCS 8726, pp. 82–97, 2014.

However, web-based education also brings with it new challenges. Students may become frustrated due to ambiguous instructions or lack of prompt feedback [25]. This triggers the need to manage the quality of online teaching material, and highlights the need for an objective system for measuring performance and for efficient resource allocation [10,43,47,36].

However, measuring the quality of teaching materials or predicting the attainment of students are challenging. Teachers who teach a similar subject are likely to have completely disjoint student cohorts, of different ability levels, backgrounds, and demographic traits. Further, students may solve different tasks or get different exams (with potentially some overlap in tasks or questions).

Many questions arise in such settings. How can we aggregate observations on outcomes in order to evaluate the abilities of students, the competence of teachers and the difficulty of exams? Can we systematically predict the attainment of students? Do demographic and personality traits correlate with performance? How can we optimize resource allocation, such as the assignment of teachers to students, so as to maximize performance?

Our Contribution: We propose a probabilistic graphical model for assessing teaching material quality and student ability, and for predicting student attainment in online education. Our model gets input data regarding past interactions between students, teachers and exams and past outcomes (whether a student succeeded in answering questions in the exam), and provides predictions regarding future interactions. We evaluate our model based on a dataset crowdsourced from Amazon's Mechanical Turk (AMT). We divided the AMT workers into "teachers" and "students". Each teacher prepared "lessons" on various topics, in the form of summaries of Wikipedia articles. For each topic we constructed a multiple choice "exam", and students were asked to solve it based on the lesson prepared by one of the teachers. We show that our model can predict outcomes in such settings and estimate the abilities of students, the competence of teachers and the difficulty of questions. We show that our model outperforms heuristic approaches for predicting outcomes. We also explore how demographic profiles and personality traits correlate with student performance in this task. Finally, given a limited pool of teachers, we propose an approach for using information from our model to optimize performance in our domain, such as the number of students passing a difficult exam. We do so by choosing the optimal assignment between teachers and students, based on our model's estimates, and evaluate the potential impact of this approach using a simulation based on our dataset.

2 Probabilistic Graphical Model for Predicting Attainment in Web-Based Education

We now describe our model for predicting performance in web-based education. Our domain consists of online *exams* given to *students* who studied various *topics* with the help of *lessons* prepared by *teachers*. We denote the student set as S, the teacher set as T, the topic set as M, and the set of questions comprising the exam on topic m as Q_m. We denote the exam on topic m as E_m. A student

$s \in S$ learns topic $m \in M$ based on the lesson prepared by teacher $t \in T$, then answers the exam E_m on topic $m \in M$. We say the *outcome* for this attempt was a success, denoted $r_{s,q} = 1$, if student s answers question q correctly, and otherwise we say it is a failure, denoted $r_{s,q} = 0$. The *raw score* of student s in the exam E_m is the number of questions she answers correctly. This raw score reflects not only the ability of the student, but also how well she was taught the topic by her teacher, and the difficulty level of the questions in the exam. Thus our dataset consists of observations of the form $z_i = (s, t, m, q, r_{s,q})$. Every student is taught each topic by a single teacher (though she may receive a different teacher for different topics).

Given our observations $Z = \{z_i\}_{i=1}^w$, we wish to predict future outcomes: how well is a student s likely to do in an exam E_m on topic m when she is taught by teacher t? We refer to our problem as the *attainment problem*. The full input data to the attainment problem potentially includes an entry for the outcome on every question for every student, so its size is $|S| \cdot |Q|$. Typically, however, the input data only includes a smaller set of observations: for example, a student may only have been taught some of the topics, or was only tested using some of the questions on a topic. Given the input data, our goal is to predict the outcomes on the missing entries, so a *query* is a tuple $u_j = (s, t, m, q)$. A query is similar to the input entries, except it is missing the outcome r, to be interpreted as requesting the model to predict whether student s would answer the question q regarding the topic m correctly when taught by teacher t.

Predicting Outcomes Using a Probabilistic Model: We propose a probabilistic graphical model for the attainment problem, called the Student-Teacher-Exam-Performance model — **STEP**. Given the input observations $Z = \{z_i\}_{i=1}^w$ and queries $U = \{u_j\}_{j=1}^l$, the model's output consists of predictions regarding the outcomes for the entries in the query set $R = (r_1, \ldots, r_l)$. STEP also outputs information regarding latent variables, such as the ability level of each student, the competence of each teacher and difficulty of each question. The outcomes in the query set U, as well as the abilities, competences and difficulties, are modeled as *unobserved random variables*. In contrast, the outcomes in the observation set Z are *observed variables*. The structure of our STEP model is governed by independence assumptions regarding the variables. Pearl discusses Bayesian Networks [42] (now referred to as directed graphical models), which represent conditional independence assumptions as a graph where each vertex corresponds to a variable and the edges capture dependencies between adjacent variables. We base STEP on a prominent extension of Bayesian Networks, called Factor Graphs (see [29]), which describes a factorial decomposition of an assumed joint probability distribution between the variables.

We first define the crux of the model in the form of a Factor Graph representation. We then set the observed variables in the graph to the values of the observations Z, consisting of the identities of the students, teachers, topics and questions, and most importantly the outcomes in our observation set. We then use approximate message passing algorithms [29] to infer marginal probability distributions of the target unknown variables: student abilities, teacher competences, question difficulties, and of course the unobserved outcomes of the query set. We thus get a posterior distribution over these unobserved variables.

The Graphical Model: Recall that the variable $r_{s,q}$ indicates whether student s answered question q correctly ($r_{s,q} = 1$ indicates the answer was correct and $r_{s,q} = 0$ indicates it was incorrect). This variable is an observed variable for every entry $z_i = (s, t, m, q, r) \in Z$ (though it is unobserved in the query set U). We model the process which causes a student $s \in S$ to either answer a question correctly or incorrectly. We assume every student $s \in S$ has an inherent *ability* $a_s \in \mathbb{R}$ reflecting how easy she finds it to learn new topics and answer questions on them, and that every teacher $t \in T$ has an inherent *competence* $c_t \in \mathbb{R}$ reflecting her ability to teach students and provide them information on a topic. We assume every question $q \in Q$ has an inherent *difficulty* $d_q \in \mathbb{R}$ determining how likely it is that a student could answer it correctly.

Our model is a joint probabilistic model with a factor graph representation given in Figure 1. The model has two parts. The first part reflects the probability that student s actually knows the correct answer to a question q, denoted by the variable $k_{s,q}$, as determined by the student ability parameter a_s, the teacher competence parameter c_t (where t is the teacher who taught s the topic of that question), and the question difficulty parameter d_q. In Figure 1, this is shown to the left and above the vertex of $k_{s,q}$. The second part determines the observed outcome, depending on $k_{s,q}$ and is shown to the right of the vertex of $k_{s,q}$.

$k_{s,q}$ is a Boolean variable. A value of 1 indicates that the student s knows the correct answer to the question q, while a value of 0 indicates she does not know the answer (but may still give the right answer to the question by making

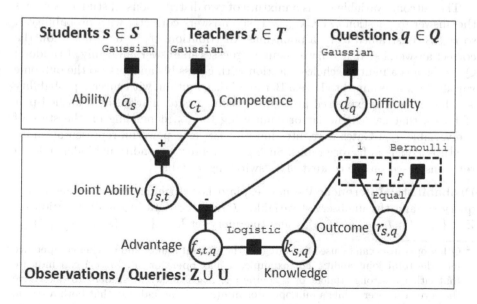

Fig. 1. Factor graph for the STEP model

a lucky guess). The probability of $k_{s,q}$ having the value 1 increases with the student ability and teacher competence and decreases with the difficulty of the question. By $f_{s,t,q}$, we denote the difference between the "total joint ability" of the student and the teacher $(a_s + c_t)$ and the difficulty of the question (d_q), so $f_{s,t,q} = (a_s + c_t) - d_q$. [2] The variable $f_{s,t,q}$ reflects the "advantage" the student has over the question after she is taught the relevant topic by the teacher.

We assume that $k_{s,q}$ depends on the advantage $f_{s,t,q}$ as follows:

$$P(k_{s,q} = 1 | f_{s,t,q}, \tau_q) := \int_{x=-\infty}^{x=\infty} \phi(\sqrt{\tau_q}(x - f_{s,t,q}))\theta(x)\, dx$$

$$= \Phi\left(\sqrt{\tau_q} f_{s,t,q}\right). \tag{1}$$

Where ϕ is the Gaussian density: $\phi(x) := \frac{1}{\sqrt{2\pi}} e^{\frac{-x^2}{2}}$, Φ is the sigmoidal cumulative Gaussian distribution: $\Phi(t) := \int_{x=-\infty}^{t} \phi(x)\, dx$, and $\theta(\cdot)$ is the Heaviside step function. The integral presentation allows for the following interpretation of this probability: this is a binary process which results from evaluating the step function θ over a variable f which is added a Gaussian noise of variance $\frac{1}{\tau}$. Another way to view this is that the data is assumed to come from a probabilistic generative process: the student's ability, teacher's competence and question's difficulty are sampled from random Gaussian distributions which reflect the distribution of those properties in the population. A random "performance noise" for each entry in the observation set Z, which may be either positive or negative, is added to the total joint ability (the sum of the student ability and teacher competence); If this number is greater than the difficulty of the question, then the student knows the correct answer so $k_{s,q} = 1$, otherwise $k_{s,q} = 0$.

The outcome variable $r_{s,q}$ is a mixture of two distributions. If student s knows the answer to question q, i.e. $k_{s,q} = 1$, she answers correctly with probability 1, so $r_{s,q}$ is constrained to be a point-mass distribution. If she does not know the correct answer, i.e. $k_{s,q} = 0$, we assume s guesses an answer uniformly at random; Question q is a multiple choice question with b possible answers, so the outcome variable $r_{s,q}$ is assumed to have a Bernoulli distribution, with success probability $\frac{1}{b}$. The mixture is expressed in Figure 1 using a *gate*, marked by a dashed pair of boxes, that switch the factor connecting to $r_{s,q}$, depending on the state of the variable $k_{s,q}$. Gates were introduced in [38] as a powerful representation for mixture models in factor graphs. Such gates represent conditional independence relations based on the context of a "switching variable".

Probabilistic Inference: We now explain how to infer the outcomes in the query set and the unobserved variables. Given the data in the observation set $\mathbf{Z} = \{z_i = (s_i, t_i, m_i, q_i, r_i^Z)\}_{i=1}^{w}$ and the query set $\mathbf{U} = \{u_j = (s_j, t_j, m_j, q_j)\}_{j=1}^{l}$

---

[2] Other operators can be used to aggregate the student ability and teacher competence into the total joint ability. For example, a max operator $\max(a_s, c_t)$ can indicate that either a strong student or a competent teach allow the student to determine the correct answer, while a min operator $\min(a_s, c_t)$ can indicate that both a strong student and a competent teacher are required. The complexity of performing the inference in such alternative graphical models depends on the operator used.

we are interested in predicting the missing outcomes for the query set $\{r_j\}_{j=1}^{l}$. We do so by simultaneously inferring several approximate posterior (marginal) distributions: the Gaussian density of the ability of each student, competence of each teacher and difficulty of each question, the Bernoulli distribution indicating whether each student knew the answer to each question for all such entries in the observation set and query set, and the Bernoulli distribution indicating whether each student gave a correct response to the question asked for all entries in the query set. The posterior distributions $\{p(r_j|\mathbf{Z}, \mathbf{U})\}_{j=1}^{l}$ can be interpreted as the probability that the outcome in the j'th entry in the query set would be a success (i.e. the probability that the student would answer the question correctly). This posterior distribution is a Bernoulli distribution, so we can simply denote the probability of a successful outcome as $p_{r_j} = p(r_j = 1|\mathbf{Z}, \mathbf{U})$. When requested to make a binary prediction rather than estimate the probability of a successful outcome, we use the *mode* of that distribution: if $p_{r_j} > \frac{1}{2}$ we predict a success, and otherwise predict a failure.

To perform the inference and compute the posterior distribution in STEP, we use Expectation-Propagation approximate message passing (see [39,29]), using Infer.NET [37], a framework for probabilistic modeling. [3]

3   Model Evaluation

We evaluated STEP using a dataset crowdsourced from Amazon's Mechanical Turk (AMT). AMT is a crowdsourcing marketplace bringing together workers interested in performing jobs remotely, and requesters interested in obtaining human labor for tasks. We constructed tasks for a remote learning experience, both on the teacher's side and the student' side. We first selected 10 Wikipedia articles covering various topics such as Chad, Saffron and DNA. We composed an "exam" on each of those topics, consisting of 5 multiple choice questions (50 questions total). We divided the worker set to two groups: "teachers" and "students". Each teacher was required to write a short (1500 character) "lesson" on each of the topics. The teachers were notified which issues to focus on when preparing the students for the exam (for example the history of Chad or the chemical structure of DNA). However, they did *not* know which specific questions were in the exam. Each student was asked to study the topic using the lesson provided by a teacher we chose, then solve the exam on that topic. The time given to solve the exam was limited to 3 minutes per topic, making it difficult (though not impossible) for students to consult external resources other than the teacher's lesson.

Data Collection: Our dataset consists of observations regarding the questions solved by students, in the form discussed in the previous section: student, teacher,

---

[3] STEP's factor graph is loopy, as we have multiple participants who respond to the same question set and share the same teacher set. Thus EP computes the posteriors by iterating until convergence. The number of iterations used in Infer.NET is constant, so the procedure runs in time linear in the input, i.e. in $O(|S| \cdot |Q|)$.

topic, question, and correctness. We sourced 237 workers for the task from AMT. We used 10 of them as teachers, and 227 as students. Each teacher had prepared a lesson on each of the 10 topics. Lessons were allocated to students as follows. Each student got 10 lessons by 10 different teachers. For each student, the teacher permutation was modified by cyclic shift, i.e. student s got the lesson by teacher $(s + m)$ mod $|T|$ on topic m, where $|T|$ is the number of teachers. Each student answered all 5 questions on each topic resulting in a total of 11,350 entries in our dataset.

The students were given a base payment of \$2 for performing the task, and a bonus of up to \$3 depending on their performance, measured by the number of questions they answered correctly. The teachers received a base payment of \$10, for writing the lessons, and engaged in a contest for an additional bonus of \$10: each teacher was randomly paired up with another teacher; The teacher with better performing students was awarded a \$10 bonus. [4] In addition to answering the questions, each student completed a demographics survey regarding their age, gender, income and education. They also completed a short personality questionnaire called TIPI [22]. TIPI follows the Five Factor Personality Model, [13,45], a generally accepted model representing the "basic structure" underlying human personality, whose ability to predict human behavior has been thoroughly investigated [15,9]. [5] The key five personality traits are Openness to experience, Conscientiousness, Extroversion, Agreeableness and Neuroticism (OCEAN for short).

Model Performance: We examined the performance of our STEP model, evaluated by randomly partitioning the data into a training set and a test set. We compared our model to heuristic approaches using two error metrics. The first error metric is the *prediction error*, which is the mean absolute difference between the actual answers (0 for an incorrect answer and 1 for a correct answer) and the model estimated probability of a correct answer. The second metric is based on a binary outcome prediction. We round the estimated probabilities of answering a question correctly to get a binary classification. The *classification error* is the proportion of entries where the model mis-classified the outcome.

We compared the performance of STEP with two heuristics. Given a target student s, our *student heuristic* examines all the entries with that student in the training set, and measures the proportion of those where the outcome was a success (i.e. the proportion of the student's entries where she gave a correct answer). This proportion is then used as the estimated probability of a successful outcome on each of that student's entries in the test set. Similarly, given a teacher

---

[4] While there is a high variance in the performance of participants in AMT [30,5], such contests are known to have good properties in terms of incentivizing the participants to exert significant effort on the task [27,3,20,52] (so long as participants are anonymous and are not colluding [35])

[5] Further, it is possible to automatically infer personality traits from peoples' social network profiles [7,32,6] or website choices [33,31], allowing such publicly available information to be used to profile students and make predictions about their performance in educational settingts.

t our *teacher heuristic* examines all the entries with that teacher in the training set and measures the proportion of those entries where the outcome was a success (i.e. the proportion of the teacher's entries where her student gave the correct answer, no matter who that student was). This proportion is then used as the estimated probability of a successful outcome on each of that teacher's entries in the test set. The student heuristic ignores information regrading who the teacher was, and the teacher heuristic ignores information regarding who the student was, while STEP uses all the available full information.

Figure 2 compares the quality of our model with the student and teacher heuristics, in terms of the classification error and prediction error metrics. The x-axis in both plots is the number of observations available in the training set. For each point in the plot we randomly selected a subset of questions, whose size was determined by the location on the x-axis, and used their entries as the training set. The remaining entries were used as a test set, with an unobserved outcome. We repeated the sampling 500 times and averaged the resulting error metrics. Figure 2 shows that the STEP produces better predictions than the heuristics, as it has a lower error for both error metrics discussed above. For both the heuristics and STEP, the error decreases as more data is given as input, but the improvement diminishes in the size of the data.

In addition to the outcome predictions regarding queries in the test set, the STEP model also returns information regarding the abilities of students, competence of teachers and difficulty of questions, captured as posterior distribution for the model parameters. These parameters allow us to rank students, teachers and questions, by their abilities, competence levels and difficulties, correspondingly. The values of these parameters are shown in Figure 3.

Figure 3 indicates high variances of the parameters. STEP sums together student ability and teacher competence and compares the sum with the question difficulty. The variability on the y axis between student abilities is larger than the variability between teacher competences, indicating that the identity of the student had stronger impact on performance than the identity of the teacher.

One simple way to "score" student abilities is by the proportion of questions they answered correctly. Correspondingly, we can score teacher competence by the proportion of questions that their students answered correctly. Similarly, we can score question difficulty by the proportion of all students who managed to correctly answer that question (here a high score means an easy question). Unsurprisingly, there is strong positive correlation ($r = 0.997$) between a student overall score in the full exam and her inferred ability, and between the average score of a teacher's students and her inferred competence ($r = 0.999$). Similarly, there is a strong negative correlation ($r = -0.946$) between the proportion of students who managed to solve a question and its inferred difficulty.

Demographics and Student Success: STEP predicts student success based on observed outcomes in previous interactions. Other sources of information regarding a student, such as demographic traits or personality traits may also help predict student performance. Previous work has already examined the correlation between a student's demographic or personality traits and success in online

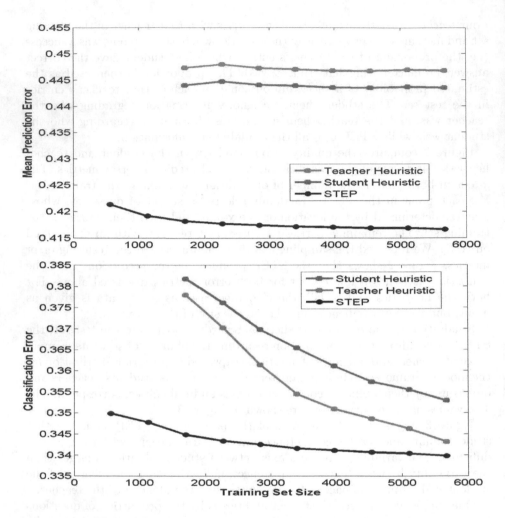

Fig. 2. Model quality - prediction and classification errors

tasks or in traditional educational settings [17,53,44,48,19]. We now examine such correlations in our web-based educational task. We measured a student's performance using the proportion of questions they answered correctly. We correlated this student performance score with other traits of the student, such as their age, level of education or personality. We found strong evidence for a positive correlation between a student's educational level and performance. We also found strong evidence of correlation between a student's personality and performance: both openness to experience and extroversion correlate positively with performance in our task; There is also some weak evidence for a positive correlation between a student's conscientiousness or agreeableness and performance. To test for the statistical significance we divided students into groups. For the educational level we used the questionnaire categories. For age and personality

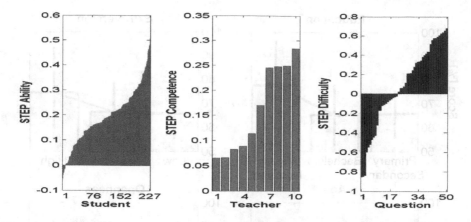

Fig. 3. STEP parameters - student ability, teacher competence and question difficulty

traits, we divided the student population into 3 equal size groups (low, medium and high) according to their responses in the questionnaire. We used a Mann-Whitney U-test (see [49]) to test the statistical significance of the differences between the low group and the high group. The statistically significant results (at a $p < 5\%$ level) are given in Table 1.

Table 1. Demographic/personality and performance

Property	Pearson Correlation	p value
Education	N/A	0.0001
Openness	0.2371	0.0001
Age	-0.1709	0.0032
Extroversion	0.2902	0.0068
Conscientiousness	0.1526	0.0405
Agreeableness	0.1867	0.0455

Table 1 shows that young or educated students had better performance. Further, those high in openess to experience or extroversion tended to do well in our task. Figure 4 visualizes these relations, showing the average performance for different groups (and showing the standard error).

Our results show a correlation between demographics or personality traits and performance in our task. Despite these correlations, there is a huge variability in performance even for workers with very similar demographic or personality profiles, highlighting the need to base predictions regarding attainment on observations regarding past performance, as done in the STEP model.

Student Teacher Matching: Our experiment used teaching materials prepared by various teachers. Online education can allow high volumes of students to access training material though the Internet. However, direct student-teacher interaction, by a phone call or a chat, allows teaching more difficult material and

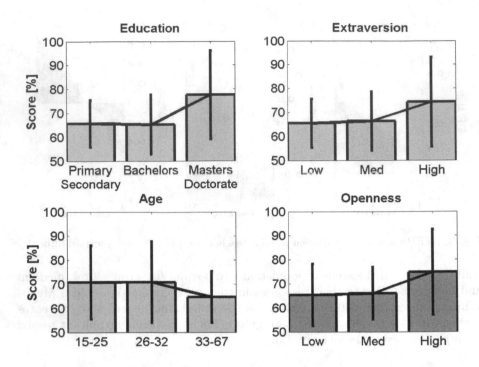

Fig. 4. Demographics and performance

achieves a higher rates of learning [41]. Such individual training requires having many teachers, as each teacher can only directly interact with few students. Nonetheless, one difference between traditional and online education systems is the flexibility in assigning teachers to students. Traditional education is constrained by physical limitations: a teacher who lives in one city cannot teach in another remote city. In online education a single student can be taught by many different teachers from across the globe, without leaving the comfort of their own home. We show that this allows us to optimize the assignment of teachers to students in order to improve the overall student performance. We use the model's estimate of a teacher t's competence in preparing teaching material as a *proxy* of how well they teach by *direct interaction*: though in our experiment the teaching materials prepared by a teacher can be used to train many students, we consider the case where a teacher can only interact with a single student.

STEP infers Gaussian posterior distributions for the competence of teachers and abilities of students. Given these parameters and a question (or exam) of a given difficulty, it infers p_c, the probability that student s would succeed in answering the question q if she is taught by the teacher t. Let $S \sim N(\mu_s, \sigma_s^2)$ be the inferred student s's ability, $T \sim N(\mu_t, \sigma_t^2)$ the inferred teacher t's competence, $D \sim N(\mu_d, \sigma_d^2)$ the question difficulty and $N \sim N(0, \sigma_n^2)$ the Gaussian noise used in the model. Let $p_c(s, t)$ be the probability that student s taught by teacher t knows the correct answer to a question of difficulty d (similar to the

Bernoulli variable $k_{s,q}$ in the previous section.) Under the assumptions of the STEP model, $p_c(s,t) = Pr(S + T + N > D)$, we can compute $p_c(s,t)$ for any student s and teacher t.

Consider a domain with an equal number n of teachers and students. Suppose every teacher has the capacity to teach a single student, and that we wish to maximize the number of students who pass an exam of difficulty d. How should we choose an assignment $A : S \to T$ between students and teachers, which respects the teacher capacity constraints (i.e. for any $t \in T$ there is only one student $s \in S$ such that $A(s) = t$), so as to maximize the expected number of passing students: $\arg\max_A \sum_{s \in S} p_c(s, A(S))$? The simplest way is a random assignment, which ignores the inferred abilities. However, when maximizing the number of passing students we only care if a student passes (rather than considering the exact score). If we have one good student and one bad student, and one good teacher and one bad teacher, we may be better off matching the good teacher to the bad student and the bad teacher to the good student, as the "returns on competence" can decrease with the ability of the student. [6] One heuristic is to sort students by increasing ability, and the teachers by decreasing competence and match them in that order. We call this the *inverse heuristic assignment*. Given an exam of difficulty d, matching a student s with teacher t has the expected return of $p_c(s,t)$. We can formulate maximizing the expected number of "passing" students as a Bipartite Maximum Weighted Matching (BMWM) problem [51]; We are given a bipartite graph of students on one side and the teachers on the other, and the edge between student s and teacher t has weight $w_{(s,t)} = p_c(s,t)$; The goal is find an assignment $A : S \to T$ matching each teacher to exactly one student so as to maximize the sum of weights of the matching. The BMWM output is the assignment A maximizing $\sum_{s \in S} w_{(s,A(s))}$. This *optimal assignment* (equivalently BMWM) can be found in polynomial time [51].

We compared the three matching algorithms (random assignment, inverse heuristic assignment and the optimal assignment) in terms of their performance, measured by the expected number of passing students. As the input data for the simulations we used the scaled output parameters of STEP on the real data discussed in the previous section. We only had 10 teachers in this dataset, so we randomly sampled a subset of 10 students many times, averaging the resulting performance under the three assignment methods. We performed the analysis on a range of question difficulty levels (matching the student abilities and teacher competences). The results are shown in Figure 5.

Figure 5 shows that for easy questions, the inverse heuristic outperforms random matching, and almost as good as the optimal assignment. However, as the difficulty increases, the inverse heuristic's performance degrades, until at some point it is even worse than random matching. For such moderate to difficult exams, there is a performance gain when switching to the optimal assignment. One possible reason for this is low ability students. If the exam is easy, such students are likely to pass when assigned a highly competent teacher, so the

---

[6] Such diminishing returns are prevalent in many resource allocation settings [12,18,8,2,40].

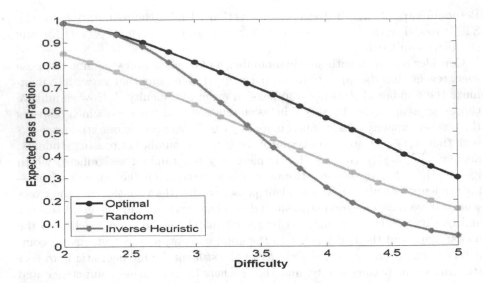

Fig. 5. Performance under assignment methods

inverse heuristic does well. However, if the exam is difficult, even a competent teacher cannot help such low ability students pass, so this heuristic "wastes" a very good teacher on a student that is very likely to fail nonetheless.

4   Related Work

Various models were proposed for assessing teacher competence [34,14,23]. To the best of our knowledge, we are the first to propose a probabilistic graphical models that simultaneously estimates student abilities, teacher competence and exam difficulties. The impact of demographics or personality on student attainment in *traditional* educational settings was studied in [17,53,44,48,19].

Our teachers' bonus was based on a competition. Such crowdsourcing contests were shown to allow the contest designer to elicit significant participant efforts [27,3].

Predicting attainment in cognitive tasks is a central topic in psychology. Psychometricians developed a framework called "test theory" to analyze outcomes in psychological testing, including intelligence and education [1]. One paradigm for designing such tests is "item-response theory" [24] (IRT for short), used to develop high-stakes adaptive tests such as the Graduate Management Admission Test (GMAT). Our STEP model relies on a probabilistic graphical model [29], and uses themes similar to the principles of IRT. A key difference is that we consider teacher competence as well, and tie the variables in the form of a factor graph. Frameworks using IRT principles and a probabilistic graphical model are [50,4,46]. However, the goal of these models is to *aggregate* multiple responses of participants to best determine the correct answers to questions, whereas our goal is to predict future performance of teachers and students in online education.

Our work ignored logical connections between questions. In many exams several questions rely on the same piece of knowledge, so a mistake regarding this information is likely to affect many responses. Frameworks such as Probabilistic Relational Models [21] combine a logical representation with probabilistic semantics, and can be used to express such structures.

5 Conclusion

We introduced the STEP model for estimating abilities and predicting outcomes in web-based education based on student abilities, teacher competences and question difficulties. We evaluated it on a crowdsourced dataset. We showed that STEP outperforms alternative approaches, and explored possible applications of this model. We have also analyzed the relation between attainment and demographics or personality traits. Finally, we have shown that the outputs of the STEP model regarding student abilities and teacher competences can be used to optimize the overall attainment of all the students by best matching teachers to students. This achieves an overall performance that is much better than a random or heuristic assignment.

Several directions remain open for future research. STEP was evaluated using data from a short experiment in AMT, which does not necessarily reflect a realistic online learning environment. Can a similar model predict outcomes in traditional education systems? Do our results generalize to real-world data from MOOCs? Can we build a dynamic model, that tracks fluctuations in student ability and teacher competence over time? How can we express dependency relations between tasks and areas of expertise?

References

1. Anastasi, A., Urbina, S., et al.: Psychological testing. Prentice Hall, Upper Saddle River (1997)
2. Anderson, T.: The theory and practice of online learning. Au Press (2008)
3. Archak, N.: Money, glory and cheap talk: analyzing strategic behavior of contestants in simultaneous crowdsourcing contests on topcoder. com. In: Proceedings of the 19th International Conference on World Wide Web, pp. 21–30. ACM (2010)
4. Bachrach, Y., Graepel, T., Minka, T., Guiver, J.: How to grade a test without knowing the answers—a bayesian graphical model for adaptive crowdsourcing and aptitude testing. In: ICML (2012)
5. Bachrach, Y., Graepel, T., Kasneci, G., Kosinski, M., Van Gael, J.: Crowd iq: aggregating opinions to boost performance. In: AAMAS (2012)
6. Bachrach, Y., Graepel, T., Kohli, P., Kosinski, M., Stillwell, D.: Your digital image: factors behind demographic and psychometric predictions from social network profiles. In: AAMAS (2014)
7. Bachrach, Y., Kosinski, M., Graepel, T., Kohli, P., Stillwell, D.: Personality and patterns of facebook usage. In: ACM WebSci (2012)
8. Bachrach, Y., Rosenschein, J.S.: Distributed multiagent resource allocation in diminishing marginal return domains. In: AAMAS (2008)

9. Barrick, M.R., Mount, M.K.: The big five personality dimensions and job performance: a meta-analysis. Personnel Psychology 44(1), 1–26 (2006)
10. Brabazon, T.: Digital hemlock: Internet education and the poisoning of teaching. University of New South Wales Press (2002)
11. Brusilovsky, P., et al.: Adaptive and intelligent technologies for web-based eduction. KI 13(4), 19–25 (1999)
12. Clearwater, S.H.: Market-based control: A paradigm for distributed resource allocation (1996)
13. Costa Jr., P.T., McCrae, R.R.: Neo personality inventory–revised (neo-pi-r) and neo five-factor inventory (neo-ffi) professional manual. Psychological Assessment Resources, Odessa (1992)
14. Darling-Hammond, L.: Evaluating teacher effectiveness: How teacher performance assessments can measure and improve teaching (2010)
15. De Raad, B., Schouwenburg, H.C.: Personality in learning and education: A review. European Journal of Personality 10(5), 303–336 (1998)
16. Downes, S.: The rise of moocs. Stephens Web (2012)
17. Dumais, S.A.: Cultural capital, gender, and school success: The role of habitus. Sociology of Education, 44–68 (2002)
18. Eichler, H.-G., Kong, S.X., Gerth, W.C., Mavros, P., Jönsson, B.: Use of cost-effectiveness analysis in health-care resource allocation decision-making: How are cost-effectiveness thresholds expected to emerge? Value in health 7(5), 518–528 (2004)
19. Engle, J., Tinto, V.: Moving beyond access: College success for low-income, first-generation students. Pell Institute for the Study of Opportunity in Higher Education (2008)
20. Gao, X.A., Bachrach, Y., Key, P., Graepel, T.: Quality expectation-variance trade-offs in crowdsourcing contests. In: AAAI (2012)
21. Getoor, L., Friedman, N., Koller, D., Pfeffer, A., Taskar, B.: 5 probabilistic relational models. Statistical Relational Learning, 129 (2007)
22. Gosling, S.D., Rentfrow, P.J., Swann, W.B.: A very brief measure of the big-five personality domains. Journal of Research in Personality 37(6), 504–528 (2003)
23. Glazerman, S., Loeb, S., Goldhaber, D., Staiger, D., Raudenbush, S., Whitehurst, G.: Evaluating teachers: The important role of value-added. Brookings Institution (2010)
24. Hambleton, R.K., Swaminathan, H., Rogers, H.J.: Fundamentals of item response theory, vol. 2 (1991)
25. Hara, N.: Student distress in a web-based distance education course. Information, Communication & Society 3(4), 557–579 (2000)
26. Harasim, L.: Shift happens: Online education as a new paradigm in learning. The Internet and Higher Education 3(1), 41–61 (2000)
27. Howe, J.: The rise of crowdsourcing. Wired Magazine 14(6), 1–4 (2006)
28. Khan, B.H.: Web-based instruction. Prentice Hall (1997)
29. Koller, D., Friedman, N.: Probabilistic Graphical Models: Principles and Techniques (2009)
30. Kosinski, M., Bachrach, Y., Kasneci, G., Van-Gael, J., Graepel, T.: Crowd iq: Measuring the intelligence of crowdsourcing platforms. In: ACM WebSci (2012)
31. Kosinski, M., Bachrach, Y., Kohli, P., Stillwell, D., Graepel, T.: Manifestations of user personality in website choice and behaviour on online social networks. Machine Learning 95(3), 357–380 (2014)
32. Kosinski, M., Stillwell, D., Graepel, T.: Private traits and attributes are predictable from digital records of human behavior. PNAS (2013)

33. Kosinski, M., Stillwell, D., Kohli, P., Bachrach, Y., Graepel, T.: Personality and website choice (2012)
34. Lavy, V.: Evaluating the effect of teachers group performance incentives on pupil achievement. Journal of Political Economy 110(6), 1286–1317 (2002)
35. Lev, O., Polukarov, M., Bachrach, Y., Rosenschein, J.S.: Mergers and collusion in all-pay auctions and crowdsourcing contests. In: AAMAS (2013)
36. Mackness, J., Mak, S., Williams, R.: The ideals and reality of participating in a MOOC. In: Networked Learing Conference (2010)
37. Minka, T., Winn, J.M., Guiver, J.P., Knowles, D.A.: Infer.NET 2.4 (2010)
38. Minka, T., Winn, J.: Gates. In: NIPS, vol. 21 (2008)
39. Minka, T.P.: A family of algorithms for approximate Bayesian inference. PhD thesis (2001)
40. Moore, M.G., Kearsley, G.: Distance education: A systems view of online learning. Wadsworth Publishing Company (2011)
41. Palloff, R.M., Pratt, K.: Lessons from the cyberspace classroom: The realities of online teaching. Wiley. Com (2002)
42. Pearl, J.: Probabilistic reasoning in intelligent systems: networks of plausible inference (1988)
43. Picciano, A.G.: Beyond student perceptions: Issues of interaction, presence, and performance in an online course. Journal of Asynchronous Learning Networks 6(1), 21–40 (2002)
44. Ridgell, S.D., Lounsbury, J.W.: Predicting academic success: General intelligence,big five personality traits, and work drive. College Student Journal 38(4), 607–618 (2004)
45. Russell, M.T., Karol, D.L.: Institute for Personality, and Ability Testing. In: The 16PF Fifth Edition Administrator's Manual, Institute for Personality and Ability Testing, Champaign (1994)
46. Salek, M., Bachrach, Y., Key, P.: Hotspotting – a probabilistic graphical model for image object localization through crowdsourcing. In: AAAI (2013)
47. Schochet, P.Z., Chiang, H.S.: Error rates in measuring teacher and school performance based on student test score gains. ncee 2010-4004. National Center for Education Evaluation and Regional Assistance (2010)
48. Scott, J.: Family, gender, and educational attainment in britain: A longitudinal study. Journal of Comparative Family Studies 35(4), 565–590 (2004)
49. Sprinthall, R.C., Fisk, S.T.: Basic statistical analysis. Prentice Hall, Englewood Cliffs (1990)
50. Welinder, P., Branson, S., Belongie, S., Perona, P.: The multidimensional wisdom of crowds. In: NIPS (2010)
51. West, D.B., et al.: Introduction to graph theory, vol. 2. Prentice Hall, Upper Saddle River (2001)
52. Witkowski, J., Bachrach, Y., Key, P., Parkes, D.C.: Dwelling on the negative: Incentivizing effort in peer prediction. In: HCOMP (2013)
53. Yorke, M., Thomas, L.: Improving the retention of students from lower socio-economic groups. Journal of Higher Education Policy and Management 25(1), 63–74 (2003)

Gaussian Process Multi-task Learning Using Joint Feature Selection

P.K. Srijith and Shirish Shevade

Computer Science and Automation, Indian Institute of Science, India
{srijith,shirish}@csa.iisc.ernet.in

Abstract. Multi-task learning involves solving multiple related learning problems by sharing some common structure for improved generalization performance. A promising idea to multi-task learning is joint feature selection where a sparsity pattern is shared across task specific feature representations. In this paper, we propose a novel Gaussian Process (GP) approach to multi-task learning based on joint feature selection. The novelty of the proposed approach is that it captures the task similarity by sharing a sparsity pattern over the kernel hyper-parameters associated with each task. This is achieved by considering a hierarchical model which imposes a multi-Laplacian prior over the kernel hyper-parameters. This leads to a flexible GP model which can handle a wide range of multi-task learning problems and can identify features relevant across all the tasks. The hyper-parameter estimation results in an optimization problem which is solved using a block co-ordinate descent algorithm. Experimental results on synthetic and real world multi-task learning data sets demonstrate that the flexibility of the proposed model is useful in getting better generalization performance.

Keywords: Gaussian process, multi-task learning, feature selection.

1 Introduction

Multi-task learning (MTL) is used in situations where one has to solve several related learning problems. MTL considers each learning problem as a separate task, but instead of learning the tasks independently, learns them together [1]. It is extremely effective when each learning problem is associated with a limited data set. It enables a task to be learnt using the data from multiple related tasks. This results in a better predictive performance of the individual tasks. It has been shown that multi-task learning performs better than learning tasks independently [2,3,4]. Multi-task learning methods have been successfully applied to applications like user preference modeling [5] and conjoint analysis [6].

Multi-task learning has recently created a lot of interest in the machine learning community. Many approaches have been proposed to effectively learn from multiple related tasks by capturing the similarity among them. Task similarity can be captured by restricting different task functions to be close to each other in some sense [4]. Bayesian approaches [5,7] capture the task similarity by sharing

T. Calders et al. (Eds.): ECML PKDD 2014, Part III, LNCS 8726, pp. 98–113, 2014.

a common prior among different tasks. Other approaches capture task similarity by sharing a common internal representation across all the tasks [2,3].

Multi-task learning using joint feature selection has been shown to improve performance in many scenarios [6,8,9,10]. These methods capture the similarity across the tasks by selecting a common subset of features or by sharing a sparsity pattern over feature representations. This is useful in situations like user preference modeling where a few product features are considered to be important by most of the users. Bayesian joint feature selection approaches [9,11,10] learn relevant features by imposing sparsity inducing priors over the feature coefficients. Regularized joint feature selection approaches [6,8] perform joint feature selection using a regularization framework in which a mixed norm regularizer is used over the feature coefficient matrix. This results in sharing the sparsity pattern over task specific feature coefficients. We propose an approach to multi-task learning based on joint feature selection using the non- parametric Bayesian framework of Gaussian process.

Gaussian process (GP) is a non-parametric model which provides a probabilistic approach to learning with kernels [12]. Being probabilistic, GP based MTL approaches provide an estimate of uncertainty over predictions. Being non-parametric, it allows the complexity of the decision function to grow with the data size. Most of the GP based approaches to MTL model task similarity by sharing a common prior across the tasks [7,13]. A task covariance matrix is learnt in [14] to model the task similarity. The semi-parametric latent factor approach [15] models each task as a linear combination of latent functions with task specific weights. We propose a general and a flexible GP based MTL approach, Gaussian process multi-task feature selection (GPMTFS), based on the idea of joint feature selection.

Gaussian process multi-task feature selection (GPMTFS) performs multi-task regression by jointly selecting features relevant across all the tasks. This is useful in many multi-task scenarios like speech recognition and handwriting recognition where some features are relevant across all the tasks while the rest are irrelevant. GPMTFS models task similarity by sharing a sparsity pattern over the feature specific parameters associated with each task. The approach considers a covariance function which implements automatic relevance determination (ARD) for each task and shares the sparsity pattern over the task specific ARD hyper-parameters. This is achieved by placing a multi-Laplacian prior over feature specific hyper-parameters across the tasks. A maximum a posteriori (MAP) estimate of the hyper-parameters is obtained using a block co-ordinate descent algorithm. The approach facilitates the selection of features which are relevant across all the tasks and leads to a better generalization performance. The proposed approach is different from Gaussian Process multi-task learning (GPMTL) [13] which can be used to perform joint feature selection by employing an ARD enabled covariance function. Fig. 1 provides a graphical model representation of the two approaches. In GPMTL, all the tasks share the same

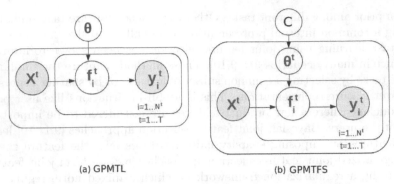

(a) GPMTL   (b) GPMTFS

Fig. 1. Graphical model for GPMTL and GPMTFS

ARD hyper- parameters. GPMTFS allows each task to have its own set of ARD hyper-parameters and shares the sparsity pattern over the task specific hyper-parameters. Hence, unlike GPMTL, GPMTFS does not restrict the functional form of the task specific functions to be the same. Moreover, it can control the degree of similarity through a regularization parameter. Such a flexible and general model enables one to handle a wide range of multi-task learning problems where the tasks are less similar. We show that the flexibility of GPMTFS leads to a better generalization performance on multi-task regression problems through experiments on synthetic and real world regression data sets.

This paper is organized as follows. We discuss the related work in section 2. Gaussian process regression is discussed in section 3. We discuss the proposed approach, Gaussian process multi-task feature selection, in section 4 and present the experimental results on synthetic and real multi-task data sets in section 5. Finally, we conclude in section 6.

Notations. We consider a multi-task regression problem with T tasks. Each task t is associated with a training data set $\mathbf{D}^t$ with N^t examples, *i.e.* $\mathbf{D}^t = (\mathbf{X}^t, \mathbf{y}^t) = \{\mathbf{x}_i^t, y_i^t\}_{i=1}^{N^t}$ and a test data set $\mathbf{D}_*^t$ with N_*^t examples, *i.e.* $\mathbf{D}_*^t = (\mathbf{X}_*^t, \mathbf{y}_*^t) = \{\mathbf{x}_{*i}^t, y_{*i}^t\}_{i=1}^{N_*^t}$. We assume that all the data sets come from the same input space $\mathcal{R}^P$ and output space $\mathcal{R}$, *i.e.* $\mathbf{x}_i^t \in \mathcal{R}^P$ and $y_i^t \in \mathcal{R}$. Let $\mathbf{D}$ be the collection of all task specific training data sets *i.e.* $\mathbf{D} = (\mathbf{X}, \mathbf{y}) = (\cup_{t=1}^{T}\mathbf{X}^t, \cup_{t=1}^{T}\mathbf{y}^t)$ and $\mathbf{D}_*$ be the collection of all task specific test data sets *i.e.* $\mathbf{D}_* = (\mathbf{X}_*, \mathbf{y}_*) = (\cup_{t=1}^{T}\mathbf{X}_*^t, \cup_{t=1}^{T}\mathbf{y}_*^t)$. Let $N = \sum_{t=1}^{T} N^t$ and $N_* = \sum_{t=1}^{T} N_*^t$ be the total number of training and test examples respectively, from all the tasks. We assume that task specific data sets are associated with a different but related sampling distributions S^t. In multi-task regression, we learn a function f^t for the task t and use it to make predictions on the test data set $\mathbf{D}_*^t$ associated with the task t. The goal in multi-task regression is to learn these functions from the training data set $\mathbf{D}$ such that they provide good generalization performance on the test data set $\mathbf{D}_*$. We denote $\|\mathbf{a}\|_2$ to represent the l_2 norm of a vector $\mathbf{a}$ and $|A|$ to denote the determinant of a matrix A.

2   Related Work

Multi-task learning using joint feature selection improves generalization performance by sharing a sparse feature pattern across the tasks. In [8], this is achieved by using the l_1/l_2 (mixed norm) regularization term. They consider the l_2 norm of coefficients associated with a feature across all the tasks. The regularization term is formed by taking the sum of this l_2 norm over all the features. The l_1/l_2 regularization term is also used in [16], but [16] uses an efficient optimization approach different from the one used in [8]. Convex multi-task feature learning (CMTFL) [6] assumes that the tasks share a small set of features and learns the feature matrix. Sparsity was induced by using the squared l_1/l_2 regularizer over the feature coefficient matrix. The resulting non-convex problem is solved using an equivalent convex formulation involving trace norm. An alternative efficient way to perform feature learning is provided in [17]. The selection of an appropriate mixed norm regularizer for performing multi-task learning is discussed in [18] and they provide a probabilistic interpretation to it. A maximum entropy discrimination framework to perform joint feature selection for multi-task learning is discussed in [19]. However, these approaches are not probabilistic in nature and cannot provide a measure of uncertainty over predictions.

A Bayesian approach in which an automatic relevance determination (ARD) prior is imposed over the feature coefficients associated with the tasks is discussed in [9]. Here, the sparsity is achieved by constraining the variance of the coefficients to a constant value. Sparse Bayesian multi-task learning [10] achieves group sparsity over the feature coefficient matrix by imposing a matrix-variate Gaussian scale mixture prior over it. In Bayesian multi-task feature selection [11], a spike and slab prior is used to enforce the selection of a common subset of features across the tasks. All these approaches are parametric and hence the model complexity of each task is limited by the parametric functional form.

Gaussian processes (GPs) provide a Bayesian non-parametric approach to multi-task learning. The GP approach to multi-task learning presented in [7] models task similarity by placing a common prior over the parameters across all the tasks. In [13], task similarity is modeled by sharing the kernel parameters across the tasks. The semi-parametric latent factor approach [15] models each task as a linear combination of latent functions with task specific weights. In [20], this is extended by putting a spike and slab prior over the task specific weights. The GP approach to multi-task regression in [14] considers the covariance matrix as a Kronecker product of the covariance matrices over the tasks and data and learns the task covariance matrix when the task specific features are not present. In [21], a mixed effect model is proposed where the task functions are assumed to be a combination of a common fixed effect and a task specific random effect. We provide an approach to perform multi-task regression using Gaussian processes which can perform joint selection of features relevant across all the tasks.

3   Gaussian Process Regression

A Gaussian Process is a collection of random variables with the property that the joint distribution of any finite subset of which is Gaussian [12]. It generalizes Gaussian distribution to infinitely many random variables. The GP is completely specified by a mean function and a covariance function. The covariance function is defined over the function values of a pair of inputs and is evaluated using a Mercer kernel function over the pair of inputs. It expresses some general properties of functions such as their smoothness, and length-scale. A commonly used covariance function is the squared exponential (SE) kernel

$$K(\mathbf{x}_i, \mathbf{x}_j) = \exp(-\frac{1}{2}\sum_{l=1}^{P}\kappa_l(x_{il} - x_{jl})^2). \tag{1}$$

Here $\kappa_1, \kappa_2, \ldots, \kappa_P$ (all non-negative) are the kernel hyper-parameters associated with the SE kernel K. The SE kernel (1) implements automatic relevance determination (ARD) through the kernel parameters $\kappa_1, \kappa_2, \ldots, \kappa_P$. A low value of κ_i implies that the dimension i is less relevant. This helps to estimate the dimensions (features) which are relevant for prediction. Let $\mathbf{K} = K(\mathbf{X}, \mathbf{X})$, $\mathbf{K}_* = K(\mathbf{X}, \mathbf{X}_*)$ and $\mathbf{K}_{**} = K(\mathbf{X}_*, \mathbf{X}_*)$. Here, $K(\mathbf{X}, \mathbf{X}_*)$ is an $N \times N_*$ matrix of covariances evaluated for all the pairs of training and test input data. The matrices $K(\mathbf{X}, \mathbf{X})$, $K(\mathbf{X}_*, \mathbf{X})$ and $K(\mathbf{X}_*, \mathbf{X}_*)$ are also defined similarly.

We consider a noisy Gaussian process regression (GPR) model where the output y lies around a latent function $f(\mathbf{x})$ with an additive, independently and identically distributed (i.i.d.) Gaussian noise ϵ with mean 0 and variance σ_n^2, i.e. $y = f(\mathbf{x}) + \epsilon$. In GPR, the likelihood is Gaussian

$$p(y|f(\mathbf{x})) = \mathcal{N}(f(\mathbf{x}), \sigma_n^2). \tag{2}$$

The GPR approach imposes a zero mean GP prior over the latent function values $\mathbf{f}$ associated with the training data and $\mathbf{f}_*$ associated with the test data. The predictive distribution on $\mathbf{f}_*$ is obtained by integrating the conditional distribution $p(\mathbf{f}_*|\mathbf{f}, \mathbf{X}_*, \mathbf{X})$ over the posterior distribution $p(\mathbf{f}|\mathbf{X}, \mathbf{y})$, i.e. $p(\mathbf{f}_*|\mathbf{X}_*, \mathbf{X}, \mathbf{y}) = \int p(\mathbf{f}_*|\mathbf{f}, \mathbf{X}_*, \mathbf{X})p(\mathbf{f}|\mathbf{X}, \mathbf{y})d\mathbf{f}$. The conditional distribution is Gaussian because of the GP prior. Due to the Gaussian form of the likelihood the posterior distribution is Gaussian. Hence, the predictive distribution over the latent function values of the test data is also Gaussian. The predictive distribution over the test outputs $\mathbf{y}_*$ is obtained as $p(\mathbf{y}_*|\mathbf{X}_*, \mathbf{X}, \mathbf{y}) = \int p(\mathbf{y}_*|\mathbf{f}_*)p(\mathbf{f}_*|\mathbf{X}_*, \mathbf{X}, \mathbf{y})d\mathbf{f}_*$, and it is also Gaussian.

The hyper-parameters $\{\kappa_1, \kappa_2, \ldots, \kappa_P, \sigma_n^2\}$ are estimated using either Bayesian techniques or cross-validation techniques [12]. Generally, the hyper-parameters are estimated by maximizing marginal likelihood $p(\mathbf{y}|\mathbf{X}) = \int p(\mathbf{y}|\mathbf{f}, \mathbf{X})p(\mathbf{f}|\mathbf{X})d\mathbf{f}$ $= \mathcal{N}(0, \mathbf{K} + \sigma_n^2\mathbf{I}_N)$ or equivalently by minimizing the negative logarithm of the marginal likelihood:

$$\underset{\theta}{\text{argmin}} \frac{1}{2}\mathbf{y}^\top(\mathbf{K} + \sigma_n^2\mathbf{I}_N)^{-1}\mathbf{y} + \frac{1}{2}log|\mathbf{K} + \sigma_n^2\mathbf{I}_N| + \frac{N}{2}log(2\pi), \tag{3}$$

4   Gaussian Process Multi-task Regression Using Feature Selection

We propose a novel approach to multi-task regression using Gaussian Processes based on the idea of joint feature selection. The proposed approach, Gaussian process multi-task feature selection (GPMTFS), improves the generalization performance by sharing the feature sparsity pattern across the tasks.

The GPMTFS approach uses a covariance function which implements automatic relevance determination (ARD). The kernel parameters in these covariance functions help in capturing the importance of each feature in the data set. In the multi-task setting, we model each task t using a latent function f^t. The latent function f^t comes from a zero mean Gaussian process with a covariance function K^t. The covariance function K^t can be any Mercer kernel implementing ARD. Thus, each task t is associated with kernel hyper-parameters $\{\kappa_1^t, \kappa_2^t, \ldots, \kappa_P^t\}$, which help in automatic feature relevance. The task specific hyper-parameters allow f^t to take distinct functional form. In this work, we consider the SE kernel (4) and the linear kernel (5).

$$K^t(\mathbf{x}_i^t, \mathbf{x}_j^t) = \exp(-\frac{1}{2}\sum_{l=1}^{P}\kappa_l^t(x_{il}^t - x_{jl}^t)^2) \tag{4}$$

$$K^t(\mathbf{x}_i^t, \mathbf{x}_j^t) = \sum_{l=1}^{P}\kappa_l^t x_{il}^t x_{jl}^t \tag{5}$$

Multi-task Marginal Likelihood. In multi-task regression, the likelihood for each task t is Gaussian as in GPR, *i.e.* $p(y_i^t|f^t(\mathbf{x}_i^t)) = \mathcal{N}(f^t(\mathbf{x}_i^t), \sigma_t^2)$, where σ_t^2 is the noise variance associated with the task t. The marginal likelihood of the examples belonging to the task t is also Gaussian, $p(\mathbf{y}^t|\mathbf{X}^t) = \mathcal{N}(0, \mathbf{K}^t + \sigma_t^2\mathbf{I}_{N^t})$, where $\mathbf{K}^t = K^t(\mathbf{X}^t, \mathbf{X}^t)$. Then, the marginal likelihood over all the tasks is obtained as $p(\mathbf{y}|\mathbf{X}) = \prod_{t=1}^{T}p(\mathbf{y}^t|\mathbf{X}^t)$. The hyper-parameters $\{\kappa_1^t, \kappa_2^t, \ldots, \kappa_P^t, \sigma_t^2\}_{t=1}^{T}$ are estimated by maximizing the marginal likelihood $p(\mathbf{y}|\mathbf{X})$ or equivalently by minimizing the negative log of the marginal likelihood,

$$- \log p(\mathbf{y}|\mathbf{X}) = \sum_{t=1}^{T} - \log p(\mathbf{y}^t|\mathbf{X}^t) \simeq$$

$$\sum_{t=1}^{T}(\frac{1}{2}\mathbf{y}^{t\top}(\mathbf{K}^t + \sigma_t^2\mathbf{I}_{N^t})^{-1}\mathbf{y}^t + \frac{1}{2}log|\mathbf{K}^t + \sigma_t^2\mathbf{I}_{N^t}|). \tag{6}$$

The objective function (6) is a sum of the negative log of the marginal likelihood for each task. Learning hyper-parameters by minimizing (6) leads to an independent learning of hyper-parameters associated with each task. This does not lead to any kind of feature sharing across the tasks and fails to model the multi-task learning situation. We model the task similarity by sharing the sparsity pattern over the kernel hyper-parameters associated with each task. This is achieved by using a hierarchical approach where we impose a sparse prior over the kernel hyper-parameters $\{\kappa_1^t, \kappa_2^t, \ldots, \kappa_P^t\}_{t=1}^T$. This results in selecting a subset of features which are relevant and common across all the tasks.

Prior Over the Hyper-parameters. We consider a Laplacian or double exponential prior as the sparsity inducing prior. It has been used as a sparsity inducing prior in various contexts [22]. It leads to a l_1 regularized function in the log space which results in sparse solutions for properly chosen regularization parameters. We collect the kernel hyper-parameters from all the tasks into a matrix $\mathbf{Q}$, where $\mathbf{Q} = [\kappa_1, \kappa_2, \ldots, \kappa_P]$ and $\kappa_i = [\kappa_i^1, \kappa_i^2, \ldots, \kappa_i^T]^\top$. Thus, the column i of the matrix $\mathbf{Q}$ denotes the kernel hyper-parameters corresponding to the dimension i for all the tasks. We denote $\kappa^t = [\kappa_1^t, \kappa_2^t, \ldots, \kappa_P^t]^\top$ as the vector of all the kernel hyper-parameters for the task t and $\sigma^2 = [\sigma_1^2, \sigma_2^2, \ldots, \sigma_T^2]^\top$ as the vector of all task specific variance hyper-parameters. To achieve our objective of sharing sparsity pattern on the kernel hyper-parameters across all the tasks, we consider imposing a sparsity inducing prior over the matrix $\mathbf{Q}$. Specifically, we impose a zero mean multi-Laplacian (ML) prior [23] over each column κ_i of the matrix $\mathbf{Q}$. The ML prior over the vector κ_i is defined as

$$
\begin{aligned}
p(\kappa_i) &= \text{Multi-Laplace}(\kappa_i | 0, C^{-1}) \\
&= C^{T/2} \exp(-C\|\kappa_i\|_2),
\end{aligned}
\tag{7}
$$

where C is the parameter associated with the ML prior. We impose independent ML priors over each column of the matrix $\mathbf{Q}$. The prior over the matrix $\mathbf{Q}$ is defined as

$$
p(\mathbf{Q}) = \prod_{i=1}^P p(\kappa_i) = C^{TP/2} \exp(-C \sum_{i=1}^P \|\kappa_i\|_2).
\tag{8}
$$

Imposing the ML prior over each column of $\mathbf{Q}$ will result in the columns becoming sparse together. Thus, the kernel hyper-parameters corresponding to an irrelevant feature for all the tasks become zero together. This will lead to sharing of sparsity pattern over features across the tasks. The variance hyper-parameter vector σ^2 is assigned independent exponential priors with the rate parameter B,

$$
p(\sigma^2) = \prod_{t=1}^T \exp(\sigma_t^2 | B) = B^T \exp(-B \sum_{t=1}^T \sigma_t^2).
\tag{9}
$$

The posterior over the hyper-parameters is given by

$$
p(\mathbf{Q}, \sigma^2 | \mathbf{y}, \mathbf{X}) \propto p(\mathbf{y} | \mathbf{X}, \mathbf{Q}, \sigma^2) p(\mathbf{Q}) p(\sigma^2).
\tag{10}
$$

Learning the Hyper-parameters. The posterior (10) cannot be obtained in closed form. The hyper-parameters are estimated using a maximum a posteriori (MAP) approach. We estimate the hyper-parameters by minimizing negative log of the posterior which results in the following optimization problem.

$$\underset{\kappa_1,\kappa_2,\ldots,\kappa_P,\sigma^2}{\text{argmin}} \sum_{t=1}^{T} (\frac{1}{2}\mathbf{y}^{t\top}(\mathbf{K}^t + \sigma_t^2\mathbf{I}_{N^t})^{-1}\mathbf{y}^t + \frac{1}{2}log|\mathbf{K}^t + \sigma_t^2\mathbf{I}_{N^t}| + B\sigma_t^2)$$

$$+C\sum_{i=1}^{P}\|\kappa_i\|_2 \quad \text{s.t.} \quad \kappa_1 \geq 0, \kappa_2 \geq 0, \ldots, \kappa_P \geq 0, \sigma^2 \geq 0 \tag{11}$$

The objective function in the optimization problem (11) consists of two terms: the first one is the loss term arising from the marginal likelihood over all the tasks and the second one is a regularization term. The regularization term penalizes the sum of the l_2 norm of kernel hyper-parameters across the tasks for a particular feature, and it helps to perform joint feature selection. It couples the kernel hyper-parameters across all the tasks and causes the task specific kernel hyper-parameters to share the sparsity pattern. The non-sparse kernel hyper-parameters correspond to the features which are relevant across all the tasks and help in feature selection. The regularization constant C controls the degree of similarity in the multi-task learning problem. When the tasks share high similarity a proper value of C results in sharing the sparsity pattern across the tasks. When the tasks are dissimilar a zero value of C results in learning the hyper-parameters independently without any sharing.

The optimization problem (11) is similar to the one used in [8] which uses the mixed norm l_1/l_2 regularizer over the task coefficients. When the number of tasks reduces to one, the l_1/l_2 regularization reduces to the l_1 regularization over the kernel hyper-parameters. In this case, GPMTFS performs Gaussian process regression with the l_1 regularization over the kernel parameters. Learning tasks independently using such l_1 regularization is not effective since the number of examples associated with a task is too small to learn the relevant features.

GPMTFS Algorithm. The optimization problem (11) is solved using the block co-ordinate descent (BCD) approach [24]. It has been applied in many multi-task learning settings with mixed norm regularizers [25]. The approach updates the parameters associated with the co-ordinates in a cyclic manner (Gauss-Seidel procedure). In the GPMTFS optimization problem (11), we consider the kernel hyper-parameters κ_i corresponding to the dimension i across all the tasks as the parameters of the co-ordinate i. We consider the variance hyper-parameter σ^2 across all the tasks as the parameters of the co-ordinate $P + 1$. Gradient based optimization approaches are used to update the hyper-parameters in each co-ordinate descent step. The hyper-parameters are updated until the relative decrease in the objective function value is small. The BCD approach for GPMTFS is summarized in Algorithm 1. Each co-ordinate descent step takes $\mathcal{O}(\sum_{t=1}^{T} N_t^3)$ time. The cubic complexity arises from the inversion of a $N_t \times N_t$ matrix in the optimization problem. However, the number of examples associated with each task is often very small. Let N_{max} denote the largest among $N_1, \ldots, N_T$. The computational complexity of the co-ordinate descent step is $\mathcal{O}(TN_{max}^3)$ and that of the GPMTFS algorithm is $\mathcal{O}(PTN_{max}^3)$. The non-smooth optimization problem in (11) is solved using a subgradient approach.

Algorithm 1. Block co-ordinate descent for GPMTFS

1: **Input** Regularization constants B and C, Data sets $\{D^t\}_{t=1}^T$
2: **Output** Matrix $\mathbf{Q}$, σ^2
3: Initialize matrix $\mathbf{Q}$ and σ^2
4: **repeat**
5:     **for** $i = 1$ **to** P **do**
6:         Update κ_i by solving the optimization problem (11) *w.r.t.* κ_i and fixing all other variables.
7:     **end for**
8:     Update σ^2 by solving the optimization problem (11) *w.r.t.* σ^2 and fixing other variables.
9: **until** relative decrease in the objective function value in (11) is not small

---

For a proper choice of the regularization constant C, the approach results in a sparse solution. In general, the sparsity increases as we increase the value of the regularization parameter C. The regularization constants B and C can be chosen using cross-validation. The estimated kernel hyper-parameters share a common sparsity pattern across the tasks and can be used to select features relevant across all the tasks.

Prediction. The estimated hyper-parameters are used to make predictions for each task t. The output predictive probability distribution for the task t on the test data $\mathbf{x}_*^t$ is Gaussian with mean $\mathbf{K}_*^{t\top}(\mathbf{K}^t + \sigma_t^2 \mathbf{I}_{N^t})^{-1}\mathbf{y}^t$ and variance $K_{**}^t - \mathbf{K}_*^{t\top}(\mathbf{K}^t + \sigma_t^2 \mathbf{I}_{N^t})^{-1}\mathbf{K}_*^t + \sigma_t^2$, where $\mathbf{K}_*^t = K^t(\mathbf{X}^t, \mathbf{x}_*^t)$ and $K_{**}^t = K^t(\mathbf{x}_*^t, \mathbf{x}_*^t)$. The mean is taken as the output predicted by the GPMTFS approach.

5   Experimental Results

We conduct experiments to study the behavior and the performance of the proposed GPMTFS approach on a synthetic and two real multi-task regression data sets, Personal Computer and School [6]. Table 1 summarizes the properties of these data sets. We compare the performance of GPMTFS against convex multi-task feature learning (CMTFL) [6] [1] and a closely related GP based multi-task learning approach (GPMTL) [13]. CMTFL is based on the idea of joint feature selection using mixed norm regularizers but is not a probabilistic approach. On the other hand, both GPMTFS and GPMTL are probabilistic models based on GP. We also compare our approach against independent task learning (ITL) and aggregate task learning (ATL). In ITL, decision functions are learnt independently for each task t from the data set $\mathbf{D}^t$ using Gaussian process regression with a regularization over hyper-parameters. In ATL, a single decision function is learnt for all the tasks from the collection $\mathbf{D}$ of the data sets $\mathbf{D}^t$ using Gaussian process regression with a regularization over hyper-parameters.

---

[1] Code is available at
http://ttic.uchicago.edu/~argyriou/code/mtl_feat/mtl_feat.tar

Table 1. Properties of the data sets

Data	Number of tasks	Dimension	Examples per task
Synthetic	10	10	25
Personal Computer	190	14	20
School	139	27	20-150

5.1 Synthetic Data

A synthetic data set is used to study the behavior of the proposed approach [6]. We assume the number of tasks to be 10 ($T = 10$) and generate a 10 dimensional synthetic data set for each task. Each task is associated with 5 training data examples and 20 test data examples. The training and test data for each task are generated randomly from a uniform distribution $[0, 1]^P$, where P is 10. We assume the first 5 features of the data set as relevant and the rest of the features as irrelevant. This is modeled by generating the coefficients ($\mathbf{w}^t$) corresponding to the first 5 dimensions from a 5 dimensional Gaussian distribution with zero mean and a diagonal covariance matrix with diagonal entries $(1, 0.5, 0.1, 0.15, 0.1)$. The coefficients corresponding to the rest of the dimensions are zero. The output y_i^t for the task t is computed as $y_i^t = \mathbf{w}^t \cdot \mathbf{x}_i^t + \nu$, where ν is Gaussian noise with mean zero and variance 0.1. The task coefficients $\mathbf{w}^t$ are generated independently for each task. All the experiments use a linear ARD kernel.

We run our approach GPMTFS over this synthetic data set and learn the task coefficient matrix. Fig. 2 denotes a color map for the generated task coefficient matrix(left) and the learnt task coefficient matrix(right). We can see that both the generated and the learnt task coefficient matrices are similar. Like the generated task coefficient matrix, the learnt task coefficient matrix also assigns zero values to the irrelevant dimensions.

We verify if the proposed approach is able to learn the dimensions relevant for all the tasks correctly. Consider the bar plot in Fig. 3 obtained using the kernel parameter values learned by our approach. For each dimension, we plot the l_2 norm of the kernel parameter values obtained for all the tasks in that dimension, i.e. $\|\kappa_i\|_2$. From the bar plot, we can observe that the first 6 dimensions are found to be relevant by GPMTFS while the rest of the dimensions are found to be irrelevant.

We study the dependence of the GPMTFS approach on the regularization parameter C in the right plot of Fig. 3. We observe that the number of selected features, root mean square error (RMSE) and the Frobenius norm difference between the actual and learnt task coefficient matrix decrease as we increase the value of the regularization parameter C from 10^{-7} to 10^{-2}. An increase in the value of C leads to sparser solutions and results in the selection of features which are most relevant across all the tasks. This improves the performance of the GPMTFS algorithm which is reflected in the RMSE values obtained. This validates the idea of using joint feature selection for multi-task learning problems.

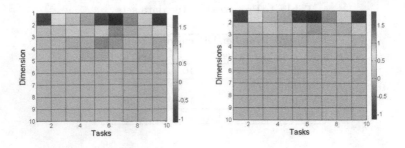

Fig. 2. Task coefficient matrix for 10 tasks and 10 input dimensions. Left : generated task coefficient matrix. Right: learnt task coefficient matrix.

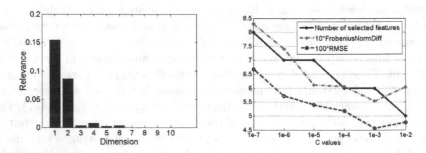

Fig. 3. Left : bar plot indicating the relevance of features in the synthetic data set using GPMTFS . Right : variation in number of features selected , RMSE and Frobenius norm difference on increasing the value of C.

We note that after a particular point, a further increase in the values of C degrades the performance as it forces the hyper-parameter values corresponding to the relevant features also to zero. Therefore the regularization parameter C needs to be chosen carefully. We use cross- validation to choose the value of C.

We compare the RMSE obtained using the proposed GPMTFS approach with CMTFL, GPMTL, ITL and ATL for 2 types of synthetic data sets in Table 2. The first one is same as the one used in the studies discussed above. It consists of highly similar tasks. The second synthetic data set consists of less similar tasks. This is obtained by considering the Gaussian generating the task coefficients $\mathbf{w}^t$ for the first 8 dimensions to have very high variance along its diagonal. The last 2 dimensions of $\mathbf{w}^t$ are considered irrelevant and are taken to be zero. We generate 10 instances of these 2 synthetic data sets and report the mean RMSE obtained for various approaches. For the highly similar synthetic data set, we observe that the performance of all the multi-task learning methods are similar and is better than ITL and ATL. In fact, the proposed approach GPMTFS gives a slightly better performance than other MTL approaches. For the less similar synthetic data set, we observe that the proposed GPMTFS approach performs better than GPMTL. GPMTFS allows each task to have its own set of

Table 2. Experimental results on the data sets for GPMTFS, CMTFL, GPMTL, ITL and ATL. Performance measure used is explained variance for the School data set and RMSE for all other data sets. The numbers in bold face style indicate the best result.

Data set	GPMTFS	CMTFL	GPMTL	ITL	ATL
Synthetic (High)	**0.041± 0.005**	0.042± 0.005	0.045± 0.006	0.071± 0.008	0.105± 0.008
Synthetic (Low)	3.201± 0.050	**3.138± 0.048**	3.532± 0.062	4.522± 0.090	6.589± 0.099
PC	**2.041 ± 0.030**	2.045 ± 0.030	2.062 ± 0.036	2.475 ± 0.060	2.283 ± 0.032
School	35.45 ± 1.36	**35.63 ± 1.25**	34.96 1.54	32.67 ± 2.17	34.23 ± 1.15

hyper-parameters which helps it to capture the variability across the tasks better. The performance of CMTFL is better than that of GP based approaches. This is possibly due to the joint feature selection in the parameter space rather than in the hyper-parameter space.

5.2 Personal Computer

Experiments are conducted on a real data set consisting of ratings of personal computers by people [26] [2]. The data set consists of ratings on 20 different personal computers by 190 people. The properties of the personal computer are represented using 14 binary features. The output consists of integer ratings on a scale of 0-10. Here, each person corresponds to a task and the ratings by the person correspond to the examples in the task. Thus, there are 190 tasks and 20 examples per task. We consider the first 8 examples in each task as the training data and the last 4 examples as the test data. The performance is measured using the root mean squared error (RMSE) averaged over each task. We report the mean RMSE values over 10 independent partitions of the training and test data sets. The experiments are conducted using the squared exponential ARD kernel.

Table 2 compares the performance of GPMTFS with CMTFL, GPMTL, ITL, and ATL on the personal computer (PC) data set. We observe that the proposed approach GPMTFS performs better than all other MTL approaches on the PC data set.

We plot the relevance of features in the PC data set using GPMTFS in Fig. 4. We observe that price (dimension 14) is the most relevant feature. We find that GPMTFS selects technical characteristics of the computer such as RAM, CPU and CDROM (dimensions 2-6) also as relevant features. We observe that for the personal computer dataset, most of the features are relevant.

---

[2] We thank Peter Lenk for kindly providing the data set.

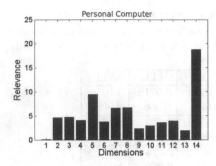 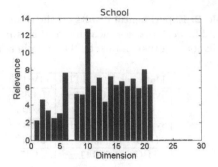

Fig. 4. Bar plot indicating the relevance of features obtained using GPMTFS. Left : Personal Computer. Right : School.

5.3   School Data

We conduct experiments on the real world school data set [4] to study the generalization performance of the proposed GPMTFS approach. The data set consists of examination records of 15362 students over 139 schools. Each student record has 27 dimensions and the number of student records associated with each school varies from 20-150. A student record consists of year of examination, student specific features and school specific features. The goal is to predict exam scores of students from each school. In order to conduct multi-task learning experiments, each school is considered as a task and the student records belonging to a school as the data corresponding to the task. Experiments are conducted on 10 partitions of data into training and test data sets with 75% of the examples from each school as the training set and the rest as the test set. All the experiments use a linear ARD kernel. We use explained variance [4] as the performance measure, which is widely used for comparing the performance of multi-task learning approaches on the School data set. Explained variance is defined as

$$\text{Explained variance} = 1 - \frac{\text{sum squared error}}{\text{total variance}}. \tag{12}$$

A high value of explained variance is preferred over a low value.

Table 2 reports the mean explained variance obtained over 10 independent training and test data instances of the school data set. We observe that the proposed approach GPMTFS performed better than GPMTL, ITL and ATL. GPMTFS captures the variability across the tasks that GPMTL fails to capture.

The features selected by GPMTFS and their relevance for the school data set are shown in Fig. 4. Relevance for a feature is obtained by using the norm of the kernel parameter values across all the tasks for that feature. It agrees well with the results obtained using CMTFL on the school data set. GPMTFS considered the dimensions 22-27 as irrelevant as these dimensions corresponding to the school specific features do not contribute much to the examination score

Table 3. Comparison of the optimal NLPD values obtained by GPMTFS, GPMTL, ITL and ATL on different data sets. The numbers in the bold face style indicate the best result.

Data set	GPMTFS	GPMTL	ITL	ATL
Synthetic (high)	-1.6521	**-1.6552**	-0.9553	1.0622
Synthetic (low)	**2.7468**	3.2681	3.4862	3.4365
PC	**2.2228**	2.3798	3.4740	6.2068
School	**3.6284**	3.7821	6.2431	5.4468

Table 4. Computed t-test statistic for different datasets. Bold face style indicates the cases for which the t-test statistic is greater than the critical value.

Synthetic (high)	Synthetic (low)	PC	School
1.514	**12.119**	**4.846**	**9.336**

of students. VR band (dimensions 10, 13-15) and ethnic background (dimensions 16-21) are the features which strongly influence the exam score of students.

5.4   Probabilistic Analysis, Statistical Significance Test and Runtime Experiments

We perform a probabilistic analysis of the proposed approach by providing the negative log predictive density (NLPD) [12] values on the test data set. Table 3 reports the optimal mean NLPD values obtained on 2 synthetic and 2 real world data sets using GPMTFS, GPMTL, ITL and ATL. CMTFL being a non-probabilistic approach, cannot be used to obtain the NLPD values. Note that low NLPD values are preferred over high NLPD values. The NLPD values clearly show the effectiveness of the proposed GPMTFS approach over the GPMTL approach.

We use the paired t-test [27] to check if the proposed GPMTFS performs significantly better than GPMTL. The null hypothesis is that both GPMTFS and GPMTL have similar performance. Under the null hypothesis, the t-test statistic follows the Students t-distribution with 9 degrees of freedom[3]. For the confidence level of 95% and 9 degrees of freedom, the critical value for the one sided t-test is 1.833. Table 4 reports the t-test statistic computed on 2 synthetic and 2 real world data sets. We find that the computed t-test statistic is greater than the critical value for all the datasets except the synthetic data set with highly similar tasks. This emphasizes the significantly better performance of the GPMTFS approach over the GPMTL approach.

The proposed GPMTFS approach and the GPMTL approach are implemented in Matlab. Publicly available CMTFL code is also implemented in Matlab. These Matlab programs are run on a 3.2 GHz Intel processor with 4GB of shared main memory in a Linux environment. Table 5 provides the runtime for different multi-task learning approaches on synthetic and real world data sets.

---

[3] We consider the results over 10 partitions of a data set.

Table 5. Runtime (in seconds) for GPMTFS, CMTFL, GPMTL, ITL and ATL on different data sets

Data set	GPMTFS	CMTFL	GPMTL	ITL	ATL
Synthetic	11.0825	11.6845	8.2543	13.1067	15.8405
PC	333.3381	243.7865	325.4205	356.0030	395.3381
School	7.5778e+03	8.5566e+03	7.1723e+03	9.8863e+03	1.5566e+04

6 Summary

We proposed a novel approach to multi-task regression using Gaussian processes and joint selection of features. The joint feature selection was done by imposing a sparse prior over the kernel hyper-parameters. This lead to a flexible model which can handle variability across the tasks. The resulting optimization problem was solved using a block co-ordinate descent algorithm. The proposed approach facilitated the selection of features relevant across all the tasks and lead to an improvement in performance. This is validated through the experiments on synthetic and real world data sets. The proposed approach performed better than other GP based approaches. Due to its Bayesian nature, it provides an estimate of uncertainty over predictions and is an useful alternative for multi-task learning. The ideas presented in this paper are general and can be easily extended to multi-task classification problems.

References

1. Caruana, R.: Multitask Learning. Machine Learning 28(1), 41–75 (1997)
2. Ando, R.K., Zhang, T.: A Framework for Learning Predictive Structures from Multiple Tasks and Unlabeled Data. JMLR 6, 1817–1853 (2005)
3. Bakker, B., Heskes, T.: Task Clustering and Gating for Bayesian Multitask Learning. JMLR 4, 83–99 (2003)
4. Evgeniou, T., Micchelli, C.A., Pontil, M.: Learning Multiple Tasks with Kernel Methods. JMLR 6, 615–637 (2005)
5. Xue, Y., Liao, X., Carin, L., Krishnapuram, B.: Multi-task Learning for Classification with Dirichlet Process Priors. JMLR 8, 35–63 (2007)
6. Argyriou, A., Evgeniou, T., Pontil, M.: Convex Multi-task Feature Learning. Machine Learning 73(3), 243–272 (2008)
7. Yu, K., Tresp, V., Schwaighofer, A.: Learning Gaussian processes from Multiple Tasks. In: ICML, pp. 1012–1019 (2005)
8. Obozinski, G., Taskar, B.: Multi-task Feature Selection. Technical report, Department of Statistics, University of California, Berkeley (2006)
9. Xiong, T., Bi, J., Rao, R.B., Cherkassky, V.: Probabilistic Joint Feature Selection for Multi-task Learning. In: SDM (2007)
10. Archembeau, C., Guo, S., Zoeter, O.: Sparse Bayesian Multi-task Learning. In: NIPS, pp. 1755–1763 (2011)
11. Hernández-Lobato, D., Hernández-Lobato, J.M., Helleputte, T., Dupont, P.: Expectation Propagation for Bayesian Multi-task Feature Selection. In: ECML-PKDD, pp. 522–537 (2010)

12. Rasmussen, C.E., Williams, C.K.I.: Gaussian Processes for Machine Learning (Adaptive Computation and Machine Learning). MIT Press (2005)
13. Lawrence, N.D., Platt, J.C.: Learning to Learn with the Informative Vector Machine. In: ICML, pp. 65–72 (2004)
14. Bonilla, E.V., Chai, K.M., Williams, C.K.I.: Multi-task Gaussian Process Prediction. In: NIPS, pp. 153–160 (2008)
15. Teh, Y.W., Seeger, M., Jordan, M.I.: Semiparametric Latent Factor Models. In: International Workshop on Artificial Intelligence and Statistics, vol. 10 (2005)
16. Liu, J., Ji, S., Ye, J.: Multi-task Feature Learning via Efficient L2,1-Norm Minimization. In: UAI, pp. 339–348 (2009)
17. Obozinski, G., Taskar, B., Jordan, M.I.: Joint Covariate Selection and Joint Subspace Selection for Multiple Classification Problems. Statistics and Computing 20(2), 231–252 (2010)
18. Zhang, Y., Yeung, D.Y., Xu, Q.: Probabilistic Multi-Task Feature Selection. In: NIPS, pp. 2559–2567 (2010)
19. Jebara, T.: Multitask Sparsity via Maximum Entropy Discrimination. JMLR 12, 75–110 (2011)
20. Titsias, M.K., Lázaro-Gredilla, M.: Spike and Slab Variational Inference for Multi-Task and Multiple Kernel Learning. In: NIPS, pp. 2339–2347 (2011)
21. Wang, Y., Khardon, R., Protopapas, P.: Shift-invariant Grouped Multi-task Learning for Gaussian Processes. In: ECML-PKDD, pp. 418–434 (2010)
22. Tibshirani, R.: Regression Shrinkage and Selection Via the Lasso. Journal of the Royal Statistical Society 58, 267–288 (1994)
23. Raman, S., Fuchs, T.J., Wild, P.J., Dahl, E., Roth, V.: Bayesian Group-Lasso for Analyzing Contingency Tables. In: ICML, pp. 881–888 (2009)
24. Bach, F.R., Jenatton, R., Mairal, J., Obozinski, G.: Optimization with Sparsity-Inducing Penalties. Foundations and Trends in Machine Learning 4(1), 1–106 (2012)
25. Liu, H., Palatucci, M., Jian, Z.: Blockwise Coordinate Descent Procedures for the Multi-task Lasso, with Applications to Neural Semantic Basis Discovery. In: ICML, pp. 649–656 (2009)
26. Lenk, P.J., DeSarbo, W.S., Green, P.E., Young, M.R.: Hierarchical Bayes Conjoint Analysis: Recovery of Partworth Heterogeneity from Reduced Experimental Designs. Marketing Science 15(2), 173–191 (1996)
27. Dietterich, T.G.: Approximate Statistical Tests for Comparing Supervised Classification Learning Algorithms. Neural Computation 10, 1895–1923 (1998)

Separating Rule Refinement and Rule Selection Heuristics in Inductive Rule Learning

Julius Stecher, Frederik Janssen, and Johannes Fürnkranz

Technische Universität Darmstadt, Knowledge Engineering
jlstecher@gmail.com, {janssen,juffi}@ke.tu-darmstadt.de

Abstract. Conventional rule learning algorithms use a single heuristic for evaluating both, rule refinements and rule selection. In this paper, we argue that these two phases should be separated. Moreover, whereas rule selection proceeds in a bottom-up specific-to-general direction, rule refinement typically operates top-down. Hence, in this paper we propose that criteria for evaluating rule refinements should reflect this by operating in an inverted coverage space. We motivate this choice by examples, and show that a suitably adapted rule learning algorithm outperforms its original counter-part on a large set of benchmark problems.

1 Introduction

Separate-and-conquer or *covering* rule learning algorithms [6,8] proceed by first learning a single rule (*conquer*) followed by the removal of all examples that are covered by this rule (*separate*). The remaining examples are then used to learn the next rule (return to the conquer step). For learning a rule, most algorithms use a *top-down hill-climbing* search that starts with the *universal rule* covering *all* examples, and subsequently add conditions that optimize a *heuristic*. Typical heuristics trade off *consistency* and *coverage*, i.e., they prefer rules that cover as few *negative* and as many positive examples as possible [7,9].

Typically, such a heuristic is used in two different places in this process: (i) for judging *rule refinements*, i.e., to select which of the refinements of the current rule will be further explored, and (ii) for *rule selection*, i.e., to finally decide which of the refinements that have been explored is added to the rule set. In this paper, we argue that these tasks should be treated separately, i.e., evaluated with separate heuristics. Moreover, we argue that the rule refinement step in a top-down search requires *inverted heuristics*, which evaluate rules from the point of view of the current base rule instead of the empty rule. We will motivate this with an example, show the derivation of such *inverted heuristics* in coverage space, and demonstrate empirically that they lead to improved performance.

We start with a brief recapitulation of separate-and-conquer rule learning, heuristics and coverage spaces (Section 2). In Section 3, we then motivate why rule refinement and rule selection should be separated, and show how the commonly used heuristics precision, Laplace, and m-estimate can be inverted to better reflect a top-down search for refinements. We will also see that other heuristics, such as weighted relative accuracy, are invariant to such inversions. Finally, in Section 4, we evaluate the use of inverted heuristics for evaluating rule refinements experimentally on 20 UCI datasets.

T. Calders et al. (Eds.): ECML PKDD 2014, Part III, LNCS 8726, pp. 114–129, 2014.

Algorithm 1. Procedure Separate-And-Conquer

Data: TrainingData
Result: theory $\mathcal{R}$

1 Start with empty theory $\mathcal{R}$
2 **while** positive examples left in TrainingData **do**
3 | Rule r = findBestRule(TrainingData)
4 | **if** positiveCovered(r) $\leq$ negativeCovered(r) **then**
5 | | **break**
6 | $\mathcal{R} = \mathcal{R} \cup$ r
7 | remove all covered examples from TrainingData
8 **return** $\mathcal{R}$

---

2 Separate-and-Conquer Rule Learning

In this section, we briefly recapitulate the necessary foundations for our contribution, the separate-and-conquer rule learning algorithm (Section 2.1), coverage spaces (Section 2.2), and rule learning heuristics (Section 2.3).

2.1 Algorithm

Most rule learning algorithms follow a so-called separate-and-conquer or covering strategy to learn from P positive and N negative training examples. This algorithm proceeds by learning one rule at a time, while removing all examples that are covered by each rule from the dataset. This is repeated until no examples remain, i.e., until all examples are covered, or until the best found rule covers more negative than positive examples. Algorithm 8 shows the basic algorithm, as it has also been described in [6,8].

In contrast to algorithms producing an unordered rule set, we consider the learned rule set as a *decision list* made up of an ordered list of rules. A decision list ends with a *default rule*, which unconditionally applies the majority class label to any example that is not covered by one of the previous rules in the list. At classification time, the ordered rule list is checked from top to bottom, assigning to each example the class label of the head of the first rule that matches the example.

For finding individual rules, we focus on the most commonly used *top-down hill-climbing* strategy, which is shown in Algorithm 11. Whenever it needs to learn a new rule the algorithm initializes it with the *universal rule* $\mathbf{r}^{\top}$, which covers *all* examples. By adding conditions to this rule, the amount of covered examples will decrease with each iteration, thereby increasing the consistency of the rule by focusing on removing more negative examples than positive examples. How much consistency is gained depends on the particular condition that is selected as a refinement of the rule in each iteration. This choice depends on a *heuristic* function h, which is applied to all possible rule refinements, choosing the refinement that scores best after applying the heuristic to all refinements. It is easy to see that the importance of a good heuristic is vital for learning a theory w.r.t. consistency and coverage as it is the only type of guidance the rule learner can make use of during the training process.

Algorithm 2. Procedure findBestRule

Data: TrainingData
Result: best rule $\mathbf{r}_{best}$

1 $\mathbf{r}_{best} = \emptyset$
2 bestValue = heuristic($\mathbf{r}_{best}$)
3 **repeat**
4   get possible refinements
5   **forall the** refinements ref **do**
6   ⌊ evaluation = heuristic(ref)
7      $\mathbf{r}_{ref}$ = best refined rule
8   **if** heuristic($\mathbf{r}_{ref}$) $\geq$ bestValue **then**
9   ⌊ $\mathbf{r}_{best} = \mathbf{r}_{ref}$
10 **until** no refinements left;
11 **return** $\mathbf{r}_{best}$

2.2 Coverage Space

Coverage spaces have been introduced as a formal framework for analyzing and visualizing the behavior of rule learning heuristics [7]. A coverage space plots the number of covered positive examples (the *true positives p*) over the number of covered negative examples (the *false positives n*), resulting in a rectangular plot with the values $\{0, 1, ..., N\}$ on the horizontal axis and $\{0, 1, ..., P\}$ on the vertical axis. This principle can then be used to both plot entire theories consisting of an ordered rule list (the decision list) as well as individual rules.

The following points of the coverage space are of special interest (cf. also Figure 1):

- $(0, 0)$ is the *empty theory*. It does not cover any examples, neither positive nor negative ones. A *bottom-up* learning algorithm would start at this point and successively add rules.
- $(0, P)$ is the *perfect theory* covering all positive, but no negative examples.
- $(N, 0)$ is the *opposite theory* covering all negative, but no positive examples.
- (N, P) is the *universal theory*. It covers all examples regardless of their label.

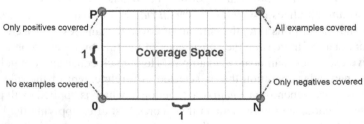

Fig. 1. Coverage space visualization with P total positive examples and N total negative examples

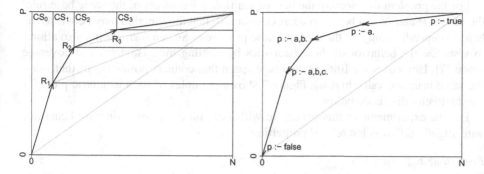

Fig. 2. Paths in coverage space for (*left*) the covering strategy of learning a rule set by adding one rule at a time and (*right*) top-down specialization of a single rule

Fürnkranz and Flach [7] have shown that learning a rule set one rule at a time may be viewed as a path through coverage space, where each point on the path corresponds to the addition of a rule to the theory. Figure 2 shows the coverage path for a theory with three rules. Each point $\mathcal{R}_i = \bigcup_{j=1}^{i}\{\mathbf{r}_j\}$ represents the rule set consisting of the first i rules. Adding a rule moves the induced rule set to the next point $\mathcal{R}_{i+1} = \mathcal{R}_i \cup \{\mathbf{r}_{i+1}\}$.

Removing the covered (positive and negative) examples has the effect of switching to a subspace of the original coverage space, using the last learned rule as the new origin. Thus the path may also be viewed as a sequence of nested coverage spaces CS_i. Each new rule is evaluated relative to the origin $(0,0)$ of this new coverage space. For example, precision would pick the rule with the steepest ascent from the origin.

The commonly used top-down strategy for rule refinement, on the other hand, successively specializes a rule by adding the most promising condition to the rule body. Just as with adding rules to a rule set, successive rule refinements describe a path through coverage space (Figure 2, right). However, in this case, the path starts at the upper right corner with the universal rule $\mathbf{r}^\top$, and successively proceeds towards the origin, which corresponds to the empty rule $\mathbf{r}_\perp$.

2.3 Rule Learning Heuristics

Any rule learning algorithm relies on some sort of measure to determine the quality of a rule; this is done with the help of a *heuristic function* h. Most heuristics implement a trade-off between consistency and coverage favoring rules that cover as many positive examples as possible (optimizing coverage) while keeping the amount of negative examples covered small (optimizing consistency). Thus, the computed value depends mostly on p (positive examples covered) and n (negative examples covered). Since for some of the examined heuristics (e.g. the m-estimate as well as the modifications suggested later) the values of P (total positive examples) and N (total negative examples) must be known, for most purposes a heuristic can be defined as a function

$$h : (p, n, P, N) \to \mathbb{R}$$

For the problem of selecting the best of multiple refinements of the same base rule, the values P and N can be regarded as constant, so that the function may be written as $h(p, n)$ depending only on the true and false positives.[1] Such a formulation also allows to visualize the behavior of these heuristics by plotting their *isometrics* in coverage space [7]. Isometrics are lines in coverage space that connect points (n, p) that share the same heuristic value $h(p, n)$. Figure 3 shows examples of such isometric plots for two heuristics discussed below.

For the experiments in this paper, we will focus on three common base heuristics with slightly different but related properties:

Precision: $h_{prec}(p, n) = \frac{p}{p+n}$
Precision prefers a rule r_1 to another rule r_2 if r_1 covers a larger *percentage* of positive examples. Note that this does not take into account coverage – a rule covering one positive and no negative examples will score the highest possible value, while a rule covering all positive and one negative example will score slightly lower. Thus a theory learned with the help of the precision heuristic is likely to overfit the training data with a bad performance when generalizing to new or noisy data. This can also be seen from its visualization in coverage space (Figure 3(a)), which shows that the isometrics of the precision heuristic rotate around the origin $(0, 0)$, and that therefore all points on the P axis receive the same evaluation.

Laplace: $h_{lap}(p, n) = \frac{p+1}{p+n+2}$
The Laplace heuristic reduces some of the overfitting drawbacks (bad generalization) of precision while following the same general intent of maximizing (mostly) consistency. Starting the p and n counts at 1 instead of 0, the origin of the isometrics shifts to $(-1, -1)$. The effects of this change is that rules on the P-axis not sharing the same value anymore. For example, if two rules r_1 and r_2 cover no negative examples, but r_1 covers 2 positives while r_2 only covers 1, the resulting heuristic values are $h_{lap}(r_1) = 0.75$ and $h_{lap}(r_2) = 0.66$, whereas evaluating both rules with precision would have yielded $h_{prec}(r_1) = h_{prec}(r_2) = 1.0$.

m-Estimate: $h_{mest}(p, n) = \frac{p+m \cdot \frac{P}{P+N}}{p+n+m}$
The m-estimate may be considered as a generalization of the Laplace heuristic. It follows the same idea, but features a parameter m that allows to shift the origin of the rotation, which is fixed at $(0, 0)$ für precision and at $(-1, -1)$ for Laplace to any place along the negative extension of the diagonal of the coverage space. Essentially, this has the effect of initializing all coverage counts with m examples, which are distributed according the overall distribution. For the special case $m = 0$, the m-estimate equals precision, and for $m \to \infty$, it approximates *weighted relative accuracy* (WRA), which means that its isometrics approach parallel lines with a slope of $\frac{P}{P+N}$ (the a priori distribution).[2] For the algorithm that we use in our experiments, an optimal value of

---

[1] Some heuristics also include additional parameters such as the length of the rule. However, this is often implicitly captured (longer rules correlate with lower coverage), and adding them does not necessarily yield increased performance [9].

[2] WRA is defined as $\frac{p+n}{P+N} \cdot (\frac{p}{p+n} - \frac{P}{P+N})$, which is equivalent to $\frac{p}{P} - \frac{n}{N}$. We will not further consider it in this paper, for reasons that will be explained in Section 3.2.

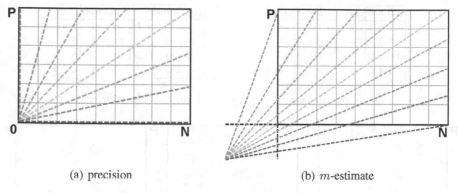

(a) precision   (b) m-estimate

Fig. 3. Visualization of the isometrics of precision and the m-estimate

$m = 22.466$ has been determined experimentally [9], and we will be use this value in our experiments as well. Figure 3(b) shows the isometrics for the m-estimate heuristic. It can be clearly seen that the isometrics rotate around a point in the negative space, which has the effect that points on the P-axis no longer receive the same evaluation.

3   Optimization via Modified Heuristics for Rule Refinement

In the standard separate-and-conquer implementation, we use the same heuristic function each time we want to evaluate an entire rule or a refinement of a rule to determine the current best rule and the best refinement w.r.t. the goals of the heuristic (usually coverage and consistency). The approach highlighted in this paper modifies this standard algorithm to use different heuristics for *rule selection* and *rule refinement*. In particular, we will propose to separate these two phases and show how to adapt the three heuristics mentioned above for top-down rule refinement.

3.1   Motivation

As we have seen in Section 2.2, top-down hill-climbing takes a path through coverage space, starting from the universal rule in its upper-right corner. Common rule learning algorithms evaluate each of the rules encountered on this path with a heuristic in the same coverage space. For example, precision would evaluate two candidate rules according to the steepest ascent from the origin, as it would do with rule selection. However, we argue that this evaluation is, in a way, irrelevant because, while it selects the best complete rule that can currently be added to the rule set, it does *not* select the best candidate for further refinement.

This illustrated in an example dataset with four binary attributes and a binary class attribute shown in Figure 4(a). The corresponding coverage statistics of all possible refinements are listed in Figure 4(b) and plotted in coverage space in Figure 4(c). According to precision h_{prec}, the refinement $a = 0$ is clearly the best choice, as is illustrated in Figure 4(d), whereas the refinement $c = 1$ would only be the third choice. However, we

a b c d	class	a b c d	class
0 1 1 1	+	0 1 1 0	+
0 1 1 1	+	0 0 1 1	+
0 0 1 0	−	1 1 1 0	−
1 1 1 0	−	1 0 1 1	+
1 0 0 1	−	1 0 0 1	−

(a) Example dataset

condition	p n	condition	p n
a = 0	**4 1**	a = 1	1 4
b = 0	2 3	b = 1	3 2
c = 0	0 2	**c = 1**	**5 3**
d = 0	1 3	d = 1	4 2

(b) Possible refinements

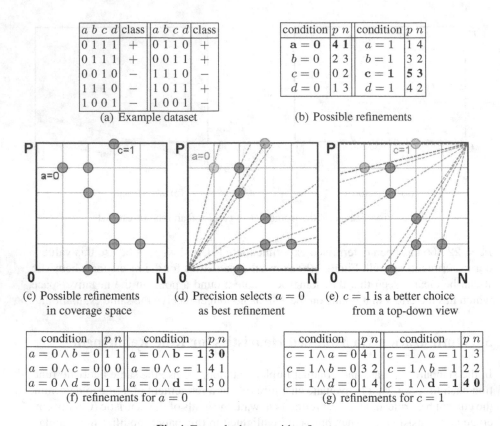

(c) Possible refinements   (d) Precision selects $a = 0$ (e) $c = 1$ is a better choice
    in coverage space   as best refinement   from a top-down view

condition	p n	condition	p n
$a = 0 \wedge b = 0$	1 1	$a = 0 \wedge b = 1$	**3** 0
$a = 0 \wedge c = 0$	0 0	$a = 0 \wedge c = 1$	4 1
$a = 0 \wedge d = 0$	1 1	$a = 0 \wedge d = 1$	**3** 0

(f) refinements for $a = 0$

condition	p n	condition	p n
$c = 1 \wedge a = 0$	4 1	$c = 1 \wedge a = 1$	1 3
$c = 1 \wedge b = 0$	3 2	$c = 1 \wedge b = 1$	2 2
$c = 1 \wedge d = 0$	1 4	$\mathbf{c = 1 \wedge d = 1}$	**4 0**

(g) refinements for $c = 1$

Fig. 4. Example dataset with refinements

argue that $c = 1$ is a better choice for a refinement, because it covers more positive and negative examples and can thus be still refined into a rule that may be better than the first refinement. As the refinement $a = 0$ already has lost one positive example, further refinements will never cover 5 positive examples as is theoretically still possible when $c = 1$ is chosen. However, this choice can be obtained if we use a precision-like heuristic, whose isometrics do not rotate around the origin, but rotate around the base rule, as sown in the right part of Figure 4(e). Indeed, as can be seen from the further possible refinements of these two rules shown in Figures 4(f) and 4(g), the best refinement from the choice $a = 0$ is a rule that covers 3 positive and no negative examples (both $b = 1$ and $d = 1$ can be selected in this case). On the other hand, the precision-like heuristic whose isometrics rotate around the best rule, would end up in the final rule $c = 1 \wedge d = 1$, which covers 4 positive and no negative examples. This rule is preferable to the previous ones but could not be found with the conventional application of precision.

In the next section, we will derive top-down versions of heuristics that correspond to precision, Laplace, and the m-estimate.

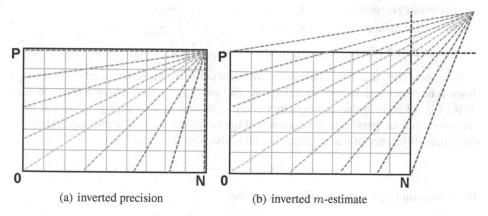

(a) inverted precision   (b) inverted m-estimate

Fig. 5. Visualization of the isometrics of the top-down versions of precision and the m-estimate

3.2 Adaptation of Heuristics to Top-Down Rule Refinement

Note that the three base heuristics (h_{prec}, h_{lap} and h_{mest}) all share similar isometrics, with the only difference being the origin (in the latter case, the location of the origin can be configured via the parameter m). As motivated in the previous section, we want to preserve this attribute, but shift the origin to the top right corner of the coverage space. The intention of this is that in our case the *rule refiner* follows the top-down strategy (starting off with the most general rule and successively adding conditions). We have to take into account that the values of P and N are not constant this time w.r.t. the heuristic function, but depend on the predecessor of the rule. This is because for our approach to work, we will want the origin of the isometrics to be placed at the point in coverage space corresponding to the base rule we want to refine, which will produce *nested* coverage spaces, and subsequently evaluate the refinements within the base rule's nested coverage space.

Figure 5 illustrates the intended behavior for the cases of precision and m-estimate. Instead of a rotation around the origin as in their original versions depicted in Figure 3, we aim for a rotation around the base rule, which is located in (N, P). Moreover, we also have to swap the positive and negative axes: While the best refinements starting from the origin lie on the P-axis, the best refinements starting from (N, P) lie on the N-axis of the coverage space. More precisely, we have to modify the heuristic in a way so that it holds that

$$ɥ(p, n) = h(N - n, P - p) \tag{1}$$

where $ɥ$ is the *inverted* or *top-down heuristic* in the coverage space with dimensions P and N, whereas h is the original heuristic, but in a coverage space with swapped dimensions N and P.

For the three heuristics discussed in Section 2.3, it is straight-forward to see that we obtain the following expressions:

- *Inverted Precision*: $\mathfrak{q}_{prec}(p,n) = \frac{N-n}{(P+N)-(p+n)}$
- *Inverted Laplace*: $\mathfrak{q}_{lap}(p,n) = \frac{N-n+1}{(P+N)-(p+n-2)}$
- *Inverted m-Estimate*: $\mathfrak{q}_{mest}(p,n) = \frac{N-n+m\cdot\frac{P}{P+N}}{(P+N)-(p+n-m)}$

Note, however, that some heuristics are insensitive to the difference between top-down refinement and bottom-up selection. For example, *weighted relative accuracy* (WRA) is a heuristic that is frequently used in subgroup discovery [11,13] and has isometrics that are parallel to the diagonal of the coverage space. It is thus equivalent to the simple difference of true positive rate and false negative rate [7]

$$h_{rdiff}(p,n) = \frac{p}{P} - \frac{n}{N}.$$

The corresponding top-down version would be

$$\mathfrak{q}_{rdiff}(p,n) = \frac{N-n}{N} - \frac{P-p}{P}.$$

Obviously $\mathfrak{q}_{rdiff}$ is equivalent to h_{rdiff} because of

$$\mathfrak{q}_{rdiff}(p,n) = \frac{N-n}{N} - \frac{P-p}{P} = \left(1 - \frac{n}{N}\right) - \left(1 - \frac{p}{P}\right) = \frac{p}{P} - \frac{n}{N} = h_{rdiff}(p,n).$$

This is also apparent from the isometric structure, which does not change if one switches from a bottom-up version with base $(0,0)$ to a top-down version with base (N,P). However, while frequently used in subgroup discovery, WRA has been shown to over-generalize in a predictive setting [16,9]. We will thus not consider it further in this paper.

3.3   Integration into the Learning Algorithm

One could now think that the new heuristics could be directly plugged into the top-down refinement algorithm of Algorithm 11. However, it is easy to see that this would not yield the desired results. For example, continuing the example of Figure 4, the algorithm would select $c = 1$ as the final refinement for $\mathfrak{q}_{prec}$, since all rules covering all positive examples share the same (maximal) heuristic value of 1.0, irrespective of the amount of negative examples they cover. The rule $c = 1 \wedge d = 1$, which covers almost all positive examples but not negative examples, would receive a worse evaluation than its predecessor. Thus, $\mathfrak{q}_{prec}$ and to a lesser extent $\mathfrak{q}_{lap}$, are not well-suited for *rule selection* because rules with high coverage are still preferred by these heuristics, whereas the rule learning process is not steered towards learning a consistent theory. In fact, in preliminary experiments which just replaced the heuristics h_x with their counter-parts $\mathfrak{q}_x$ so that the latter was used for both rule selection *and* rule refinement, the resulting classifiers were sometimes unable to label *any* new testing example correctly.

Thus, we would like to maintain the conventional heuristics for rule selection, and need to adapt the learning algorithm so that it can use separate heuristics for rule selection and for rule refinement. To realize this, we need to adapt top-down hill-climbing so that different heuristics can be used for rule refinement and rule selection, as marked in the comments of the pseudo-code of Algorithm 11. Algorithm 11 shows the resulting algorithm; lines 2, 6 and 8 have changed.

Algorithm 3. Procedure findBestRule
Data: TrainingData
Result: best rule r_{best}
1 $r_{best} = \emptyset$
2 bestValue = selection_heuristic(r_{best})
3 **repeat**
4   get possible refinements
5   **forall the** refinements ref **do**
6   evaluation = refinement_heuristic(ref)
7   r_{ref} = best refinement
8   **if** selection_heuristic(r_{ref}) $\geq$ bestValue **then**
9           $r_{best} = r_{ref}$
10 **until** no refinements left;
11 **return** r_{best}

4 Experiments

In our experimental evaluation, we intend to answer the question whether the proposed separation of rule selection and rule refinement heuristics does indeed yield an improved performance over the standard technique that uses the same heuristic for both tasks.

4.1 Experimental Setup

For our experiments, we use the top-down hill-climbing algorithm implemented in the SECO-library [10] that has also been used in [9]. This is a simple and straight-forward rule learner that solely relies on heuristic rule evaluation to learn a classifier that generalizes well to new data. In particular, no additional procedures, such as pruning or rule optimization, are used to avoid overfitting on the training data. Multiple classes are handled using an ordered one-against-all strategy, as originally proposed for the RIPPER rule learning algorithm [3].

We modified the algorithm as described in Section 3.3, so that it allows for separate criteria for rule selection and rule refinement. We can thus denote an algorithm by a pair $(h_{selection}, h_{refinement})$. For the experiments we use each of the three standard heuristics for rule selection, and evaluate it with four different heuristics for rule refinement, yielding a total of $3 + 3 \times 3 = 12$ different algorithms. The four heuristics for rule refinement are to use the same heuristic as for rule selection (yielding a standard rule learning algorithm), and to use each of the three inverted heuristics.

For all experiments with h_{mest} and q_{mest}, we used a value of $m = 22.446$ which has been experimentally determined in [9] for the same learning algorithm that forms the basis of our experiments. Thus, this setting is optimized for the use of the m-estimate for guiding both rule selection and refinement. It is most likely a suboptimal value for q_{mest}. However, our main purpose in this paper was not to achieve optimal performance, but to investigate the general properties of different top-down rule refinement heuristics. As we have discussed in Section 2.3 the m-estimate provides a trade-off

Table 1. Number of classes (C), examples (E), and attributes(A) of the 20 datasets used in the experiments

Dataset	C	E	A	Dataset	C	E	A
breast-cancer	2	286	10	car	4	1728	7
futebol	2	14	5	contact-lenses	3	24	5
hepatitis	2	155	20	glass	7	214	10
hypothyroid	2	3163	26	idh	3	29	5
horse-colic	2	368	23	iris	3	150	5
ionosphere	2	351	35	lymphography	4	148	19
labor	2	57	17	primary-tumor	22	339	18
mushroom	2	8124	23	monk3	2	122	7
soybean	19	683	36	tic-tac-toe	2	958	10
vote	2	435	17	zoo	7	101	18

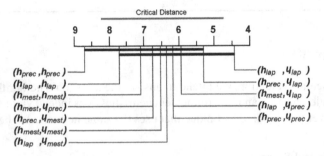

Fig. 6. Nemenyi Test with a significance level of 0.1

between precision and weighted relative accuracy, where larger values of m approach the behavior of WRA, which is insensitive to inversion. In this sense, the m-estimate with $m = 22.446$ nicely complements precision and Laplace in that it is much closer to WRA than the others.

We will evaluate the twelve combinations listed above on 20 datasets by the means of estimated average accuracy. The evaluation method is ten-fold cross-validation to reduce bias and increase the quality of the resulting performance estimate. As can be seen from Table 1, the chosen datasets range from very small datasets (where we feel that good selection heuristics are particularly important) to datasets with several thousand examples. For checking for statistical differences we use Friedman rank tests with a post-hoc Nemenyi test, as recommended by [4].

4.2 Comparison of Average Accuracies

Table 2 shows the detailed results with respect to accuracy. There are three main columns, each corresponding to one of the three rule selection strategies h_{prec}, h_{lap}, and h_{mest}. Each of them has four subcolumns, each corresponding to a rule refinement strategy. The left-most is the standard strategy, and the three others are all three inverted strategies u_{prec}, u_{lap}, and u_{mest}. The best results in each line and each group are underlined.

Table 2. Average accuracies obtained via ten-fold cross-validation on 20 datasets. The best result for each rule selection heuristic is underlined. The bottom line shows the average rank of each rule refinement strategy for each rule selection heuristic.

Dataset	$(h_{prec}, \cdot)$				$(h_{lap}, \cdot)$				$(h_{mest}, \cdot)$			
	h_{prec}	q_{prec}	q_{lap}	q_{mest}	h_{lap}	q_{prec}	q_{lap}	q_{mest}	h_{mest}	q_{prec}	q_{lap}	q_{mest}
breast-cancer	68.53	72.38	72.03	<u>73.43</u>	69.58	70.63	71.33	<u>72.73</u>	71.33	72.03	72.38	<u>73.78</u>
car	90.10	90.34	<u>90.51</u>	88.66	90.45	91.20	<u>91.73</u>	91.20	89.64	<u>90.45</u>	90.28	87.91
contact-lenses	79.17	<u>87.50</u>	<u>87.50</u>	83.33	79.17	<u>87.50</u>	<u>87.50</u>	83.33	<u>87.50</u>	<u>87.50</u>	<u>87.50</u>	83.33
futebol	28.57	<u>64.29</u>	57.14	42.88	28.57	<u>64.29</u>	57.14	42.88	50.00	<u>64.29</u>	57.14	42.86
glass	56.54	65.89	<u>68.69</u>	62.15	61.22	65.89	<u>68.69</u>	62.15	69.16	67.29	<u>71.50</u>	63.55
hepatitis	78.07	79.36	<u>80.00</u>	76.77	78.71	79.36	<u>80.00</u>	76.74	78.07	79.36	<u>80.00</u>	76.77
hypothyroid	98.23	98.61	98.74	<u>98.83</u>	98.39	98.61	98.74	<u>98.83</u>	98.80	98.61	98.74	<u>98.83</u>
horse-colic	72.01	<u>79.35</u>	<u>79.35</u>	77.99	70.65	79.35	<u>80.16</u>	77.99	77.45	<u>79.35</u>	78.80	77.99
idh	62.07	<u>82.76</u>	75.86	75.86	62.07	<u>82.76</u>	75.86	75.86	68.97	<u>82.76</u>	75.86	75.86
iris	92.67	93.33	<u>95.33</u>	94.67	94.00	93.33	<u>95.33</u>	94.67	94.00	93.33	<u>95.33</u>	94.67
ionosphere	<u>95.16</u>	82.62	83.19	89.46	<u>94.87</u>	82.62	93.19	89.46	<u>91.74</u>	82.91	83.19	91.17
labor	<u>91.23</u>	80.70	82.46	89.47	<u>91.23</u>	80.70	82.46	89.47	85.97	80.70	82.46	<u>89.47</u>
lymphography	83.78	77.70	<u>84.46</u>	83.11	<u>85.14</u>	77.70	84.46	83.11	75.00	76.35	81.08	<u>83.78</u>
mushroom	100.0	100.0	100.0	100.0	100.0	100.0	100.0	100.0	100.0	100.0	100.0	100.0
monk3	<u>87.71</u>	82.79	82.79	84.43	<u>88.53</u>	85.25	84.43	86.89	81.15	79.51	81.15	<u>82.79</u>
primary-tumor	33.63	<u>39.23</u>	35.10	30.97	32.45	<u>39.23</u>	35.99	30.38	33.92	<u>37.76</u>	34.51	30.68
soybean	90.04	91.51	<u>92.24</u>	91.36	90.34	91.80	<u>92.39</u>	90.63	<u>91.51</u>	90.92	90.48	91.36
tic-tac-toe	97.39	<u>98.02</u>	97.60	97.81	97.60	<u>98.02</u>	97.60	97.91	<u>98.12</u>	98.02	97.60	97.81
vote	<u>94.94</u>	93.56	94.25	94.48	<u>95.40</u>	94.25	94.25	94.94	93.33	93.56	94.71	<u>96.09</u>
zoo	84.16	88.12	<u>92.08</u>	90.01	86.14	88.12	<u>92.08</u>	90.10	89.11	88.12	<u>92.08</u>	90.10
average rank	3.075	2.400	<u>1.975</u>	2.550	3.000	2.500	<u>1.975</u>	2.525	2.700	2.625	<u>2.225</u>	2.450

Not surprisingly, we can see that each of the combinations works best in some cases, and that the differences can be quite large in some cases (mostly for rather small datasets). In order to get a better overall impression, we show the average ranks within each group in the last line of the table. This gives a fairly consistent picture in that the standard strategy always performs worst, i.e., on average all three inverted rule refinement heuristics perform better than the case where rule refinements are evaluated with the same heuristic as rule refinements. Thus, our expectation that top-down refinement heuristics work better than conventional heuristics has been confirmed.

The results are also consistent in that the inverted Laplace-heuristic always performs best for all rule selection strategies, whereas the m-estimate and precision are about equal on ranks 2 and 3. We can also see that the differences between the methods are much smaller for the m-estimate than for the others. In fact, a Friedman test reveals that the results within the h_{prec} rule selection group are statistically significant at a 5% level, the results within the h_{lap} group at the 10% level, whereas the results in the m-estimate are only weakly different. However, this is not surprising, because as we noted above, for larger values of m, the behavior of the m-estimate approaches the behavior of WRA, which is insensitive to inversion. Thus, with increasing values of m, the results must become more and more similar to each other.

Table 3. Comparison of the number of rules (R) and conditions (L) for regular and inverted Laplace heuristics

Dataset	(h_{lap}, h_{lap})		(h_{lap}, h'_{lap})		Dataset	(h_{lap}, h_{lap})		(h_{lap}, h'_{lap})	
	R	L	R	L		R	L	R	L
breast-cancer	25	67	38	173	ionosphere	17	25	8	42
car	107	495	107	506	labor	5	7	3	12
contact-lenses	5	14	5	15	lymphography	18	42	11	47
futebol	4	7	2	5	monk3	13	38	11	32
glass	50	103	14	83	mushroom	11	13	7	35
hepatitis	13	26	7	46	primary-tumor	80	319	72	518
horse-colic	44	114	19	111	soybean	62	134	45	195
hypothyroid	27	65	9	69	tic-tac-toe	22	84	16	69
iris	7	15	5	17	vote	13	48	12	58
idh	4	5	2	5	zoo	19	19	6	14
averages						27.3	82.0	20.0	102.6

4.3   Validation and Algorithm Comparison

Overall, the combination $(h_{lap}, ꓕ_{lap})$ outperforms other combinations on seven datasets (namely *car, contact-lenses, hepatitis, horse-colic, iris, soybean* and *zoo*). As such, this combination in particular becomes interesting for further validation. We will now conduct statistical tests to try and prove the assumption that the combination $(h_{lap}, ꓕ_{lap})$ is superior w.r.t. accuracy.

Using $N = 20$ datasets with $k = 12$ algorithms, we obtain a chi-square value of 19.792 and a corresponding F_F statistic of 1.878. The corresponding critical value based on a significance level of 0.05 with 11 and 209 degrees of freedom is 1.834, resulting in a passed Friedman test (failure at level 0.01). The ranks of the algorithms as well as the critical distance for the post-hoc Nemenyi test are shown in Figure 6. Although the results only show that the combination $(h_{lap}, ꓕ_{lap})$ is significantly better than the algorithm (h_{prec}, h_{prec}), which is known to overfit the data, it is still remarkable that all combinations that involve an inverted heuristic are higher-ranked than all three original heuristics, including (h_{mest}, h_{mest}), which was one of the best-performing algorithms in a previous study [9].

4.4   Number of Rules and Conditions

Using inverted heuristics also has an effect on the nature of conditions that are selected. In short, whereas regular heuristics focus mostly on consistency, inverted heuristics tend to add conditions that maintain completeness. For example, if at any point, both heuristics are faced with the choice of adding an incomplete but consistent rule r_1 (a point on the P-axis) and a complete but inconsistent rule r_2 (a point (P, n) for some value $0 < n < N$), regular precision would give a maximum evaluation of $h_{prec}(r_1) = 1.0$ to r_1, whereas inverted precision gives a maximum score $ꓕ_{prec}(r_2) = 1.0$ to r_2. This has the effect that inverted heuristics bias the learner towards conditions that do not add additional discriminative power (but are nevertheless informative).

```
2160  p :- odor = f.
1152  p :- gill-color = b.
 256  p :- odor = p.
 192  p :- odor = c.
  72  p :- spore-print-color = r.
  36  p :- stalk-color-below-ring = c.
  24  p :- stalk-color-below-ring = y.
   4  p :- cap-surface = g.
   1  p :- cap-shape = c.
  16  p :- stalk-color-below-ring = n, stalk-surface-above-ring = k.
   3  p :- habitat = l, stalk-color-below-ring = w.
```

(a) using h_{lap} for refinement

```
2192  p :- veil-color = w, gill-spacing = c, bruises? = f, ring-number = o,
            stalk-surface-above-ring = k.
 864  p :- veil-color = w, gill-spacing = c, gill-size = n, population = v,
            stalk-shape = t.
 336  p :- stalk-color-below-ring = w, ring-type = p, stalk-color-above-ring = w,
            ring-number = o, cap-surface = s, stalk-root = b, gill-spacing = c.
 264  p :- stalk-surface-below-ring = s, stalk-surface-above-ring = s,
            ring-type=p, stalk-shape=e, veil-color=w, gill-size=n, bruises?=t.
 144  p :- stalk-shape = e, stalk-root = b, stalk-color-below-ring = w,
            ring-number = o.
  72  p :- stalk-shape = e, gill-spacing = c, veil-color = w, gill-size = b,
            spore-print-color = r.
  44  p :- stalk-surface-below-ring = y, stalk-root = c.
```

(b) using q_{lap} for refinement

Fig. 7. Decision lists learned for the class poisonous in the *mushroom* dataset, along with the number of positive examples covered by each rule (no rule covers any negative examples)

In practical terms, inverted heuristics tend to learn longer rules, which will nevertheless, somewhat counter-intuitively, have a higher coverage than those learned with regular heuristics. As an illustration, Table 3 compares the number of rules and conditions induced with h_{lap} to those induced with $qlap$, the latter being the the configuration that achieved the best results in our experiments. On 17 out of 20 datasets the inverted version learns a lower number of rules, on two an equal number of rules, and only on one dataset a higher number of rules, which clearly confirms that the learned rules on average tend to have a higher coverage. Moreover, on 13 datasets q_{lap} has a higher number of conditions, on one dataset it is equal and on 6 the number is smaller, which confirms that the rules learned by inverted heuristics tend to be longer. Both findings are also confirmed by the averages shown in the last line of Table 3.

Note, however, that this does not necessarily reduces the comprehensibility of the learned rules. In a way, in the terminology of Michalski [14], inverted heuristics tend to find *characteristic* descriptions, whereas standard heuristics tend to find *discriminative* descriptions. As an illustration, Figure 7 shows the rule sets learned for the *mushroom* dataset. Both rule sets cover all 3196 examples of poisonous mushrooms. However, while the first rule set, learned with a traditional heuristic, focuses on single characteristics such as the odor of the mushroom, the second rule set contains much more descriptive rules. Interestingly, the used attributes are quite different (e.g., odor does not appear at all in the latter rule set). Another interesting observation is that the former rule set contains some rules with only very low coverage: the last six rules all cover fewer examples than the last rule of the rule set learned with the inverted heuristic. One reason for this is that because the previous rules are somewhat less general, they also leave more examples to be classified for subsequent rules. For example, the last rule of Figure 7 (b) still classifies 44 examples, whereas a similar rule that consisting only of the first condition has only 24 examples left to classifier in Figure 7 (a).

We are not expert enough to judge the plausibility of the rules of Figure 7, but in general, we think that more detailed rules can be more convincing and are certainly no less comprehensible than the general discriminative rules. In fact, we think that this property of inverted heuristics is also of particular interest to subgroup discovery [12], although we leave this as subject for future work.

5   Conclusions and Open Questions

In this paper, we made two contributions to heuristic inductive rule learning. First, we argued that it may be beneficial to separate the evaluation of candidates for rule refinement and the selection of rules for the final theory. Accordingly, we suggest to use different criteria for both. Second, we showed that conventional precision-based heuristics can be inverted in the sense that they do not evaluate candidate refinements from the point of view of the origin of the coverage space, but from the point of view of their predecessor rule in a top-down search. Our experiments showed that the use of such inverted heuristics for evaluating rule refinements leads to better results than the use of the original versions. Interestingly, inverted heuristics also have the tendency to learn rule sets with longer but fewer rules.

Our results are so far confined to top-down covering rule learning algorithms. While we do not expect that bottom-up algorithms would profit from inverted heuristics, which reflect a top-down search strategy, it remains an open question whether other top-down algorithms may benefit from their use. In particular, the fact that inverted heuristics tend to learn longer, characteristic rules may be of interest for subgroup discovery.

We have also only considered precision-like heuristics in this work, mainly because the m-estimate has delivered a state-of-the-art performance in a large comparative study [9], so that it seemed a natural point of departure for our experiments. While we have shown that other linear heuristics such as WRA, which is popular in subgroup discovery, cannot be inverted, we still need to look at heuristics with non-linear isometrics. In particular, the proposed separation of rule refinement and rule selection criteria is also closely related to the use of pruning criteria, which filter out unpromising rules. It remains to be seen whether conventional rule pruning criteria, such as the significance test of CN2 [2,1] may also be used favorably as rule selection criteria. Furthermore, we also deliberately refrained from optimizing the m-parameter of the inverted heuristics in any way because we wanted to avoid to obtain good results for the inverted heuristics that are only due to an extensive search for optimal parameter values. However, such an evaluation is planned as the next step in our work.

Finally, we note that the use of precision for rule selection may be viewed as a simple, greedy maximization of the area under the ROC curve (AUC) [7]. The inverted precision heuristic introduced in this paper may be viewed as a counter-part that maximizes the AUC for individual rules. Interestingly, in preliminary experiments we could not demonstrate that improving the AUC maximization for individual rules also leads to a better AUC for the entire theory. However, this needs a deeper investigation and needs to be put into perspective with alternative approaches to maximize the AUC in inductive rule learning [15,5].

Acknowledgements. The authors would like to thank the anonymous reviewers for their comments, which helped to improve this paper.

References

1. Clark, P., Boswell, R.: Rule induction with CN2: Some recent improvements. In: Proceedings of the 5th European Working Session on Learning (EWSL 1991), Porto, Portugal, pp. 151–163. Springer, Heidelberg (1991)
2. Clark, P., Niblett, T.: The CN2 induction algorithm. Machine Learning 3(4), 261–283 (1989)
3. Cohen, W.W.: Fast effective rule induction. In: Prieditis, A., Russell, S. (eds.) Proceedings of the 12th International Conference on Machine Learning, Tahoe City, CA, July 9-12, vol. 123, pp. 115–123. Morgan Kaufmann (1995)
4. Demšar, J.: Statistical comparisons of classifiers over multiple data sets. Journal of Machine Learning Research 7, 1–30 (2006)
5. Fawcett, T.E.: PRIE: A system for generating rulelists to maximize ROC performance. Data Mining and Knowledge Discovery 17(2), 207–224 (2008)
6. Fürnkranz, J.: Separate-and-conquer rule learning. Artificial Intelligence Review 13(1), 3–54 (1999)
7. Fürnkranz, J., Flach, P.A.: ROC 'n' rule learning – Towards a better understanding of covering algorithms. Machine Learning 58(1), 39–77 (2005)
8. Fürnkranz, J., Gamberger, D., Lavrač, N.: Foundations of Rule Learning. Springer, Heidelberg (2012)
9. Janssen, F., Fürnkranz, J.: On the quest for optimal rule learning heuristics. Machine Learning 78(3), 343–379 (2010)
10. Janssen, F., Zopf, M.: The SeCo-framework for rule learning. In: Proceedings of the German Workshop on Lernen, Wissen, Adaptivität - LWA (2012)
11. Klösgen, W.: Explora: A multipattern and multistrategy discovery assistant. In: Fayyad, U.M., Piatetsky-Shapiro, G., Smyth, P., Uthurusamy, R. (eds.) Advances in Knowledge Discovery and Data Mining, pp. 249–271. AAAI Press (1996)
12. Kralj Novak, P., Lavrač, N., Webb, G.I.: Supervised descriptive rule discovery: A unifying survey of contrast set, emerging pattern and subgroup mining. Journal of Machine Learning Research 10, 377–403 (2009)
13. Lavrač, N., Kavšek, B., Flach, P., Todorovski, L.: Subgroup discovery with CN2-SD. Journal of Machine Learning Research 5, 153–188 (2004)
14. Michalski, R.S.: A theory and methodology of inductive learning. Artificial Intelligence 20(2), 111–162 (1983)
15. Prati, R.C., Flach, P.A.: Roccer: An algorithm for rule learning based on ROC analysis. In: Kaelbling, L.P., Saffiotti, A. (eds.) Proceedings of the 19th International Joint Conference on Artificial Intelligence (IJCAI 2005), Edinburgh, Scotland, pp. 823–828. Professional Book Center (2005)
16. Todorovski, L., Flach, P.A., Lavrač, N.: Predictive performance of weighted relative accuracy. In: Zighed, D.A., Komorowski, J., Żytkow, J.M. (eds.) PKDD 2000. LNCS (LNAI), vol. 1910, pp. 255–264. Springer, Heidelberg (2000)

Scalable Information Flow Mining in Networks

Karthik Subbian[1], Chidananda Sridhar[1], Charu C. Aggarwal[2], and Jaideep Srivastava[1]

[1] University of Minnesota, Minneapolis, MN, USA
{karthik,sridh050,srivasta}@cs.umn.edu
[2] IBM T. J. Watson Research Center, Yorktown Heights, NY, USA
charu@us.ibm.com

Abstract. The problem of understanding user activities and their patterns of communication is extremely important in social and collaboration networks. This can be achieved by tracking the dominant content flow trends and their interactions between users in the network. Our approach tracks all possible paths of information flow using its network structure, content propagated and the time of propagation. We also show that the complexity class of this problem is #P-complete. Because most social networks have many activities and interactions, it is inevitable the proposed method will be computationally intensive. Therefore, we propose an efficient method for mining information flow patterns, especially in large networks, using distributed vertex-centric computational models. We use the Gather-Apply-Scatter (GAS) paradigm to implement our approach. We experimentally show that our approach achieves over *three orders of magnitude* advantage over the state-of-the-art, with an increasing advantage with a greater number of cores. We also study the effectiveness of the discovered content flow patterns by using it in the context of an influence analysis application.

Keywords: Information Flow Mining, Vertex-centric models, Influence Analysis Network-centric approach, Scalable Influence Analysis.

1 Introduction

The problem of finding dominant content flow trends in networks is an important problem in the context of online social and collaboration networks. In social networks, such as *Twitter* and *Facebook*, every user posts messages, photos and comments to exchange information with their neighbors in the network. The daily volume of content propagation in these networks is in the order of hundreds of millions of posts per day[1]. These posts typically propagate as short phrases [12], topics [1], *hashtags* [2], or *URLs* [7] in specific patterns on the underlying friend or follower network. Some of these posts may go *viral* and reach millions of users within a few hours. The massive reach of these flows may result in significant influence in online user behavior [21]. Therefore, it is desirable to understand such viral information flows for online marketing, advertisement and a variety of other applications. There are several recent works that attempt to understand these viral information flows in terms of memes [11,12], cascades [13], and events [2].

---

[1] http://www.digitalbuzzblog.com/
facebook-statistics-facts-figures-for-2010/

T. Calders et al. (Eds.): ECML PKDD 2014, Part III, LNCS 8726, pp. 130–146, 2014.
© Springer-Verlag Berlin Heidelberg 2014

The existing literature on understanding information flows using cascades [11,12,13] and memes [11,12] analyze a stream or a corpus of text documents where there is no explicit network structure used for communication. For instance, the work in [13] determines the cascade patterns from the temporal sequence of blog posts across multiple blogs. Here, the term cascade, refers to a phenomenon in which a topic is adopted by a blog and further propagated through the creation of hyperlinks. Note that there is no underlying network structure between the bloggers. On the other hand, online social networks use an explicit network structure, such as follower or friend network, to propagate the information. Therefore, understanding the patterns of propagation in a network structure is likely to yield superior insights than observing patterns from general population. In several other related works, the flow of information is typically analyzed in the absence of the underlying network structure [18,19].

Another disadvantage of existing approaches [7] is that they ignore the life-span of influence due to information flows, which is very relevant in the social context. For example, a message posted on a *Facebook* wall may not even be available on the first page after a day elapses. If a receiver of that message re-posts the same message after a week, then it is less likely that the user was influenced by the original post sent to his wall. As the life span is not considered, the existing methods produce a large number of cascades as opposed to more active and meaningful ones.

Most of the existing approaches for mining information flows [18,19,11,12,13] cannot handle large amounts of data, as their processing is centralized in a single server. We propose a distributed approach using vertex-centric computational models [8]. In these models, each vertex is a separate computational unit (available in a core or a machine), and it result in a high level of parallelism. As we show in our experiments, our approach is *three orders of magnitude* faster than existing state-of-the-art approaches. To the best of our knowledge, our approach is the first work in this area and we are not aware of any distributed or parallel information flow mining algorithms.

In this paper, we propose an efficient information FLOWExtractoR algorithm, called *FLOWER*, to discover these information flow patterns. We establish the complexity class of this problem, by showing the counting problem of all maximal information flow patterns is #P-complete. In order to scale up to large networks, we propose a parallel version called *pFLOWER* that runs on vertex-centric graph computational models [8]. In the experimental section, we show that our parallel method *pFLOWER* is faster than the state-of-the art algorithms by up to *three orders of magnitude*. We also study the effectiveness of the discovered information flows in the context of an influence analysis application. Our approach consistently outperforms the popular baselines in terms of precision, recall and F_1 measure.

The paper is organized as follows. The remainder of this section discusses related work. In the next section, we introduce the preliminaries for the problem of flow mining in networks. Then we describe the flow mining algorithm and propose a parallel version using a vertex-centric computation model. Finally, we demonstrate the efficiency and effectiveness of our approach using multiple real-life data sets.

1.1 Related Work

The problem of analyzing information flow has been studied using content influence cascades [12,13,1]. Most of these work analyze the blogosphere, and there is *no explicit network* structure over which users exchange information in the blogosphere. However, in general, for social [2] and biological networks [10], there is an explicit network over which the flow of information occurs. Using this structure is important, because it enhances the discoverability of the relevant information flow patterns. In addition, information cascades have a limited life span, and therefore the use of the information life-time of the cascade helps in finding more active cascades caused by intrinsic social network influence rather than external sources [16]. In some recent works [15,4] an instance of the Independent-Cascade (IC) model is built using the log of past propagation in the network. The aim of these techniques are to sparsify the network for scalability, while our intent is to extract the dominant information flow patterns. There are several other works that use only a network structure to compute social centrality of the users [22].

Content propagation in online media is usually tracked as short and distinct phrases, referred to as memes [12,1]. Memes tend to have a broader stable vocabulary, that mimic a slowly evolving genetic signature over time. The key idea of this work [12] is to understand how the short phrases evolve over time while the several words of the phrase are intact during the entire period of propagation. However, there is no notion of tracking content flows in a network, while it does track how content evolves over time. There are several other papers that tend to capture such bursty topic behavior over time, based on different notions of topic identification [23]. A more detailed survey of evolution of content in network structures can be found in [3].

In a recent experimental study [11], the diffusion of stories in social networks, such as *Twitter* and *Digg*, are analyzed using the evolution of the number of fan votes in general population and in a network structure. More specifically, this work confirms the importance of using network structure in such studies. There are other recent works that study the distribution of URL cascades [7] in *Twitter* and propose a prediction model to predict the number of mentions of an URL after its posting. However, none of these related work track the dominant content flow information in a network structure over time.

2   Information Flow Mining Model

We define the information flow mining problem and related information flow properties in this section.

Let $G = (V, E)$ be the relationship network containing the node set V and the edge set E. Each actor $a_i \in V$ performs a number of *content-based actions* such as sending tweets or posting wall posts. We denote a content posted as U_j, where j is index of the message, and the time of its posting as t_k. The time points are ordered using their index $k = 1 \ldots T$, such that $t_k < t_{k+1}$. A message U_j can be propagated by different nodes at the same time in different parts of the network, and not all nodes may necessarily propagate all messages. Also, all nodes need not propagate a message at every time point. Consider Figure 1, where an example network G and a table of different messages propagated are shown. The node C does not propagate message U_4, and none of the nodes propagate a message at time point t_4.

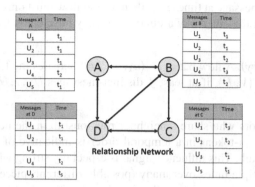

Fig. 1. An illustrative example showing different messages $U_1, \ldots, U_5$ propagated at different times $t_1, \ldots, t_5$ from different actors (A, B, C, and D) in a small example network

Definition 1 (Flow Path). *A flow path s in the network $G = (V, E)$ is a sequence of distinct actors $\langle a_1, \ldots, a_k \rangle$, where each actor $a_i \in V$ propagated the same content U_j at least once.*

Each information flow path s is a sequence of distinct *actors* $a_1, \ldots, a_k$, who are involved in multiple content interactions over a period of observation. Note that there are no cycles in a single flow path as it contains a distinct set of actors. The purpose of our approach is to determine such frequent information flow paths, where certain content flows may occur frequently in specific paths of the network. In this paper, we assume that flow patterns are sequential paths in the network, and a general cascade can be constructed by overlaying multiple such sequential paths. In order to distinguish the interesting flow patterns, we define several flow properties. These flow properties ensure the interestingness in terms of *network structure, causality, frequency* and *life-time.*

Property 1 (Network Structure). A flow path $s = \langle a_1, \ldots, a_k \rangle$ satisfies the network structure property in the network $G = (V, E)$, if for each $r \in \{1 \ldots k - 1\}$, an edge exists between a_r and a_{r+1} in E.

By Property 1, we consider only the information flows that adhere to the network structure, which also has the effect of focussing on relevant patterns. As in social networks, similar nodes are related to each other by a neighboring relationship, such as a friend or a follower, and the content recommendation in a social network is often guided by such relationships.

Property 2 (Causality). A flow path $s = \langle a_1, \ldots, a_k \rangle$ satisfies the causality property in the network $G = (V, E)$, if the actors in the flow path propagate a message U_j at time points $t_1 \ldots t_k$ where $t_m < t_{m+1}, \forall m = 1 \ldots k - 1$.

The causality[2] property defines the interestingness of a flow pattern in terms of the time of propagation. An actor a_i broadcasts a message at time t_k and the neighbor a_j

---

[2] The notion of "causality" in this paper is only based on temporal ordering, and no explicit mechanisms of cause and effect are assumed. Clearly, such temporal orderings might also occur by chance.

broadcasts the same message at time t_{k+1}, then a valid flow must only consider the path from a_i to a_j and not the other way around, as a_i may have caused a_j to transmit the message.

Property 3 (Frequency). A flow path $s = \langle a_1, \ldots, a_k \rangle$ satisfies the frequency property in the network $G = (V, E)$, if at least f distinct messages $U_1 \ldots U_f$ are propagated over the flow path s.

A sequence of actors who share a neighbor relationship in the network and post a particular set of content-tokens in a temporal order is indicative of a signal of influence along the flow path. This influence signal is especially strong when actors behave in a similar way multiple times over many (possibly different) pieces of content. The frequency property captures the strength of such influences in a flow path through a pre-specified frequency level. The frequency parameter f is the count of the number of such repeated flows.

Property 4 (Life-time). A flow path $s = \langle a_1, \ldots, a_k \rangle$ has a *valid life-time period* τ, if the message propagated U_j at time points $t_1 \ldots t_k$ is such that $t_i - t_{i-1} \leq \tau, i = 2 \ldots k$.

A post on a *Facebook* wall is not available forever for further propagation due to new incoming posts [11], or due to limited user attention span [24]. The notion of life-time is designed to model such real-life situations. All the aforementioned definitions can be generalized to multiple messages.

The problem of information flow mining is to extract all valid flow paths s that satisfy these flow properties. In the following problem definition, we denote the set of all messages U_j and the corresponding time stamps t_j sent by each actor a_i as T_i. The size of T_i is denoted as m_i. For example, in Figure 1, message table T_C of actor C contains four messages and time-stamp pairs: $T_C = \{(U_1, t_3), (U_2, t_2), (U_3, t_1), (U_5, t_1)\}$.

Problem 1 (Information Flow Mining). Given a graph $G = (V, E)$, a set of m_i messages propagated by each actor a_i, and their corresponding time-stamps denoted by $T_i = \{(U_j, t_j)_{j=1\ldots m_i}\}$, the problem of information flow mining is to extract the set of all valid flow paths $F = \{s_1, \ldots, s_p\}$ that satisfy the network, causality, frequency, and life-time properties (Properties 1-4).

Our approach provides a generic framework to analyze information flows in a variety of domains such as social networks, fMRI brain networks, or Internet networks. The messages propagated in these networks correspond to user posts, molecular interactions or data packets, respectively. With appropriate functions to compare and track similar signals across multiple nodes, our approach can be easily generalized to other domains [10]. However, in this paper, we restrict our attention to propagation of textual content as information signals over discrete time points.

3   Information Flow Mining Algorithm

A major challenge in information flow mining is that of incorporating the impact of network structure directly into the flow mining process. The key issue here is that a set of users who propagate the same message at approximately the same time period provide us with very little knowledge about their actual path through the network. Typically, when a message is popular, it might be independently propagated by users in many different regions of the network. Therefore, how does one "connect up" the propagation of the different users over the entire network? It is here that the linkage information between the users comes in handy. A message is assumed to have been propagated from one user to the other, only if two neighboring users propagate the same message within the life-time constraint. Therefore, an efficient algorithm for mining the information flow patterns needs to integrate the sequences of user posts, the network structure, and the temporal aspects in a holistic way, to extract the relevant flow patterns over the network structure.

One way of mining the patterns is to extract the flow paths in a *content-centric* fashion. Consider a message U_j sent by a set of actors $a_1, a_2, ..., a_k$. These actors might have sent these messages at different time points. One can order the actors in increasing temporal order and extract all subsequences of actors that appear in at least f such messages. We can then eliminate all flow paths that do not have a valid edge in the network. In this approach, we first use the content to create the flow paths and finally we apply the network validity property. The main disadvantage of this approach is that the number of possible subsequences of actors in general population is very large. Therefore, it is prudent to use the network structure to eliminate such unnecessary candidates directly *during* the mining process. This idea is the key ingredient of the *network-centric approach*. In this approach, the message table T_i for the actor a_i is sent to each of its neighbor a_j iteratively. Each neighbor checks for validity and lifetime constraints by comparing the table T_i and T_j. If there are at least f messages that survive after the validation, the message is sent to the neighbors of a_j and so on. The advantage of this approach is that the sparsity of the network reduces the number of candidate flow paths dramatically [20]. An added advantage of the network-centric approach is that it can be easily parallelized using vertex-centric computational models. This aspect will be addressed in Section 4.

The pseudocode for our algorithm is shown in Algorithm 1. We refer to our approach as *FLOWER*, which stands for FLOW ExtractOR. The algorithm first extract the actors that have at least f distinct messages in their respective message tables (lines 2-4). Then, the information flow paths originating at each actor a_i are extracted by calling a recursive procedure *FLOWPROP*. This procedure extracts the messages that support causality and life-time property, compared to the incoming message table T_{seq}. If the number of messages in new supporting message table is at least f, then all its neighbors are iteratively explored (lines 4-5). If there are no neighbors for the actor a_i, the message is added to set F. This *if* condition ensures that the flow paths added to F are maximal in nature; in other words, a flow path is added only when none of its sub-sequence flow paths are already in F. One can ignore lines 6 and 7 in FLOWPROP to extract all paths that are not only maximal.

3.1 Complexity Analysis

In this section, we show that the counting version of the information flow mining problem is #P-complete. For this purpose, we reduce the maximal frequent sequence mining problem that is #P-complete [25], in polynomial time to the problem of mining information flow patterns (Problem 1). The notion of maximal information flow patterns is

---

Algorithm 1. *FLOWER*

Input: $G = (V, E)$: Relationship network; T_i: message table for actor a_i; f: frequency;
and τ: life-time;

Output: F: Set of flows satisfying all the properties

1 Initialize V' and S to empty set
2 **for** *each* $a_i \in V$ **do**
3   **if** *number of distinct messages in* $T_i \geq f$ **then**
4   Add a_i to V' and F

5 **for** *each* $a_i \in V'$ **do**
6   FLOWPROP($\{\phi\}, T_i, a_i, T_i, \tau, f, V', F$)

7 **return** F;

Algorithm 2. FLOWPROP

Input: seq: current flow path; T_{seq}: message table supporting the current flow path; a_i:
current actor; T_i: message table for actor a_i; τ: life-time; f: frequency; V':
frequent actors set; F: frequent flows discovered;

1 T_{new} = Extract messages that satisfy causality and life-time property from T_i and T_{seq}
2 **if** *number of messages in* $T_{new} \geq f$ **then**
3   Γ_i = Get neighbors of a_i not in current flow path seq and in V'
4   **for** *each* $a_j \in \Gamma_i$ **do**
5   FLOWPROP($\langle seq \cup \{a_i\}\rangle, T_{new}, a_j, T_j, \tau, f, V', F$)

6   **if** Γ_i *is empty* **then**
7   Add $\langle seq, \{a_i\}\rangle$ to F

---

to retain only the longest frequent flow paths in the set F of Algorithm 1, beyond which the flow path does not satisfy one of the frequency, network, or lifetime constraints.

Theorem 1. *[25] Let $\mathcal{D}$ be a database of sequences with m transactions. The problem of counting the number of maximal f-frequent subsequences in $\mathcal{D}$, where $1 \leq f \leq m$ is #P-complete.*

Let T_i be the message table of the actor a_i in network $G = (V, E)$. We define a function Q that converts the database $\mathcal{D}$ to individual user message-tables $T_1, \ldots, T_{|V|}$.

Definition 2 (Function Q). *Let $Q : \mathcal{D} \rightarrow \{T_1, \ldots, T_n\}$, where T_i is the message table of actor a_i. The ith transaction maps to a unique message id U_i. For each item k in the ith transaction of $\mathcal{D}$, a corresponding set $(U_i, \gamma(k, i))$ is added to T_k by Q, where $\gamma(k, i)$ denotes the first occurrence of actor k in the ith transaction of $\mathcal{D}$.*

The running time of function Q for a database D with m transactions and n items is $O(mn)$. Let $S_{\mathcal{D}}(f)$ be the set of maximal frequent sequences for support f for database $\mathcal{D}$. Let $F(f, G, \tau)$ be the set of all maximal frequent flows discovered for the information flow mining problem, as described in Problem 1, for support f, graph G and lifetime τ.

Lemma 1. *All maximal frequent sequences of set* $S_\mathcal{D}(f)$ *are present in* $F(f, G, \tau)$, *when G is complete and* $\tau = 0$.

Proof: Consider a path P that is a valid maximal frequent flow for $\tau = 0$ and the underlying graph G is complete. When $\tau = 0$ all valid paths in the graph G satisfy the lifetime constraint, because all lifetimes are non-negative. Also, given a complete graph any permutation of n nodes in the graph is a valid path. As the path P is frequent, there are at least f messages flowing along path P. Hence, the actors in that path must be appearing in that order in at least f such transactions in database $\mathcal{D}$, as per the function Q. Thus, the path P must be a frequent sequence in the database $\mathcal{D}$. So every f-frequent flow path P that is a valid for $\tau = 0$ and for a complete graph G is present in $S_\mathcal{D}(f)$. Because the path P is maximal, there cannot exist a longer path in set F that contains some actor a_r after P. If this is the case, per function Q, there cannot also exist a minimum of f transactions where P followed a_r is present.   ∎

From Lemma 1, it is evident that set $F(f, G, \tau)$ can be extracted from set $S_\mathcal{D}(f)$ by pruning the frequent sequences with lifetime lower than τ. For sparser graphs, several paths are invalid and hence several sequences are removed. Because the pruning process results in a much smaller set F compared to the original set $S_\mathcal{D}$, the complexity of mining maximal frequent sequences acts as an upper bound on mining sets of maximal frequent information flows.

Theorem 2. *The problem of counting all maximal flow patterns in the information flow mining problem is #P-complete.*

Proof: The maximal sequence mining problem can be reduced to an equivalent maximal information flow mining problem in two steps: (a) converting the database $\mathcal{D}$ using function Q (see Definition 2) into actor level message tables and (b) create a complete graph G. The computational complexity of function Q is $O(mn)$ and the complete graph creation is polynomial in n and the total time required is $O(mn + n^2)$. When $n \gg m$, the total complexity is polynomial in n and when $m \gg n$ it is polynomial in m. In either case, the maximal frequent sequence mining problem can be reduced to maximal information flow mining in polynomial time in the size of the sequence database $\mathcal{D}$ and hence it is #P-complete.   ∎

4   Accelerating FLOWER

There are several computational challenges associated with the flow mining problem, which can affect the performance of the *FLOWER* algorithm presented in Section 3. While *FLOWER* is designed to be inherently efficient because of careful network-centric pruning, the problem itself can sometimes be fundamentally intractable for large networks. For example, in a completely connected graph, traversing every possible actor sequence from every source vertex has $O(|V|!)$ complexity. This is, of course, not true in most real networks, where the linkage structure is sparse and not all actors send the same set of messages at the same time. Nevertheless, it is still possible to envision scenarios, where the *FLOWER* approach might be undesirably slow for certain parameter settings (such as low values of f and high values of τ). Because these challenges are *inherent* to the problem at hand, it is natural to explore whether parallelization can be used to accelerate *FLOWER*.

There are several ways to parallelize the *FLOWER* algorithm. In Algorithm 1, line 6 executes the subroutine *Flowprop* for each vertex v. This can be executed in parallel, as each call is independent of the other. Similarly, each of the flow paths in the recursion tree from root to leaf is independent of one another and is therefore easy to parallelize. In these approaches, the parallelism is performed at a path-level, where each path can be treated as a independent computational unit. In path-wise parallelization, however, care must be taken to reduce redundant computations at the parent nodes, as they form common prefixes in different flow paths.

The highest level of parallelism, however, can be achieved if we can parallelize at the vertex level, where each vertex can be treated as a separate computational unit that can be executed in parallel. This is typical in vertex-centric computational models, such as GraphLab [14] or Pregel. As seen earlier, our sequential approach propagates messages between the neighboring vertices, and it naturally fits into this framework. We discuss a brief overview of the vertex-centric computational models in the next couple of paragraphs.

In vertex-centric computational models, any *vertex* needs to perform three main operations: *Gather*, *Apply*, and *Scatter*. The *Gather* operation receives messages through the incoming edges of a vertex, the *Apply* operation processes the incoming messages and the *Scatter* operation distributes the processed messages to the neighbors via outgoing edges. Due to these three operations, vertex-centric computational abstractions are popularly referred to as the GAS framework.

The main problem with the GAS framework is in scenarios, where they deal with natural graphs having power-law degree distribution. Such graphs have very few nodes with extremely high degree and the remaining nodes have very small degree [8]. Hence balanced distribution of computational load, storage and communication is extremely challenging in this framework and to address this issue new frameworks, such as Power-Graph [8], have been developed. For a more detailed review of the PowerGraph, please refer to [8].

Algorithm 3. Scatter

Input: *icontext_type*: context, *vertex_type*: current_vertex, *edge_type*: edge

1 **if** *edge.destination_node does not have f words in its message table* **then**
2 | **return**

3 **for** *each sequence_to_send in current_vertex* **do**
4 | **if** *sequence_to_send in current_vertex has the destination vertex Id* **then**
5 | | continue
6 | **if** *edge.destination_node satisfies all properties (1)-(4)* **then**
7 | | add the destination id to sequences_to_send and copy it to edge data

---

In the distributed version of our algorithm, each vertex contains three pieces of metadata: its message table $(U_j, t_l)_{j=1...m}$, frequent flows ending at that vertex, and messages to forward to neighbors in the next iteration. Each edge acts as a channel that carries the message from a source to a destination vertex. The message carried by each

Algorithm 4. Apply

Input: *icontext_type*: context, *vertex_type*: current_vertex, *gather_type*:
      incoming_object
1 **for** *each sequence in the incoming_object.sequences* **do**
2 | Add *sequence* to the current_vertex saved_sequences list
3 |_ Add *sequence* in to sequences_to_send array for scatter method to pick up
4 **if** *(context.iteration <= max_iterations) && (number of sequences_to_send in
   current_vertex > 0)* **then**
5 |_ schedule current_vertex for next iteration

Algorithm 5. Gather

Input: *icontext_type*: context, *vertex_type*: current_vertex, *edge_type*: edge
Output: *gather_type*: gathered_obj
1 **return** *received sequences from edge data*

Algorithm 6. Gather Operator+=

Input: *gather_type*: that
Output: *gather_type*: ret_obj
1 **if** *this.incoming_sequences has no sequence* **then**
2 | copy that.incoming_sequences to this.incoming_sequences
3 |_ **return** *this*
4 **else if** *that.incoming_sequences has no sequence* **then**
5 |_ **return** *that*
6 **else**
7 | **for** *each seq in that.incoming_sequences* **do**
8 | |_ this.incoming_sequences.push_back(seq)
9 |_ **return** *this*

edge has a set of flow objects, where each flow object contains a flow sequence and a set of word messages that support the sequence. Note that we do not need to carry any temporal information along the edges, because it can be reconstructed at each vertex based on the set of words that support the flow sequence. This approach provides significant savings in time and space. We also optimize the computation by initializing only the vertices that have message table of length at least f.

During the *Scatter* phase, each edge is invoked to scatter a message from source to the destination vertex. Each source vertex does an advanced lookup of the destination vertex message table, to verify the possibility of extending the flow by adding the destination node. If any of the properties (1)-(4) fail, then the message is not scattered to the destination. The pseudocode for the *Scatter* subroutine is listed in Algorithm 3.

In the *Gather* operation, each vertex is invoked to gather the messages from the incoming edges. This step eventually appends all the incoming flows, one after the other, from different edges into a single incoming flow object containing several flow sequences and corresponding word signals. In GraphLab the *operator+=* appends all the sequences from each edge and the *Gather* function merely copies the reference of data from each edge and passes it to the operator. The pseudocode for the *Gather* subroutine and *operator+=* are provided in Algorithm 5 and Algorithm 6, respectively.

Each vertex during the apply phase saves the incoming sequences (from the *Gather* operation) in its own frequent sequence table. As the *Scatter* phase does advance lookup and scatters only valid sequences, the apply phase can save these sequences with no additional validations. Each vertex then schedules itself for the next iteration, as there could be potential extensions of recently added frequent flow sequences. Also, if one is interested in sequences of length not more than L, then the vertex can stop scheduling, if the current iteration number is greater than L. The pseudocode for the *Apply* routine is listed in Algorithm 4.

The main advantage of the GraphLab framework is that it is a unified framework for multi-core and distributed computation. Graphlab can use multiple cores on a single multi-core server, and if that is not sufficient it can scale to multiple servers. There is no additional coding or algorithmic changes required to switch from one infrastructure to another. We refer to our parallel version of FLOWER as *pFLOWER*.

5   Experimental Results

We evaluate the efficiency of our algorithm in terms of runtime of the algorithm. The implementation of the algorithm was done using C++ and the runtime was evaluated on a Linux server with Ubuntu 10.04 OS, 24GB RAM, 24 cores with each running 2.67GHz Intel Xeon processor. We used Graphlab version 2.1 [8] for our parallel *pFLOWER* evaluation.

5.1   Data Sets

We used two data sets: the *DataBase List of Publications* (DBLP) and the *US Patent Office* (USPTO) database. These data sets are described below in detail. We are interested in extracting the information flow patterns in the co-authorship network of both these data sets. In DBLP, for instance, the "mining" keyword may propagate across a sequence of authors forming an information flow path. We use all the words in the abstract of the papers and patents to generate the messages. The time-stamp of the document was used to generate the message time-stamp. The co-authorship network was used as the underlying network of communication.

DBLP Data Set: We downloaded the publicly available DBLP data set[3], and extracted the year of publication, abstract and authors for each of the published documents. We removed entities with multiple identities using the data available in the DBLP website [4]. The cleaned data set had 444,406 authors and 1,572,277 papers. We stemmed the words, removed stop words, and stripped off punctuations in the abstract. The resulting

---

[3] http://arnetminer.org/citation
[4] http://dblp.uni-trier.de/xml/

dictionary was 600,718 words. All the publications were between the time-period of 1945 to 2011. We used publication abstracts to generate the content tokens. The network was constructed using the co-authorship relationship with 444,406 authors (nodes) and 1,280,168 edges.

US Patent Data Set: The United States (US) patent database is publicly available for access from the US Patent Office (USPTO)[5]. We downloaded the following set of attributes for all patents granted from June 21, 1977 to December 28, 1999: *Patent Number, Granted Date, Abstract, Inventors, Assignee, Legal Representative,* and *Application Number.* After cleaning the data set of documents containing missing meta-information, a total of 1,813,616 patents remained in the patent database. We used the patent abstract to generate the content tokens. The co-authorship network for the US Patent database contained 1,310,057 nodes and 2,444,474 edges.

5.2 Evaluation Approach

We measured the efficiency in terms of the running time of the algorithm. We evaluate the scalability of the distributed approach by varying the number of cores used for *pFLOWER*. We used the *PrefixSpan*[6] [17] sequence mining algorithm followed by post-processing of the output sequences to apply the network and life-time properties. The input to *PrefixSpan* is a set of transactions, where each transaction corresponds to a message U_j and the temporal order of the actors a_i who propagated that message (as ordered singleton itemsets). The output of *PrefixSpan* is a set of author sequences (corresponding to information flow paths), except that they do not satisfy the network validity and lifetime constraints. This is checked explicitly by using a constant time look-up table for each author and message pair. The resulting output of *PrefixSpan*, after post-processing yields the same output as our algorithm. Therefore, the running times of the methods can be meaningfully compared. We also compared the running time of our sequential version of the algorithm (*FLOWER*) against our parallel version (*pFLOWER*).

5.3 Results

We compared the running time of *PrefixSpan*, *FLOWER* and *pFLOWER*. However, in Figure 2, we could not plot *PrefixSpan* running times as they were extremely large. Therefore, we list the running times of *PrefixSpan* separately in Table 1. Furthermore, we are unable to show the results for several values of $f \leq 450$, because *PrefixSpan* did not complete within a day. On the other hand, as evident from Figure 2, both *FLOWER* and *pFLOWER* completed in less than a couple of minutes over most parameter settings.

We compared the running times of *FLOWER* and *pFLOWER* algorithm, in Figures 2(a) and (b). As the number of words (f) required for the frequency property (Property 3) decreases, the number of possible flow paths increases exponentially. The *FLOWER* approach explores each of these paths sequentially, resulting in an exponential complexity with path length. On the other hand, the parallel algorithm *pFLOWER*

---

[5] http://uspto.org/
[6] http://www.cs.uiuc.edu/homes/hanj/software/prefixspan.htm

Table 1. The running time (seconds) for the *PrefixSpan* baseline for DBLP and USPTO data sets. For an *f* value smaller than 450, *PrefixSpan* ran for more than a day (>86400 seconds) and did not complete.

DBLP		USPTO	
f	Runtime (secs.)	f	Runtime (secs.)
480	36185.52	470	20905.63
540	16284.14	530	12311.91

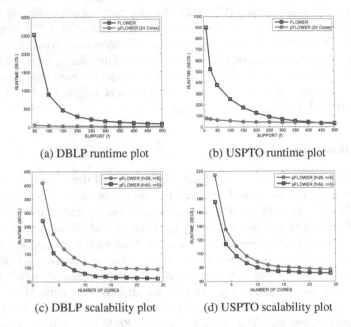

(a) DBLP runtime plot   (b) USPTO runtime plot

(c) DBLP scalability plot   (d) USPTO scalability plot

Fig. 2. The two plots in the *top row* show the running time measurements for the DBLP and USPTO dataset by varying *f*. The two plots in the *bottom row* show the scalability analysis for the DBLP and USPTO data sets by varying the number of cores.

scales extremely well at very low *f* values and the running time remains extremely small throughout the entire range of *f* values. The *pFLOWER* algorithm performs up to *three orders of magnitude faster* than PrefixSpan, and *two orders of magnitude faster* than *FLOWER* at low *f* values. These observations are consistent in both DBLP and USPTO data sets as shown in Figures 2(a) and (b), respectively. These observations also highlight the importance of a *network-centric approach* for computing information flow paths.

We evaluated the scalability of the *pFLOWER* algorithm in terms of the number of cores in Figures 2(c) and (d). The figure shows that the running time is roughly inversely proportional to the number of cores used for computation. In other words, linear speed-up is achieved in terms of the number of cores. It also demonstrates the efficiency of vertex-centric computational models in scaling up scenarios where sequential approaches are computationally infeasible. In this case, 14 cores were sufficient to complete the flow mining algorithm in less than a minute for low values of *f*, whereas straightforward sequential approaches do not terminate in reasonable running times (see Table 1). Thus, the proposed approach can be used to find information flows in networks with large number of activities and interactions, which may otherwise be computationally intractable using a single core.

6 Influence Analysis: An Application

Information flow patterns are sequences of actors who propagate at least f messages repeatedly preserving the temporal order in each propagation. These flow patterns denote the flow of influence along the network paths. The nature of influence depends on the nature of underlying network relationship or interactions. In DBLP and USPTO data set, we considered the co-authorship network and the nature of influence in these data sets are through co-authorship interactions. For instance, a flow path $\langle a, b, c \rangle$ denotes a word w used by author a, followed by b, and then c. When a used the word w because a and b have a co-authorship relationship, b may have been influenced by the word w through a and propagated it further to its neighbors. Similarly, c may have been influenced from b and propagated the word w to its neighbors. In a sense, for the example sequence, a is the leader and b and c are its followers. Similarly, b is the leader of the sub-sequence $\langle b, c \rangle$ with c as its follower. For each actor, we can compute the total number of followers (in this way) across all the flow patterns and we refer to this as the (co-authorship) influence score of that actor in the (co-authorship) network. The actor with the highest influence score in this DBLP or USPTO co-authorship network denotes the most influential co-author.

One might argue that using centrality measures or popular influence mining algorithms (such as PMIA [5], DegreeDiscountIC [6]) in a static co-authorship network are sufficient to measure the influence. We evaluate this hypothesis by comparing the influential co-authors found using the popular influence analysis algorithms such as degree-centrality, PageRank, PMIA [5] and DegreeDiscountIC [6] against the influencers found using the flow patterns. As the notion of influence has *no absolute ground truth* (similar to intelligence or trust), we use the author *citation counts as a proxy* for author influence. Here, we assume that an author has very high citation count if the author has considerable influence in the area. We computed the precision-at-K (P@K), precision-recall, and the F_1 score for the top-500 influencers found by each method (compared against the ground truth).

6.1 Evaluation Baselines

Let us now describe the baselines we used for evaluating our hypothesis. PMIA [5] is the prefix excluded extension of Maximum Influence Arborescense model. We used the weighted cascade model proposed in [9] to compute the edge probabilities for this approach. The degree-centrality approach uses the maximum total out-degree and DegreeDiscountIC [6] heuristic developed for the uniform IC [9] model with propagation probability $p = 0.01$. For PageRank, the restart probability was set to 0.15 and the stopping criterion, which is based on the L_1 norm difference between two successive iterations, was set to 10^{-7}. We use *FLOWER* to denote the influencers found using the information flow-based approach.

6.2   Evaluation Results

Our evaluation results are shown in Figure 3. The figure clearly shows that the order of baselines are not consistent in both data sets. The first- or second-order centrality measures might work in some data sets, while the information diffusion based method might work in others. But the flow-based techniques (like FLOWER) capture the lead authors whose ideas propagate dominantly and later gets picked up by other highly cited authors, resulting in high precision and recall compared to baselines. Moreover, our approach works consistently well in both data sets. In Figures 3(a) and (d), the precision gradually reduces as the top-K increases. This is because the number of authors in the ground truth reduces significantly as K increases. However, our method does not suddenly drop unlike the baseline methods, such as PMIA. Our approach is very stable and decreases gradually. As evident from Figures 3(b) and (e), the precision and recall of our methods are considerably better than the baselines. In terms of the F_1 measure (see Figures 3(c) and (f)), our approach performs better than baselines over all values of K.

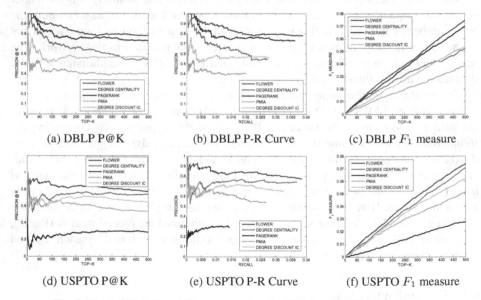

(a) DBLP P@K   (b) DBLP P-R Curve   (c) DBLP F_1 measure

(d) USPTO P@K   (e) USPTO P-R Curve   (f) USPTO F_1 measure

Fig. 3. The P@K, P-R and F_1 measure plots for DBLP and USPTO data sets

7   Conclusions

In this paper, we proposed an information flow mining problem with several desired properties. We developed a sequential version of the algorithm and established that the computational complexity of this problem is #P-complete. In order to scale for large networks, we described a parallel algorithm using vertex-centric computational models.

Our parallel algorithm provides *three orders of magnitude* scale up over the state-of-the-art and with an increasing advantage with greater number of cores. Finally, we showed the effectiveness of the discovered flow patterns using an influence analysis application.

Acknowledgments. This research was sponsored by the Defense Advanced Research Project Agency (DARPA) agreement number W911NF-12-C-0028 and Army Research Laboratory (ARL) cooperative agreement number W911NF-09-2-0053.

References

1. Adar, E., Adamic, L.: Tracking information epidemics in blogspace. In: Web Intelligence, pp. 207–214 (2005)
2. Aggarwal, C., Subbian, K.: Event detection in social streams. In: SDM, pp. 624–635 (2012)
3. Aggarwal, C., Subbian, K.: Evolutionary network analysis: A survey. ACM Comput. Surv. 47(1), 10 (2014)
4. Bonchi, F., De Francisci Morales, G., Gionis, A., Ukkonen, A.: Activity preserving graph simplification. Data Mining and Knowledge Discovery 27(3), 321–343 (2013)
5. Chen, W., Wang, C., Wang, Y.: Scalable influence maximization for prevalent viral marketing in large-scale social networks. In: KDD, pp. 1029–1038 (2010)
6. Chen, W., Wang, Y., Yang, S.: Efficient influence maximization in social networks. In: KDD, pp. 199–208 (2009)
7. Galuba, W., Aberer, K., Chakraborty, D., Despotovic, Z., Kellerer, W.: Outtweeting the twitterers-predicting information cascades in microblogs. In: WOSN (2010)
8. Gonzalez, J., Low, Y., Gu, H., Bickson, D., Guestrin, C.: Powergraph: Distributed graph-parallel computation on natural graphs. In: USENIX (2012)
9. Kempe, D., Kleinberg, J.M., Tardos, E.: Maximizing the spread of influence through a social network. In: KDD, pp. 137–146 (2003)
10. Kim, Y.A., Przytycki, J.H., Wuchty, S., Przytycka, T.M.: Modeling information flow in biological networks. Physical Biology 8(3), 035012 (2011)
11. Lerman, K., Ghosh, R.: Information contagion: An empirical study of the spread of news on digg and twitter social networks. In: ICWSM (2010)
12. Leskovec, J., Backstrom, L., Kleinberg, J.M.: Meme-tracking and the dynamics of the news cycle. In: KDD, pp. 497–506 (2009)
13. Leskovec, J., McGlohon, M., Faloutsos, C., Glance, N.S., Hurst, M.: Cascading behavior in large blog graphs. In: SDM (2007)
14. Low, Y., Gonzalez, J., Kyrola, A., Bickson, D., Guestrin, C., Hellerstein, J.M.: Graphlab: A new framework for parallel machine learning. arXiv:1006.4990 (2010)
15. Mathioudakis, M., Bonchi, F., Castillo, C., Gionis, A., Ukkonen, A.: Sparsification of influence networks. In: KDD, pp. 529–537 (2011)
16. Myers, S.A., Zhu, C., Leskovec, J.: Information diffusion and external influence in networks. In: KDD, pp. 33–41 (2012)
17. Pei, J., Pinto, H., Chen, Q., Han, J., Mortazavi-Asl, B., Dayal, U., Hsu, M.-C.: Prefixspan: Mining sequential patterns efficiently by prefix-projected pattern growth. In: ICDE, pp. 215–215 (2001)
18. Rodriguez, M.G., Leskovec, J., Krause, A.: Inferring networks of diffusion and influence. In: KDD, pp. 1019–1028 (2010)
19. Rodriguez, M.G., Leskovec, J., Schölkopf, B.: Structure and dynamics of information pathways in online media. In: WSDM, pp. 23–32 (2013)

20. Subbian, K., Aggarwal, C., Srivastava, J.: Content-centric flow mining for influence analysis in social streams. In: CIKM (2013)
21. Subbian, K., Melville, P.: Supervised rank aggregation for predicting influencers in twitter. In: SocialCom, pp. 661–665 (2011)
22. Subbian, K., Sharma, D., Wen, Z., Srivastava, J.: Social capital: the power of influencers in networks. In: AAMAS, pp. 1243–1244 (2013)
23. Wang, X., Zhai, C., Hu, X., Sproat, R.: Mining correlated bursty topic patterns from coordinated text streams. In: KDD, pp. 784–793 (2007)
24. Weng, L., Flammini, A., Vespignani, A., Menczer, F.: Competition among memes in a world with limited attention. Scientific Reports 2 (2012)
25. Yang, G.: The complexity of mining maximal frequent itemsets and maximal frequent patterns. In: SIGKDD, pp. 344–353 (2004)

Link Prediction in Multi-modal Social Networks

Panagiotis Symeonidis[1] and Christos Perentis[2],[*]

[1] Aristotle University, Department of Informatics, Thessaloniki 54124, Greece
symeon@csd.auth.gr
[2] Fondazione Bruno Kessler, Trento 38123, Italy
perentis@fbk.eu

Abstract. Online social networks like Facebook recommend new friends to users based on an *explicit* social network that users build by adding each other as friends. The majority of earlier work in link prediction infers new interactions between users by mainly focusing on a single network type. However, users also form several *implicit* social networks through their daily interactions like commenting on people's posts or rating similarly the same products. Prior work primarily exploited both explicit and implicit social networks to tackle the group/item recommendation problem that recommends to users groups to join or items to buy. In this paper, we show that auxiliary information from the user-item network fruitfully combines with the friendship network to enhance friend recommendations. We transform the well-known Katz algorithm to utilize a multi-modal network and provide friend recommendations. We experimentally show that the proposed method is more accurate in recommending friends when compared with two single source path-based algorithms using both synthetic and real data sets.

Keywords: link prediction, friend recommendation.

1 Introduction

Web 2.0 technologies and especially social networking services have gradually allowed users to form different types of interactions, like sharing and rating online items, but primarily to form online friendship networks. For example, online social networks (OSNs) such as Facebook have become popular, since they enable users to share digital content and expand their social circle by recommending new friends, based on their explicit friendship network. Moreover, social rating networks (SRNs) like Epinions and Flixter mainly focus on enabling users to share opinions and rate online items (e.g. posts and movies, respectively), but also to articulate an explicit network of trust. Both OSNs and SRNs constitute multi-modal social networks (MSNs) since they allow people to form simultaneously more than one type of explicit and/or implicit networks. In Figure 1c, we demonstrate an example of an MSN, where thick black edges connect users in an explicit friendship social network and thin edges connect users with items in an implicit user-item social network. In MSNs, explicit social relationships

---

[*] This work has been conducted during co-author's affiliation with Aristotle University.

T. Calders et al. (Eds.): ECML PKDD 2014, Part III, LNCS 8726, pp. 147–162, 2014.

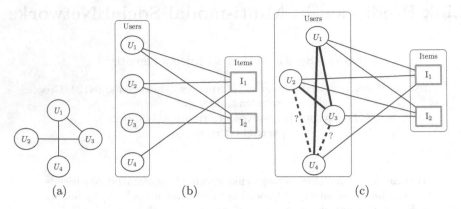

Fig. 1. Example of (a) Unipartite, (b) Bipartite and (c) Multi-modal Social Network

among users co-evolve simultaneously with their interactions with several digital items (e.g. co-participating in groups, co-commenting on posts, co-rating on products etc.). MSNs have recently attracted a lot of research attention. For example, an interesting research question is how to recommend new friends to users by combining their existing social circle with the auxiliary information derived from their user-item network. The main goal is to enhance the accuracy of the future friendship prediction by using also the user-item network. Notice that available information from the bipartite user-item network is crucial due to possible absence of information from the friendship network.

There has been extended research [1,11,12] addressing the link prediction problem within the OSNs, by only exploiting single-source information (i.e. the unipartite user-user friendship network). However, little research has focused on exploiting multiple sources of information in predicting links within MSNs. Lu et al. [15] proposed a supervised framework, by incorporating three real implicit networks (i.e. co-author, co-citation and co-reference) to predict links in the co-author network. Vasuki et al. [22] exploited available information derived from both explicit and implicit social networks such as Orkut and Youtube to provide users with group recommendations. They have tackled the group/affiliation recommendation problem by employing both latent factor and graph proximity models, whereas the latter turned out to be the most effective.

In this paper, we propose a framework that aims to boost the friend recommendation task. Unlike previous works that primarily focused on recommending affiliated groups to users [22], we recommend new friends to users. But to do this, we look simultaneously into the user's explicit friendship and user-item implicit network. Our approach, elaborates one combined form of Katz algorithm [11] into an MSN context. We first utilize the unipartite friendship network and consider human chains of varying lengths corresponding to paths of $user_i \rightarrow user_j$ and $user_i \rightarrow user_j \rightarrow user_k$ forms. Then, we expand our approach to an auxiliary bipartite user-item network where we consider paths of this type $user_i \rightarrow item_j \rightarrow user_k$. This combined Katz approach allows us to provide recommendations in a unified level, traversing new paths for users to connect between and through two discrete networks: user-user and user-item. Our experimental evaluation provides evidence that the usage of auxiliary information from the bipartite user-item network succeeds in enhancing the friend recommendation task.

The rest of this paper is organized as follows. Section 2 summarizes the related work, whereas Section 3 briefly reviews preliminaries in graphs and presents a motivating example of the proposed approach. In Section 4, we present the experimental protocol and our results. Finally, in Section 5 we further discuss the proposed approach and possible directions, while Section 6 concludes this paper.

2 Related Work

The research area of link prediction in social networks tries to infer which new interactions among members of a social network are likely to occur in the near future. There are two main approaches [12] that handle the *link prediction* problem. The first approach is based on local topological features of a network, focusing mainly on the structure of the nodes. There is a variety of local similarity measures such as common neighbors, Jaccard's coefficient, Adamic/Adar index [2], Friend of a Friend (FOAF) algorithm [4] and Preferential Attachment [12], which compute the proximity between a potential pair of nodes. These similarity measures employ local features of the network like the number of common neighbors or the total number of connections and several other combinations.

The second approach is based on global features, detecting the overall path structure in a network. There is a variety of global approaches, such as Random Walk with Restart algorithm [18] and Katz status index [11], SimRank and PageRank [12], which have been used to compute the similarity between a pair of nodes. The Katz status index is a proximity measure that directly sums over the collection of all different length paths that connect two users. An attenuation factor weights the contribution of the paths to the overall similarity according to their length. Symeonidis et al. [20] proposed the FriendTNS algorithm to provide more accurate friend recommendations. They defined a transitive node similarity measure in OSNs by taking into account local and global features of a social graph. Finally, Scholz et al. [19] performed unsupervised random walks for predicting links in user-user networks (i.e. co-author in DBLP).

Besides the aforementioned link prediction algorithms that are based solely on graph structure, there are also other methods that exploit other data sources such as messages among users, co-authored paper and common tagging. For instance, Ido Guy et al. [10], proposed a novel user interface widget for providing users with recommendations of people. Their people recommendations were based on aggregated information collected from various sources across IBM (e.g. common tagging, common link structure, common co-authored papers). Chen et al. [4] evaluated four recommender algorithms (Content Matching, Content-plus-Link, the FOAF algorithm and, SONAR) to help users discover new friends on IBM's OSN. Lo and Lin [13] proposed two algorithms, denoted as *weighted minimum message ratio* (WMR) and *weighted information ratio* (WIR), respectively, which generate a friend list based on real-time message interaction among members of an OSN. Cha et al. [3] collected and analyzed large-scale traces of information dissemination in the Flickr social network. They experimentally derived that over 50% of users find their favorite pictures (i.e., pictures they bookmark) from their friends in an OSN. TidalTrust [9] and MoleTrust [16] are also hybrid approaches that combine the rating data of collaborative filtering systems with the link data of trust-based social networks (i.e. Epinions.com) in order to improve the recommendation accuracy.

There has also been research work that uses supervised approaches to address the link prediction problem in multiple data sources. For instance, Lu et al. [15] exploited topological features from four networks and applied a probabilistic model to learn the network dynamics. They showed that supervised approaches can improve link prediction tasks, suggesting that independency assumptions and scaling issues should be further investigated. In addition, Davis et al. [5] introduced a probabilistically weighted extension of the local-based Adamic/Adar measure for heterogenous networks and showed that a supervised approach based on topological features enhances prediction performance. Finally, maximum-likelihood methods have been proposed to deal with the link prediction problem providing insights about network organization that are difficult to obtain from similarity-based approaches [14]. However, these methods presume specific organizing principles of the network structure and suffer from scalability and accuracy issues.

3   Preliminaries in Multi-modal Graphs

In this section, we present the most important notations with the corresponding definitions and a motivating example based on Figure 1 that will be used throughout the rest of the paper. The multi-modal graph of Figure 1c consists of (i) friendships among users of an OSN and (ii) users' affiliations with items shown in Figure 1a and 1b, respectively. For our calculations, we will use well-known representations, such as the adjacency matrix $\mathbf{A}^{u \times u}$ of friendship network, and the user-item matrix $\mathbf{R}^{u \times w}$ of the affiliation network.

3.1   Link Prediction Based on User-User Unipartite Graph

Let $\mathcal{G}$ be a graph with a set of nodes $\mathcal{V}$ and a set of edges $\mathcal{E}$. Every edge is defined by a specific pair of graph nodes (v_i, v_j), where $v_i, v_j \in \mathcal{V}$. We assume that the graph $\mathcal{G}$ is undirected and unweighted, thus the graph edges do not have any weights, plus the order of nodes in an edge is not important. Therefore, (v_i, v_j) and (v_j, v_i) denote the same edge on $\mathcal{G}$. We also assume that the graph $\mathcal{G}$ can not have multiple edges that connect two nodes, thus if two nodes v_i, v_j are connected with an edge of $\mathcal{E}$, then there can not exist another edge in $\mathcal{E}$ also connecting them. Finally, we assume that there can not be self loop edges on $\mathcal{G}$ (i.e. a node can not be connected to itself). A common graph representation is the *adjacency matrix* $\mathbf{A}^{n \times n}$, where $n=|\mathcal{V}|$ is the number of nodes in $\mathcal{G}$. Therefore, it has n rows and n columns labelled by the graph nodes. For an unweighted non-multiple graph (such as $\mathcal{G}$), the adjacency matrix values are set as $\mathbf{A}_{ij}=1$ if $(v_i, v_j) \in \mathcal{E}$ and $\mathbf{A}_{ij}=0$ otherwise. Following all previous assumptions and definitions, the adjacency matrix of an undirected and unweighted graph such as $\mathcal{G}$, is a symmetric matrix with values 0 and 1, if two nodes are neighbors or not, respectively. In addition, as there are no self loop edges, the main matrix diagonal has zero values. The adjacency matrix of the friendship network for our running example is depicted in Figure 2a. As we want to investigate the relations with ?, we can assume that initially are equal to 0 (i.e. there are no connections between the corresponding users). It is obvious from Figure 1a and its corresponding adjacency matrix $\mathbf{A}$ of Figure 2a that U_1 is connected with U_3 and U_4, while U_2 only to U_3. In terms of social networks, U_1 and U_2 have a "mutual" friend U_3, since they are both connected to this user. Let's assume in our running example, that we want to propose new friends to user U_4. There are several global similarity measures [12] (i.e Katz status index, RWR algorithm, SimRank algorithm, etc.) for capturing similarity of nodes in a network, which are path-dependent. We apply the Katz status index, which defines the similarity score between two nodes V_x and V_y, by summing over paths of varying length ℓ connecting V_x to V_y given by Equation 1:

$$Katz_\beta = \sum_{\ell=1}^{\infty} \beta^\ell |paths_{V_x,V_y}^\ell| \qquad (1)$$

where $paths_{V_x,V_y}^\ell$ is the set of all length-ℓ paths from node V_x to V_y, which are computed by the adjacency matrix $\mathbf{A}$. Note that the algorithm can also handle directed graphs, but this not the case for friendship relationships. Katz status index exploits that raising the adjacency matrix in the power of n produces the number of n-paths connecting one pair of nodes. An attenuation factor β is introduced to efficiently weight the contribution of different lengths of paths to the final similarity score between node pairs. Very low values of β force long paths connecting a pair of nodes to contribute much less to the final similarity score. Thus, it is possible to limit the reach of the similarity measure by weighting higher the shorter paths from node's neighborhood. Both analytical and factorized forms of Katz are given by Equation 2 when applied to the adjacency matrix

	U_1	U_2	U_3	U_4
U_1	0	0	1	1
U_2	0	0	1	?
U_3	1	1	0	?
U_4	1	?	?	0

(a)

	U_1	U_2	U_3	U_4
U_1	0	0.16	0.49	0.43
U_2	0.16	0	0.43	0.05
U_3	0.49	0.43	0	0.16
U_4	0.43	**0.05**	**0.16**	0

(b)

Fig. 2. Running Example: (a) Adjacency $\mathbf{A}$ and (b) Similarity Matrix of User-User Unipartite Social Network

A of Figure 2a:

$$Katz(\mathbf{A}; \beta) = \beta \mathbf{A} + \beta^2 \mathbf{A}^2 + \beta^3 \mathbf{A}^3 + ... = (I - \beta \mathbf{A})^{-1} - I \qquad (2)$$

The identity matrix $\mathbf{I}_n$ is a $n \times n$ matrix of size n holding ones on the main diagonal and being of the same size n as the adjacency matrix $\mathbf{A}$. The attenuation factor β should take values that can ensure series convergence and allow the computation of the $\mathbf{A}^{-1}$ inverse matrix. Therefore, the β attenuation factor can take values $\beta < 1/\lambda$, where λ is the largest absolute value among any eigenvalue of matrix $\mathbf{A}$ [8,11]. We choose β equal to $1/(1+K)$, as L.Katz originally introduced [11] and Foster et al. [8] employed for the fast approximation implementation, where K is the maximum row/column sum of $\mathbf{A}$. This choice satisfies the sufficient condition for the computations to fulfill (i.e. series convergence) and secondly, allows the factor to adopt to each matrix, thus, to each dataset. Back to our running example, we want to recommend new friends to U_4. Thus, we apply Katz algorithm to the unipartite friendship graph $\mathcal{G}$, in order to provide recommendations based on an induced similarity matrix. We compute the Katz status index by applying Equation 2 to the adjacency matrix $\mathbf{A}$ of Figure 2a. The attenuation factor β for matrix $\mathbf{A}$ is $\beta = 1/(1+2)$, equal to 0.33. Notice that Katz calculates similarity between two nodes taking into account paths of length $\ell > 1$.

Firstly, the similarity between U_4 and U_2 is computed based on the unique path that connects them $4 \to 1 \to 3 \to 2$, shown in Figure 1a. This path of length-3 contributes a similarity score of 0.05 given in matrix of Figure 2b. For the similarity between U_4 and U_3, there is only one path of length-2 ($4 \to 1 \to 3$) corresponding to a score of 0.16. The user-user similarity matrix entries of Figure 2b capture the friendship relationships in the unipartite social network and its rows show the "proximity" among users. There is a clear indication from the above similarity matrix that U_3 should be recommended as friend to U_4 instead of U_2, with similarity value 0.16>0.05. Notice that the similarity score of (U_1, U_4) pair is the highest observed matrix entry, but we do not recommend U_1 to U_4, since they are already "friends" and it is not a new link.

3.2   Link Prediction Based on User-Item Bipartite Graph

Users can also form several implicit social networks through their daily interactions like co-commenting on people's posts, co-rating products, and co-tagging people's photos [22]. These implicit relations contain edges between two types of entities (vertices in a graph), such as a user-item bipartite graph. Let $\mathcal{G}' = (\mathcal{V} + \mathcal{W}, \mathcal{E})$ be a bipartite graph with two sets of nodes $\mathcal{V}$ and $\mathcal{W}$, and a set of edges $\mathcal{E}$. Every edge is defined by a specific pair of graph nodes (v_i, w_j), where $v_i \in \mathcal{V}$ denotes users set and $w_j \in \mathcal{W}$ items set. Following the unipartite adjacency matrix notation, we define the biadjacency matrix $\mathbf{R}$ corresponding to bipartite user-item network as a new matrix $\mathbf{B} = \begin{bmatrix} \mathbf{B}_{11} & \mathbf{B}_{12} \\ \mathbf{B}_{21} & \mathbf{B}_{22} \end{bmatrix}$ equal to $\begin{bmatrix} 0 & \mathbf{R} \\ \mathbf{R}^T & 0 \end{bmatrix}$, where $R_{v_i, w_j} = 1$ if $(v_i, w_j) \in \mathcal{E}$ and $\mathbf{R}_{v_i, w_j} = 0$ otherwise.

We extend our running example by affiliating users with items, as depicted in Figure 1b and the corresponding biadjacency matrix $\mathbf{R}$ of Figure 3a. Our main task remains the friend recommendation for U_4 by using this time only the bipartite user-item $\mathbf{R}$. Edges of $\mathbf{R}$ represent length-1 paths of from a user U_i ending to an item I_j. By multiplying matrix $\mathbf{R}$ with its transpose $\mathbf{R}^T$, we derive all length-2 paths of this form $U_i \rightarrow I_j \rightarrow U_k$, where users are connected through items. We employ the $\mathbf{B}^{n \times n}$ adjacency matrix of Figure 3b where block $\mathbf{B}_{11}^2(U_i, U_j) = \mathbf{R}(U_i, I_j) \times \mathbf{R}^T(I_j, U_i)$. If $B_{11}^2(U_i, U_j) > 1$, these two users are connected with an *implicit* (i.e co-share, co-like, etc.) relationship with a potential item. Katz algorithm is next applied to adjacency matrix $\mathbf{B}$ using Equation 3 to obtain a new similarity matrix derived only from the bipartite user-item network.

	I_1	I_2
U_1	1	1
U_2	1	1
U_3	0	1
U_4	1	0

(a)

	U_1	U_2	U_3	U_4	I_1	I_2
U_1	0	0	0	0	1	1
U_2	0	0	0	0	1	1
U_3	0	0	0	0	0	1
U_4	0	0	0	0	1	0
I_1	1	1	0	1	0	0
I_2	1	1	1	0	0	0

(b)

	U_1	U_2	U_3	U_4	I_1	I_2
U_1	0	0.18	0.09	0.09	0.36	0.36
U_2	0.18	0	0.09	0.09	0.36	0.36
U_3	0.09	0.09	0	0.01	0.04	0.31
U_4	0.09	**0.09**	**0.01**	0	0.31	0.04
I_1	0.36	0.36	0.04	0.31	0	0.19
I_2	0.36	0.36	0.31	0.04	0.19	0

(c)

Fig. 3. Running Example: (a) User-Item $\mathbf{R}$, (b) Adjacency $\mathbf{B}$ and (c) Similarity Matrix of Bipartite Social Network

$$Katz(\mathbf{B}; \beta) = \beta\mathbf{B} + \beta^2\mathbf{B}^2 + \beta^3\mathbf{B}^3 + \beta^4\mathbf{B}^4 \cdots = \sum_{\ell=1}^{\infty} \beta^\ell \mathbf{B}^\ell \qquad (3)$$

The odd factors of Equation 3 do not contribute to the similarity among users denoted in $\mathbf{B}_{11}$ block, because they represent paths ending to items (we could exclude them from the equation). Back to the running example, we aim to recommend friends to U_4, thus we calculate its similarity with U_2 and U_3. We apply Katz algorithm to the bipartite graph $\mathcal{G}'$ by applying Equation 3 to the adjacency matrix $\mathbf{B}$ of Figure 3b. The computed similarities are summarized in the matrix of Figure 3c and the attenuation factor for the bipartite network is $\beta=1/(1+3)$, equal to 0.25.

In the 4th row of similarity matrix of Figure 3c is clearly indicated that user U_2 should be recommended to user U_4 as a friend instead of U_3, with similarity value 0.09>0.01. There is a difference between the produced recommendations when using different information sources, since previously we recommended U_3 to U_4 using only the user-user unipartite social network. The information from user-item bipartite network suggests that we should recommend U_2 to U_4, since more paths through the items connect these two users. Specifically, U_4 and U_2 are connected through one path of length-2 ($U_4 \rightarrow I_1 \rightarrow U_2$) and two paths of length-4 ($U_4 \rightarrow I_1 \rightarrow U_1 \rightarrow I_1 \rightarrow U_2$ and $U_4 \rightarrow I_1 \rightarrow U_1 \rightarrow I_2 \rightarrow U_2$). In contrast, U_4 and U_3 are connected through two paths of length-4 ($U_4 \rightarrow I_1 \rightarrow U_1 \rightarrow I_2 \rightarrow U_3$ and $U_4 \rightarrow I_1 \rightarrow U_2 \rightarrow I_2 \rightarrow U_3$).

In our running example, we produced all the possible similarity scores concerning both the user-user and the user-item relationships, by using the adjacency matrix $\mathbf{B}$ of Figure 3b. We exploit only the information from $\mathbf{B}_{12}$ and $\mathbf{B}_{21}$ blocks of matrix $\mathbf{B}$ that correspond to the user-item network, in order to capture similarities concerning block $\mathbf{B}_{11}$. We also produced the similarities for the auxiliary item-item network given by block $\mathbf{B}_{22}$ that is not currently used here. In the future this block of the matrix could reveal semantic relationships between items for other recommendation tasks, like cross-domain.

3.3   Proposed Approach: Link Prediction in Multi-modal Graphs

In this section, the approach of combining the heterogeneous multiple sources of the unipartite user-user and the bipartite user-item graphs, is presented. These two graphs are combined in a multi-modal graph of Figure 1c. This approach enables recommendations to be made in a unified way by opening new paths for users to connect among two distinct sets: users and items. Similarity among users results from both the *explicit* user-user friendship and the *implicit* user-item networks. Therefore, in case the friendship network fails to capture similarity between two users, the auxiliary user-item network could be used for this task, and vice versa. The combined adjacency matrix $\mathbf{C}$ of Figure 4a is introduced in the following form of four blocks: $\begin{bmatrix} \mathbf{A} & \mathbf{R} \\ \mathbf{R}^T & 0 \end{bmatrix}$. To obtain the combined similarity matrix of Figure 4b, which uses information from both user-user $\mathbf{A}$ and $\mathbf{RR^T}$, we apply Equation 4 to $\mathbf{C}$:

$$Katz(\mathbf{C}; \beta) = \beta\mathbf{C} + \beta^2\mathbf{C}^2 + \beta^3\mathbf{C}^3 + \beta^4\mathbf{C}^4 \cdots = \sum_{\ell=1}^{\infty} \beta^\ell \mathbf{C}^\ell \qquad (4)$$

The computed attenuation factor for the multi-modal network is $\beta=1/(1+4)$, equal to 0.2. Unlike we did previously in the bipartite network where we used only the $\mathbf{B}_{12}$ and $\mathbf{B}_{21}$ blocks of the bipartite network, for the multi-modal we exploit information from blocks $\mathbf{C}_{11}$, $\mathbf{C}_{12}$ and $\mathbf{C}_{21}$. Block $\mathbf{C}_{22}$ holds also for the multi-modal network non observed values. The combined version of Katz constructs

	U_1	U_2	U_3	U_4	I_1	I_2
U_1	0	0	1	1	1	1
U_2	0	0	1	0	1	1
U_3	1	1	0	0	0	1
U_4	1	0	0	0	1	0
I_1	1	1	0	1	0	0
I_2	1	1	1	0	0	0

(a)

	U_1	U_2	U_3	U_4	I_1	I_2
U_1	0	0.225	0.379	0.332	0.370	0.379
U_2	0.225	0	0.357	0.106	0.307	0.357
U_3	0.379	0.357	0	0.109	0.169	0.392
U_4	0.332	**0.106**	**0.109**	0	0.313	0.1098
I_1	0.370	0.307	0.169	0.313	0	0.169
I_2	0.379	0.357	0.392	0.109	0.169	0

(b)

Fig. 4. Running Example: (a) Adjacency $\mathbf{C}$ and (b) Similarity Matrix of Multi-modal Social Network

multiple paths using both unipartite friendship and bipartite user-item networks by traversing previously unreached paths between users. Generalization of Katz for $\mathbf{C_{11}}$ user-user block is given by Equation 5 showing such form of paths:

$$Katz(\mathbf{C}; \beta)_{11} = \beta\mathbf{A} + \beta^2(\mathbf{A}^2 + \mathbf{RR}^T) + \beta^3(\mathbf{A}^3 + \mathbf{ARR}^T + \mathbf{RR}^T\mathbf{A}) +$$

$$\beta^4(\mathbf{A}^4 + \mathbf{A}^2\mathbf{RR}^T + \mathbf{RR}^T\mathbf{A}^2 + \mathbf{ARR}^T\mathbf{A} + \mathbf{RR}^T\mathbf{RR}^T)\cdots = \sum_{\ell=1}^{\infty} \beta^\ell \mathbf{C}_{11}^\ell \quad (5)$$

For instance, the $\mathbf{ARR}^T$ factor shown in Equation 5 contains new traversable length-3 paths of this form: $U_i \xrightarrow{A} U_j \xrightarrow{RR^T} U_k$. Finally, the 4th row of the similarity matrix of Figure 4b indicates that U_3 should be recommended to U_4 as a new friend and not U_2, with similarity value 0.109>0.106. One can observe that both unipartite and multi-modal approaches resulted in the same recommendation, but with much smaller difference after the bipartite network was also considered.

4   Experimental Evaluation

In this section, we experimentally compare the performance of the multi-modal link prediction approach with two other single network algorithms. We want to discover in what extent an auxiliary user-item bipartite network contributes to predicting links in the friendship network. Firstly, we evaluate the combined (cKatz) Katz utility for handling more networks, one *user-user* friendship and one *user-item* network. Then, we employ RWR [18,21] and Katz algorithm [11] for predicting links in single social networks as comparison partners:

RWR is the well-known Random Walk with Restart algorithm [18,21] taking into account only one single friendship social network for providing recommendations. In general, RWR considers one random walker starting from an initial node V_x and randomly choosing among the available edges with a probability α. Every time, before random walker makes a choice returns back to the initial node with a probability $1 - \alpha$. Similarity among nodes is computed by Equation 6:

$$\mathbf{RWR}(\mathbf{P}; \alpha) = (1 - \alpha)(I - \alpha\mathbf{P})^{-1} \quad (6)$$

where $\mathbf{I}_n$ is the identity and $\mathbf{P}$ the transition-probability matrix.

sKatz is the model proposed in [11], which takes into account only the single friendship social network, and analyzed in Section 3.1. The proposed approach of this paper *cKatz* considers both the unipartite friendship and the bipartite *user-item* auxiliary network, discussed in Section 3.3.

Parameter's values were tuned as described in [8] and Section 3, therefore α and β for both single network algorithms RWR and sKatz, is set at 0.0008 and 0.0003 for xSocial synthetic and Epinions 49K real data set, respectively. For cKatz parameter β is set at 0.0005 and 0.0003. We employ a fast approximate method of Katz introduced by [8] reducing the computational cost to $O(n+m)$, where n is the number of nodes and m the number of edges, since matrix operations require $O(n^3)$ used by the original Katz algorithm. In this implementation, adjacency matrix is normalized by dividing each entry by the row/column degree. Concerning the maximum length of paths that Katz algorithm employs, we denote ℓ equal to infinite in Equation 2, considering all paths until series convergence. Our experiments were performed on a Core 2 Duo processor with 4 GB of memory. All algorithms were implemented in C. To evaluate the examined algorithms, we have generated synthetic data set using the xSocial generator [7] and chosen one real data set from Epinions web site.

4.1   Real World Networks and Data Sets

Recognizing real-network evolution patterns enables us to better understand the human social behavior and capture similarities among people or about their preferences, detect network intrusions or virus propagation and highlight anomalies [6]. There is a range of patterns that have been identified in real life networks, such as power law distributions [7], six degrees [17](small worlds), scale-free and other log-normal distributions [6], which are powerful tools to mimic observed behaviors. Faloutsos et.al [7] classify graph generators models into emergent (e.g. small-world), where the macro network properties emerge from the micro interactions, and generative graph models, which facilitate a utility function performing recursive iterations until the generated networks meet real network properties.

xSocial Synthetic Data. *xSocial Generator* proposed by [7], is a multi-modal graph generator that mimics real social networking sites to produce simultaneously a network of friends and a network of their co-participation. In particular, xSocial builds a network with N nodes performing three independent actions at each step (i.e. write a message, add a friend and comment on a message). A node chooses his friends either by their popularity of by the number of messages on which they have commented together, which is determined by a unique preference value. A node can also follow the updated status of his friends by putting comments on the corresponding new written messages. In our experiments we use xSocial generator to produce simultaneously one explicit friendship and one implicit network of co-comments. In particular, we generated a MSN data set[1] with 100K users and 384K edges among pair of users, in which users contributed 233K messages and 467K comments. The derived MSN for xSocial data set consists of 330K user and item nodes with 852K edges. In Figure 5a we calculated several topological properties for xSocial data set revealing a large clustering coefficient (LCC) equal to 0.2 and small average shortest path length value (ASD) equal to 2.1 discovered mostly in small-worlds networks [17]. Such networks hold sub-networks with connections between most pairs of nodes (i.e. high LLC) which are connected by at least one short path (i.e. small ASD).

---

[1] http://delab.csd.auth.gr/~symeon/

TOPOLOGICAL PROPERTIES OF FRIENDSHIP NETWORKS:
N = total number of nodes
E = total number of edges
ASD = average shortest path distance between node pairs
ADEG = average node degree
LCC = average local clustering coefficient
GD = graph diameter (maximum shortest path distance)
GGS = global graph sparsity (number of zeros in adjacency matrix/ N^2)

PROPERTIES OF USER-ITEM BIPARTITE NETWORKS:
N = total number of Nodes (users)
R = total number of Ratings
I = total number of Items
MINR = minimum rating value
MAXR = maximum rating value
AVGR = average rating value
GGS = global graph sparsity (zeros in matrix / existing users x items)

Data Set	Type	N	E	ASD	ADEG	LCC	GD	GGS
xSocial 100K	undirected	100000	384458	2.10	6.06	0.20	7	99.99%
Epinions 49K	Directed	49288	487183	4.00	19.77	0.26	14	99.96%

Data-Set	N	R	I	MINR	MAXR	GGS
xSocial 100K	100000	467640	233820	0	0	99.99%
Epinions 49K	49288	664824	139738	1	5	99.98%

(a)   (b)

Fig. 5. Topological properties of (a) friendship and (b) bipartite user-item networks

Real Data. We employ the Epinions 49K[2] data set, which is a who-trusts-whom social network. In particular, users of Epinions.com express their Web of Trust, i.e. reviewers whose reviews and ratings they have found to be valuable. In addition, users are enabled to rate a variety of online items (e.g. books, computers, movies, toys) using a 5 star rating scale. Epinions data set contains 49K users and 487K edges among pair of users, constituting one single friendship social network. Apart from that, it offers a *user-item* network with 140K items and 664K ratings as shown in Figure 5b. In our experiments, we use the whole single network and we keep from the *user-item* network only items rated by users with $r \geq 3$, positively affiliating users with items. Keeping all edges is meaningful in rating prediction tasks, but for friend recommendation this binarization process supports the intuition that we should not recommend users who rated differently similar items. After this, the number of ratings, i.e. edges in *user-item* network, is 570K. The MSN for Epinions 49K data set, when combining the trust and rating network, has 189K nodes of users and items with more than 1M edges. The calculated topological features of the Epinions 49K data set shown in Figure 5a characterize also Epinions 49K as a small-world network with LCC equal to 0.26 and ASD equal to 4. Our evaluation considers the division of friends of each target user into two sets: (i) the training set $\mathcal{E}^T$ is treated as known information and, (ii) the probe set $\mathcal{E}^P$ is used for testing and no information in the probe set is allowed to be used for prediction. It is obvious that, $\mathcal{E} = \mathcal{E}^T \cup \mathcal{E}^P$ and $\mathcal{E}^T \cap \mathcal{E}^P = \varnothing$. Therefore, for a target user we generate the recommendations based only on the friends in $\mathcal{E}^T$. Each experiment has been repeated 30 times (each time a different training set is selected at random) and the presented measurements, based on two-tailed t-test, are statistically significant at the 0.05 level. All algorithms predict the friends of the target users in the probe set. We use the classic precision/recall metric as performance measure for friend recommendations. For a test user receiving a list of k recommended friends (top-k list): *precision* is the ratio of the number of relevant users in the top-k list (i.e. those in the top-k list that belong in the probe set $\mathcal{E}^P$ of friends of the target user) to k. *Recall* is the ratio of the number of relevant users in the top-k list to the total number of relevant users (all friends in the probe set $\mathcal{E}^P$ of the target user). *F1-measure* is the normalized harmonic mean of precision and recall, providing the overall performance metric.

---

[2] http://www.trustlet.org/wiki/Downloaded_Epinions_dataset

4.2   Combined Katz Sensitivity Analysis

In this section, we examine the sensitivity of the combined and single Katz in terms of accuracy performance when we set different density degree of observed items in the user-item network. We want to identify under which circumstances and to what extend the recommendation task is enhanced when we gradually use auxiliary information from an implicit user-item network.

In particular, we test how the performance of cKatz, a multi-modal network approach, is affected when we keep the fraction of observed friend nodes fixed and gradually increase the fraction of observed items as we select user-items edges randomly. We test both in synthetic and real data sets. Firstly, for the synthetic 100K xSocial data set, we set 5 different density cases (i.e. 0.2, 0.4, 0.6, 0.8, 1) by varying the fraction of observed co-comments, as depicted in Figure 6a, while y-axis holds F1-measure at top-1, which is the average performance of the algorithm in terms of both precision and recall when we recommend only one user. Since, sKatz exploits only the friendship network to provide recommendations, increasing the density of the user-item network has no effect in its performance. The fraction of observed friend nodes in the friendship network is fixed to 0.5, where sKatz achieves its best performance. However, cKatz constantly improves its predicting performance as more items from the user-item networks are being observed. We further verify our results in the Epinions 49K real data set shown in Figure 6b. As expected, cKatz improves its overall predicting performance when the fraction of observed items increases. The auxiliary information derived from the affiliation of users with the positively rated items boosts the overall performance, showing that there is fruitful information in the bipartite network. The best performance that sKatz achieves is in 0.5 fraction of observed users, since it does not exploit any auxiliary information and after a certain fraction of friend edges the prediction space of new possible links in the friendship network decreases. Henceforth, we tune the fraction of user nodes observed in 0.5 and this of items observed in 1, for the rest of the experimental evaluation.

Next, we focus on the combined Katz algorithm and we further investigate its performance sensitivity when we vary the number of k recommended friends

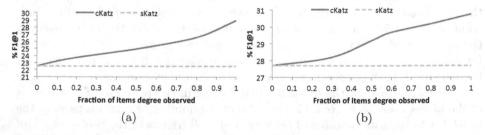

(a)   (b)

Fig. 6. Comparing cKatz with sKatz Performance in terms of F1-measure at Top-1 vs. fraction of items degree for (a) 100K xSocial synthetic and (b) Epinions 49K data set

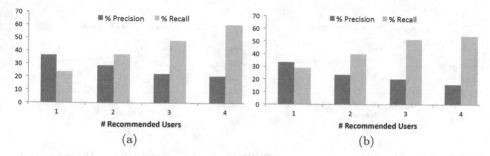

Fig. 7. cKatz Performance in terms of Precision and Recall vs. Top-k for (a) 100K xSocial synthetic and (b) Epinions 49K data set

in the top-k list. We depict the cKatz precision and recall scores versus the varying number of recommended users when applied to the synthetic xSocial and Epinions 49K data sets in Figure 7a and 7b, respectively. In both synthetic and real data sets, cKatz achieves the most accurate scores when we recommend top-1 user. The precision accuracy of cKatz, as expected, gradually decays when we ask for a higher number k of predictions while recall scores increase. Recall is the ratio of the number of correct predictions to the number of all the actual friends in the test set. Each user has a different number of actual friends and this indicates the difficulty of getting better predictions as we increase the number of requested recommendations. The average number of friends (ADEG) for xSocial is 6 and for Epinions 49K data set is 19, depicted in Figure 5a for both data sets. Thus, it is more possible that we return correct recommendations in the Epinions 49K data set as we increase k in the top-k list. In Figure 7a and 7b the recall scores versus top-k diagram are depicted with k varying from 1, 2, 3 and 4 for the xSocial and Epinions 49K data set, respectively. In both data sets we observe, as expected, that we get more correct predictions when we ask for more recommendations. When we produce the top-4 list we achieve the best results for both xSocial and Epinions 49K with recall equal to 59,7% and 54,2%, respectively. We would expect that we get better recall scores in the Epinions 49K data set but the average shortest path distance (ASD) is 2 for xSocial and 4 for Epinions 49K, meaning that it is easier to produce more predictions localized in node's neighborhood since we use small values of β.

4.3   Comparison with other Methods

In this section, we conducted the comparison of our multi-modal proposed combined Katz approach with the two other single network comparison partners i.e. sKatz and RWR algorithms, in terms of precision and recall. As the number k of the list varies starting from the top-1 user to top-4, we examine the precision and recall scores. Achieving high recall scores while precision follows with the minimum decline indicates the robustness of the examined algorithm.

For the xSocial synthetic data set, in Figure 8a we visualize the precision vs. recall curve for all three algorithms. As k increases, precision falls while recall

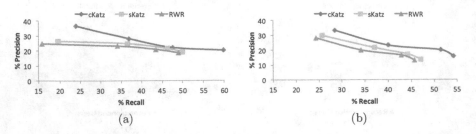

Fig. 8. Comparing cKatz, sKatz, and RWR Performamce in terms of Precision and Recall at Top-k for (a) 100K xSocial synthetic and (b) Epinions 49K data set

increases as expected for all algorithms. cKatz attains the best results achieving the highest precision, outperforming both single network algorithms. This is due the fact that cKatz exploits information from both friendship and the user-item networks. We conduct the same experimental configuration for the Epinions 49K, shown in Figure 8b real data set to confirm our initial results in the synthetic one. It is clear that cKatz outperforms the two single network partners in terms of both precision and recall, exploiting the user-item auxiliary network. Between the two single network algorithms sKatz performs again better than RWR.

5   Discussion

In this section we discuss several issues concerning the multi-modal network context and our approach. We based our method on path-dependent approaches since they capture the overall structure of the network and can limit their reach to node neighborhood level by using attenuation factors. Furthermore, we understand that weighting strategies are essential to effectively control the contribution of various social networks to the final similarity among users. For us, the main task is to recommend new friends to users by exploiting both explicit and implicit social networks. Therefore, we promote the information derived from the unipartite friendship network and control the contribution of the auxiliary information from the user-item network. In this context, the combined adjacency matrix C takes the following form $C = \begin{bmatrix} A & w\mathbf{R} \\ w\mathbf{R}^T & 0 \end{bmatrix}$, where $w \geq 0$ is the weighting parameter controlling the user-item network contribution to the final similarity.

When $w=0$, the bipartite social network does not offer any information in the similarity between users. In this case, the combined Katz behaves like the single version of Katz, sKatz. Earlier in our running example, we observed from the similarity matrix of Figure 2b that when we use information only from the friendship network we recommend U_3 to U_4. The same result is acquired when using the MSN and matrix of Figure 4b, where U_3 is again recommended to U_4, but with much smaller similarity difference from U_2. However, when we exploit information only from the user-item network, U_2 is recommended to U_4 as seen in matrix of Figure 3c. Therefore, we understand that the friendship network is in any case important for providing friend recommendations within the friendship domain. However, the contribution of the user-item network could be proven both fruitful, but in some cases also noisy. Parameter w is a factor that could be tuned by either learning the dynamics of the network, or following a specific range according to the recommendation domain, or being adjusted by the user.

Concerning computational issues, our approach is based on a fast approximation of Katz algorithm introduced by [8], who reduce the computational cost to $O(n+m)$ where n is the number of nodes and m the number of edges, since matrix operations require $O(n^3)$ used by the original Katz algorithm [11]. Concerning the maximum length of paths that Katz algorithm employs, we set ℓ equal to infinite taking into account all the paths until the convergence of the series. Nevertheless, wisely tuning ℓ could potentially improve the proposed approach in terms of efficiency by not traversing very long paths. Truncated versions of Katz can reduce the computational cost, but can also improve the efficacy of the recommendations by learning how to avoid uninformative paths [15].

6 Conclusions and Future Work

In this paper, we presented an extended framework exploiting multi-modal social networks to provide friend recommendations. We experimentally showed that implicit information can be proven fruitful for the friend recommendation task. In the future, MSNs will allow us to perform more cross-domain recommendation tasks, but will also raise challenges like scaling, the effective weighting of multiple information sources and the exploitation of semantic information.

References

1. Adamic, L., Adar, E.: Friends and neighbors on the web. Social Networks 25(3), 211–230 (2003)
2. Adamic, L., Adar, E.: How to search a social network. Social Networks 27(3), 187–203 (2005)
3. Cha, M., Mislove, A., Gummadi, K.: A measurement-driven analysis of information propagation in the flickr social network. In: 18th International World Wide Web Conference, pp. 721–730 (2009)
4. Chen, J., Geyer, W., Dugan, C., Muller, M., Guy, I.: Make new friends, but keep the old: recommending people on social networking sites. In: 27th International Conference on Human Factors in Computing Systems, pp. 201–210 (2009)
5. Davis, D., Lichtenwalter, R., Chawla, N.V.: Multi-relational link prediction in heterogeneous information networks. In: IEEE International Conference Advances in Social Networks Analysis and Mining, pp. 281–288 (2011)
6. Du, N., Faloutsos, C., Wang, B., Akoglu, L.: Large human communication networks: Patterns and a utility-driven generator. In: 15th ACM SIGKDD International Conference on Knowledge Discovery and Data Mining, pp. 269–278 (2009)
7. Du, N., Wang, H., Faloutsos, C.: Analysis of large multi-modal social networks: Patterns and a generator. In: Balcázar, J.L., Bonchi, F., Gionis, A., Sebag, M. (eds.) ECML PKDD 2010, Part I. LNCS, vol. 6321, pp. 393–408. Springer, Heidelberg (2010)

8. Foster, K., Muth, S., Potterat, J., Rothenberg, R.: A faster katz status score algorithm. Computational & Mathematical Organization Theory 7(4), 275–285 (2001)
9. Golbeck, J.: Personalizing applications through integration of inferred trust values in semantic web-based social networks. In: Semantic Network Analysis Workshop at the 4th International Semantic Web Conference, vol. 16, p. 30 (2005)
10. Guy, I., Ronen, I., Wilcox, E.: Do you know?: Recommending people to invite into your social network. In: 14th ACM International Conference on Intelligent User Interfaces, pp. 77–86 (2009)
11. Katz, L.: A new status index derived from sociometric analysis. Psychometrika 18(1), 39–43 (1953)
12. Liben-Nowell, D., Kleinberg, J.: The link-prediction problem for social networks. In: Conference on Information and Knowledge Management, pp. 556–559 (2003)
13. Lo, S., Lin, C.: Wmr: a graph-based algorithm for friend recommendation. In: IEEE/ACM International Conference on Web Intelligence, pp. 121–128 (2006)
14. Lü, L., Zhou, T.: Link prediction in complex networks: A survey. Physica A: Statistical Mechanics and its Applications 390(6), 1150–1170 (2011)
15. Lu, Z., Savas, B., Tang, W., Dhillon, I.S.: Supervised link prediction using multiple sources. In: 10th International Conference on Data Mining, pp. 923–928 (2010)
16. Massa, P., Avesani, P.: Trust-aware collaborative filtering for recommender systems. In: Meersman, R. (ed.) OTM 2004. LNCS, vol. 3290, pp. 492–508. Springer, Heidelberg (2004)
17. Milgram, S.: The small world problem. Psychology Today 22, 61–67 (1967)
18. Pan, J., Yang, H., Faloutsos, C., Duygulu, P.: Automatic multimedia cross-modal correlation discovery. In: 10th ACM SIGKDD International Conference on Knowledge Discovery and Data Mining, pp. 653–658 (2004)
19. Scholz, C., Atzmueller, M., Barrat, A., Cattuto, C., Stumme, G.: New insights and methods for predicting face-to-face contacts. In: 7th International AAAI Conference on Weblogs and Social Media, pp. 281–288 (2013)
20. Symeonidis, P., Tiakas, E., Manolopoulos, Y.: Transitive node similarity for link prediction in social networks with positive and negative links. In: 4th ACM Conference on Recommender systems, pp. 183–190 (2010)
21. Tong, H., Faloutsos, C., Pan, J.Y.: Fast random walk with restart and its applications. In: 6th International Conference on Data Mining, pp. 613–622 (2006)
22. Vasuki, V., Natarajan, N., Lu, Z., Savas, B., Dhillon, I.: Scalable affiliation recommendation using auxiliary networks. ACM Trans. Intell. Syst. Technol. 3, 3:1–3:20 (2011)

Faster Way to Agony
Discovering Hierarchies in Directed Graphs

Nikolaj Tatti

Helsinki Institute for Information Technology,
Department of Information and Computer Science,
Aalto University, Finland
nikolaj.tatti@aalto.fi

Abstract. Many real-world phenomena exhibit strong hierarchical structure. Consequently, in many real-world directed social networks vertices do not play equal role. Instead, vertices form a hierarchy such that the edges appear mainly from upper levels to lower levels. Discovering hierarchies from such graphs is a challenging problem that has gained attention. Formally, given a directed graph, we want to partition vertices into levels such that ideally there are only edges from upper levels to lower levels. From computational point of view, the ideal case is when the underlying directed graph is acyclic. In such case, we can partition the vertices into a hierarchy such that there are only edges from upper levels to lower edges. In practice, graphs are rarely acyclic, hence we need to penalize the edges that violate the hierarchy. One practical approach is agony, where each violating edge is penalized based on the severity of the violation. The fastest algorithm for computing agony requires $O(nm^2)$ time. In the paper we present an algorithm for computing agony that has better theoretical bound, namely $O(m^2)$. We also show that in practice the obtained bound is pessimistic and that we can use our algorithm to compute agony for large datasets. Moreover, our algorithm can be used as any-time algorithm.

Keywords: Graph mining, agony, hierarchy discovery, maximum eulerian subgraph.

1 Introduction

Many real-world phenomena exhibit strong hierarchical structure [2, 5, 9, 10, 11]. For example, it is more likely that a manager in a large company will write emails to the her subordinates than an employee writes an email to his manager. As another example, in a tournament, it is more likely that a better team will win a second-tear team.

Discovering hierarchy in the context of directed networks can be viewed as the following optimization problem. Given a directed graph, partition vertices into levels such that there are only edges from upper levels to lower levels. For example, consider an email communication network of a large institute, directed edge $x \rightarrow y$ is created if x has written an email to y. We should expect that the

T. Calders et al. (Eds.): ECML PKDD 2014, Part III, LNCS 8726, pp. 163–178, 2014.

upper level of the hierarchy consists of top-level managers and each level consists of subordinates of the previous level.

Unfortunately, such a partition is only possible when the graph does not have cycles, a rare case in practice. Instead a more fruitful approach is to find a hierarchy that minimizes some cost function. One possible cost function is to penalize every edge that violates the hierarchy with a constant cost. Unfortunately, this problem leads to FEEDBACK ARC SET problem, where we are asked to discover a maximal directed acyclic subgraph. This problem is a classic **NP**-hard problem [4].

A practical variant of discovering hierarchies that was introduced recently by Gupte et al. [7] is to weight the edges based on the severity of the violation of hierarchy. Unlike the constant weights, this problem can be solved in $O(nm^2)$, polynomial time, where n is the number of vertices and m is the number of edges.

In this paper we introduce a new algorithm for computing a hierarchy that minimizes agony. Our algorithm achieves computational complexity of $O(m^2)$ which is significantly better than $O(nm^2)$, the computational complexity of the currently best approach. We also demonstrate empirically that $O(m^2)$ is in fact pessimistic and that we can compute agony using our approach for large networks.

Our approach is based on a primal-dual technique. Minimizing agony has an interpretable dual problem, finding eulerian subgraph, a graph where the in-degree is equal to the out-degree for each vertex, with the maximum number of edges. This relation implies that the agony will always be at least as large as any eulerian subgraph. We are able to exploit this relation by designing an iterative algorithm. At each iteration we decrease the gap between the current agony and the current eulerian subgraph by either modifying the hierarchy or modifying the eulerian subgraph. We show that each iteration requires only $O(m)$ time and we need at most m steps.

The rest of the paper is organized as follows. We introduce the notation and state the optimization problem in Section 2. In Section 3 we review the connection between agony and eulerian subgraphs. In Section 4 we introduce our optimization algorithm. We discuss the related work in Section 5 and present experimental evaluation in Section 6. Finally, we conclude the paper with remarks in Section 7.

2   Preliminaries and Problem Statement

Throughout the whole paper we assume that we are given a directed graph $G = (V, E)$. We will denote the number of vertices by $n = |V|$ and the number of edges by $m = |E|$. All graphs in this paper are directed and have the same vertices V. Given a graph $H = (V, F)$, we will write $E(H) = F$.

In this paper our goal is to discover a hierarchy among vertices in a graph G. That is, assume that we are given a graph $G = (V, E)$ and our goal is to discover a partition of vertices $\mathcal{P} = P_1, \ldots, P_k$, such that $P_i \cap P_j = \emptyset$ and $\bigcup_{i=1}^{k} P_i = V$, optimizing a certain quality score which we will define later. It will be more convenient to express this partition using a rank function, that is, our goal is to construct a function $r : V \to \mathbb{Z}$ mapping each vertex to an integer. We can easily construct a partition from this rank function by grouping the nodes mapping to the same value together.

Our next step is to define the quality score.

Given a rank function r, we say that an edge $e = (u, v) \in E$ is *forward* if $r(u) < r(v)$. Similarly, we say that $e = (u, v) \in E$ is *backward* if $r(u) \geq r(v)$. Note that the inequality is strict for forward edges.

As our goal is to discover hierarchy in G, in an ideal partition all edges are forward. This is only possible if G is a DAG which is rarely the case in practice. Consequently, we need a quality score that would penalize the backward edges. Given a rank r we define the *agony* of an edge $(u, v) \in E$ to be

$$q((u, v), r) = \max(r(u) - r(v) + 1, 0) \ .$$

The agony for forward edges is 0 while the agony for backward edges is the difference between ranks plus 1. Note that the edges within the same block are penalized by 1.

Given a graph G and a rank r we define the agony of the whole graph to be the sum of individual edges,

$$q(G, r) = \sum_{e \in E} q(e, r) \ .$$

Example 1. The agony of the left graph given in Figure 1 is equal to

$$q((b, a)) + q((d, b)) + q((e, g)) + q((g, f)) = 2 + 3 + 1 + 2 = 8 \ .$$

The agony of the right graph is equal to

$$q((b, a)) + q((d, b)) + q((c, d)) + q((e, g)) + q((g, f)) = 1 + 3 + 1 + 2 + 1 = 8 \ .$$

We can now state the main optimization problem of this paper.

Problem 1. Given a graph G find a rank function r minimizing agony $q(G, r)$.

Fig. 1. Toy graphs. Dotted edges represent the eulerian subgraph. Ranks are represented by dashed grey horizontal lines.

Graph $H = (V, F)$ is called *eulerian* if the out-degree of each vertex is equal to its in-degree,

$$|\{u \in V; (v, u) \in F\}| = |\{w \in V; (v, w) \in F\}| \quad .$$

In the literature, H is sometimes required to be connected but here we do not impose this constraint.

Example 2. An example of eulerian subgraph in the left graph of Figure 1 consists of (b, a), (a, c), (c, d), (d, b), (f, e), (e, g), and (g, f).

An example of eulerian subgraph in the right graph of Figure 1 consists of (b, a), (a, e), (e, c), (c, d), (d, b), (f, e), (e, g), and (g, f).

Given a graph G we say that H is a *maximum* eulerian subgraph G if H is an eulerian subgraph of G and has the highest number of edges among all eulerian subgraphs of G. This graph is not necessarily unique. For notational simplicity, we require that G and H have the same vertices, $V(H) = V(G) = V$. This restriction does not impose any difficulties since we can always add missing vertices as singletons to H.

Given a graph G we say that H is a *maximal* eulerian subgraph G if H is an eulerian subgraph of G and we cannot increase H by adding new edges without making it non-eulerian. Note that maximum eulerian subgraph is necessarily maximal but not the other way around. It is easy to see that H is maximal if and only if the remaining edges in G form a DAG.

As we see in the next section, the following optimization problem, that is, finding the maximum eulerian subgraph is closely related to optimizing agony.

Problem 2. Given a graph G find an eulerian subgraph H maximizing $|E(H)|$, the number of edges.

3   Agony and Eulerian Subgraphs

In this section we review the connection between agony and discovering maximum eulerian subgraph. In fact, they are dual problems. This connection allows us to develop our algorithm in the next sections.

To see the connection let us first write the agony optimization problem as an integer linear program, that is, our goal is to solve the following program.

$$\min \sum_{(u,v)\in E} p(u, v) \qquad \text{such that} \qquad (1)$$

$$p(u, v) \geq r(v) - r(u) + 1 \qquad \text{for all } (u, v) \in E,$$
$$p(u, v) \geq 0 \qquad \text{for all } (u, v) \in E,$$
$$p(u, v), \; r(w) \in \mathbb{Z} \qquad \text{for all } (u, v) \in E, \; w \in V \quad .$$

The solution for Eq. 1 will contain the optimal rank function r and agony for individual edges $p(u, v)$.

Let us relax the program by dropping the integrality conditions, thus transforming the program into a standard linear program. The dual of this program is equal to

$$
\max \sum_{(u,v)\in E} c(u,v) \qquad\qquad \text{such that} \qquad\qquad (2)
$$

$$
c(u,v) \leq 1 \qquad\qquad \text{for all } (u,v) \in E,
$$

$$
\sum_{(u,v)\in E} c(u,v) = \sum_{(v,w)\in E} c(v,w) \qquad\qquad \text{for all } v \in V,
$$

$$
c(u,v) \geq 0 \qquad\qquad \text{for all } (u,v) \in E .
$$

$$
(3)
$$

Assume that we are given a feasible solution to a dual problem such that $c(u,v)$ are integral. The conditions imply that $c(u,v)$ is either 0 or 1. If we form a subgraph H by taking the edges for which $c(u,v) = 1$, then the equality condition implies immediately that H is eulerian. Consequently, the solution for the dual problem is at least as large as the number of edges in the maximum eulerian graph.

Since the primal solution is always larger than the dual solution we have the following proposition.

Proposition 1. *Assume that we are given a graph G. Let r be a rank function and let H be an eulerian subgraph. Then $|E(H)| \leq q(G,r)$. Moreover, if $|E(H)| = q(G,r)$, then r minimizes agony and H has the maximum number of edges.*

Proof. Let P be the solution of Eq. 1 and let D be the solution of Eq. 2. Primal-dual theory (see, for example, [13]) states that $D = P$. We now have $q(G,r) \geq P = D \geq |E(H)|$. If $|E(H)| = q(G,r)$, then this immediately implies $|E(H)| = q(G,r) = P = D$, proving the optimality of r and H. □

The previous result only proves that *if* there is a rank function r whose agony corresponds to the number of edges in the eulerian subgraph H, then r and H are optimal. It does not guarantee that such solution exists. Gupte et al. [7] showed that such solution always exists. However, we do not need this result. Instead, in the next section we introduce an algorithm that finds r and H satisfying the conditions of Proposition 1 which immediately implies the optimality of r.

4 Algorithm for Discovering Agony

In this section we present our algorithm based on the results of previous section. As our first step, we characterize the difference between the agony of the current rank function and the number of edges in the eulerian subgraph. We then present an algorithm that minimizes this difference and by doing so leads to the optimal solution. Finally, we present a fast algorithm for discovering a maximal eulerian subgraph, an initialization step that is needed for our main algorithm.

4.1   Gap between Agony and Eulerian Subgraphs

In order to characterize the gap between the scores we need several concepts.

Assume that we are given a graph G and let H be a maximal eulerian subgraph G. We say that a rank function r *conforms* H if all backward edges with respect to r are in H. Note that this is possible only if H is maximal, otherwise there will be at least one backward edge in $E(G) \setminus E(H)$.

We will express the gap as a sum of slacks. More formally, given a rank r we define the *slack* of an edge as

$$slack((u,v),r) = \max(r(v) - r(u) - 1, 0) \quad .$$

Slack of (u,v) will be positive only if the edge is forward and the rank $r(v)$ is at least $r(u) + 2$.

We saw in the previous section that the agony is always larger than the number of edges in an eulerian graph. We can express this difference under certain conditions using slacks.

Proposition 2. *Assume that we are given a graph $G = (V, E)$ and let $H = (V, F)$ be a maximal eulerian subgraph. Let r be a rank function of V conforming H. Then*

$$q(G, r) = |F| + \sum_{e \in F} slack(e, r) \quad .$$

Moreover, if the sum of slacks is 0, then r has the lowest possible agony.

Proof. Since H is an eulerian graph, we can partition H into s edge-disjoint cycles $C_1, \ldots, C_s$. Since backward edges are only in H we can write agony as

$$q(G, r) = \sum_{e \in F} q(e, r) = \sum_{i=1}^{s} \sum_{e \in C_i} q(e, r) \quad .$$

The agony of a single edge $e = (u, v)$ can be written as

$$\begin{aligned} q(e, r) &= \max(r(v) - r(u) + 1, 0) = r(v) - r(u) + 1 - \min(r(v) - r(u) + 1, 0) \\ &= r(v) - r(u) + 1 + \max(r(u) - r(v) - 1, 0) \\ &= r(v) - r(u) + 1 + slack(e, r) \quad . \end{aligned}$$

Summing the edges in a single cycle gives us

$$\sum_{e \in C_i} q(e, r) = \sum_{e=(u,v) \in C_i} r(v) - r(u) + 1 + slack(e, r) = |C_i| + \sum_{e \in C_i} slack(e, r) \quad .$$

Since the cycles are edge-disjoint, we get the first result of the proposition. If the sum of slacks is 0, then the the agony $q(G, r)$ is equal to the number of edges in eulerian subgraph. Proposition 1 now implies that r is optimal and H is in fact a maximum eulerian subgraph.   □

Example 3. Consider the left graph in Figure 1. The current agony is equal to 8 and the size of the current eulerian subgraph is equal to 7. There is one slack edge, namely $slack((a, c), r) = 1$. On the other hand, the right graph in Figure 1 has agony of 8 which is equivalent to the number of edges in the eulerian subgraph. There are no slack edges.

4.2 Algorithm for Computing Agony

We are ready to describe the algorithm. Assume that we are given a graph G and assume that we have obtained a maximal eulerian subgraph and a rank r that conforms H. We will describe later how to obtain the initial H and r.

Proposition 2 states that r is optimal if there are no edges with slack in H. Assume there is one, say (p, s). We begin the algorithm by increasing the rank of p so that (p, s) has no slack. This may result that some of the edges outside H become backward, hence we will increase the rank of the end point of each new backward edge to make sure that there are no new backward edges. In addition, some of edges H may obtain more slack, hence we will also increase those vertices. These increases may require additional increases for other vertices and we keep doing this until either there are no more increases needed. If we do not encounter s during this algorithm, then we have successfully reduced agony by the $slack((p, s), r)$. Otherwise, we will show that we can modify H such that the number of edges in increased.

The visiting order of vertices is important in order to guarantee that the algorithm runs in $O(m)$ time. We will show that we can guarantee the running time if we keep the vertices in a priority queue based on how much we need to increase their rank, larger increases first.

The pseudo-code for the algorithm is given in Algorithm 1. The algorithm takes as an input the underlying graph G, current maximal eulerian subgraph H and conforming r, and an edge $(p, s) \in E(H)$ with positive slack. The algorithm outputs a new subgraph H' and a new rank function r'.

Case 1: we can increase $r(a)$ without increasing $r(c)$

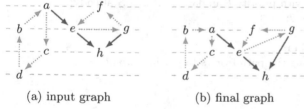

(a) input graph   (b) final graph

Case 2: we cannot increase $r(a)$ without increasing $r(c)$

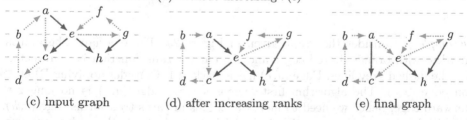

(c) input graph   (d) after increasing ranks   (e) final graph

Fig. 2. Two examples of applying RELIEF for (a, c). Dotted edges represent the eulerian subgraph. Ranks are represented by dashed grey horizontal lines.

Algorithm 1. RELIEF, given an maximal eulerian subgraph H and a conforming rank r, computes a new subgraph H' and a new rank function r' such that the agony or r is closer to the number of edges in the subgraph.

---

input : underlying graph G, current maximal eulerian subgraph H, current rank function r, $(p, s) \in E(H)$ an edge with positive slack

output : updated maximal eulerian subgraph and new rank function

```
 1  F ← E(H);
 2  r' ← r;
 3  t(v) ← 0 for all v ∈ V {how much we need to increase v}
 4  t(p) ← r(s) − r(p) − 1;
 5  add p to S with priority t(p);
 6  while S is not empty do
 7      u ← pop first element from S;
 8      r'(u) ← r'(u) + t(u);
 9      foreach (u, v) ∈ E \ F do
10          if r'(v) ≤ r'(u) then
11              t ← r'(u) + 1 − r'(v);
12              if t > t(v) then
13                  t(v) ← t;
14                  add v to S with priority t, update v if v ∈ S already;
15                  parent(v) ← u;
16      foreach e = (w, u) ∈ F do
17          if slack(e, r') > slack(e, r) then
18              t ← slack(e, r') − slack(e, r);
19              if t > t(w) then
20                  t(w) ← t;
21                  add w to S with priority t, update w, if w ∈ S already;
22                  parent(w) ← u;
23      if slack((p, s), r') > 0 then
24          O ← edges in E along the path from s to p using parent;
25          F ← (F \ O) ∪ (O \ F);
26          delete (p, s) from F;
27      return (V, F), r';
```

Example 4. Consider the graph given in Figure 2a. The eulerian subgraph is marked with orange dotted edges and the current rank function is represented by the dashed grey lines. Edge (a, c) has a slack of 1. Consider applying RELIEF on edge (a, c). The algorithm first increases $r(a)$. Edge (a, e) is no longer a forward edge, hence we need to increase e. This in turns transforms edge (e, h) into backward and increases the slack of (f, e). Ranks for both vertices are also increased. No other modifications are needed and the final graph is given in Figure 2b.

Now consider the graph given in Figure 2c and apply RELIEF on edge (a, c). As in previous case, e, f, and h are increased, but in addition c. Note that we did not manage to reduce the slack between a and c. However, if travel back along the *parent* links, $parent(c) = e$ and $parent(e) = a$ we obtain a path from c to a. By replacing (a, c) with these edges in the eulerian subgraph we obtain a new subgraph that has more edges.

The previous example showed the two possible outcomes for RELIEF, in both cases we reduce the slack. The following proposition states that this holds in general, that is, the new H and r are valid and that the difference between the costs is smaller.

Proposition 3. *Assume that we are given a graph G. Let $H = (V, F)$ be a maximal eulerian subgraph of G and let r be a rank function conforming H. Assume that there is an edge $(p, s) \in E(H)$ such that $slack((p, s), r) > 0$. Let $H', r' = \text{RELIEF}(H, r, G, p, s)$. Then H' is a maximal eulerian subgraph of G, r' is conforming H', $\max_{e \in E(H')} slack(e, r') \leq \max_{e \in E(H)} slack(e, r)$, and*

$$|\{e \in E(H') \mid slack(e, r') > 0\}| < |\{e \in E(H) \mid slack(e, r) > 0\}| \ .$$

In order to prove this result we need the following lemma.

Lemma 1. *Each vertex visited at most once during RELIEF. Order the visited vertices based on their visiting order, say u_k. Let $t_k = t(u_k)$, where $t(u_k)$ is the priority of u_k at the time when u_k is visited. Then $t_{k+1} \leq t_k$.*

Proof. We will prove by induction over the iteration of RELIEF that once a vertex u has been removed from S it will never be added again to S and the priorities of newly added vertices into S during processing u is at most $t(u)$.

Assume that this holds for $k - 1$ first iterations, and let $u = u_k$ be a vertex that is visited during the kth iteration. Since S selects elements with the highest priorities, the induction assumption implies that $t_{i+1} \leq t_i$ for $i = 1, \ldots, k - 2$.

Let $(u, v) \in E \setminus F$. Let $t = r'(u) + 1 - r'(v)$. Since u is visited for the first time, we must have $r'(u) = r(u) + t(u)$ which implies that $t = t(u) + r(u) + 1 - r'(v) \leq t(u)$, where the inequality holds since (u, v) is a forward edge w.r.t. r. If v is added in S, then its priority is at most $t(u)$. This proves the second part of the induction step. On the other hand, if v has been already visited, say $v = u_j$, then $t_j = t(v) \geq t_k$ and v will not be added into S.

A similar argument can be made for the edges in F.

This proves the induction step and the lemma as the first step is trivial.   □

Proof (of Proposition 3). Let us consider two separate cases. In Case 1, s remains unvisited while in Case 2 we visit s.

Case 1: Assume that we do not visit s. In such case, $H' = H$, hence H' is maximal.

We need to first show that edges in $E \setminus F$ remain forward. Whenever we increase the rank of u we check that none of the edges in $E \setminus F$ are backward. Assume there is one, say (u, v). If v is already visited, then Lemma 1 states that $t(v) \geq t(u)$. Since each vertex is visited only once, this implies that $r'(v) = r(v) + t(v)$ and $r'(u) = r(u) + t(u)$. This is a contradiction since (u, v) is a forward edge w.r.t. r. Hence, either v is not visited or is in S. Either way, we will increase $r'(v)$ so that v will become a forward edge at some point.

Using similar argument, we see that the slack of edges in F is not increased. Since the edge (p, s) is no longer a slack edge and we do not increase slack of any other edges, we have proved the proposition for Case 1.

Case 2: Assume that we have visited s. Write $F' = E(H')$.

Let us first argue that we can reach p by using the *parent* links. Lemma 1 implies that each vertex is visited only once which guarantees that *parent* links form a tree whose root is p.

Using the same argument as in Case 1, we see that forward edges in $E \setminus F$ remain forward edges and the slackness of edges in F is not increased. Moreover, one can easily show that (p, s) and the edges in $O \cap F$ are also forward edges. This means that $E \setminus F'$ contains only forward edges. This means that r' conforms H' and $E \setminus F'$ form a DAG which is only possible when H' is maximal. It is easy to see that H' is also eulerian.

Let $O_1 = O \cap (E \setminus F)$. For any edge $(u, v) \in O_1$, we must have $r'(v) = r'(u) + 1$, otherwise $parent(v) \neq u$. This shows that the slack of the new edges is 0. Since (p, s) is removed from H' and we do not increase slack of any other edges, we have proved the proposition for Case 2.   □

Our next step that this single iteration is linear in the number of edges.

Proposition 4. *The running time of* RELIEF *is* $O(m + slack((p, s), r))$.

Proof. Since each vertex is visited only once (Lemma 1) we will consider each edge only once. Hence, the inner for-loops are executed m times at most. Since the priorities of vertices are integers, we can implement the priority queue by storing each vertex into an array of $slack((p, s), r) - 1$ linked lists. Inserting or updating a vertex will take a constant time. Since the new priorities will always be smaller or equal, obtaining the maximum element takes $O(slack((p, s), r))$ of *total* time due to the fact that we need to possibly check some empty linked lists. This proves the proposition.   □

Alternatively, we can implement the priority queue as a heap which gives the running time to be $O(m \log n)$.

In practice, we also apply the following speed-up. We monitor $t(s)$ constantly and we visit only those vertices that have larger priority. Since, Lemma 1 states that the priorities are non-increasing, we simply stop the main loop once we encounter a vertex with the same priority as $t(s)$. In addition, once we are done we backtrack rank of each visited vertex by $t(s)$. If s is not visited, then $t(s)$ remains 0 and this speed-up has no effect. However, if s is inserted in the stack S, we will prune vertices that have the same or lower priority than S. We ignore any vertex that should be lowered by at most $t(s)$. This may transform some forward edges into backward edges but we counter this by lowering the rank of the already visited vertices by $t(s)$. This implies that the forward edges remain forward and the arguments done in proof of Proposition 3 are valid.

We are now ready to state the main loop, given in Algorithm 2, which applies RELIEF to the edge with the largest slack.

Algorithm 2. MINAGONY, given a graph G, a maximal eulerian subgraph H and a rank function r conforming H, finds a rank function optimizing agony

---

   input   : underlying graph G, maximal eulerian subgraph H, rank function r
   output : optimal maximal eulerian subgraph and rank function
1 **while** $q(G, r) > |E(H)|$ **do**
2 | $(p, s) \leftarrow$ an edge in $E(H)$ with largest slack;
3 | $H, r \leftarrow$ RELIEF(H, r, G, p, s);
4 **return** H, r;

Our next step is to show that we need to call RELIEF at most $O(m)$.

Proposition 5. *Assume a graph G, a maximal eulerian subgraph H and a rank function conforming H such that $slack(e, r) \leq m$ for any edge $e \in H$. Then MINAGONY(G, H, r) takes $O(m^2)$ time.*

Proof. Proposition 3 states that each call reduces the number of slack edges by at least 1. There can be at most m slack edges. Hence, the number of RELIEF calls is at most m. Since $slack(e, r) \leq m$ at the beginning and Proposition 3 states that slack is never increased, Proposition 4 implies that calling RELIEF takes $O(m)$ time. This completes the proof. $\square$

Assume that we are given H, a maximal eulerian subgraph of G. Then $E(G) \setminus E(H)$ is a DAG, and any topological order will provide a rank function that is conforming with H. In this paper, we use a rank function, where we first remove all source vertices simultaneously from the DAG and assign them the same rank. We continue this until DAG is empty. The largest rank in this case is at most n, this also bounds the slack and consequently the conditions in Proposition 5 are satisfied.

4.3 Discovering Maximal Eulerian Subgraph

Our final step is to discover a maximal eulerian subgraph. This can be done naively by running a DFS, finding and a removing a cycle and repeating until no cycles are left. This gives us running time of $O(m^2)$. A more sophisticated approach can be done with a single DFS, given in Algorithm 3.

CYCLEDFS starts with DFS and the moment it discovers a back edge, it finds a corresponding cycle. The algorithm proceeds by deleting the cycle and backtracking to the first vertex of visited cycle. The following proposition shows that the algorithm indeed finds a maximal eulerian subgraph.

Algorithm 3. CYCLEDFS, discovers a maximal eulerian subgraph

---

input : G, directed graph
output : F, edges corresponding to a maximal eulerian subgraph

```
1  while V ≠ ∅ do
2  │   S ← any vertex in V;
3  │   while S ≠ ∅ do
4  │   │   u ← first vertex in S;
5  │   │   if there is (u, v) ∈ E then
6  │   │   │   if v ∈ S then
7  │   │   │   │   O ← (u, v) and the path from v to u along S;
8  │   │   │   │   F ← F ∪ O;
9  │   │   │   │   delete O from G;
10 │   │   │   │   pop vertices from S until the last vertex is v;
11 │   │   │   else
12 │   │   │   │   push v to S;
13 │   │   else
14 │   │   │   pop u from S;
15 │   │   │   remove u from G;
16 return F;
```

Proposition 6. CYCLEDFS *discovers maximal eulerian subgraph.*

Proof. F consists of edge-disjoint cycles, and by definition is eulerian. Assume that F is not maximal, that is, there is a cycle C. Let u be the first vertex in C that is deleted from G. Let e be the outgoing edge from u in C. By definition, e is not added in F. This implies that when we delete u from G, e is still present in G which is a contradiction since we only delete vertices with no outgoing edges.   □

As a final step we show that CYCLEDFS runs in linear time.

Proposition 7. CYCLEDFS *executes in* $O(m)$ *time.*

Proof. During a single iteration of the inner while-loop we either delete x edges or push a vertex into a stack. Hence, the total running time is bounded by the number of edges deleted plus the number of pushes. Since each edge can be deleted only once, the first term is bounded by m. The number of times we will push a vertex u into S is bounded by the in-degree of u plus 1. Consequently, the number of pushes we will do in total is $O(n + m)$, which proves the result.   □

5   Related Work

From algorithmic point of view, the relation between our approach and the algorithm given by Gupte et al. [7] is intriguing. Both methods are based on primal-dual techniques, that is, they rely on the relationship between the primal problem, minimizing agony, and the dual problem, maximizing the eulerian subgraph. Gupte's algorithm is essentially an instance of the primal-dual algorithm, where one tries to improve the dual problem, in this case discovering maximum eulerian subgraph, until no improvement is possible. This improvement correspond to finding the negative cycle in a certain weighted graph, that is, a cycle whose sum of weights is negative. Currently the best algorithm for discovering negative cycle needs $O(nm)$ time [1] and this can be achieved with a Bellman-Ford algorithm [3]. Since we need m iterations at most, the computational complexity of this approach is $O(nm^2)$.

On the other hand, our approach is also an instance of the primal-dual algorithm. Especially, both algorithms improve the current eulerian subgraph. The difference is that while Gupte's algorithm searches the improvement by transforming the problem into discovering negative cycles, we discover the improvement in several calls of RELIEF. During each call of RELIEF if we have not able to find a new improvement for the eulerian subgraph, then we are able to improve the primal problem, that is, minimizing agony. In other words, while searching for improvement for the eulerian subgraph, we are able to use intermediate calculations to minimize the agony. This allows us to achieve a better computational complexity of $O(m^2)$.

The agony of a single edge is chosen very carefully. For example, if we choose agony to be 1 for every backward edge, then the problem is related to FEEDBACK ARC SET, where the goal is to discover a directed acyclic graph H. from a given directed graph G such that $E(G) \setminus E(H)$ is minimized. This problem is not only **NP**-hard, it is also **APX**-hard with a coefficient of $c = 1.3606$ [4]. There is no known constant-ratio approximation algorithm for FAS and the best known approximation algorithm has ratio $O(\log n \log \log n)$ [6].

Next, we highlight some of the existing methods for discovering hierarchies. Maiya and Berger-Wolf [11] suggested a statistical model where the probability of an edge is high between a parent and a child. To find the hierarchy they employ a greedy heuristic. Clauset et al. [2] studied discovering hierarchy in undirected graphs, where given a dendrogram, the probability of an edge between two vertices is based on Erdős-Rényi model, with a probability depending on the lowest common ancestor in the dendrogram. The authors then sample dendrograms using MCMC techniques. Macchia et al. [10] used agony to discover summaries of propagations based on traces. Jameson et al. [9] applied a model, where the likelihood of the the vertex dominating other is based on the difference of their ranks, to animal dominance data. Similar ideas has been used for ranking chess players by Elo [5]. Finally, hierarchy partitions vertices into groups, the top-level vertices having very different role than the bottom-level vertices. Assigning different roles to vertices have received some attention. Henderson et al. [8] consider assigning roles to vertices based on features while McCallum et al. [12] assigned topic distributions to individual vertices. An interesting direction for future work would be to study how hierarchy can be used for role mining in graphs.

6 Experimental Evaluation

While we were able to improve the computational complexity of computing agony from $O(nm^2)$ to $O(m^2)$, the bound is still impractical even for graphs of modest size. Our next goal is to demonstrate empirically that this bound is in fact pessimistic and that we can compute the agony for large graphs.

In order to do so, we applied our algorithm for several large directed graphs, downloaded from Stanford Large Network Dataset Collection (SNAP).[1] We removed any edges of form (u, u) as they have no effect on the rank. The characteristics of the datasets are given in the first 2 columns in Table 1. In addition, to our algorithm we applied a baseline algorithm of Gupte et al. [7]. The algorithm requires a subroutine for detecting a negative cycle. We used Bellman-Ford algorithm with an additional speed-up, where after each iteration over the edges we check whether a cycle has been discovered. We implement both algorithms in C++ and performed experiments using a Linux-desktop equipped with a Opteron 2220 SE processor. The running times and detailed statistics are given in Table 1.

Table 1. Basic characteristics of the datasets and statistics from experiments. The 3rd column indicates the number of iterations, the 4th column indicates the slack of the starting point, the 5th depicts the final score. The running times for MINAGONY and the baseline are given in 6th and 7th columns, respectively.

Dataset	$|V|$	$|E|$	iterations	gap	agony	time	baseline
Amazon	403 394	3 387 388	89 046	911 095	1 973 965	4h27m	–
Gnutella	62 586	147 892	1 907	150 851	18 964	45s	20m
EmailEU	265 214	418 956	27 679	500 177	120 874	2m	3h45m
Epinions	75 879	508 837	18 652	922 817	264 995	20m	1h40m
Slashdot	82 168	870 161	37 858	1 891 586	748 582	1h5m	7h3m
WebGoogle	875 713	5 105 039	164 708	4 110 696	1 841 215	2h32m	–
WikiVote	7 115	103 689	865	76 149	17 676	7s	1m

Our first observation is that the theoretical bound is indeed pessimistic. We are able to compute the agony for large networks in reasonable time. We spend 2.5 hours for computing agony for *WebGoogle*, a graph with over 5 million edges, and 4 hours for *Amazon*, a graph with over 3 million edges. For smaller graphs, the running time is significantly faster, either seconds or minutes.

The reason for this scalability is two-fold. First of all, the number of iterations, given in 4th column, is significantly lower than the number of edges. Secondly, RELIEF typically affects less than m vertices.

Our method performs better than the baseline for all datasets. We interrupted the baseline calculation after 24 hours for *Amazon* and *WebGoogle*.

---

[1] The datasets and their detailed descriptions are available at
http://snap.stanford.edu/data/index.html

Finally, let us consider behaviour of agony and the size of the eulerian graph as a function of iterations. In order to do so we plot the evolution of scores, normalized by the final agony, in Figure 3.

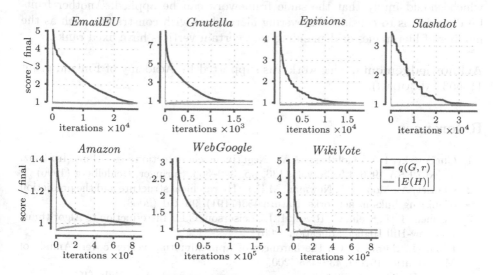

Fig. 3. Scores as a function of iteration. Each plot represents a single dataset. The upper line depicts the current agony score, normalized by the final score, as a function of a current iteration. The lower line depicts the current number of edges in the eulerian subgraph, normalized by the final score, as a function of a current iteration. Note that the x-axis is scaled.

We see that the initial agony is significantly larger than the final agony. For most datasets the agony drops quickly. For the largest datasets the algorithm achieves approximation ratio of 2 relatively quickly: for *WebGoogle* the algorithm achieves approximation ratio of 2 during the first 8% of iterations. This suggests that we can use the algorithm as any-time algorithm, stopping iterations early once we achieved acceptable approximation ratio. Note that since the optimal solution is at least as large as the current eulerian subgraph, we can at any time bound the approximation ratio of the current agony.

7   Concluding Remarks

In this paper we introduced an algorithm for discovering hierarchy among vertices in a given directed graph. The hierarchy should minimize agony, the edges that violate the hierarchical structure. We show that our algorithm achieves computational complexity of $O(m^2)$ which is significantly better than the current bound of $O(nm^2)$. We also demonstrate that $O(m^2)$ is a pessimistic estimate of the running time and in practice the algorithm scales up for large networks.

There are several interesting directions for future work. An obvious and practical extension is to make edges weighted. Weighting edges will change the definition of the dual problem as we no longer are looking for maximum eulerian subgraph. On the other hand, integer weights can be viewed as multiple edges which should imply that the same framework can be applied. Another fruitful direction is to consider discovering hierarchies with constraints, such as the number of hierarchies or demanding that certain vertices have fixed ranks.

Acknowledgements. This work was supported by Academy of Finland grant 118653 (ALGODAN).

References

[1] Cherkassky, B.V., Goldberg, A.V.: Negative-cycle detection algorithms. In: Díaz, J. (ed.) ESA 1996. LNCS, vol. 1136, pp. 349–363. Springer, Heidelberg (1996)

[2] Clauset, A., Moore, C., Newman, M.E.J.: Hierarchical structure and the prediction of missing links in networks. Nature 453(7191), 98–101 (2008)

[3] Cormen, T.H., Stein, C., Rivest, R.L., Leiserson, C.E.: Introduction to Algorithms. McGraw-Hill Higher Education (2001)

[4] Dinur, I., Safra, S.: On the hardness of approximating vertex cover. Annals of Mathematics 162(1), 439–485 (2005)

[5] Elo, A.E.: The rating of chessplayers, past and present. Arco Pub. (1978)

[6] Even, G., Naor, J.S., Schieber, B., Sudan, M.: Approximating minimum feedback sets and multicuts in directed graphs. Algorithmica 20(2), 151–174 (1998)

[7] Gupte, M., Shankar, P., Li, J., Muthukrishnan, S., Iftode, L.: Finding hierarchy in directed online social networks. In: Proceedings of the 20th International Conference on World Wide Web, pp. 557–566 (2011)

[8] Henderson, K., Gallagher, B., Eliassi-Rad, T., Tong, H., Basu, S., Akoglu, L., Koutra, D., Faloutsos, C., Li, L.: Rolx: Structural role extraction & mining in large graphs. In: Proceedings of the 18th ACM SIGKDD International Conference on Knowledge Discovery and Data Mining, pp. 1231–1239 (2012)

[9] Jameson, K.A., Appleby, M.C., Freeman, L.C.: Finding an appropriate order for a hierarchy based on probabilistic dominance. Animal Behaviour 57, 991–998 (1999)

[10] Macchia, L., Bonchi, F., Gullo, F., Chiarandini, L.: Mining summaries of propagations. In: IEEE 13th International Conference on Data Mining, pp. 498–507 (2013)

[11] Maiya, A.S., Berger-Wolf, T.Y.: Inferring the maximum likelihood hierarchy in social networks. In: Proceedings IEEE CSE 2009, 12th IEEE International Conference on Computational Science and Engineering, pp. 245–250 (2009)

[12] McCallum, A., Wang, X., Corrada-Emmanuel, A.: Topic and role discovery in social networks with experiments on enron and academic email. J. Artif. Int. Res. 30(1), 249–272 (2007)

[13] Papadimitriou, C.H., Steiglitz, K.: Combinatorial Optimization: Algorithms and Complexity. Prentice-Hall, Inc. (1982)

Speeding Up Recovery from Concept Drifts

Silas Garrido Teixeira de Carvalho Santos[1], Paulo Mauricio Gonçalves Júnior[2], Geyson Daniel dos Santos Silva[1], and Roberto Souto Maior de Barros[1]

[1] Centro de Informática, Universidade Federal de Pernambuco, Brazil
{sgtcs,gdss,roberto}@cin.ufpe.br
[2] Instituto Federal de Educação, Ciência e Tecnologia de Pernambuco, Brazil
paulogoncalves@recife.ifpe.edu.br

Abstract. The extraction of knowledge from data streams is an activity that has progressively been receiving an increased demand. However, in this type of environment, changes in data distribution, or concept drift, can occur constantly and is a challenge. This paper proposes the *Adaptable Diversity-based Online Boosting (ADOB)*, a modified version of the online boosting, as proposed by Oza and Russell, which is aimed at speeding up the experts recovery after concept drifts. We performed experiments to compare the accuracy as well as the execution time and memory use of ADOB against a number of other methods using several artificial and real-world datasets, chosen from the most used ones in the area. Results suggest that, in many different situations, the proposed approach maintains a high accuracy, outperforming the other tested methods in regularity, with no significant change in the execution time and memory use. In particular, ADOB was specially efficient in situations where frequent and abrupt concept drifts occur.

Keywords: data stream, concept drift, ensemble classifier, online boosting, diversity.

1 Introduction

Nowadays, several applications need the use of mechanisms that enable the extraction of knowledge in real time. Examples of such applications include monitoring the purchase history of customers, the movement data from sensors, or water temperatures. Thus, the algorithms used for this purpose must be constantly updated, trying to adapt to new instances and taking into account the computational constraints.

When working in environments with a continuous flow of data, it is not possible to guarantee that the distribution of the data will remain stationary. On the contrary, several changes may occur over time, triggering situations commonly known as concept drift. The speed with which the changes occur may be classified as abrupt, when the transition from an old to a new concept occurs suddenly, or gradual, when such a transition is smooth [1].

There are several approaches proposed to detect changes in concepts. Some of these approaches provide for the adaptation of the internal structure of a

T. Calders et al. (Eds.): ECML PKDD 2014, Part III, LNCS 8726, pp. 179–194, 2014.

classifier to deal with the changes of concept. Others try to identify when a concept drift has occurred, and then drop the old classifier and create a new one for the most actual concept [1–6]. Other existing methods use ensemble classifiers with some weighting policy applied to their members as well as for dropping the worst classifiers and adding new ones [7–11]. Finally, there are also approaches that seek to increase the efficiency of change detection when dealing with concepts that have previously occurred [12–14].

In situations where many changes of concepts occur, learning algorithms constantly need to adapt to the new distribution. In such scenarios, it is common to observe a delay before the complete adaptation occurs, i.e., a period that is used for learning a satisfactory generalization. The longer this period, the greater the number of incorrect predictions.

Based on these observations and using an ensemble of classifiers, we decided to try modifying the online version of boosting, originally proposed by Oza and Russell [15], aiming at a more rapid recovery of the experts accuracy in environments with frequent changes of concepts. More specifically, we have changed the way diversity is distributed during training.

This paper describes the Adaptable Diversity-based Online Boosting (ADOB) and compares the performance of our proposal with some of the major existing ensemble methods for dealing with data streams and concept drifts using the Massive Online Analysis (MOA) framework [16]. The results indicate that ADOB maintains good accuracies in different situations, surpassing the other tested methods in regularity.

The rest of this paper is organized as follows: Section 2 surveys the related work; Section 3 introduces the datasets used in the experiments; Section 4 describes in detail the operation of our proposal; Section 5 compares the proposed method to other existing ensemble methods; and, finally, Section 6 presents our conclusions.

2 Related Work

A large number of methods have been proposed to learn from data streams containing concept drifts. Examples of older methods include [12, 17, 18]. Nowadays, several methods have been proposed using the concept of ensemble classifiers.

Bagging [19] and Boosting [20] are techniques that use a set of classifiers trained on the original data by aggregating the responses of each classifier to get a better prediction. They use different strategies both to manipulate the data during the training of experts and also to combine their predictions. Online Bagging and Boosting [15] are adapted versions of these techniques to data stream environments. They both make use of the Poisson distribution to simulate a behavior similar to their offline versions.

Adwin Bagging (ADWINBAG) [8] and Leveraging Bagging (LEVERAGING) [9] both make use of the online version of bagging, as it was defined by Oza and Russell [15], adding Adaptive Windowing (ADWIN) [3] as their concept drift detector. In addition, Leveraging Bagging makes two changes to the original proposal: the first is to increase the value of diversity (λ) which, as a consequence, leads to an increase in the probability that an expert trains on the same instance. The second is to change the way the experts predict instances in order to increase diversity and reduce the correlation.

The Dynamic Weighted Majority (DWM) [7] extends the Weighted Majority Algorithm [21] and implements a weighted ensemble classifier specifically designed to identify concept drifts. This method adds and removes classifiers according to the ensemble global performance. If the ensemble commits an error, then a classifier is added. If one classifier commits an error, its weight is reduced. If after many examples a classifier continues with a low accuracy, indicated by a low weight, it is removed from the ensemble.

Diversity for Dealing with Drifts (DDD) [10] uses four ensemble classifiers with high and low diversity, before and after a concept drift is detected. A previous study [22] analyzed how these ensembles behaved in data sets suffering from abrupt and gradual concept drifts with several speeds of change, right after the drift and longer after. With the results obtained, Diversity for Dealing with Driftsd was proposed, trying to select the best ensemble (or weighted majority of ensembles) before and after drifts, detected by the use of a drift detection method.

The Accuracy Updated Ensemble (AUE2) [11], recently proposed, maintains a set of classifiers and its strategy is to, at every n instances (called chunks), remove the expert with the worst accuracy and replace it with a new one. The weight of the experts are also defined according to their accuracy, making the most accurate one have a greater influence on the prediction. The way the experts are updated makes the method sensitive to changes in concepts.

3 Datasets

This section describes the datasets that were selected for the experiments used to analyze the performance of ADOB against those of other recent methods. We chose both real-world and artificial datasets. In the artificial ones, it is possible to define the position of the concept drifts as well as its quantity and size. Thus, several situations can be simulated. In the real-world datasets, the unpredictability and volume of data makes their use interesting, complementing the scenarios provided by the artificial data. All the datasets used are available, most of them in the MOA website at the address http://moa.cs.waikato.ac.nz/.

3.1 Artificial Datasets

For the experiments described in Section 5, we chose four artificial datasets, two of them with gradual concept drifts and two with abrupt concept drifts. These are: LED [23, 24], RBF [8, 25], Sine [1, 2, 4], and Stagger [7, 17].

The LED dataset is composed of 24 categorical attributes, 17 of which are irrelevant, and one categorical class with ten possible values. It represents the problem of predicting the digit shown by a seven-segment LED display, where each attribute has 10% probability of being inverted (noise). We used a version of LED available at MOA that includes concept drifts to the data sets by simply changing the attributes positions. This dataset was used in our experiments to test gradual concept drifts.

RBF (Radial Basis Function) creates complex concept drifts that are not straightforward to approximate with a decision tree model. It works as follows: a fixed number of random centroids are generated. Each center has a random position, a single standard deviation, a class label, and a weight. New examples are generated by selecting a center at random, taking weights into consideration so that centers with higher weight are more likely to be chosen. A random direction is chosen to offset the attribute values from the central point. The length of the displacement is randomly drawn from a Gaussian distribution with standard deviation determined by the chosen centroid. The chosen centroid also determines the class label of the example. This effectively creates a normally distributed hypersphere of examples surrounding each central point with varying densities. Only numeric attributes are generated. Drift is introduced by moving the centroids with constant speed. This speed is initialized by a drift parameter. This dataset is composed of six classes and 20 attributes, and was also used to test gradual concept drifts.

Sine presents the problem of identifying the position of coordinates, represented by two attributes, in relation to the curve $y = sin(x)$. In the first context, points below the curve are classified as positive. After each concept drift, the classification is reversed. Each coordinate has values uniformly distributed in the [0,1] interval. It is possible to include two other attributes, filled with random data in the same interval, with no influence on the classification function (irrelevant data). Gama et al. [1] named these data sets as Sine1 and Sinirrel1, respectively. They also described Sine2, similar to Sine1 but using a different curve: $y < 0.5 + 0.3sin(3\pi x)$. Positive and negative examples are interchanged to ensure a stable learning environment. This dataset was used to test abrupt concept changes and is available at https://sites.google.com/site/moaextensions/.

In Stagger, each example consists of the following attributes: $color \in \{green, blue, red\}$, $shape \in \{triangle, circle, rectangle\}$, and $size \in \{small, medium, large\}$. According to the Stagger original paper [17], there are three kinds of different concepts: in concept 1, $color = red \wedge size = small$; in concept 2, $color = green \vee shape = circle$; and in concept 3, $size = medium \vee size = large$. This data set is usually used to simulate abrupt concept drifts and is fairly simple to learn – it has few attributes and concepts, and concepts 2 and 3 overlap.

3.2   Real Datasets

In addition to the artificial datasets, we chose three real-world datasets, from the most used ones in the area of data streams, all with very different number of instances and complexity. In these datasets, the number and position of concept drifts (if existent) are unknown.

The Electricity dataset [1, 2, 7, 8, 10, 25], composed of 45,312 instances and eight attributes, presents data collected from the Australian New South Wales Electricity Market. In that market the prices are not fixed, varying based on market demand and supply. The prices are set every five minutes and the class label identifies the change of the price related to a moving average of the last 24 hours. The goal of the problem is to predict if the price will increase or decrease.

The Forest Covertype dataset [8, 15, 24, 25] contains the forest cover type for 30 x 30 meter cells obtained from US Forest Service (USFS) Region 2 Resource Information System (RIS) data. The goal is to predict the forest cover type from cartographic variables. It contains 581,012 instances and 54 attributes, including numeric and categoric ones.

The Poker Hand data set [8, 25] represents the problem of identifying the value of five cards in the Poker game. It is constituted of five categoric and five numeric attributes and one categoric class with 10 possible values informing the value of the hand; for example, one pair, two pairs, a sequence, a street flush, etc. "In the Poker hand data set, the cards are not ordered, i.e., a hand can be represented by any permutation, which makes it very hard for propositional learners, especially for linear ones" [25]. Even though a simpler modified version exists (where the cards are sorted by rank and suit, and duplicates are removed), we decided to use the original harder version in our experiments. The used data set contains 1,000,000 instances.

4   Adaptable Diversity-Based Online Boosting

The Adaptable Diversity-based Online Boosting (ADOB) is a variation of the Online Boosting [15] method, which proposes to distribute instances more efficiently among experts, aiming to more quickly adapt to the situation where concept drifts occur frequently, specially if they are abrupt. This distribution is performed by controlling the diversity (λ) through the accuracy of each expert.

When a new distribution starts, the accuracy of the experts can be used to reduce the initial error, increasing the focus on instances of difficult classification and accelerating the diversity. As the experts have a high degree of similarity in the beginning, due to the reduced number of known instances, the accuracy can be used to define their behaviors. However, because more instances of the same distribution are coming and experts are diversifying, the use of the experts accuracies will tend to have little influence in their behaviors.

Algorithm 1 shows the ADOB pseudo-code, which is our modified version of the Online AdaBoost algorithm [15]. Initially, the ensemble of classifiers (h) is sorted by accuracy in ascending order. Before, several variables are initialized, including $minPos$ and $maxPos$ with values that represent the classifier with the worst and best accuracy, respectively, as well as λ, λ^{sc}, and λ^{sw} (lines 1 to 4).

When an instance d arrives, initially the expert with less accuracy will be selected. If the instance is correctly classified, we assume that probably the other experts, which are more accurate, will also have good chances of correctly classifying it. However, the correct classification of the worst expert does not guarantee that the others will do it properly too, even if an error is unlikely. Accordingly, we will refer to the error of another expert with better accuracy as an unlikely error.

Algorithm 1. Adaptable Diversity-based Online Boosting

Input: ensemble size M, instance d, ensemble h
1 $minPos \leftarrow 1$; $maxPos \leftarrow M$;
2 $correct \leftarrow$ false;
3 $\lambda \leftarrow 1.0$; $\lambda^{sc} \leftarrow 0.0$; $\lambda^{sw} \leftarrow 0.0$;
4 **sort** h by accuracy in ascending order;
5 **for** $m \leftarrow 1$ **to** M **do**
6 | **if** $correct$ **then**
7 | | $pos \leftarrow maxPos$; $maxPos \leftarrow maxPos$ - 1;
8 | **else**
9 | | $pos \leftarrow minPos$; $minPos \leftarrow minPos$ + 1;
10 | **end**
11 | $K \leftarrow$ Poisson(λ);
12 | **for** $k \leftarrow 1$ **to** K **do**
13 | | $h_{pos} \leftarrow$ Learning(h_{pos}, d);
14 | **end**
15 | **if** $h_{pos}(d)$ *was correctly classified* **then**
16 | | $\lambda^{sc}_m \leftarrow \lambda^{sc}_m + \lambda$;
17 | | $\lambda \leftarrow \lambda \left(\frac{N}{2\lambda^{sc}_m} \right)$;
18 | | $correct \leftarrow$ true;
19 | **else**
20 | | $\lambda^{sw}_m \leftarrow \lambda^{sw}_m + \lambda$;
21 | | $\lambda \leftarrow \lambda \left(\frac{N}{2\lambda^{sw}_m} \right)$;
22 | | $correct \leftarrow$ false;
23 | **end**
24 **end**
25 **return** h;

Looking into lines 15 to 23 of Algorithm 1, it is possible to observe that the value of λ will be reduced when the classification is done correctly and increased when it is incorrect. In this way, if an unlikely error occurs, the later it occurs, the smaller the influence on λ it will have. To minimize the consequences of an unlikely error, the next expert selected to do the classification will be the one with the best accuracy, followed by the second best, and so on (lines 6 to 7). Using this procedure, experts with the worst accuracies, and most likely to make mistakes, will only be selected at the end.

Another possible scenario is the case where the expert with the worst accuracy incorrectly classifies the instance. In this situation, we distribute the greatest possible λ for the next experts which are more likely to make mistakes in the classification. Therefore, the next experts will be selected according to their performances, from the worst to the best (lines 8 to 9). Assuming that experts with the lowest performances have higher probabilities of making mistakes in the classification, we force them to be selected earlier and maximize λ.

As previously mentioned, after more instances of the same distribution are presented, the lower the influence of this procedure in the experts accuracy will be. At this stage, the experts tend to have a low correlation, and the accuracy of the worst expert will have low importance to the others. Thus, ADOB will now have similar behavior to the original online boosting [15], except for the fact that experts begin to be selected unpredictably, varying for each instance. Thus, these changes are especially valid for situations in which the concept changes often and abruptly, as a consequence of their rapid recovery.

To help understanding how ADOB is used, Algorithm 2 presents a simplified version of MOA's singleClassifierDrift. The classification result is monitored by a concept drift detection method – we used ADWIN. To classify new instances, $ADOB_{classifier}$ (line 4) behaves the same as defined in [15] and the return is:

$$h(x) = argmax_{y \in Y} \sum m : h_m(x) = y^{log \frac{1}{\beta_m}}, \text{ where}$$

$$\beta_m = \frac{\epsilon_m}{1-\epsilon_m}, \ \epsilon_m = \left(\frac{\lambda_m^{sw}}{\lambda_m^{sc}+\lambda_m^{sw}} \right),$$

and $m \in [1..M]$ is limited by the number of experts. If ADWIN returns a warning, a new ensemble h_2 immediately starts to be trained using ADOB alongside the existing one (lines 6-7). When the drift is confirmed, the newly created ensemble is used and the old one, representing the last distribution, is removed (lines 8-9).

Algorithm 2. Simplified code of MOA's singleClassifierDrift with ADWIN

Input: ensemble size M, data stream D, base learner b
1 $ADWIN \leftarrow$ new ADWIN method;
2 $h, h_2 \leftarrow$ new ensemble using M times b;
3 **foreach** *instance d in D* **do**
4 $\quad ADWIN \leftarrow ADOB_{classifier}(M, d, h)$;
5 $\quad$ **switch** *ADWIN* **do**
6 $\quad\quad$ **case** *detect a warning level*
7 $\quad\quad\quad | \quad h_2 \leftarrow ADOB(M, d, h_2)$;
8 $\quad\quad$ **case** *detect a drift*
9 $\quad\quad\quad | \quad h \leftarrow h_2$; h2 $\leftarrow$ reset ensemble;
10 $\quad$ **endsw**
11 $\quad h \leftarrow ADOB(M, d, h)$;
12 **end**

It is worth pointing out that, in the real code, h_2 is also reset when a warning is *not* confirmed. This and other less important details were omitted here.

5 Experiments Configuration and Results

This section describes the set up and results of the experiments used to evaluate ADOB against other implementations of online bagging and boosting, as well as other recent ensemble methods aimed at detecting concept drifts in data streams. The chosen methods are: ADWINBAG, LEVERAGING, DWM, DDD, and AUE2. The original online version of AdaBoost (OzaBoost) [15] was included in the tests for comparative purposes, also using ADWIN to detect concept drifts. This method will be called Adwin Boosting (ADWINBOOST). All these methods were compared in terms of accuracy, execution time, and memory used.

The choice of methods was also based on the following additional criteria: ADWINBAG and ADWINBOOST were selected because they implement the original version of online bagging and boosting, respectively; LEVERAGING and DDD, because they use modified versions of the online bagging; whereas DWM and AUE2, because they set their own training strategies to detect changes.

To compute the precision, memory usage, and execution time of the methods, experiments were repeated 40 times in the artificial datasets. Average was computed alongside with a 95% confidence interval. In addition, each artificial dataset was composed of 10,000 instances.

Finally, all the tests were performed in a Core i3 350M processor, 2GB of main memory, running the Ubuntu 12.04 64 bits operating system.

5.1 Drift Configuration in the Artificial Datasets

In the configuration of the artificial datasets, we inserted abrupt and gradual concept drifts. Noise was also inserted in some datasets in order to check the behavior of the methods in these situations.

Two versions of the LED dataset were used, both with gradual concept drifts. In one dataset, drifts occur at instances 3,000 and 6,000. In the other, four gradual changes were inserted at instances 2,000, 4,000, 6,000, and 8,000. In both versions, every time a change occurs, 10% of noise was added.

Other gradual changes were tested, this time making use of the Random RBF dataset. The position of the concept drifts in the two versions of RBF were the same used in the LED data sets.

The two versions of both Sine and Stagger datasets have four and eight abrupt changes, respectively. In their first configurations, the changes were inserted at instances 2,000, 4,000, 6,000, and 8,000. In their second versions, eight changes occur at instances 2,000, 3,000, 4,000, ..., 8,000, and 9,000, respectively.

5.2 Ensemble Methods Configuration

To perform a fair comparison between the methods, common parameters were all set similarly: the base learner was a Hoeffding Tree [18] and the number of experts was set to ten. To set individual parameters, each method was executed with ten different configurations for each dataset used, to check if their default values were the ones which produced the best accuracy. In most cases this was indeed the case, but there were exceptions. In these few cases, we adopted a different parametrization – specific values are given below.

To detect concept drifts, ADWINBAG, ADWINBOOST, LEVERAGING, and ADOB all make use of ADWIN. In all of them, the δ parameter of ADWIN, that corresponds to the maximum global error, was set to 0.1. This value influences the hypothesis test used to check for any change in the distribution [3].

DDD originally uses the Early Drift Detection Method (EDDM) [2] to detect changes. The parameter values used for EDDM were their defaults, i.e., $n = 30$, $w = 0.95$, and $d = 0.99$. These represent, respectively, the number of instances before starting to detect changes, the confidence level to activate the warning level, and the confidence level to detect a change.

Regarding the parameters of the methods, LEVERAGING uses λ, which controls the weight of resampling. The higher the value of λ, the greater the probability that a given instance is repeated for each expert ensemble. This probability is defined according to the Poisson distribution [9]. In our tests, we used $\lambda = 6$.

DWM uses three parameters: p, which corresponds to the time needed to verify if any expert will be removed or added as well as to update their weights if any classifier incorrectly classifies the actual instance; β, the value that will be decremented by the expert every time it makes a mistake; and, finally, θ, which is the minimum value that an expert can have without being removed [7]. We used $p = 100$ (artificial datasets), $p = 250$ (real datasets), $\beta = 0.5$, and $\theta = 0.01$.

The parameters of DDD are: W, responsible for controlling the robustness of the method to false alarms; λ_l, useful to define the value that will represent an ensemble with low diversity; and λ_h, a parameter that will represent an ensemble with high diversity [10]. In our experiments we used $W = 1$ (except in the LED datasets), $W = 3$ (LED datasets), $\lambda_l = 1$, and $\lambda_h = 0.05$.

Finally, AUE2 has two parameters, which control the memory usage (m) and the chunk size (c). The first is responsible for limiting the maximum amount of memory that each component of the ensemble may have. The latter defines the number of instances needed to check the accuracy and memory usage of the ensemble members [11]. After preliminary tests, their values were defined as $m = 32MB$, $c = 50$ (except in the Forest Covertype and Poker Hand datasets), and $c = 500$ (Forest Covertype and Poker Hand datasets).

5.3   Accuracy Analysis

Table 1 presents the accuracies obtained for each method on the artificial and real-world datasets. Bold values identify the best results. The average rank is the average of the positions that each method achieved in different datasets.

In the LED dataset with two concepts drifts, the ADWINBAG method had the best accuracy, closely followed by leveraging. Following, there are DDD, ADOB, and AUE2. In general, these methods have similar performance, with differences ranging from 0.1% to 3%, approximately. On the other hand, ADWINBOOST and DWM obtained the worst results. As can be seen in the first graphic of Figure 1, the slow recovery of ADWINBOOST negatively affects its accuracy at every concept drift. In addition, DWM had the worst performance throughout the dataset.

With the addition of two more drifts in the LED dataset, as might be expected, the performance of all methods deteriorated. The most affected method was AUE2, with a drop of about 3%. The less impacted was DWM, with a worsening of only 0.74%. However, DWM's performance remained well below those of the others. In descending order of accuracy, the result is LEVERAGING, ADWINBAG, DDD, ADOB, AUE2, ADWINBOOST, and DWM.

Table 1. Average accuracy in percentage (%) with 95% confidence in artificial datasets

	ADOB	AdwinBag	AdwinBoost	Leveraging	DWM	DDD	AUE2
LED_2	59.30±0.14	**61.18±0.08**	55.52±0.53	61.08±0.09	45.48±0.78	60.62±0.16	58.18±0.23
LED_4	56.86±0.10	58.57±0.09	53.69±0.52	**59.01±0.06**	44.74±0.89	57.92±0.14	55.13±0.29
RBF_2	53.08±0.26	**54.59±0.15**	47.13±0.45	39.20±0.15	40.12±0.62	53.41±0.21	53.26±0.27
RBF_4	56.77±0.40	**58.29±0.12**	51.13±0.36	43.58±0.20	42.75±0.59	57.58±0.22	57.19±0.28
$Sine_4$	90.28±0.10	81.84±0.79	**90.68±0.13**	89.90±0.14	87.60±0.43	88.97±0.31	87.57±0.21
$Sine_8$	88.42±0.10	80.14±0.53	**88.91±0.12**	87.42±0.14	86.90±0.36	87.55±0.33	86.06±0.23
$Stagger_4$	**99.02±0.02**	92.73±0.21	96.00±1.21	97.80±0.05	96.45±0.32	97.47±0.15	94.27±0.17
$Stagger_8$	**98.74±0.02**	90.32±0.18	96.52±0.82	94.70±0.13	95.34±0.29	96.59±0.13	88.29±0.19
Elec	87.98	86.44	87.09	**89.71**	88.17	85.72	80.98
Cov	82.79	84.44	81.31	**88.16**	87.34	83.96	65.95
Poker	**53.74**	53.20	52.42	52.18	46.60	53.19	48.84
Rank	**2.82**	3.82	4.18	3.36	5.09	3.18	5.55

Analyzing the RBF dataset with two concept drifts, ADWINBAG again had the best accuracy, closely followed by DDD, AUE2, and ADOB. ADWINBOOST, DWM, and LEVERAGING had the worst results. Unlike the previous dataset, with the insertion of two more concept drifts, the performance of the methods improved. This can be explained by how the changes were defined. In RBF_2, the gradual changes take twice as long to fully occur than in RBF_4. Thus, methods spend more time to detect if there was a change and, consequently, take longer to recover. Although all methods increased their accuracies in this version, the order remained the same, except for the fact that DWM assumed the worst position, swapping places with LEVERAGING.

Up to this point, the presented datasets included gradual concept drifts. Making a general analysis of this type of change, the methods that had the best accuracies were: ADWINBAG, DDD, ADOB, and AUE2, with differences ranging from 0.77% to 2.21%. Right after, with a considerable distance, follows ADWINBOOST, LEVERAGING, and DWM occupying the last positions, respectively.

Differently from the other datasets, at Sine, ADWINBAG returned the worst results. Its differences to AUE2 are of 5.73% and 5.92%, in the datasets with four and eight drifts, respectively. These are significant differences, given that the differences from the first (ADWINBOOST) to the second to last (AUE2, in both cases) are of only 3.11% and 2.85%, respectively.

Finally, on Stagger the results were somewhat different from those on Sine. AUE2 and ADWINBAG were again the worst methods, but the order was different in the two versions used. The best accuracies were achieved by ADOB and DDD.

Analyzing the overall accuracy of the methods in all datasets with abrupt concept drifts, ADOB had the best performance, followed by ADWINBOOST, DDD, LEVERAGING, DWM, AUE2, and ADWINBAG. Comparing these results with the results of the datasets with gradual concept drifts, ADWINBAG had the highest drop in performance, from first to last. The methods that maintained a better balance in different situations were ADOB and DDD, respectively.

In the real datasets, LEVERAGING was the method with the best overall performance, followed by ADOB, ADWINBAG, DDD, ADWINBOOST, DWM, and AUE2, respectively. It is worth noting that, despite its poor overall performance in these datasets, DWM was the second best in both Electricity and Forest Covertype. This is explained by its very bad performance in the Poker Hand dataset.

Finally, observe that, in the first two graphics of Figure 1, it is possible to visualize the gradual and abrupt changes, respectively. However, in the third, referring to the Electricity dataset, apparently no drastic change occurs.

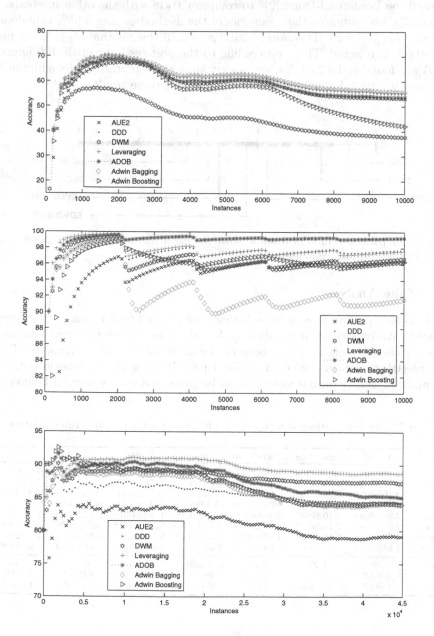

Fig. 1. Accuracy results in the LED$_2$, Stagger$_4$ and Electricity datasets

Complementing the accuracy analysis, a statistic based on the nonparametric Friedman test [26] was used. For this test, the null hypothesis states that all methods are statistically equal. If this hypothesis is rejected, the test indicates that there is a statistical difference in any of the methods, but it does not specify which method(s). For this task, the use of a post-test is required. In our case, we used the Bonferroni-Dunn [27] to compare ADOB with the other methods.

Initially, we compared the accuracies of the methods, using a 95% confidence interval: $F_{6;60} = 2.25$. Therefore, with $F_F = 2.79$ (bigger than $F_{6;60}$), the null hypothesis is rejected. Then, proceeding to the post-test, the critical difference (CD) was found to be 2.44. So, we can say that ADOB is statistically superior in accuracy to AUE2. Figure 2 graphically represents these results.

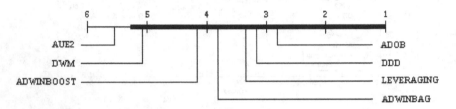

Fig. 2. Comparison results of ADOB against the other methods using the Bonferroni-Dunn test with a 95% confidence interval

5.4 Time Analysis

Table 2 displays the values in seconds that each method took to rank the different datasets. An important observation is that the times of the real-world datasets were significantly higher than those of the artificial datasets, except for RBF and Electricity, which presented similar times. Because the real-world datasets are much bigger, requiring more time to be processed is an expected behavior.

Table 2. Average runtime in seconds (s) with 95% confidence in artificial datasets

	ADOB	AdwinBag	AdwinBoost	Leveraging	DWM	DDD	AUE2
LED$_2$	3.34±0.04	3.05±0.03	2.86±0.04	3.74±0.02	**2.74±0.05**	4.87±0.14	4.09±0.02
LED$_4$	3.62±0.05	3.23±0.03	3.05±0.04	3.90±0.03	**2.72±0.04**	5.74±0.11	4.13±0.02
RBF$_2$	8.10±0.07	8.79±0.10	**7.76±0.10**	12.63±0.06	9.63±0.11	13.80±0.50	11.65±0.11
RBF$_4$	9.34±0.14	9.46±0.09	**7.79±0.09**	13.43±0.10	9.42±0.11	16.28±0.61	13.16±0.14
Sine$_4$	2.30±0.03	2.10±0.02	2.36±0.04	3.79±0.04	**1.48±0.04**	2.97±0.06	2.49±0.03
Sine$_8$	2.59±0.04	2.14±0.03	2.58±0.05	3.80±0.04	**1.58±0.03**	3.37±0.06	2.53±0.04
Stagger$_4$	1.67±0.03	1.66±0.03	1.68±0.03	2.09±0.02	**0.70±0.01**	1.59±0.02	1.99±0.02
Stagger$_8$	1.81±0.03	1.59±0.03	1.84±0.03	2.22±0.02	**0.69±0.01**	1.78±0.02	2.93±0.07
Elec	8.98	7.22	8.27	16.65	**4.73**	15.25	8.18
Cov	291.52	228.08	229.42	468.86	**165.49**	545.17	312.90
Poker	469.67	358.27	374.05	3574.45	**155.88**	358.56	152.66

Analyzing these results, the slowest artificial dataset was RBF, with an average of 11.27s and 10.34s in the versions with four and two concept drifts, respectively. All the others were much faster. The fastest was Stagger, with an average time of 1.84s and 1.63s in the versions with 8 and 4 drifts, respectively.

In the real-world datasets, the average times were somewhat proportional to the number of instances: Poker Hand was the one that took longer (777.65s), followed by Forest Covertype (320.21s), and Electricity (9.90s).

A fact that possibly influences the time is diversity. The higher the diversity, the higher the probability of the same instance be repeatedly distributed to a different expert and hence more time be used. ADOB and ADWINBOOST are methods that tend to increase their diversities in proportion to the error. Thus, the lower the accuracy of the method – which suggests more errors – the greater the diversity, causing a longer running time.

For example, observing the RBF and Stagger results, the ones with the lowest and highest accuracies, were the ones with the highest and lowest times in the artificial datasets, respectively. The same idea might be true in the real-world datasets, but these should be compared separately because they possess very different numbers of instances. The differences in execution times between ADOB and ADWINBOOST demonstrate the first attempt to maximize/minimize diversity, as discussed in Section 4, in favor of a better accuracy in different situations.

On the other hand, one of the characteristics of LEVERAGING is to maintain a higher diversity during the processing of the instances. As a result, it is the slowest method, followed by DDD. Although ADWINBAG and LEVERAGING use a static diversity, Minku et al. [22] show that different diversities in different situations of concept drifts contribute to improve the accuracy of the method.

5.5 Memory Analysis

We decided to monitor the memory usage to confirm that, despite using more memory than ADWINBOOST, ADOB is *not* memory intensive in absolute terms.

The evaluation was based on a metric that computes the amount of memory used by the methods per hour (in KB) and the results are presented in Table 3. Again, in the real-world datasets, the methods used much more memory than in the artificial datasets as a consequence of the greater number of instances.

The method with the lowest memory usage was DWM, followed by AUE2, ADWINBOOST, ADWINBAG, ADOB, DDD, and LEVERAGING. A possible explanation for the high memory usage of DDD is that it stores four times more classifiers than the value set by the user. But not all of them are used to classify each instance: the best ones are chosen within the ensemble according to each situation.

Memory usage by methods using both the original and the modified online bagging/boosting versions can be explained similarly to the explanation made to the execution time. The greater the diversity, the greater the likelihood of repetition of an instance among experts involving more training. As a direct consequence, it increases the used memory. The observation of the relationship with the precision, made earlier, apparently also applies to the memory usage. However, it is important to notice that there are exceptions in both cases.

Table 3. Average memory usage in (KB/h) with 95% confidence in artificial datasets

	ADOB	AdwinBag	AdwinBoost	Leveraging	DWM	DDD	AUE2
LED_2	0.34±0.00	0.21±0.00	0.29±0.00	0.26±0.00	**0.02±0.00**	1.29±0.05	0.26±0.00
LED_4	0.36±0.01	0.22±0.00	0.31±0.00	0.26±0.00	**0.02±0.00**	1.56±0.04	0.27±0.00
RBF_2	1.14±0.01	0.94±0.01	1.09±0.01	2.18±0.05	**0.10±0.00**	6.23±0.29	1.20±0.01
RBF_4	1.31±0.02	1.00±0.01	1.09±0.01	2.89±0.06	**0.10±0.00**	7.36±0.37	1.36±0.01
$Sine_4$	0.18±0.00	0.10±0.00	0.18±0.00	0.46±0.01	**0.01±0.00**	0.44±0.01	0.07±0.00
$Sine_8$	0.20±0.00	0.10±0.00	0.19±0.00	0.39±0.01	**0.01±0.00**	0.49±0.01	0.07±0.00
$Stagger_4$	0.10±0.00	0.07±0.00	0.10±0.00	0.11±0.00	**0.00±0.00**	0.18±0.00	0.04±0.00
$Stagger_8$	0.11±0.00	0.06±0.00	0.11±0.00	0.12±0.00	**0.00±0.00**	0.20±0.00	0.29±0.01
Elec	0.85	0.47	0.86	3.43	**0.05**	2.40	0.29
Cov	37.54	25.75	31.79	290.38	**3.76**	168.50	27.41
Poker	293.80	208.34	195.37	30816.18	**0.67**	266.08	6.17

6   Conclusion

Dealing with concept drifts in data streams is a challenging topic, given that such drifts can be abrupt or gradual, slow or fast, rare or frequent, cyclical or not, etc. Thus, the single classifier approach is unlikely to achieve good results in general and, so, the ensemble of classifiers methods are becoming more popular.

This paper presented ADOB, an ensemble algorithm based on the Online Boosting [15] method specially built to deal more efficiently with *frequent* and *abrupt* concept drifts on on-line learning environments. More specifically, ADOB proposes to distribute instances more efficiently among experts, by controlling the diversity (λ) through the accuracy of each expert, aiming at recovering faster from the situations where concept drifts occur frequently.

We run experiments to compare ADOB to six different online ensemble methods, including other variations of Online Bagging and Boosting [15], all of them using ADWIN [3] as their drift detector, namely Adwin Bagging [8], Leveraging Bagging [9], and ADWINBOOST, as well as other well known and/or recent ensembles such as DWM [7], DDD [10], and AUE2 [11].

To perform the comparison, we used two different versions of four selected artificial datasets (eight in total), with both abrupt and gradual concept drifts, as well as three real-world datasets, all of them chosen from the most used ones in the concept drift research area.

It is important emphasizing that our main subject of interest in these experiments was the performance evaluation of Algorithm 1 – more specifically we wanted to compare it to the ADWINBOOST version that inspired it. Mainly for this reason, ADOB, ADWINBOOST, ADWINBAG, and LEVERAGING all used very similar versions of Algorithm 2 – MOA's singleClassifierDrift using ADWIN as drift detection method, as well as the same parametrization. DDD also used a similar version of singleClassifierDrift but the selected drift detection method was EDDM, as in its original reference [10].

The tested ADOB configuration presented good precision in several situations and, in particular, it was specially efficient in the Stagger [17] and Sine [1] datasets, which had abrupt concept drifts. It is worth pointing out that ADOB presented the best overall accuracy considering all tested datasets. In addition, according to a statistic based on the non-parametric Friedman test, ADOB presented statistically superior accuracy, when compared to AUE2, and comparable performance to the other methods in the tested data sets.

Even so, we believe the efficiency of ADOB can be further improved, both by optimizations in the algorithms and by using different drift detection methods in different types of datasets. These might be investigated in the near future. Another possible future work is a deeper investigation of the relationship between diversity and accuracy of the methods with the run time and memory usage.

Finally, it is worth pointing out that both ADOB and DDD were implemented as part of this work. They have been added to the MOA framework and are freely available at https://sites.google.com/site/moamethods. The implementation of ADWINBOOST was a mere parametrization of code previously available in MOA. DWM was already available at https://sites.google.com/site/moaextensions.

Acknowledgements. Silas Santos is supported by postgraduate grant number 0837-1.03/12 from FACEPE. We also thank the comments and suggestions from the anonymous referees which helped to improve this final version of the paper.

References

1. Gama, J., Medas, P., Castillo, G., Rodrigues, P.: Learning with drift detection. In: Bazzan, A.L.C., Labidi, S. (eds.) SBIA 2004. LNCS (LNAI), vol. 3171, pp. 286–295. Springer, Heidelberg (2004)
2. Baena-García, M., Del Campo-Ávila, J., Fidalgo, R., Bifet, A., Gavaldá, R., Morales-Bueno, R.: Early drift detection method. In: International Workshop on Knowledge Discovery from Data Streams, IWKDDS 2006, pp. 77–86 (2006)
3. Bifet, A.: Learning from time-changing data with adaptive windowing. In: Proceedings of the Seventh SIAM International Conference on Data Mining, SDM 2007, Lake Buena Vista, Florida, USA, pp. 443–448. SIAM (2007)
4. Ross, G.J., Adams, N.M., Tasoulis, D.K., Hand, D.J.: Exponentially weighted moving average charts for detecting concept drift. Pattern Recognition Letters 33(2), 191–198 (2012)
5. Nishida, K., Yamauchi, K.: Detecting concept drift using statistical testing. In: Corruble, V., Takeda, M., Suzuki, E. (eds.) DS 2007. LNCS (LNAI), vol. 4755, pp. 264–269. Springer, Heidelberg (2007)
6. Page, E.S.: Continuous inspection schemes. Biometrika 41(1/2), 100–115 (1954)
7. Kolter, J.Z., Maloof, M.A.: Dynamic weighted majority: An ensemble method for drifting concepts. Journal of Machine Learning Research 8, 2755–2790 (2007)
8. Bifet, A., Holmes, G., Pfahringer, B., Kirkby, R., Gavaldà, R.: New ensemble methods for evolving data streams. In: Proceedings of the 15th ACM SIGKDD International Conference on Knowledge Discovery and Data Mining, pp. 139–148. ACM, New York (2009)
9. Bifet, A., Holmes, G., Pfahringer, B.: Leveraging bagging for evolving data streams. In: Balcázar, J.L., Bonchi, F., Gionis, A., Sebag, M. (eds.) ECML PKDD 2010, Part I. LNCS, vol. 6321, pp. 135–150. Springer, Heidelberg (2010)

10. Minku, L.L., Yao, X.: DDD: A new ensemble approach for dealing with concept drift. IEEE Transactions on Knowledge and Data Engineering 24(4), 619–633 (2012)
11. Brzezinski, D., Stefanowski, J.: Reacting to different types of concept drift: The accuracy updated ensemble algorithm. IEEE Transactions on Neural Networks and Learning Systems 25(1), 81–94 (2013)
12. Widmer, G., Kubat, M.: Learning in the presence of concept drift and hidden contexts. Machine Learning 23(1), 69–101 (1996)
13. Ramamurthy, S., Bhatnagar, R.: Tracking recurrent concept drift in streaming data using ensemble classifiers. In: Proceedings of the 6th International Conference on Machine Learning and Applications, ICMLA 2007, pp. 404–409. IEEE Computer Society, Los Alamitos (2007)
14. Gonçalves, Jr. P.M., Barros, R.S.M.: RCD: A recurring concept drift framework. Pattern Recognition Letters 34(9), 1018–1025 (2013)
15. Oza, N.C., Russell, S.: Online bagging and boosting. In: Artificial Intelligence and Statistics 2001, pp. 105–112. Morgan Kaufmann (2001)
16. Bifet, A., Holmes, G., Kirkby, R., Pfahringer, B.: MOA: Massive online analysis. Journal of Machine Learning Research 11, 1601–1604 (2010)
17. Schlimmer, J.C., Granger, R.H.: Incremental learning from noisy data. Machine Learning 1(3), 317–354 (1986)
18. Hulten, G., Spencer, L., Domingos, P.: Mining time-changing data streams. In: Proceedings of the Seventh ACM SIGKDD International Conference on Knowledge Discovery and Data Mining, KDD 2001, pp. 97–106. ACM, New York (2001)
19. Breiman, L.: Bias, variance, and arcing classifiers. Technical report, Statistics Department, University of California, Berkeley, CA, USA (1996)
20. Freund, Y., Schapire, R.E.: Experiments with a new boosting algorithm. In: International Conference on Machine Learning, vol. 96, pp. 148–156 (1996)
21. Blum, A.: Empirical support for winnow and weighted-majority algorithms: Results on a calendar scheduling domain. Machine Learning 26(1), 5–23 (1997)
22. Minku, L.L., White, A.P., Yao, X.: The impact of diversity on online ensemble learning in the presence of concept drift. IEEE Transactions on Knowledge and Data Engineering 22(5), 730–742 (2010)
23. Breiman, L., Friedman, J.H., Olshen, R.A., Stone, C.J.: Classification and Regression Trees. In: Wadsworth Statistics / Probability series. Wadsworth International Group, Belmont (1984)
24. Gama, J., Rocha, R., Medas, P.: Accurate decision trees for mining high-speed data streams. In: Proceedings of the Ninth ACM SIGKDD International Conference on Knowledge Discovery and Data Mining, KDD 2003, pp. 523–528. ACM Press, New York (2003)
25. Bifet, A., Holmes, G., Pfahringer, B., Frank, E.: Fast perceptron decision tree learning from evolving data streams. In: Zaki, M.J., Yu, J.X., Ravindran, B., Pudi, V. (eds.) PAKDD 2010. LNCS, vol. 6119, pp. 299–310. Springer, Heidelberg (2010)
26. Demšar, J.: Statistical comparisons of classifiers over multiple data sets. Journal of Machine Learning Research 7, 1–30 (2006)
27. Dunn, O.J.: Multiple comparisons among means. Journal of the American Statistical Association 56(293), 52–64 (1961)

Training Restricted Boltzmann Machines with Overlapping Partitions

Hasari Tosun and John W. Sheppard

Montana State University,
Department of Computer Science, Bozeman, Montana, USA

Abstract. Restricted Boltzmann Machines (RBM) are energy-based models that are successfully used as generative learning models as well as crucial components of Deep Belief Networks (DBN). The most successful training method to date for RBMs is the Contrastive Divergence method. However, Contrastive Divergence is inefficient when the number of features is very high and the mixing rate of the Gibbs chain is slow. We propose a new training method that partitions a single RBM into multiple overlapping small RBMs. The final RBM is learned by layers of partitions. We show that this method is not only fast, it is also more accurate in terms of its generative power.

Keywords: Restricted Boltzmann Machine, Machine Learning.

1 Introduction

The Restricted Boltzmann Machine was introduced by Hinton *et al.* as a parallel network for constraint satisfaction [1]. Since computing the partition function in the Boltzmann distribution is not tractable, training was initially inefficient, and RBMs did not gain popularity for seventeen years until Hinton *et al.* developed Contrastive Divergence, a method based on Gibbs Sampling [8]. Since then, RBMs are used as basic components of deep learning algorithms [3,7,9]. RBMs have also been successfully applied to classification tasks [5,10,12]. Moreover, RBMs have been applied to many other learning tasks including Collaborative Filtering [14].

As RBMs became popular, research on training them efficiently increased. Tieleman modified the Contrastive Divergence method by making Markov chains persistent [15]. Thus, the Markov chain is not reset for each training example. This has been shown to outperform Contrastive Divergence with one step, *CD-1*, with respect to classification accuracy. However, it does not address the problem of training speed. Brekal *et al.* introduced an algorithm to parallelize training RBMs using parallel Markov chains [4]. Resulting Markov chains need to share messages and the gradient is estimated by averaging chains.

In the context of Deep Belief Networks (DBN), the DistBelief model and data parallelization framework was developed by [6]. Here, the DBN model is partitioned into parts. Overlapping parts then exchange messages. Moreover, models are replicated in different computation nodes and trained on different subsets of data to provide data parallelization.

T. Calders et al. (Eds.): ECML PKDD 2014, Part III, LNCS 8726, pp. 195–208, 2014.

In this paper, we propose a novel algorithm, *RBM-Partition*, for training RBMs that splits a single RBM into multiple partitions. Each partition then trains on a subsection of the data instance. We explore the effects of permitting these partitions to overlap to improve training across the boundaries of the partitions. We then investigate the generative power of model and find that this training process improves training performance of CD-1 in terms of both generative power and speed.

The rest of this paper is organized as follows. In Section 2 we briefly introduce the Boltzmann Distribution and RBMs. We describe our training method in Section 3 and show experimental results in Section 4. Finally, we discuss future work in Section 5.

2   Restricted Boltzmann Machines

In statistical mechanics, the Boltzmann distribution is the probability of a random variable that realizes a particular energy level (Equation 1) [11]. Here $\beta = \frac{1}{kT}$ where T is temperature and k is the Boltzmann constant. In machine learning, β is usually set to 1, except in the context of algorithms such as simulated annealing. Z is the partition function, which is generally intractable to compute. However, when Z is computable, all other properties of the system such as entropy, temperature, etc. can be calculated. The equation for Z, which summarizes over the micro states of the system, is shown in Equation 2.

$$p(\mathbf{x}) = \frac{e^{-\beta E(\mathbf{x})}}{Z} \tag{1}$$

$$Z = \sum_i \left(e^{-\beta E(x_i)} \right) \tag{2}$$

As a type of Hopfield Network, an RBM is a generative model with visible and hidden nodes as shown in Figure 1. There are no dependencies between hidden nodes, or between visible nodes, thus an RBM forms a bipartite graph. The model represents a Boltzmann energy distribution [11], where the probability distribution of the RBM with visible (x) and hidden node (h) is given in Equation 3.

$$p(\mathbf{x}, \mathbf{h}) = \frac{e^{-E(\mathbf{x}, \mathbf{h})}}{Z} \tag{3}$$

If we marginalize over the hidden variables, we obtain the probability of the visible variables $p(\mathbf{x}) = \sum_h \left(\frac{e^{-E(\mathbf{x},h)}}{Z} \right)$. Inspired from statistical mechanics, we write $p(\mathbf{x})$ in terms of Free Energy (A) as follows:

$$A(\mathbf{x}) = -\log \left(\sum_h e^{-E(\mathbf{x},h)} \right) \tag{4}$$

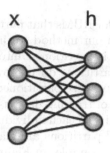

Fig. 1. Restricted Boltzmann Machine

Thus, rewriting $p(\mathbf{x})$ results in

$$p(\mathbf{x}) = \frac{e^{-A(\mathbf{x})}}{Z} \tag{5}$$

The partition function, $Z = \sum_x e^{-A(x)}$. The energy function of an RBM is given in the following equation.

$$E(\mathbf{x}, \mathbf{h}) = -b\mathbf{x} - c\mathbf{h} - \mathbf{h}\mathbf{W}\mathbf{x}.$$

If θ represents the model parameters, then the gradient of the log-likelihood is calculated as in Equation 6. The gradient contains two terms that are referred as the *positive* and the *negative* terms respectively. The first term increases the probability of the training data by decreasing free energy while the second term decreases the probability of a sample generated by the model. Computing the expectation over the first term is tractable; however, for the second term it is not. Thus, Hinton introduced the Contrastive Divergence algorithm that uses Gibbs sampling to estimate the second term [8].

$$\frac{-\partial \log(p(x))}{\partial \theta} = \frac{\partial A(x)}{\partial \theta} - \sum_{\tilde{x}} p(\tilde{x}) \frac{\partial A(\tilde{x})}{\partial \theta} \tag{6}$$

We provide a more detailed description of the CD algorithm in Section 4. An alternative description of Contrastive Divergence algorithm is given by Bengio [2].

CD provides a reasonable approximation to the likelihood gradient. The CD-1 algorithm (i.e, Contrastive Divergence with one step) is usually sufficient for many applications; for CD-k, resetting the Markov chain after each parameter update is inefficient because the model has already changed [15,2].

3 Partitioned Restricted Boltzmann Machines

We propose a training method for RBMs that partitions the network into several overlapping subnetworks. With our method, training involves several partition steps. In each step, the RBM is partitioned into multiple RBMs as shown in Figure 2. In this figure, the partitions do not overlap; we discuss the version with overlap later in this section These partitioned RBMs are trained in parallel with a corresponding partition of training data using CD-1. In other words, the feature vector is also partitioned, and each individual RBM is trained on a section of that feature vector. Once all partitions are trained, we generate another set of RBMs with fewer splits. For example, in Figure 2, we initially generate four RBMS. In the second step, we generate two, and final training occurs on the full RBM. It should be noted that the training process in all steps is over the same weight vector.

The motivation behind our approach is that when RBMs are small, they can be trained with more training epochs. However, as we reduce the number of splits, training requires fewer epochs and therefore less time to train. The pseudo-code for our training procedure is given in Algorithm 1. Since the details for overlapping partitions is omitted, we added notes where overlapping partitions will need different logic.

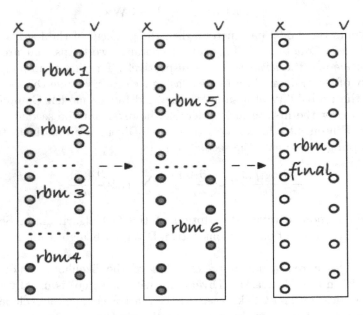

Fig. 2. RBM Partitions

Algorithm 1. Training with Partitions

1: *partition_configurations*: a list of configurations that describe splits and training instances for each training step
2: $W \leftarrow$ Create and initialize weight vector
3: *vbias* $\leftarrow$ Create and initialize bias vector for visible layer
4: *hbias* $\leftarrow$ Create and initialize bias vector for hidden layer
5: **for** each *configuration* in *partition_configurations*:
6:    *train* $\leftarrow$ Training instances
7:    **train_partitions**(*configuration, train, visible, hidden, W, vbias, hbias*)

Algorithm 2. create_rbm_partitions(configuration, W, vbias, hbias)

1: //for overlapping, *visible_nodes* and *hidden_nodes* will increase based
2:    on percentage of overlap
3: *visible_nodes* $\leftarrow$ *configuration.visible/configuration.splits*
4: *hidden_nodes* $\leftarrow$ *configuration.hidden/configuration.splits*
5: *rbms*: RBM list
6: **for** *i* in *configuration.splits*:
7:    //Each RBM will operate on a region of the visible vector
8:       and hidden vector.
9:    //For overlapping partitions, *vbase* and *hbase* will change based
10:       on overlap percentage
11:    *vbase* $\leftarrow$ base index in visible vector
12:    *hbase* $\leftarrow$ base index in hidden vector
13:    *rbm(i)* $\leftarrow$ RBM(*W, vbias, hbias, vbase, hbase, visible_nodes, hidden_nodes*)
14: **return** rbms

Algorithm 3. train_partitions(configuration, train, visible, hidden, W, vbias, hbias)

1: *rbm_list* $\leftarrow$ **create_rbm_partitions**(*configuration, W, vbias, hbias*)
2: for each instance in *train*:
3:    //for overlapping, partition will change according to
4:       *configuration.overlap* percentage
5:    *splits* $\leftarrow$ split instance into number of configuration.splits partitions
6:    for *rbm* in *rbms_list*
7:       *rbm(i).contrastive_divergence(splits(i))*

Based on the intuition that neighboring RBMs may share some features (nodes), for overlapping partitions, we define similar partitions as described above. However, in this model, each partition has some percent of its nodes overlap with its neighboring partitions. As shown in Figure 3, the RBMs are

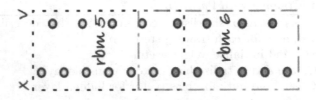

Fig. 3. RBM With Overlapping Partitions

sharing two hidden and two visible nodes. Since nodes are shared, partitioned RBMs cannot be trained concurrently without some kind of message passing. It should be noted that when trained sequentially, message passing between the partitions is not required.

4   Experimental Results

The MNIST dataset is used for our experiments due to its wider association with RBMs. MNIST has 60,000 training samples and 10,000 test samples of images. Each image is 28×28 pixels corresponding to handwritten digits from 0 to 9. Some sample images are presented in Figure 4. We measure performance of our method using reconstruction error, which is defined to be the average pixel differences between the original and reconstructed images (Equation 7). For each epoch, we use a batch size of 10 images from the training samples. Thus, for 6,000 epochs, 60,000 samples are used for training. Unless stated otherwise, we use CD-1 for all training steps. The unpartitioned RBM has 500 hidden nodes and $28 \times 28 = 784$ visible nodes. To have a fair comparison in terms of performance, rather using CPU time, we applied the following method. CD-1 training for one sample is carried out as follows:

- For all hidden nodes, find the probability of hidden node h_i as $\sigma(c_i + \sum_j W_{ij} x_j)$ and sample h_{i1} from a binomial distribution given h_i.
- For all visible nodes, find the probability of visible node x_j as $\sigma(b_j + \sum_i W_{ij} h_{i1})$ and sample x_{j1} from a binomial distribution given x_j.
- For all hidden nodes, find the probability of hidden node h_{i2} as $\sigma(c_i + \sum_j W_{ij} x_{j1})$.
- Calculate the gradient:
  - $W = W + \epsilon(h_{i1} x_j - h_{2i} x_{j1})$ where ϵ is the learning rate.
  - $b = b + \epsilon(x_j - x_{j1})$
  - $c = c + \epsilon(h_{i1} - h_{i2})$

where $\sigma(x) = \frac{1}{1+e^{-x}}$. Since operations at each step of CD involves *visible_nodes×hidden_nodes* updates, we estimate that the total number of Markov chain calculations is

$$ChainOperations = visible\ nodes \times hidden\ nodes \times samples$$

Fewer chain operations translate into less CPU time.

We used reconstruction error to compare our algorithms. For reconstruction error, we first obtain the binary representation of the original image and the reconstructed image. 30 is chosen as the threshold for converting pixel values [0-255] to binary 0 or 1. Thus, pixel values greater than or equal to 30 are set to 1 while values less than 30 are set to 0. Then, the reconstruction error is calculated as in Equation 7.

$$error = \frac{\sum_i (image(i) \neq reconstructedImage(i))}{total\ pixels} \tag{7}$$

Table 1 shows the results of our first experiment where the learning rate is set to 0.1 for all RBMs. *Single RBM* represents a fully connected RBM that is used as a baseline for comparison. The training sample for each RBM is equal to the number of epochs times the batch size (10). For instance, the Single RBM algorithm is trained on 60,000 images. Moreover, we use each image sample once. Unless stated otherwise, for following experiments, we ran training algorithms on samples for one iteration only—at the most, each sample is used only once. Each RBM-X represents a step with X partitions. Samples chosen for RBM-X are always from first N samples of total images. RBM-1 represents the final model. For these experiments, partitions are trained sequentially. Thus, if we train them concurrently, the total *ChainOperations* will be lower. As compared to Single RBM, RBM-1 has significantly lower reconstruction error. The total *ChainOperations* for partitioned RBMs is also less than Single RBM. In the table, using a t-test, significant results with 99% confidence are shown in bold. RBM-Partition after training on 20 partitions, significantly outperformed the Single RBM. Furthermore, the total number of chain operations for RBM-Partition is substantially less than for Single RBM.

Since we want fast convergence in the first step, in the following experiment we varied the learning rate to enable this. Results are shown in Table 2.

Table 1. Training Characteristics

Configuration	Number of RBMs	Epochs (batch=10)	Learning Rate	Reconstruction Error (%)	Chain Operations (10^9)
Single RBM	1	6000	0.1	3.85	23.52
RBM-28	28	6000	0.1	4.76	0.84
RBM-20	20	5000	0.1	4.03	0.98
RBM-15	15	4000	0.1	**3.19**	1.05
RBM-10	10	3000	0.1	**2.67**	1.18
RBM-5	5	2500	0.1	**2.29**	1.96
RBM-2	2	2000	0.1	**2.33**	3.92
RBM-1	1	2000	0.1	**2.36**	7.84
Total					**17.77**

Table 2. Training Characteristics wrt Learning Rate

Configuration	Number of RBMs	Epochs (batch=10)	Learning Rate	Reconstruction Error (%)	Chain Operations (10^9)
RBM-28	28	6000	0.3	**3.23**	0.84
RBM-20	20	5000	0.3	**2.93**	0.98
RBM-15	15	4000	0.3	**2.66**	1.05
RBM-10	10	3000	0.25	**2.30**	1.18
RBM-5	5	2500	0.20	**2.08**	1.96
RBM-2	2	2000	0.10	**2.10**	3.92
RBM-1	1	2000	0.10	**2.10**	7.84
Total					**17.77**

Table 3. Training Characteristics wrt Learning Rate

	$lr = 0.3$	$lr = 0.1$	$lr = 0.05$	$lr = 0.005$	$lr = 0.0005$
Single RBM	4.43	3.85	4.22	9.40	23.15
RBM-1	**3.45**	**2.10**	**1.95**	**1.83**	**1.92**

Table 4. Overlapping Partitions

Configuration	Number of RBMs	Epochs (batch=10)	Learning Rate	Reconstruction Error (%)	Chain Operations (10^9)
RBM-28	28	6000	0.3	**3.11**	0.90
RBM-20	20	5000	0.3	**2.76**	1.10
RBM-15	15	4000	0.3	**2.50**	1.18
RBM-10	10	3000	0.25	**2.19**	1.35
RBM-5	5	2500	0.20	**1.95**	2.27
RBM-2	2	2000	0.10	**1.92**	4.30
RBM-1	1	2000	0.10	**2.08**	7.84
Total					**18.94**

The RBM-Partition with 99% confidence outperforms the Single RBM in all steps including the first partition, RBM-28 (with 28 partitions). Moreover, reconstruction errors are even lower compared to our previous experiment.

Using the same configuration above, we varied the learning rate (denoted lr in the results) for the Single RBM and RBM-1. Learning rates for other RBM-X are fixed as in the configuration given in Table 2. Reconstruction errors for different learning rates are given in Table 3. Results demonstrate that RBM-Partition is less sensitive to different learning rates as compared to the Single RBM with 99% confidence.

We also wanted to determine if overlapping partitions would affect the results. We ran our experiment with 5% overlap, which means that each RBM shares 5% of its neighbor's nodes (5% from the left neighbor and 5% from the right neighbor). We ran overlapping partitions sequentially. As shown in Table 4, reconstruction errors are even lower with only a modest increase in overhead in terms of *ChainOperations*.

Comparing overlapping with non-overlapping RBM-Partition algorithms using the t-test, results show that the overlapping algorithm outperforms the non-overlapping algorithm with 99% confidence in almost every stage. However, in the last stage, the results were not significantly different, as shown in Table 5. We hypothesize that since overlapping partitions have more connections in each partition, they will require more training samples.

Table 5. Non-overlapping vs. Overlapping Partitions

Configuration	Number of RBMs	Epochs (batch=10)	Learning Rate	Overlapping Reconstruction Error(%)	NonOverlapping Reconstruction Error(%)	Overlapping Chain Operations (10^9)	NonOverlapping Chain Operations (10^9)
RBM-28	28	6000	0.3	3.11	3.23	0.90	0.84
RBM-20	20	5000	0.3	2.76	2.93	1.10	0.98
RBM-15	15	4000	0.3	2.50	2.66	1.18	1.05
RBM-10	10	3000	0.25	2.19	2.30	1.35	1.18
RBM-5	5	2500	0.20	1.95	2.08	2.27	1.96
RBM-2	2	2000	0.10	1.92	2.10	4.30	3.92
RBM-1	1	2000	0.10	2.08	2.10	7.84	7.84
Total						18.94	17.77

Table 6. 10-Fold Cross Validation Results

Configuration	Number of RBMs	Samples	Learning Rate	No Overlap: Average Reconstruction Error (%)	Overlap: Average Reconstruction Error(%)	Chain Operations per fold (10^9) No-Overlap/Overlap
Single RBM	1	60000	0.1	4.06		21.12
RBM-28	28	60000	0.3	3.32	3.32	0.76/0.81
RBM-20	20	50000	0.3	3.07	2.92	0.88/1.00
RBM-15	15	40000	0.3	2.68	2.62	0.94/1.06
RBM-10	10	30000	0.25	2.35	2.29	1.06/1.21
RBM-5	5	25000	0.20	2.12	2.09	1.76/2.04
RBM-2	2	20000	0.10	2.15	2.08	3.53/3.88
RBM-1	1	20000	0.10	2.18	2.14	7.06/7.06
Total						16.00/17.06

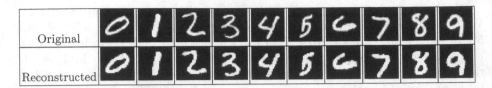

Original	
Reconstructed	

Fig. 4. Original vs. Reconstructed Images

Finally, 10-fold cross validation results are given in Table 6. Rather than using the provided training and test data sets. we pooled all of the data and split samples into 10 equal size subsamples. One subsample was used as the validation data for testing and the remaining 9 subsamples were used for training. We repeated this process 10 times. It should be noted that the numbers of samples for partitioned RBMs are not equal (Table 6) because we wanted to keep the total time complexity of RBM-Partition to be no worse than the Single RBM. RBM-Partition outperforms Single RBM with 99% confidence. Moreover, overlapping RBMs have lower average reconstruction error as compared to non-overlapping ones.

To visually compare the original images with the some of our reconstructed images, we present some examples in Figure 4.

Learning behavior with respect to the number of training samples is given in Figure 5. We compare RBM-10 with RBM Single. After each training cycle where we add 10,000 more images, we tested the algorithms on 10,000 images. RBM-10 outperforms RBM Single with 99% confidence on all training steps.

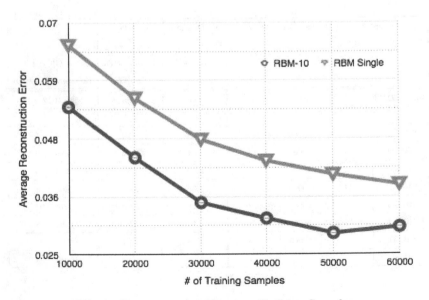

Fig. 5. Reconstruction Error vs. Training Samples

As we described at the begining of the Section 4, so far we ran these experiments for one iteration only. To see how our learning method will behave with additional iterations, we ran RBM-Partition and Single RBM for 15 iterations. Results are shown in Table 7. Starting with RBM-10, RBM-Partition significantly outperforms Single RBM with 99% confidence. For RBM-Partition, on average, the error is approximately 5 pixels out of 28 × 28 pixels, whereas it is 10 pixels for Single RBM.

The *Special Database 19* dataset from the National Institute of Standards and Technology (NIST) is the official training dataset for handprinted document and character recognition from 3600 writers, including 810K character images and 402K handwritten digits. Unlike the MNIST dataset, images are 128 by 128 pixels. We selected 62K images for training and testing. The dataset consists of 62 types of images for lowercase and uppercase letters, and numbers. Thus, in our dataset each type has 1,000 images. We used 10% for testing and 90% for training. 1-fold validation results are shown in Table 8. Based on the

Table 7. Training Iterations

Configuration	Number of RBMs	Samples	Learning Rate	Reconstruction Error (%)
Single RBM	1	60000	0.05	1.29
RBM-28	28	60000	0.3	1.75
RBM-20	20	30000	0.3	1.45
RBM-15	15	20000	0.3	1.35
RBM-10	10	20000	0.25	**1.08**
RBM-5	5	20000	0.2	**0.94**
RBM-2	2	20000	0.1	**0.75**
RBM-1	1	30000	0.05	**0.67**

Table 8. Training Characteristics with NIST dataset

Configuration	Number of RBMs	Training Samples	Learning Rate	Reconstruction Error (%)	Chain Operations (10^9)
Single RBM	1	62000	0.1	4.82	507.90
RBM-28	28	62000	0.3	**3.74**	19.92
RBM-20	20	50000	0.3	**3.65**	24.09
RBM-15	15	40000	0.3	**3.70**	25.19
RBM-10	10	30000	0.25	**3.69**	28.70
RBM-5	5	25000	0.20	**3.63**	47.78
RBM-2	2	20000	0.10	**3.67**	90.14
RBM-1	1	20000	0.10	**3.74**	163.8
Total					**399.62**

Table 9. Reconstruction Error per Character

Uppercase	Error	Lowercase	Error	Number	Error
A	3.93	a	3.55	0	2.95
B	**6.24**	b	4.14	1	*1.95*
C	2.8	c	2.68	2	3.55
D	5.22	d	4.49	3	4.01
E	3.34	e	2.79	4	3.67
F	3.61	f	4.14	5	4.07
G	5.68	g	5.15	6	3.35
H	4.25	h	3.35	7	3.2
I	*1.47*	i	*1.93*	8	4.49
J	4.47	j	3.2	9	3.97
K	4.89	k	4.15		
L	3.08	l	1.97		
M	4.26	m	4.69		
N	3.64	n	2.83		
O	2.58	o	2.69		
P	4.62	p	3.78		
Q	**6.62**	q	4.5		
R	3.53	r	2.37		
S	3.24	s	2.98		
T	3.08	t	3.48		
U	3.56	u	3.09		
V	3.18	v	2.87		
W	**6.92**	w	4.17		
X	4.38	x	3.19		
Y	3.75	y	3.37		
Z	5.04	z	3.57		

t-test results, RBM-Partition significantly outperforms Single RBM, again with substantially fewer chain operations.

Finally, RBM-Partitioned reconstruction error for each character is given in Table 9. The average reconstruction error is lowest for I, i, and 1 and it is highest for W,Q and B.

5   Conclusions and Future Work

We showed that our RBM-Partition training algorithm with small RBM partitions outperforms training full RBMs using CD-1. In addition to having superior results in terms of reconstruction error, RBM-Partition is also faster as compared to the single, full RBM. The reason that RBM-Partition is faster is due to having fewer connections in each training step. However, the reasons for the superior generative characteristics in terms of reconstruction error is not that obvious. We hypothesize that it is because in each training step, fewer nodes are involved and a small partition RBM settles in a low energy configuration more rapidly. As we move to other stages with less partitions, fewer training instances are needed to modify the energy configuration to obtain lower energy in the full network. Furthermore, in spatial data like an image, only neighboring nodes are involved in representing a feature. Therefore, a fully connected RBM is not optimal for training spatial datasets.

Our algorithm also has similarities to transfer learning. Since in each stage we learn some weights and those weights are used as a base configuration for the next stage, in way it corresponds to feature representation transfer [13]. One interesting direction of future work is to investigate whether other methods of transfer learning can be used during training or not.

Moreover, our approach opens the door to many potential applications. Since training is done on partitioned small RBMs, we believe the method will learn multi-model data, that is data from multiple sources, more accurately. Thus, other directions for future work include: 1) carrying out additional experiments to demonstrate that this training method can be applied to other domains with a high volume of features; 2) investigating if the training layers can be used in the form of a Deep Belief Network (i.e., the process will still require partitions as we described; however, instead of training each layer independently, a layer-wise training may produce more accurate results); and 3) investigating the discriminative power of the model by running it on classification tasks.

References

1. Ackley, D.H., Hinton, G.E., Sejnowski, T.J.: A learning algorithm for boltzmann machines. Cognitive science 9(1), 147–169 (1985)
2. Bengio, Y.: Learning deep architectures for ai. Foundations and trends in Machine Learning 2(1), 1–127 (2009)
3. Bengio, Y., Lamblin, P., Popovici, D., Larochelle, H., et al.: Greedy layer-wise training of deep networks. In: Advances in Neural Information Processing Systems, vol. 19, p. 153 (2007)
4. Brakel, P., Dieleman, S., Schrauwen, B.: Training restricted boltzmann machines with multi-tempering: Harnessing parallelization. In: Villa, A.E.P., Duch, W., Érdi, P., Masulli, F., Palm, G. (eds.) ICANN 2012, Part II. LNCS, vol. 7553, pp. 92–99. Springer, Heidelberg (2012)
5. Dahl, G.E., Adams, R.P., Larochelle, H.: Training restricted boltzmann machines on word observations. In: Proceedings of the 29th International Conference on Machine Learning, pp. 679–686. ACM (2012)
6. Dean, J., Corrado, G., Monga, R., Chen, K., Devin, M., Le, Q.V., Mao, M.Z., Ranzato, M., Senior, A.W., Tucker, P.A., et al.: Large scale distributed deep networks. In: Advances in Neural Information Processing Systems, pp. 1232–1240 (2012)
7. Hinton, G., Salakhutdinov, R.: Discovering binary codes for documents by learning deep generative models. Topics in Cognitive Science 3(1), 74–91 (2011)

8. Geoffrey, E.: Hinton. Training products of experts by minimizing contrastive divergence. Neural Computation 14(8), 1771–1800 (2002)
9. Hinton, G.E., Salakhutdinov, R.R.: Reducing the dimensionality of data with neural networks. Science 313(5786), 504–507 (2006)
10. Larochelle, H., Bengio, Y.: Classification using discriminative restricted boltzmann machines. In: Proceedings of the 25th International Conference on Machine Learning, pp. 536–543. ACM (2008)
11. Lemons, D.S.: A student's guide to entropy. Cambridge University Press (2013)
12. Louradour, J., Larochelle, H.: Classification of sets using restricted boltzmann machines, pp. 463–470. AUAI (2011)
13. Pan, S.J., Yang, Q.: A survey on transfer learning. IEEE Transactions on Knowledge and Data Engineering 22(10), 1345–1359 (2010)
14. Salakhutdinov, R., Mnih, A., Hinton, G.: Restricted boltzmann machines for collaborative filtering. In: Proceedings of the 24th International Conference on Machine Learning, pp. 791–798. ACM (2007)
15. Tieleman, T.: Training restricted boltzmann machines using approximations to the likelihood gradient. In: Proceedings of the 25th International Conference on Machine Learning, pp. 1064–1071. ACM (2008)

Integer Bayesian Network Classifiers

Sebastian Tschiatschek*, Karin Paul, and Franz Pernkopf

Signal Processing and Speech Communication Laboratory,
Graz University of Technology, Graz, Austria
tschiatschek@tugraz.at, karin.paul@student.tugraz.at, pernkopf@tugraz.at

Abstract. This paper introduces integer Bayesian network classifiers (BNCs), i.e. BNCs with discrete valued nodes where parameters are stored as integer numbers. These networks allow for efficient implementation in hardware while maintaining a (partial) probabilistic interpretation under scaling. An algorithm for the computation of margin maximizing integer parameters is presented and its efficiency is demonstrated. The resulting parameters have superior classification performance compared to parameters obtained by simple rounding of double-precision parameters, particularly for very low number of bits.

Keywords: Bayesian networks, Bayesian network classifiers, custom precision analysis, parameter learning.

1 Introduction

Bayesian networks (BNs) are probabilistic graphical models used to represent probability distributions. They are widely used for data modeling, e.g. in medicine, bioinformatics, image processing and pattern recognition. Their applications include inference and classification tasks, i.e. BNs are used to answer probabilistic queries on certain random variables.

Inference and classification with BNs are typically performed on computers with high numerical precision, i.e. using double-precision floating-point calculations. However, because of energy and computational constraints, low-power and integrated applications implemented on embedded systems require low complexity algorithms. Such applications are, for example, auditory scene classification in hearing aids and on-satellite computations[1]. In these kinds of applications, a trade-off between accuracy and algorithm complexity is essential.

In this paper, we argue that maximum-margin (MM) BNs, i.e. discriminatively optimized BNs, achieve a good trade-off in this respect and that careful algorithm design for the resource-constrained destination platform is advantageous.

---

* This work was supported by the Austrian Science Fund (project number P25244-N15).

[1] Computational capabilities on satellites are still severely limited due to power constraints and restricted availability of hardware satisfying the demanding requirements with respect to radiation tolerance.

T. Calders et al. (Eds.): ECML PKDD 2014, Part III, LNCS 8726, pp. 209–224, 2014.

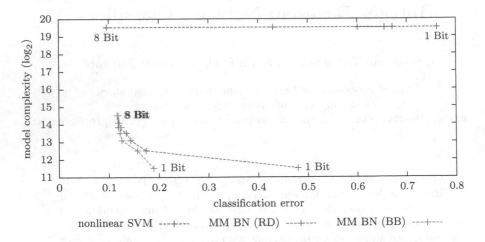

Fig. 1. Model complexities versus achieved classification errors. *Nonlinear SVM* refers to an SVM with radial-basis-function kernel, *MM BN (RD)* refers to an MM BN with parameters obtained by rounding, and *MM BN (BB)* refers to an MM BN with parameters obtained by the method proposed in this paper.

This is substantiated in Figure 1 for the *satimage* dataset from the UCI repository [5]. The model complexity in terms of bits required to store the classifier parameters versus the achieved classification error for SVMs with radial-basis-function kernels and for MM BNs is shown. In case of MM BNs, the performance of conventionally full-precision optimized and subsequently rounded parameters (RD) and that of parameters optimized for resource constraint environments using branch and bound (BB) techniques is presented — details on both parameter learning approaches are provided in the forthcoming sections. Note that the model complexity of the SVM is significantly higher than that of MM BNs, while classification performance is only slightly worse. Thus, if the application of interest allows to trade-off (slightly) reduced classification performance for tremendous savings in model complexity, MM BNs are obviously a good choice. If very low complexity models are desired, then MM BNs using BB are the best choice.

In this paper, we devise algorithms for efficiently learning such high performance low complexity models. While in [18], the authors already showed that parameters in Bayesian network classifiers (BNCs) can be mapped to the integer domain without considerable loss in classification rate (CR) performance, we take the analysis further: A principled approach for BNC parameter learning of margin-maximizing parameters over a discrete search space, i.e. maximum-margin (BB) parameters, is considered. This includes BNs with fixed-point parameters and (by proper scaling) integer parameters. In the sequel, also fixed-point parameters are termed as *integer parameters* because scaling enables the representation as integers. An algorithm for parameter optimization based on BB techniques is presented. For low bit-widths, the obtained parameters lead to significantly better performance than parameters obtained by rounding double-precision maximum-margin parameters.

Our main contributions can be summarized as follows:

- An efficient algorithm for computing margin maximizing integer parameters. The algorithm is based on the branch and bound algorithm and a set of greedy heuristics. This offers a gain in computation time and makes learning tractable.
- Experiments demonstrating that integer BNs with small bit-widths can be widely applied. We especially show that a very low number of integer bits is often sufficient to obtain classification performance close to full-precision MM BNCs and SVMs. This offers considerable advantages when implementing BNs on embedded systems, i.e. data storage and bandwidth requirements are minimized.
- A brief theoretical analysis of BNs with rounded parameters.

This paper is structured as follows: In Section 2 we summarize related work. Section 3 introduces our notation and provides background on BNs, BNCs and parameter learning. In Section 4, integer Bayesian network classifiers (iBNCs) are introduced and an efficient algorithm for learning margin maximizing integer parameters is provided. Experimental results are provided in Section 5. We conclude the paper in Section 6 and provide an outlook on future work.

2   Related Work

Literature on BNs with reduced precision parameters is scarce. Directly related work investigates the effect of parameter quantization in BNCs with focus on comparing the robustness of BNCs with generatively and discriminatively optimized parameters [18], and investigates bounds on the performance loss due to parameter quantization [17]. In the former work, the authors use bit-width reduced floating-point parameters, while they resort to fixed-point parameters in the latter.

Indirectly related work deals with (a) *sensitivity analysis* of Bayesian networks [3,4], stating essentially that classification using BNCs is insensitive to parameter deviations whenever either of these parameters are not close to zero or the class posteriors differ significantly, (b) credal networks, i.e. generalizations of Bayesian networks that associate a whole set of conditional probability densities (CPDs) with every node in the directed acyclic graph (DAG) [19], allowing for robust classification and supporting imprecisely specified CPDs.

In terms of undirected graphical networks, an interesting work on approximating undirected graphical models using integer parameters has been published recently [16]. The authors propose methods to perform inference *and* learning entirely in the integer domain. While undirected graphical models are more amenable to an integer approximation, there are domains where directed graphical models are more desirable and describe the probability distributions of interest more naturally, e.g. expert systems in the medical domain.

In terms of parameter learning using a BB scheme, there is related work for integer parameter learning of SVMs in the dual [1]. While some of the ideas presented by the authors are similar, classification with non-linear SVMs is computationally more demanding than classification using BNs[2]. Furthermore, when memory consumption is an issue, non-linear SVMs are disadvantageous because all support-vectors must be stored for classification.

3 Background

3.1 Notation and Bayesian Networks

We assume a set of random variables (RVs) $X_0, \ldots, X_L$. These RVs are related by a joint probability distribution $P^*(\mathbf{X})$, where $\mathbf{X} = (X_0, \ldots, X_L)$ is a random vector. BNs [12,9,15] are used to represent such joint probability distributions in a compact and intuitive way. A BN $\mathcal{B} = (\mathcal{G}, \mathcal{P_G})$ consists of a DAG $\mathcal{G} = (\mathbf{V}, \mathbf{E})$, where $\mathbf{V} = \{X_0, \ldots, X_L\}$ is the set of nodes and $\mathbf{E}$ the set of edges of the graph, and a set of local conditional probability distributions $\mathcal{P_G} = \{P(X_0|\mathbf{Pa}(X_0)), \ldots, P(X_L|\mathbf{Pa}(X_L))\}$. The terms $\mathbf{Pa}(X_0), \ldots, \mathbf{Pa}(X_L)$ denote the set of parents of $X_0, \ldots, X_L$ in $\mathcal{G}$, respectively. Assuming discrete valued nodes, we abbreviate the conditional probability $P(X_i = j|\mathbf{Pa}(X_i) = \mathbf{h})$ as $\theta^i_{j|\mathbf{h}}$ and the corresponding logarithmic probability as $w^i_{j|\mathbf{h}} = \log(\theta^i_{j|\mathbf{h}})$. Without loss of generality, we further assume that $X_i \in \{1, \ldots, |\mathrm{sp}(X_i)|\}$, where $\mathrm{sp}(X_i)$ is the set of possible values of RV X_i. Each node of the graph corresponds to an RV and the edges of the graph determine dependencies between these RVs. A BN induces a joint probability $P^{\mathcal{B}}(\mathbf{X})$ according to

$$P^{\mathcal{B}}(\mathbf{X}) = \prod_{i=0}^{L} P(X_i|\mathbf{Pa}(X_i)). \tag{1}$$

To represent $P^*(\mathbf{X})$ by the BN $\mathcal{B} = (\mathcal{G}, \mathcal{P_G})$, the graph $\mathcal{G}$ and the conditional probabilities in $\mathcal{P_G}$ must be selected such that $P^{\mathcal{B}}(\mathbf{X})$ *matches* $P^*(\mathbf{X})$. In typical settings, however, the joint distribution $P^*(\mathbf{X})$ and its properties are assumed to be unknown and only a limited number of samples drawn from this distribution, i.e. a training set $\mathcal{D}$, is available. This set $\mathcal{D}$ consists of N i.i.d. samples, i.e. $\mathcal{D} = \{\mathbf{x}^{(n)}|1 \leq n \leq N\}$, where $\mathbf{x}^{(n)}$ is the nth training sample and denotes an instantiation of $\mathbf{X}$. From this training set, the graph structure $\mathcal{G}$ of the BN as well as its parameters $\mathcal{P_G}$ have to be derived. Selecting the graph structure is known as *structure learning* and selecting $\mathcal{P_G}$ is known as *parameter learning*. The structures considered throughout this paper are fairly simple. In detail, we used naive Bayes (NB) and tree augmented network (TAN) structures [6]. Details on selecting the parameters are provided in Section 3.3.

---

[2] For classification with non-linear SVMs, the kernel must be evaluated for all support-vectors and a weighted summation must be performed. Classification using BNs with naive Bayes (NB) or tree augmented network (TAN) structures [6] corresponds to a simple summation of log-probabilities followed by an arg-max operation. Classification using linear SVMs is similar to classification using BNs.

3.2   Probabilistic Classification

In probabilistic classifiers, one RV in $X_0, \ldots, X_L$ takes the role of the class variable. Without loss of generality, we assume that X_0 corresponds to this class variable and denote it as C. The remaining RVs $X_1, \ldots, X_L$ represent the attributes/features of the classifier and are collected in the random vector $\widetilde{\mathbf{X}} = [X_1, \ldots, X_L]$. The aim is to induce *good* classifiers provided the training set, i.e. classifiers with high CR. Formally, a classifier h is a mapping

$$h: \mathrm{sp}(\widetilde{\mathbf{X}}) \to \mathrm{sp}(C), \tag{2}$$
$$\widetilde{\mathbf{x}} \mapsto h(\widetilde{\mathbf{x}}),$$

where $\widetilde{\mathbf{x}}$ denotes an instantiation of $\widetilde{\mathbf{X}}$, $\mathrm{sp}(\widetilde{\mathbf{X}})$ denotes the set of all assignments of $\widetilde{\mathbf{X}}$ and $\mathrm{sp}(C)$ is the set of classes. The CR of this classifier is

$$\mathrm{CR}(\mathrm{h}) := \mathbb{E}_{\mathrm{P}^*(C, \widetilde{\mathbf{X}})} \left[\mathbf{1}\{C = h(\widetilde{\mathbf{X}})\} \right], \tag{3}$$

where $\mathbf{1}\{A\}$ denotes the indicator function and $\mathbb{E}_{\mathrm{P}^*(C, \widetilde{\mathbf{X}})} [\cdot]$ is the expectation operator with respect to the distribution $\mathrm{P}^*(C, \widetilde{\mathbf{X}})$. Typically, the CR cannot be evaluated because $\mathrm{P}^*(C, \widetilde{\mathbf{X}})$ is unknown. It is rather estimated using cross-validation [2]. To determine BNCs, the training set $\mathcal{D}$ is assumed to consist of N i.i.d. labeled samples, i.e. $\mathcal{D} = \{(c^{(n)}, \widetilde{\mathbf{x}}^{(n)}) | 1 \leq n \leq N\}$, where $c^{(n)}$ denotes the instantiation of the RV C and $\widetilde{\mathbf{x}}^{(n)}$ the instantiation of $\widetilde{\mathbf{X}}$ in the nth training sample.

Any probability distribution, hence also any BN $\mathcal{B}$, induces a classifier $h_{\mathrm{P}^{\mathcal{B}}(C, \widetilde{\mathbf{X}})}$ according to

$$h_{\mathrm{P}^{\mathcal{B}}(C, \widetilde{\mathbf{X}})}: \mathrm{sp}(\widetilde{\mathbf{X}}) \to \mathrm{sp}(C), \tag{4}$$
$$\widetilde{\mathbf{x}} \mapsto \arg \max_{c \in C} \mathrm{P}^{\mathcal{B}}(C = c | \widetilde{\mathbf{X}} = \widetilde{\mathbf{x}}).$$

In this way, each instantiation $\widetilde{\mathbf{x}}$ of $\widetilde{\mathbf{X}}$ is classified as the maximum a-posteriori (MAP) estimate of C given $\widetilde{\mathbf{x}}$ under $\mathrm{P}^{\mathcal{B}}(C, \widetilde{\mathbf{X}})$.

3.3   Parameter Learning for Bayesian Networks

The parameters of a BN $\mathcal{B}$ can be optimized either generatively or discriminatively [14]. Discriminative parameter learning is suitable for classification tasks, while in generative parameter learning one aims at identifying parameters representing the generative process of the considered data. In this paper, we advocate a hybrid generative-discriminative parameter optimization according to [13]. The objective is the joint maximization of the data likelihood and the margin on the data. Formally, MM parameters $\mathcal{P}_\mathcal{G}^{\mathrm{MM}}$ are learned as

$$\mathcal{P}_\mathcal{G}^{\mathrm{MM}} = \arg\max_{\mathcal{P}_\mathcal{G}} \left[\sum_{n=1}^{N} \log \mathrm{P}^\mathcal{B}(\mathbf{x}^{(n)}) \right. \tag{5}$$

$$\left. + \lambda \sum_{n=1}^{N} \min\left(\gamma, \log \mathrm{P}^\mathcal{B}(\mathbf{x}^{(n)}) - \max_{c \neq c^{(n)}} \mathrm{P}^\mathcal{B}([c, \tilde{\mathbf{x}}^{(n)}])\right) \right],$$

where $\mathrm{P}^\mathcal{B}(\mathbf{X})$ is the joint distribution in (1) induced by the BN $(\mathcal{G}, \mathcal{P}_\mathcal{G})$, λ is a trade-off parameter between likelihood and margin, i.e. generative and discriminative optimization, and γ is the desired margin. The margin of sample n is defined as the difference in log-likelihood of the sample belonging to the correct class to belonging to the most likely alternative class, i.e. $\log \mathrm{P}^\mathcal{B}(\mathbf{x}^{(n)}) - \max_{c \neq c^{(n)}} \mathrm{P}^\mathcal{B}([c, \tilde{\mathbf{x}}^{(n)}])$. Consequently, a sample is classified correctly iff it has positive margin and incorrectly otherwise. Both parameters λ and γ are typically set using cross-validation.

4   Integer Bayesian Network Classifiers

In this section, we introduce iBNCs, i.e. BNCs with integer parameters. Further, we present an algorithm for determining margin maximizing parameters for iBNCs.

4.1   Definition

According to (1), the BN $\mathcal{B}$ assigns the probability

$$\mathrm{P}^\mathcal{B}(\mathbf{x}) = \prod_{i=0}^{L} \mathrm{P}(X_i = \mathbf{x}_{X_i} | \mathbf{Pa}(X_i) = \mathbf{x}_{\mathbf{Pa}(X_i)}), \tag{6}$$

to an instantiation $\mathbf{x}$ of $\mathbf{X}$, where $\mathbf{x}_{X_k}$ denotes the instantiation of X_k and $\mathbf{x}_{\mathbf{Pa}(X_k)}$ the instantiation of the parents of X_k according to $\mathbf{x}$, respectively. The above equation can be equivalently stated in the logarithmic domain, i.e.

$$\log \mathrm{P}^\mathcal{B}(\mathbf{x}) = \sum_{i=0}^{L} \log \mathrm{P}(X_i = \mathbf{x}_{X_i} | \mathbf{Pa}(X_i) = \mathbf{x}_{\mathbf{Pa}(X_i)}). \tag{7}$$

Hence, computing the likelihood of a sample $\mathbf{x}$ corresponds to a summation of log-probabilities. Assuming all log-probabilities are represented using B_I integer bits and B_F fractional bits, they can be written as

$$w_{j|\mathbf{h}}^i = \log \mathrm{P}(X_i = j | \mathbf{Pa}(X_i) = \mathbf{h}) = - \sum_{k=-B_F}^{B_I - 1} b_{j|\mathbf{h}}^{i,k} \cdot 2^k, \tag{8}$$

where $b_{j|\mathbf{h}}^{i,k} \in \{0,1\}$ denotes the kth bit of the binary representation of $w_{j|\mathbf{h}}^i$. Hence, ignoring the possibility of underflows, all $w_{j|\mathbf{h}}^i$ are in the set of negative fixed-point numbers $-\mathbb{B}_{B_F}^{B_I}$ with B_I integer bits and B_F fractional bits, i.e.

$$w_{j|\mathbf{h}}^i \in -\mathbb{B}_{B_F}^{B_I} = - \left\{ \sum_{k=-B_F}^{B_I-1} d_k \cdot 2^k : d_k \in \{0,1\} \right\}. \tag{9}$$

Introducing this in (7) and scaling by 2^{B_F} results in

$$2^{B_F} \log P^{\mathcal{B}}(\mathbf{x}) = - \sum_{i=0}^{L} \sum_{k=-B_F}^{B_I-1} b_{\mathbf{x}_{X_i}|\mathbf{x}_{\mathbf{Pa}(X_i)}}^{i,k} \cdot 2^{k+B_F}, \tag{10}$$

i.e. all summands are integer valued. The largest summand is at most $2^{B_I+B_F}-1$. The summation is over $L+1$ (scaled) log probabilities, i.e. the number of nodes in $\mathcal{B}$. Hence, in total at most

$$\log_2(L+1) + B_I + B_F \tag{11}$$

bits are required to calculate the joint probability. This transformation to the integer domain is advantageous in several aspects: (1) no floating-point rounding errors of any kind are introduced when working purely in the integer domain, (2) computations using integer arithmetic are typically faster and more efficient, (3) the need for a floating point processing unit is eliminated which encourages usage in many embedded systems, and (4) the integer parameters require less memory for storage.

When used for classification, we call BNs parametrized as above iBNCs. Note that iBNCs could also be formulated considering probabilities instead of log probabilities. Then, the sum-to-one constraint on the parameters can always be achieved. However, representing the log probabilities has the advantage that a large dynamic range is achieved and that classification essentially resorts to evaluating sums of log probabilities (more generally, max-sum message-passing can be easily performed).

4.2   Learning iBNCs

In principle, parameters for iBNCs can be determined by first learning BNC parameters using full-precision floating-point computations and subsequent rounding (and scaling) to the desired number format — a brief analysis of this approach is provided at the end of this section. However, such parameters are in general not optimal in the sense of the MM criterion (5) and we aim at a more principled approach.

Our approach is based on the branch and bound procedure [10], exploiting convexity of (5) under suitable parametrization. The implications will become clear immediately. Optimization of the MM criterion can be represented as

$$\underset{\mathbf{w}}{\text{maximize}} \quad \sum_{n=1}^{N} \boldsymbol{\phi}(\mathbf{x}^{(n)})^T \mathbf{w} + \lambda \sum_{n=1}^{N} \min\left(\gamma, \boldsymbol{\phi}(\mathbf{x}^{(n)})^T \mathbf{w} - \max_{c \neq c^{(n)}} \boldsymbol{\phi}([c, \tilde{\mathbf{x}}^{(n)}])^T \mathbf{w}\right)$$

$$(12)$$

$$\text{s.t.} \quad \sum_{j=1}^{|\text{sp}(X_i)|} \exp(w_{j|\mathbf{h}}^i) = 1 \qquad \forall i, \mathbf{h},$$

where we exploit that any log probability $\log \mathrm{P}(\mathbf{x})$ can be written as

$$\log \mathrm{P}(\mathbf{x}) = \sum_{i=0}^{L} \sum_{\mathbf{h} \in \text{sp}(\mathbf{Pa}(X_i))} \sum_{j \in \text{sp}(X_i)} w_{j|\mathbf{h}}^i \cdot \mathbf{1}(\mathbf{x}_{X_i} = j, \mathbf{x}_{\mathbf{Pa}(X_i)} = \mathbf{h}) \qquad (13)$$

$$= \boldsymbol{\phi}(\mathbf{x})^T \mathbf{w} \qquad (14)$$

by collecting for a given instantiation $\mathbf{x}$ the values of the indicator functions $\mathbf{1}(\mathbf{x}_{X_i} = j, \mathbf{x}_{\mathbf{Pa}(X_i)} = \mathbf{h})$ in vector $\boldsymbol{\phi}(\mathbf{x})$ and the corresponding $w_{j|\mathbf{h}}^i$ in vector $\mathbf{w}$. The above problem in (12) is nonconvex and hard to solve. However, when relaxing normalization constraints to

$$\sum_{j=1}^{|\text{sp}(X_i)|} \exp(w_{j|\mathbf{h}}^i) \leq 1, \qquad (15)$$

the problem becomes convex and can hence be solved efficiently. If all components of $\sum_{n=1}^{N} \boldsymbol{\phi}(\mathbf{x}^{(n)})$ are positive, e.g. when applying Laplace smoothing, then (15) is automatically satisfied with equality by any optimal solution of the relaxed problem, i.e. the original constraints are recovered [13].

For learning integer parameters, we restrict the parameters $\mathbf{w}$ to $-\mathbb{B}_{B_F}^{B_I}$ and further relax the normalization constraints to

$$\sum_{j=1}^{|\text{sp}(X_i)|} \exp(w_{j|\mathbf{h}}^i) \leq 1 + \xi(|\text{sp}(X_i)|, B_I, B_F), \qquad \forall i, \mathbf{h} \qquad (16)$$

where $\xi(|\text{sp}(X_i)|, B_I, B_F)$ is an additive constant depending on $|\text{sp}(X_i)|, B_I$ and B_F. This further relaxation of the normalization constraints is necessary, as in general reduced precision parameters do not correspond to correctly normalized parameters. The additive constant is required, as for very small bit-widths there are no parameters that are sub-normalized, i.e. $\sum_j \exp(M) > 1$, where $M = -2^{B_I} + 2^{-B_F}$ is the smallest value that can be represented. Therefore, without this additional constant, our optimization problem would be infeasible. In our experiments, we set

$$\xi(|\text{sp}(X_i)|, B_I, B_F) = \max\left(0, \frac{|\text{sp}(X_i)|}{2}[\exp(M) + \exp(M + 2^{-B_F})] - 1\right),$$

$$(17)$$

allowing at least half of the parameters of every conditional probability distribution $P(X_i|\mathbf{Pa}(X_i))$ to take on values larger than M. Note that $\xi(|\mathrm{sp}(X_i)|, B_I, B_F)$ quickly goes down to zero with increasing B_I. Thus, our final optimization problem is

$$
\begin{aligned}
\underset{\mathbf{w}}{\text{maximize}} \quad & \sum_{n=1}^{N} \phi(\mathbf{x}^{(n)})^T \mathbf{w} + \lambda \sum_{n=1}^{N} \min\left(\gamma, \phi(\mathbf{x}^{(n)})^T \mathbf{w} - \max_{c \neq c^{(n)}} \phi([c, \tilde{\mathbf{x}}^{(n)}])^T \mathbf{w}\right) \\
& \tag{18}
\end{aligned}
$$

$$
\text{s.t.} \quad \sum_{j=1}^{|\mathrm{sp}(X_i)|} \exp(w_{j|\mathbf{h}}^i) \leq 1 + \xi(|\mathrm{sp}(X_i)|, B_I, B_F) \qquad \forall i, \mathbf{h},
$$

$$
w_{j|\mathbf{h}}^i \in -\mathbb{B}_{B_F}^{B_I} \qquad \forall i, j, \mathbf{h}.
$$

For efficiently finding (global) minimizers of (18), we propose to use a BB algorithm [8] and greedy heuristics for creating candidate solutions and branching orders:

Branch and Bound Algorithm. The optimal iBNC parameters have to be searched in a discrete solution space, i.e. $w_{j|\mathbf{h}}^i \in -\mathbb{B}_{B_F}^{B_I}$. For optimization, the BB algorithm is used. BB searches the solution space by creating a tree of subproblems and dynamically adding (*branch*) and discarding (*bound*, also referred to as pruning) branches. The algorithm iteratively solves (18) using upper and lower bounds for $w_{j|\mathbf{h}}^i$ depending on the considered leaf of the search tree, i.e. the subproblem corresponding to the kth leaf is given as

$$
\underset{\mathbf{w}}{\text{maximize}} \quad \sum_{n=1}^{N} \phi(\mathbf{x}^{(n)})^T \mathbf{w} + \lambda \sum_{n=1}^{N} \min\left(\gamma, \phi(\mathbf{x}^{(n)})^T \mathbf{w} - \max_{c \neq c^{(n)}} \phi([c, \tilde{\mathbf{x}}^{(n)}])^T \mathbf{w}\right)
$$

$$
\text{s.t.} \quad \sum_{j=1}^{|\mathrm{sp}(X_i)|} \exp(w_{j|\mathbf{h}}^i) \leq 1 + \xi(|\mathrm{sp}(X_i)|, B_I, B_F) \qquad \forall i, \mathbf{h},
$$

$$
l_{j|\mathbf{h}}^{i,(k)} \leq w_{j|\mathbf{h}}^i \leq u_{j|\mathbf{h}}^{i,(k)} \qquad \forall i, j, \mathbf{h},
$$

where $l_{j|\mathbf{h}}^{i,(k)}, l_{j|\mathbf{h}}^{i,(k)} \in \mathbb{R}$ are the lower and upper bounds on the parameters, respectively. These subproblems are convex and can be exactly and efficiently solved. If the determined solution does not fit the required precision for all parameters, the algorithm performs one of the following options:

(a) *Bound.* If no global maximizer is to be found, the algorithm prunes the whole subtree (this happens if the best feasible solution found so far is better (has larger objective) than the optimal solution of the subproblem of the current leaf).

(b) *Branch.* Alternatively, the algorithm creates two new problems by adding new lower and upper bounds to one of the parameters (branching variable) which does not satisfy the desired precision. If multiple parameters do not satisfy the desired precision, i.e. different branching variables are possible, we use the *branching heuristic* described below to select the branching variable. Furthermore, to efficiently prune subtrees, it is important to generate good lower bounds for the objective at this stage, cf. the paragraph on *rounding heuristics* below.

Subproblems in the search tree are processed in order of their highest achievable objective value (the achievable objective value of subproblem k is upper bounded by the objective value of the relaxed problem of the parent of k according to the search tree). In this way, the subproblem of the search tree with highest upper-bound is processed next. The BB algorithm terminates either after a specified amount of time, returning the best solution found so far (anytime solution), or after there are no more open subproblems. In the latter case, the found solution is the global optimizer of (18).

Rounding Heuristic. To efficiently apply the BB algorithm, it is important to prune large parts of the search space at an early stage. Therefore, we need to obtain good lower bounds for the objective every time a problem corresponding to a leaf in the search tree has been solved. We try to achieve this using two simple rounding heuristics. If any of these heuristics yields a better feasible solution for (18) than the best solution found so far, the best solution is updated.

Let $\hat{\mathbf{w}}$ correspond to the intermediate solution. Then, the candidate solutions **a** and **b** are generated as follows:

- *Rounding*: Set

$$\hat{a}_{j|\mathbf{h}}^i = \max\left(M, \left[\frac{\hat{w}_{j|\mathbf{h}}^i}{q} \right]_R q \right), \tag{19}$$

where $[\cdot]_R$ denotes rounding to the closest integer, $q = 2^{-B_F}$ is the quantization interval and $M = -2^{B_I} + 2^{-B_F}$ the minimum value that can be represented. Set $\mathbf{a} = \Pi(\hat{\mathbf{a}})$, where Π is a projection-like operator ensuring that **b** is feasible for (18).
- *Gradient Guided Rounding*: Let **g** be the gradient of the objective at $\hat{\mathbf{w}}$. Then,

$$\hat{b}_{j|\mathbf{h}}^i = \begin{cases} \left\lceil \frac{\hat{w}_{j|\mathbf{h}}^i}{q} \right\rceil q & \text{if } g_{j|\mathbf{h}}^i > 0, \text{and} \\ \max\left(M, \left\lfloor \frac{\hat{w}_{j|\mathbf{h}}^i}{q} \right\rfloor q \right) & \text{if } g_{j|\mathbf{h}}^i \leq 0, \end{cases} \tag{20}$$

where $\lfloor \cdot \rfloor$ and $\lceil \cdot \rceil$ denote the floor and ceil function, respectively. Set $\mathbf{b} = \Pi(\hat{\mathbf{b}})$, where Π is as above.

Branching Heuristics. After solving one of the subproblems of the search tree, we check the obtained solution $\hat{\mathbf{w}}$ for optimality. If the solution is not optimal, we branch on the entry $\hat{w}^i_{j|\mathbf{h}}$ that has the largest deviation from the desired precision, i.e.

$$(i', j', \mathbf{h}') = \arg\max_{i,j,\mathbf{h}} \left| \hat{w}^i_{j|\mathbf{h}} - \left[\frac{\hat{w}^i_{j|\mathbf{h}}}{q} \right]_R q \right|. \tag{21}$$

4.3 Approximate iBNCs by Rounding

In this section, we provide a short analysis of the effect of rounding log parameters to their closest fixed-point representation. This reveals interesting insights into why classification performance of BNCs with rounded parameters is better than one might anticipate. Performing a similar analysis for iBNCs using BB is much more difficult because the objective for learning margin maximizing parameters does not decompose as a product of conditional probabilities.

We start by analyzing the Kullback-Leibler (KL)-divergence introduced by rounding, i.e. the KL-divergence between an *optimal* distribution, e.g. the original full-precision distribution, and its *approximation* obtained by rounding of the log-probabilities. Clearly, the approximate distribution is not necessarily properly normalized. Therefore, we compare the KL-divergence of the optimal distribution and the renormalized approximate distribution. This yields the following lemma:

Lemma 1 (KL-divergence). *Let $\mathbf{w}^i_{\cdot|\mathbf{h}}$ be a vector of normalized log probabilities (optimal distribution), i.e. $\sum_j \exp(w^i_{j|\mathbf{h}}) = 1$, and let $\widetilde{\mathbf{w}}^i_{\cdot|\mathbf{h}}$ (approximate distribution) be such that*

$$\widetilde{w}^i_{j|\mathbf{h}} = \left[\frac{w^i_{j|\mathbf{h}}}{q} \right]_R q, \tag{22}$$

where $q = 2^{-B_F}$ is the quantization interval. Then the KL-divergence between the optimal and the renormalized approximate distribution is bounded by q, i.e.

$$\mathcal{D}(\mathbf{w}^i_{\cdot|\mathbf{h}} \| \log \alpha + \widetilde{\mathbf{w}}^i_{\cdot|\mathbf{h}}) \le q, \tag{23}$$

where $\alpha = (\sum_j \exp(\widetilde{w}^i_{j|\mathbf{h}}))^{-1}$ ensures renormalization such that $\sum_j \exp(\log \alpha + \widetilde{w}^i_{j|\mathbf{h}}) = 1$.

Proof. We calculate

$$\mathcal{D}(\mathbf{w}^i_{\cdot|\mathbf{h}} \| \log \alpha + \widetilde{\mathbf{w}}^i_{\cdot|\mathbf{h}}) = \sum_j \exp(w^i_{j|\mathbf{h}}) \log \frac{\exp(w^i_{j|\mathbf{h}})}{\alpha \exp(\widetilde{w}^i_{j|\mathbf{h}})} \tag{24}$$

$$= \sum_j \exp(w^i_{j|\mathbf{h}}) \left[(w^i_{j|\mathbf{h}} - \widetilde{w}^i_{j|\mathbf{h}}) - \log \alpha \right] \tag{25}$$

$$\overset{(a)}{\le} \sum_j \exp(w^i_{j|\mathbf{h}}) \left[\frac{q}{2} - \log \alpha \right] \tag{26}$$

$$= \frac{q}{2} - \log \alpha, \tag{27}$$

where (a) is because $\widetilde{\mathbf{w}}^i_{\cdot|\mathbf{h}}$ is derived from $\mathbf{w}^i_{\cdot|\mathbf{h}}$ by rounding the parameters, i.e. $(w^i_{j|\mathbf{h}} - \widetilde{w}^i_{j|\mathbf{h}}) \leq \frac{q}{2}$. It remains to upper bound $-\log\alpha$. Straightforward calculation yields

$$-\log\alpha = \log\sum_j \exp(\widetilde{w}^i_{j|\mathbf{h}}) \leq \log\sum_j \exp(w^i_{j|\mathbf{h}} + \frac{q}{2}) = \frac{q}{2}. \qquad (28)$$

Hence, the statement follows. $\qquad\qquad\qquad\qquad\qquad\qquad\qquad\qquad\qquad\qquad$ □

This bound is tight. Assuming that sufficient integer bits are used so that no log-probabilities have to be truncated, $q = 2^{-B_F}$ (truncation must be performed if some $w^i_{j|\mathbf{h}} \leq -2^{B_I} + 2^{-B_F}$, i.e. $w^i_{j|\mathbf{h}}$ is smaller than the smallest value that can be represented using the chosen number format). Hence, the KL-divergence decays rapidly with increasing B_F.

When using only a finite number of bits for the integer part, log-probabilities may be truncated. Still, a bound on the KL-divergence can be derived:

Lemma 2 (KL-divergence). *Let $\mathbf{w}^i_{\cdot|\mathbf{h}}$ be a vector of normalized log probabilities (optimal distribution), and let $\widetilde{\mathbf{w}}^i_{\cdot|\mathbf{h}}$ (approximate distribution) be such that*

$$\widetilde{w}^i_{j|\mathbf{h}} = \max\left(M, \left[\frac{w^i_{j|\mathbf{h}}}{q}\right]_R q\right), \qquad (29)$$

where $q = 2^{-B_F}$ is the quantization interval and where M is the minimal representable log-probability. Then the KL-divergence between the optimal and the renormalized approximate distribution is bounded as

$$\mathcal{D}(\mathbf{w}^i_{\cdot|\mathbf{h}}\|\log\alpha + \widetilde{\mathbf{w}}^i_{\cdot|\mathbf{h}}) \leq \frac{3q}{2} + |\mathrm{sp}(X_i)|\exp(M), \qquad (30)$$

where $\alpha = (\sum_j \exp(\widetilde{w}^i_{j|\mathbf{h}}))^{-1}$ ensures renormalization.

Typically, $M = -2^{B_I} + 2^{-B_F}$. Hence, also in this case the bound decays with an increasing number of bits. One can further observe a dependency on the size of individual conditional probability tables (CPTs).

Both, Lemma 1 and 2, guarantee that simply rounding the log-probabilities of an optimal distribution does yield a good approximation in terms of KL-divergence. Therefore, it is not surprising that BNCs with parameters obtained by rounding achieve good performance. Furthermore, this justifies the rounding heuristics for obtaining good candidate solutions in the BB algorithm.

5   Experimental Results

In the following, we present classification experiments. In particular, we use BNCs with parameters determined as follows:

- *branch and bound* (BB): These integer parameters are obtained using the branch and bound algorithm presented in Section 4.2.

- *rounded* (RD): Integer parameters are obtained by rounding double-precision log parameters. If necessary, parameters are clipped to the considered number of integer bits.
- *double precision* (DP): Double precision parameters are obtained by solving (12) using methods proposed in [13].

5.1   Classification Experiments

We consider classification experiments for four real world datasets:

- **USPS** [7]. This dataset contains 11000 uniformly distributed handwritten digit images from zip codes of mail envelopes, of which 8000 are used for training and 3000 for testing. Each digit is represented as a 16×16 grayscale image, where each pixel is considered as feature.
- **MNIST** [11]. The MNIST dataset contains a training set of 60000 size-normalized and centered images of handwritten digits of size 16×16, accompanied by a test set of 10000 samples.
- **satimage/letter** [5]. From the UCI repository, we considered the *satimage* and the *letter* dataset. The satimage dataset consists of multi-spectral satellite images. Given a 3×3 multi-spectral pixel image patch, the task is to classify the central pixel as either red soil, cotton crop, grey soil, damp grey soil, soil with vegetation stubble, mixture class (all types present), or very damp grey soil. In total there are 6435 samples with 36 attributes. Performance is evaluated using 5-fold cross-valdiation. The letter dataset consists of 20000 samples, where two third of the data are used for training and one third for testing. Each sample is a character from the English alphabet and described by 16 numerical attributes, i.e. statistical moments and edge counts. The task is, based on these attributes, to classify each character as the represented English letter.

On these datasets, we compare the CR performance of BNCs (and iBNCs) with BB, RD, and DP parameters [3].

For RD parameters and a specific number of bits $B = B_I + B_F$, we determine the splitting into integer bits B_I and fractional bits B_F such that the classification rate on the training data is maximized. The same splitting is used for learning BB parameters with B bits. The hyper-parameters λ and γ in (18) are set using 5-fold cross-validation. For learning BB parameters with B bits, we allowed for up to five hours CPU time on a 3 GHz personal computer. If the parameter learning did not finish within this time, the best solution found so far was returned, cf. Section 4.2.

The observed CRs are shown in Figures 2, 3 and 4, for satimage, letter, USPS, and MNIST data using BNCs with NB structures, respectively. In case of USPS

---

[3] As mentioned in Section 4, up to $\log_2(L+1) + B_I + B_F$ bits are necessary for classification using BNCs with reduced precision parameters. In the presented experiments, we assume that these additional bits are available, i.e. summation does not cause overflows.

data, also CR performance for BNCs with TAN structures is shown. Only 5 to 6 integer bits for RD parameters are necessary to achieve CRs close to DP CRs. BNCs with BB parameters achieve better CRs than BNCs with RD parameters. Especially for low number of bits, BNCs with BB parameters are significantly better in terms of CR performance. This suggests that parameter learning under precision constraints is advantageous over full-precision parameter learning followed by subsequent rounding.

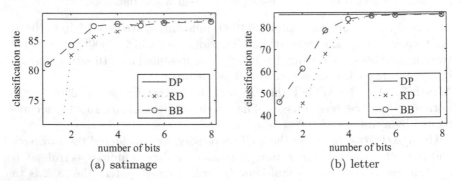

Fig. 2. CRs for *satimage* and *letter* data of BNCs with BB, RD and DP parameters for NB structures

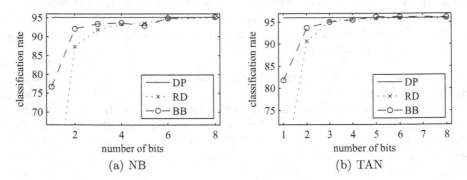

Fig. 3. CRs for USPS data of BNCs with BB, RD and DP parameters for NB and TAN structures

One important aspect of integer/reduced precision parameters, is their lower memory usage. This is exemplarily shown for USPS and MNIST data and NB and TAN structures in Table 1. The reduction in storage requirements by a factor of ~ 10 can positively influence the memory access when implementing iBNCs on embedded hardware.

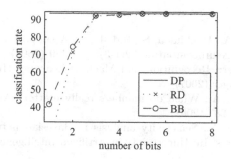

Fig. 4. CRs for MNIST of BNCs using NB structure with DP, RD and BB parameters

Table 1. Memory usage for parameter storage of DP, BB and RD parameters in reduced precision

Dataset	Structure	# Parameters	# bits	Storage [kB]	
				DP	BB/RD
USPS	NB	8650	6	67.6	6.3
	TAN	32970	6	257.6	24.1
MNIST	NB	25800	3	201.6	9.4

6   Conclusions and Future Work

In this paper we considered BNs with discrete valued nodes and parameters represented by integer numbers. We presented an efficient algorithm for computing margin maximizing integer parameters, where subproblems are convex and can be solved efficiently.

In experiments, we showed that a low number of bits is sufficient to achieve good performance in classification scenarios. Furthermore, we showed that parameter learning under precision constraints is advantageous over full-precision parameter learning followed by subsequent rounding to the desired precision. The presented results support to understand the implications of implementing BNCs on embedded hardware and can greatly reduce the storage requirements and thus the time required for memory access.

Future work aims at a sample implementation of iBNCs on embedded hardware for speed comparison. Furthermore, we want to derive methods for parameter learning using integer/reduced precision computations only. Another interesting direction for future work is to incorporate reduced precision constraints into the task of structure learning, e.g. learning BN structures such that rounding of parameters degrades classification performance as little as possible.

References

1. Anguita, D., Ghio, A., Pischiutta, S., Ridella, S.: A support vector machine with integer parameters. Neurocomputing 72(1-3), 480–489 (2008)
2. Bishop, C.M.: Pattern Recognition and Machine Learning (Information Science and Statistics). Springer (2007)
3. Chan, H., Darwiche, A.: When do numbers really matter? Artificial Intelligence Research 17(1), 265–287 (2002)
4. Chan, H., Darwiche, A.: Sensitivity analysis in Bayesian networks: From single to multiple parameters. In: Uncertainty in Artificial Intelligence (UAI), pp. 67–75 (2004)
5. Frank, A., Asuncion, A.: UCI machine learning repository. University of California, Irvine, School of Information and Computer Sciences (2010), http://archive.ics.uci.edu/ml
6. Friedman, N., Geiger, D., Goldszmidt, M.: Bayesian network classifiers. In: Machine Learning, pp. 131–163 (1997)
7. Hastie, T., Tibshirani, R., Friedman, J.: The Elements of Statistical Learning: Data Mining, Inference, and Prediction. Springer (August 2003)
8. Horst, R., Tuy, H.: Global Optimization: Deterministic Approaches. Springer, Heidelberg (1996)
9. Koller, D., Friedman, N.: Probabilistic Graphical Models: Principles and Techniques. MIT Press (2009)
10. Land, A.H., Doig, A.G.: An automatic method of solving discrete programming problems. Econometrica 28(3), 497–520 (1960)
11. LeCun, Y., Bottou, L., Bengio, Y., Haffner, P.: Gradient-based learning applied to document recognition. Proceedings of the IEEE 86(11), 2278–2324 (1998)
12. Pearl, J.: Probabilistic Reasoning in Intelligent Systems: Networks of Plausible Inference. Morgan Kaufmann Publishers Inc., San Francisco (1988)
13. Peharz, R., Tschiatschek, S., Pernkopf, F.: The most generative maximum margin Bayesian networks. In: Proceedings of the 30th International Conference on Machine Learning (ICML), vol. 28, pp. 235–243 (2013)
14. Pernkopf, F., Wohlmayr, M., Tschiatschek, S.: Maximum margin Bayesian network classifiers. IEEE Transactions on Pattern Analysis and Machine Intelligence (TPAMI) 34(3), 521–531 (2012)
15. Pernkopf, F., Peharz, R., Tschiatschek, S.: Introduction to probabilistic graphical models. In: Academic Press Library in Signal Processing, vol. 1, ch. 18, pp. 989–1064 (2014)
16. Piatkowski, N., Sangkyun, L., Morik, K.: The integer approximation of undirected graphical models. In: 3rd International Conference on Pattern Recognition Applications and Methods, ICPRAM (2014)
17. Tschiatschek, S., Cancino Chacòn, C.E., Pernkopf, F.: Bounds for Bayesian network classifiers with reduced precision parameters. In: International Conference on Acoustics, Speech, and Signal Processing (ICASSP), pp. 3357–3361 (2013)
18. Tschiatschek, S., Reinprecht, P., Mücke, M., Pernkopf, F.: Bayesian network classifiers with reduced precision parameters. In: Flach, P.A., De Bie, T., Cristianini, N. (eds.) ECML PKDD 2012, Part I. LNCS, vol. 7523, pp. 74–89. Springer, Heidelberg (2012)
19. Zaffalon, M.: A credal approach to naive classification. In: ISIPTA (1999)

Multi-target Regression via
Random Linear Target Combinations

Grigorios Tsoumakas, Eleftherios Spyromitros-Xioufis,
Aikaterini Vrekou, and Ioannis Vlahavas

Department of Informatics, Aristotle University of Thessaloniki, Thessaloniki, Greece
{greg,espyromi,agvrekou,vlahavas}@csd.auth.gr

Abstract. Multi-target regression is concerned with the simultaneous prediction of multiple continuous target variables based on the same set of input variables. It arises in several interesting industrial and environmental application domains, such as ecological modelling and energy forecasting. This paper presents an ensemble method for multi-target regression that constructs new target variables via random linear combinations of existing targets. We discuss the connection of our approach with multi-label classification algorithms, in particular RAkEL, which originally inspired this work, and a family of recent multi-label classification algorithms that involve output coding. Experimental results on 12 multi-target datasets show that it performs significantly better than a strong baseline that learns a single model for each target using gradient boosting and compares favourably to multi-objective random forest approach, which is a state-of-the-art approach. The experiments further show that our approach improves more when stronger unconditional dependencies exist among the targets.

Keywords: multi-target regression, multi-output regression, multivariate regression, multi-label classification, output coding, random linear combinations.

1 Introduction

Multi-target regression, also known as multivariate or multi-output regression, aims at simultaneously predicting multiple continuous target variables based on the same set of input variables. Such a learning task arises in several interesting application domains, such as predicting the wind noise of vehicle components [1], ecological modelling [2], water quality monitoring [3], forest monitoring [4] and more recently energy-related forecasting[1], such as wind and solar energy production forecasting and load/price forecasting.

Multi-target regression can be considered as a sibling of multi-label classification [5,6], the latter dealing with multiple *binary* target variables, instead of *continuous* ones. Recent work [7] stressed the close connection among these two tasks and argued that ideas from the more popular and developed area of

---

[1] http://www.gefcom.org

T. Calders et al. (Eds.): ECML PKDD 2014, Part III, LNCS 8726, pp. 225–240, 2014.
© Springer-Verlag Berlin Heidelberg 2014

multi-label learning could potentially be transferred to multi-target regression. Following up this argument, we present here a multi-target regression algorithm that was conceived as analogous to the RAkEL [8] multi-label classication algorithm. In particular, the proposed method creates new target variables by considering random linear combinations of k original target variables. Experiments on 12 multi-target datasets show that our approach is significantly better than a strong baseline that learns a single model for each target using gradient boosting [9] and compares favourably to the state-of-the-art multi-objective random forest approach [10]. The experiments further show that our approach improves more when stronger unconditional dependencies exist among the targets.

The rest of this paper is organized as follows. Section 2 discusses related work on multi-target regression, as well as on output coding, a family of multi-label learning algorithm of similar nature to our approach, which is presented in Section 3. Section 4 presents the setup of our empirical study (methods and their parameters, implementation details, evaluation process, datasets) and Section 5 discusses our experimental results. Finally, section 6 summarizes the conclusions of this work and points to future work directions.

2   Related Work

2.1   Multi-target Regression

Multivariate regression was studied many years ago by statisticians and two of the earliest methods were reduced-rank regression [11] and C&W [12]. A large number of methods for multi-target regression are derived from the predictive clustering tree (PCT) framework [13]. These are presented in more detail in subsequent paragraphs. An approach for learning multi-target model trees was proposed in [14]. One can also find methods that deal with multi-target regression problems in the literature of the related topics of transfer learning [15] and multi-task learning [16]. Undoubtedly, the simplest approach to multi-target regression is to independently construct one regression model for each of the target variables.

The main difference between the PCT algorithm and a standard decision tree is that the variance and the prototype functions are treated as parameters that can be instantiated to fit the given learning task. Such an instantiation for multi-target prediction tasks are the multi-objective decision trees (MODTs), where the variance function is computed as the sum of the variances of the targets, and the prototype function is the vector mean of the target vectors of the training examples falling in each leaf [13,17]. Bagging and random forest ensembles of MODTs were developed in [10] and found significantly more accurate than MODTs and equally good or better than ensembles of single-objective decision trees for both regression and classification tasks. In particular, multi-objective random forest (MORF) yielded better performance than multi-objective bagging.

Motivated by the interpretability of rule learning algorithms, other researchers developed multi-target rule learning algorithms that again fall in the PCT framework. Focusing on multi-label classification problems, [18] proposed the predictive clustering rules (PCR) method that extends the PCT framework by combining a rule learning algorithm with a search heuristic that derives from clustering. PCR yielded comparable accuracy to using multiple single-target rule learners using a much smaller and interpretable collection of rules. Later, the FIRE rule ensemble algorithm [19] was proposed, specifically designed for multi-target regression. FIRE works by first transforming an ensemble of decision trees into a collection of rules and then using an optimization procedure that assigns proper weights to individual rules in order to prune the initial rule set without compromising its accuracy. The connection of this method to the PCT framework lies in the fact that the ensemble of trees comes from the MORF method of [10]. Recently, [20] presented FIRE++, an improved version of FIRE, which among other optimizations, offers the ability to combine rules with simple linear functions. FIRE++ was found better than FIRE, but slightly worse than the less interpretable MORF.

2.2 Output Coding

Linear combinations of targets have been recently used by a number of output coding approaches [21,22,23,24] for the related task of multi-label classification [5,6]. The motivation of the methods in [21] and [24] was the reduction of large output spaces for improving computational complexity, which goes towards the opposite direction of our approach. The methods in [22] and [23] on the other hand, aimed at improving the prediction accuracy similarly to our approach.

The approach most similar to ours is the chronologically first one [21], which is based on the technique of compressed sensing and considers random linear combinations of the labels. This is also the only output coding method from the ones mentioned here, where the dimensionality of the new output space is allowed to be larger than the original output space, as in our case. Besides the opposite motivation (compression of output space) compared to our approach, [21] starts from the concept of output sparsity (sparsity of the output conditioned on the input), while in multi-target data, the output space is generally non-sparse. The encoding step of [21] is therefore based on compression matrices that satisfy a restricted isometry property, based on a sparsity level defined by the user and the decoding step is based on sparse approximation algorithms. In contrast, our approach uses uniform non-zero random weights for a user-defined number of targets in the encoding step, and standard unregularized least squares in the decoding step.

3   Random Linear Target Combinations

Consider a set of p input variables $\mathbf{x} \in R^p$ and a set of q target variables $\mathbf{y} \in \mathcal{R}^q$. We have a set of m training examples: $\mathbf{D} = (\mathbf{X}, \mathbf{Y}) = \{(\mathbf{x}^{(i)}, \mathbf{y}^{(i)})\}_{i=1}^m$, where $\mathbf{X}$ and $\mathbf{Y}$ are matrices of size $m \times p$ and $m \times q$, respectively.

Our approach constructs $r >> q$ new target variables via corresponding random linear combinations of $\mathbf{y}$. To achieve this, we define a coefficient matrix $\mathbf{C}$

of size $q \times r$ filled with random values uniformly chosen from $[0..1]$. Each column of this matrix contains the coefficients of a linear combination of the target variables. Multiplying $\mathbf{Y}$ with $\mathbf{C}$ leads to a transformed multi-target training set $\mathbf{D}' = (\mathbf{X}, \mathbf{Z})$, where $\mathbf{Z} = \mathbf{YC}$ is a matrix of size $m \times r$ with the values of the new target variables. A user-specified multi-target regression learning algorithm is then applied to $\mathbf{D}'$ in order to build a corresponding model.

Note that our approach expects that the original target variables take values from the same domain, as otherwise their linear combinations could be dominated by the values of targets with a much wider domain than the others. To ensure this, it applies 0-1 normalization in order to bring the values of all targets into the range $[0..1]$.

We consider an additional parameter $k \in \{2, \ldots, q\}$ for specifying the number of original target variables involved in each random linear combination, by setting the coefficients for the rest of the target variables to zero. Higher k means that potential correlations among more targets are being considered. However, at the same time, it means that the new targets are more difficult to predict, especially in the absence of actual correlations among the targets. We therefore hypothesize that low k values will lead to the best results. In practice, when $k < q$, for each linear combination our approach selects k targets at random, but with priority to targets with the lowest frequency of participation to previously considered linear combinations. This ensures that all targets will participate in $\mathbf{C}$ as equivallently (i.e. with similar frequency) as possible.

Given a new test instance, $\mathbf{x}'$, the multi-target regression model is first invoked to obtain a vector $\mathbf{z}'$ with r predictions. The estimates $\hat{\mathbf{y}}'$ for the original target variables are then obtained by solving for $\hat{\mathbf{y}}'$ the following overdetermined (as $r >> q$) system of linear equations: $\mathbf{C}^{\top}\hat{\mathbf{y}}' = \mathbf{z}'$.

As an example of our approach, consider a multi-target training set with $q = 6$ targets and $m = 10$ training examples. Figure 1(a) shows the normalized targets, $\mathbf{Y}$ of such a dataset, based on the first 10 training examples of the atp1d dataset (see Section 4.4 for a description of this dataset). Figure 1(b) shows a potential coefficient matrix $\mathbf{C}$ for $r = 8$ and $k = 2$. Finally, Figure 1(c) shows the values of the new targets $\mathbf{Z}$.

(a) A sample **Y**

1	1,00	0,99	0,76	0,82	0,99	0,71
2	1,00	0,99	0,76	0,82	0,99	0,71
3	1,00	0,99	0,76	0,82	0,99	0,71
4	0,23	0,37	0,51	0,43	0,14	0,38
5	0,23	0,37	0,51	0,43	0,14	0,38
6	0,23	0,37	0,51	0,43	0,14	0,38
7	0,57	0,84	0,61	0,62	0,52	0,55
8	0,21	0,84	0,61	0,62	0,11	0,55
9	0,21	0,84	0,61	0,62	0,11	0,55
10	0,20	0,84	0,61	0,62	0,10	0,55

(b) A sample **C** for k=2

0	0,80	0	0	0,69	0	0	0,49
0	0	0,96	0	0	0,41	0	0
0,94	0	0	0,06	0	0	0,50	0
0	0	0,36	0	0	0,27	0	0
0,53	0	0	0,16	0	0	0,36	0
0	0,88	0	0	0,20	0	0	0,25

(c) Corresponding **Z**

1	1,24	1,42	1,25	0,20	0,84	0,63	0,74	0,67
2	1,24	1,42	1,25	0,20	0,84	0,63	0,74	0,67
3	1,24	1,42	1,25	0,20	0,84	0,63	0,74	0,67
4	0,55	0,52	0,51	0,05	0,24	0,27	0,30	0,21
5	0,55	0,52	0,51	0,05	0,24	0,27	0,30	0,21
6	0,55	0,52	0,51	0,05	0,24	0,27	0,30	0,21
7	0,85	0,94	1,04	0,12	0,51	0,52	0,49	0,42
8	0,63	0,65	1,04	0,05	0,25	0,52	0,34	0,24
9	0,63	0,65	1,04	0,05	0,25	0,52	0,34	0,24
10	0,63	0,64	1,04	0,05	0,25	0,52	0,34	0,24

Fig. 1. An example of our approach. The $q = 6$ targets of a multi-target regression dataset with $m = 10$ examples is shown in (a). A coefficient matrix for $k = 2$ and $r = 8$ is shown in (b). The values of the new targets is shown in (c).

Our approach was inspired from recent work on drawing parallels between multi-label classification and multi-target regression [7] and conceived as the twin of the multi-label classification algorithm RA*k*EL [8] for multi-target regression tasks. Similarly to RA*k*EL, our approach aims to exploit correlations among target variables on one hand and to achieve the error-correction effect of ensemble methods on the other hand, as it implicitly pools multiple estimates for each original target variable (one for each linear combination that it participates in). We therefore expect that the larger r is, the better the estimate of the original target variables. Our approach follows the *randomness injection* paradigm of ensemble construction [25] at a larger degree than RA*k*EL, as it may combine the same target variables twice, but with different random coefficients. Randomness is a key component for improving supervised learning methods [26,27].

After inventing our approach, we realized that linear target combination approaches have been used for multi-label data in the past. From this viewpoint, our approach could also be considered as a sibling of multi-label compressed sensing [21], if we set aside the different goal and the technical differences among the two approaches discussed in Section 2.2.

4   Experimental Setup

This section offers details on the setup of the experiments that we conducted. We first present the participating methods and their parameters, then provide implementation details, followed by a description of the evaluation measure and process that was followed. We conclude this section by presenting the datasets that were used, their main statistics, as well as statistics of the pairwise correlations among their target variables.

4.1   Methods and Parameters

Our approach (dubbed RLC) is parameterized by the number of new target variables, r, the number of original target variables to combine, k, the multi-target regression algorithm that is used to learn from the transformed multi-target training set $\mathbf{D}'$ and the approach used to solve the overdetermined system of linear equations during prediction. The first two we discuss together with the results in Section 5. The multi-target regression algorithm we employ is to learn a single independent regression model for each target (dubbed ST). Each regression model is built using gradient boosting [9] with a 4-terminal node regression tree as the base learner, a learning rate of 0.1 and 100 boosting iterations. The system of linear equations is solved by the unregularized least squares approach.

The multi-target regression algorithm employed by our approach, ST with gradient boosting, is also directly used on the original target variables as a baseline. We further compare our approach against the state-of-the-art multi-objective random forest algorithm [10] (dubbed MORF). We used an ensemble size of 100 trees and the values suggested in [10] for the rest of the parameters.

4.2 Implementation

The proposed method was implemented within the open-source multi-label learning Java library Mulan[2] [28], which has been recently expanded to handle multi-target prediction tasks and includes an implementation of ST too, as well as a wrapper of the CLUS software[3], including support for MORF. Mulan is built on top of Weka[4] [29], which includes an implementation of gradient boosting. Therefore, the comparative evaluation of all methods was achieved using a single Java-based software framework.

In support of open science, Mulan includes a package called *experiments*, which contains experimental setups of various algorithms based on the corresponding papers. To ease replication of the experimental results of this paper, we have included a class called *ExperimentRLC* in that package.

4.3 Evaluation

We use the average Relative Root Mean Squared Error (aRRMSE) as evaluation measure. The RRMSE for a target is equal to the Root Mean Squared Error (RMSE) for that target divided by the RMSE of predicting the average value of that target in the training set. This standardization facilitates performance averaging across non-homogeneous targets.

The aRRMSE of a multi-target model h that has been induced from a train set $\mathbf{D_{train}}$ is estimated based on a test set $\mathbf{D_{test}}$ according to the following equation:

$$aRRMSE(h, \mathbf{D_{test}}) = \frac{1}{q} \sum_{j=1}^{q} RRMSE = \frac{1}{q} \sum_{j=1}^{q} \sqrt{\frac{\sum_{(\mathbf{x},\mathbf{y}) \in \mathbf{D_{test}}} (h(\mathbf{x})_j - y_j)^2}{\sum_{(\mathbf{x},\mathbf{y}) \in \mathbf{D_{test}}} (\bar{y}_j - y_j)^2}}$$

where $\bar{y}_j$ is the mean value of target variable y_j within $\mathbf{D_{train}}$ and $h(\mathbf{x})_j$ is the output of h for target variable y_j.

The aRRMSE measure is estimated using the hold-out approach for large datasets, while 10-fold cross-validation is employed for small datasets.

4.4 Datasets

Our experiments are based on 12 datasets[5]. Table 1 reports the name (1st column), abbreviation (2nd column) and source (3rd column) of these datasets, the number of instances of the train and test sets or the total number of instances if cross-validation was used (4th column), the number, p, of input variables (5th column) and the number, q, of output variables (6th column).

One of the motivations of our approach is the exploitation of potential dependencies among the targets. We hypothesize that our approach will do better in

---

[2] http://mulan.sourceforge.net
[3] http://dtai.cs.kuleuven.be/clus/
[4] http://www.cs.waikato.ac.nz/ml/weka
[5] http://users.auth.gr/espyromi/datasets.html

Table 1. Name, abbreviation, source, number of train and test examples or total number of examples in the case of cross-validation, number of input variables and number of output variables per dataset used in our empirical study

Name	Abbreviation	Source	Examples	p	q
Airline Ticket Price 1	atp1d	[7]	337	411	6
Airline Ticket Price 2	atp7d	[7]	296	411	6
Electrical Discharge Machining	edm	[30]	154	16	2
Occupational Employment Survey 1	oes1997	[7]	334	263	16
Occupational Employment Survey 2	oes2010	[7]	403	298	16
River Flow 1	rf1	[7]	4165/5065	64	8
River Flow 2	rf2	[7]	4165/5065	576	8
Solar Flare 1	sf1969	[31]	323	26	3
Solar Flare 2	sf1978	[31]	1066	27	3
Supply Chain Management 1	scm1d	[7]	8145/1658	280	16
Supply Chain Management 2	scm20d	[7]	7463/1503	61	16
Water Quality	wq	[3]	1060	16	14

datasets where target dependencies exist. To facilitate the discussion of results in this context, Figure 2 shows box-plots summarizing the distribution of the correlations among all pairs of targets for all datasets, while Figure 3 shows a heat-map of the pairwise target correlations for a sample dataset with a relatively large number of targets (*scm20d*). The rest of this section provides a short description for each of the datasets.

Airline Ticket Price. The *airline ticket price* dataset [7] was constructed for the prediction of airline ticket prices for a specific departure date. There are two versions of this datasets. The target attributes are the next day price (atp1d) or the minimum price within the next 7 days (atp7d) for 6 characteristics: any airline with any number of stops, any airline non-stop only, Delta Airlines, Continental Airlines, Airtran Airlines and United Airlines. The input attributes are the number of days between the observation and departure date, 7 binary attributes that refer to the day-of-the-week of the observation date and the complete enumeration of: 1) the minimum price, mean price and number of quotes from, 2) all airlines and from each airline quoting more than 50% of the observation days, 3) for non-stop, one-stop and two-stop flights, 4) for the current day, previous day and two days before. There are 411 input attributes in total.

Electrical Discharge Machining. The *electrical discharge machining* dataset [30] represents a two-target regression problem. The task is to shorten the machining time by reproducing the behavior of a human operator which controls the values of two variables. Each of the target variables takes 3 distinct numeric values (1,0,1) and there are 16 continuous input variables.

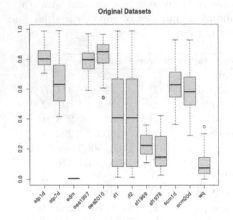

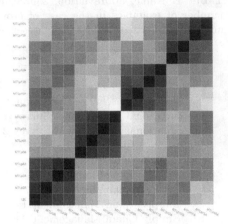

Fig. 2. Box-plots summarizing the distribution of all pairwise target correlations for all datasets

Fig. 3. Heat-map of the pairwise target correlations for the *scm20d* dataset

Occupational Employment Survey. The *occupational employment survey* dataset [7] was obtained from the annual occupational employment survey that is performed by the US Bureau of Labor Statistics. Every instance contains the aproximate number of full-time equivalent employees of different employment positions for a specific city. There are two versions of this datasets, one with data for 334 cities in the year 1997 (oes1997) and one with data for 403 cities in the year 2010 (oes2010). The employment types that were present in at least 50% of the cities were considered as variables. From these, the targets are 16 randomly selected variables, while the rest constitute the input variables.

River Flow. The *river flow* dataset [7] was constructed for the prediction of the flow in a river network at 8 specific sites, 48 hours in the future. Those sites are located in the Mississippi River in the USA. There are two versions of this dataset. River Flow 1 (rf1) contains 64 input variables that refer to the most recent observations of the 8 sites and the observations from 6, 12, 18, 24, 36, 48 and 60 hours in the past. River Flow 2 (rf2) contains additional input variables that refer to precipitation forecasts for 6 hour windows up to 48 hours in the future for each gauge site. The target attributes are 8, each one corresponding to each of the 8 sites. The data were collected from September 2011 to September 2012.

Solar Flare. The *solar flare* dataset [31] has 3 target variables that correspond to the number of times 3 types of solar flare (common, moderate, severe) are observed within 24 hours. There are two versions of this dataset. Solar Flare 1 (sf1969) contains data from year 1969 and Solar Flare 2 (sf1978) from year 1978.

Water Quality. The *water quality* dataset [3] has 14 target attributes that refer to the relative representation of plant and animal species in Slovenian rivers and 16 input attributes that refer to physical and chemical water quality parameters.

Supply Chain Management. The *supply chain management* dataset [7] is obtained from the Trading Agent Competition in Supply Chain Management (TAC SCM) tournament from 2010. The precise methods for data preprocessing and normalization are described in detail in [32]. Some benchmark values for prediction accuracy in this domain are available from the TAC SCM Prediction Challenge [33]. These data sets correspond only to the *Product Future* prediction type. The input attributes contain the observed prices for a specific day in the tournament for each game. Moreover, 4 time-delayed observations for each observed product and component (1, 2, 4 and 8 days delayed). The target attributes are 16 and refer to the next day mean price (scm1d dataset) or the mean price within the next 20 days (scm20d dataset).

5 Results

5.1 Investigation of Parameters

We first investigate the behaviour of our method with respect to its two main parameters: the number of models, r, which we vary from q to 500 and the number of targets that are being combined, k, which we vary from 2 to q.

Figure 4 shows the aRRMSE of our method (y-axis) at the *atp1d* dataset with respect to r (x-axis) for $k \in \{2, 3, 4, 5, 6\}$. We notice that the curves have logarithmic shape, steeply decreasing with approximately the first 50 models and converging after approximately 250 models. The addition of models has the typical error-correction behaviour exhibited by ensemble methods, in accordance with our expectations. We further notice, again as we expected, that low numbers of k (2 and 3) lead to the best results.

The behaviour of our approach with respect to r is similar in all datasets. Figure 5 shows the average aRRMSE of our method (y-axis) with respect to r (x-axis) across all datasets and all k values. Averages of performance estimates across datasets are not appropriate for summarizing and comparing the accuracy of different methods [34] and averages across different values of a parameter may hide salient effects of this parameter. However, we believe that this average serves well our purpose of summarizing a large number of results in a concise way in order to highlight the general behaviour of our method, which is consistent across all datasets and k values. The number of participating models starts from 16, to ensure that the displayed average values are based on all datasets (recall that the minimum number of models in our approach is q and that the maximum number of labels across our datasets is 16). We again see that the error follows the shape of a logarithmic curve, steeply decreasing with the first approximately 75 models and converging after approximately 280 models.

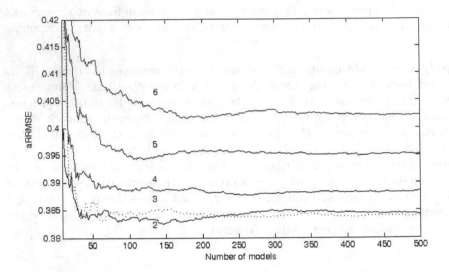

Fig. 4. aRRMSE of our method (y-axis) for $k \in \{2, 3, 4, 5, 6\}$ with respect to the number of participating regression models (x-axis) at the *atp1d* dataset. The line corresponding to $k = 3$ is dotted instead of solid, so as to contrast it with the overlapping line of $k = 2$.

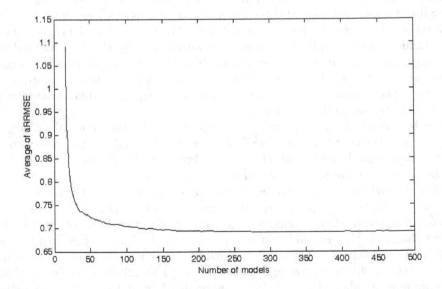

Fig. 5. Average aRRMSE of our method (y-axis) with respect to r (x-axis) across all datasets and all different k values

Table 2. aRRMSE of our method in each dataset for $r = 500$ and all possible k values. The best result of our approach in each dataset is underlined. The last two rows show the aRRMSE of ST and MORF.

k	atp1d	atp7d	edm	sf1969	sf1978	oes10	oes97	rf1	rf2	scm1d	scm20d	wq
2	0.3842	0.4614	0.6996	1.2312	1.5746	0.5026	0.5593	0.7265	0.7036	0.4572	0.7469	0.9100
3	0.3840	0.4653		1.2172	1.5675	0.5084	0.5588	0.7878	0.7584	0.4610	0.7467	0.9080
4	0.3884	0.4796				0.5232	0.5730	0.8204	0.7922	0.4663	0.7472	0.9085
5	0.3952	0.4917				0.5359	0.5837	0.8584	0.8327	0.4699	0.7477	0.9086
6	0.4022	0.5029				0.5472	0.5889	0.8515	0.8257	0.4775	0.7490	0.9089
7						0.5551	0.5958	0.8446	0.8106	0.4820	0.7513	0.9090
8						0.5734	0.6076	0.8868	0.8655	0.4855	0.7536	0.9107
9						0.5911	0.6153			0.4889	0.7548	0.9122
10						0.6031	0.6229			0.4932	0.7537	0.9128
11						0.6154	0.6348			0.4978	0.7573	0.9150
12						0.6285	0.6449			0.5020	0.7571	0.9163
13						0.6354	0.6590			0.5057	0.7619	0.9188
14						0.6428	0.6682			0.5133	0.7640	0.9217
15						0.6525	0.6860			0.5155	0.7681	
16						0.6652	0.6916			0.5218	0.7704	
ST	0.3980	0.4735	0.7316	1.2777	1.6158	0.5421	0.5727	0.7171	0.6897	0.4625	0.7571	0.9200
MORF	0.4223	0.5508	0.7338	1.2620	1.4020	0.4528	0.5490	0.8488	0.9189	0.5635	0.7775	0.8994

The performance of our approach with respect to k is similar in all datasets too. The first 16 rows of Table 2 shows the aRRMSE of our method for 500 models. We notice that the best results of our approach, which are underlined in the table, are obtained for $k \in \{2, 3\}$, while the error is in most cases monotonically increasing with higher values of k.

5.2   Comparative Evaluation

The last two rows of Table 2 shows the aRRMSE of the ST strong baseline and the MORF state-of-the-art approach. To compare our approach with ST and MORF, we follow the recommendations of [34]. We first discuss the number of datasets where each of the methods is better than each of the others based on Table 3. We see that RLC with $r = 500$ is better than ST in 10/12 datasets and better than MORF in 8/12 datasets, both for $k = 2$ and for $k = 3$. The strength of the baseline is demonstrated by the fact that it is better than MORF in 7/12 datasets.

Table 3. Number of datasets where a method is better than another method (wins:losses) for each pair of methods

	RLC	ST	MORF
RLC	-	10:2	8:4
ST	2:10	-	7:5
MORF	4:8	5:7	-

The mean rank of RLC with $r = 500$ and $k = 2$ or $k = 3$ (same k for all datasets), ST and MORF are 1.5, 2.25 and 2.25 respectively. The variation of the Friedman test described in [34] to compare the three algorithms rejects the null hypothesis for a p-value of 0.0828 (i.e. requires $a = 0.1$). Proceeding to a post-hoc Nemenyi test with $a = 0.1$, the critical difference is 0.8377, slightly more than the 0.75 difference among the mean rank of RLC and that of ST and MORF. So, these differences should not be considered statistically significant based on this test.

We also applied the Wilcoxon signed-ranks test between RLC with $r = 500$ and $k = 2$ and the other two algorithms. While multiple tests are involved in this process, these are limited to just 2, and therefore a small bias will be introduced if any due to this multiple testing process. For the comparison with ST the p-value is 0.0210 suggesting that the differences are statistically significant for $a = 0.05$, while for the comparison with MORF the p-value is 0.1763 suggesting that the differences are statistically insignificant even for $a = 0.1$.

One could argue that a fairer comparison between RLC and MORF should have setup MORF to use 500 trees instead of 100. The answer to such critique is that each target is involved in rk/q regression models in RLC and thus in datasets such as *oes*, *scm* and *wq*, RLC is actually at disadvantage. Three of the wins of MORF over RLC actually occur in the *oes* and *wq* datasets. Perhaps a fairer experiment would set $r = 100q/k$, assuming 100 trees in MORF. Selecting the number of models in RLC and MORF via cross-validation would perhaps be even fairer. Such experiments will be considered in future work.

Summarizing the comparative results, we argue that the proposed approach is worthy of being considered by a practitioner for a multi-target regression domain, as there is a high chance that it could give the best results compared to state-of-the-art methods. Futhermore, being algorithm independent, it has the flexibility and potential of doing better in a specific application, by being instantiated with a different base learner whose hypothesis representation is more suited to the given problem (e.g. a support vector regression algorithm), in contrast to MORF (and other variants of the predictive clustering trees framework), whose representation is fixed to trees.

5.3   Error with Respect to Average Pairwise Target Correlation

No clear conclusion can be drawn on whether the intensity of pairwise correlations affects the improvement that our approach can give over the baseline. The correlation among the median of the absolute value of pairwise target correlations and the gain in performance over ST is 0.15.

Noticing that the high variance of pairwise correlations in the river-flow datasets co-occurs with the failure of our approach to improve upon ST, we also calculated the correlation between the standard deviation of the pairwise target correlations and the gain in performance over ST, which is -0.68 (*edm* was excluded in this computation as it only has two targets). This apparently suggests that low variance of absolute value of pairwise target correlations leads to improved gains. However, we do not have a theory to explain this correlation.

Pairwise target correlations do not take the input features into account, so they do not measure potential conditional dependencies among targets given the inputs [35]. We do however notice that in the three pairs of datasets with similar nature and amount of features (the two versions of *atp*, *oes* and *sf* datasets), higher median of absolute value of pairwise target correlations does lead to improved performance. We simplistically assume here that similar nature and amount of features introduce similar conditional dependencies of the targets given the features, even though the aforementioned pairs of datasets have different, yet of similar nature, targets.

Table 4 presents the data, upon which the discussion of this subsection is based. In specific, the 1st row shows the percentage of improvement of our approach compared to ST, while the next two rows show the median and standard deviation respectively of the absolute value of pairwise target correlations.

Table 4. For each dataset, the 1st row shows the percentage of accuracy gain of our method compared to ST, and the next two rows show the median and standard deviation respectively of the absolute value of pairwise target correlations

	atp1d	atp7d	edm	sf1969	sf1978	oes10	oes97	rf1	rf2	scm1d	scm20d	wq
gain (%)	3.6	2.6	4.6	5.0	3.1	7.9	2.5	-1.3	-2.0	1.6	1.4	1.3
median	0.8013	0.6306	0.0051	0.2242	0.1484	0.8479	0.7952	0.4077	0.4077	0.6526	0.5785	0.0751
stdev	0.0788	0.1602	-	1.1247	1.2006	0.0972	0.0785	0.3125	0.3125	0.1316	0.1483	0.0717

To the best of our knowledge, a discussion of accuracy with respect to target dependencies has not been attempted in past multi-target regression work. We believe such an analysis is quite interesting both theoretically and practically and might be good on one hand to be adopted by future work in this area, and on another hand to be studied more elaborately by itself.

6 Conclusions and Future Work

Multi-target regression is a learning task with interesting practical applications. We expect its popularity to rise in the near future with the proliferation of multiple sensors in our everyday life (Internet of Things) recording multiple values that we might want to predict simultaneously.

Motivated from the practical interest of multi-target regression and recent work on drawing parallels between multi-label classification and multi-target regression, we developed an ensemble method that constructs new target variables by forming random linear combinations of existing targets, as a twin of the RA*k*EL multi-label classification algorithm. At the same time, we highlighted an additional connection of the proposed approach with recent multi-label classification algorithms based on output coding.

The proposed approach was found significantly better than a strong baseline that learns a single model per target using gradient boosting and compares favourably against the state-of-the-art ensemble method MORF, based on experiments on 12 multi-target regression datasets. Furthermore, the empirical study reveals a relation among the pairwise correlation of targets and the gains of the proposed approach given similar input features, suggesting succesful exploitation of existing unconditional target dependencies by the proposed approach.

The proposed approach has the potential to be further improved in the future. Towards that direction, we intend to investigate alternative randomization injection processes (e.g. normal instead of uniform coefficients) and constructing ensembles of our approach using different coefficient matrices. For example, instead of constructing 500 models with one matrix, we could construct 100 models with 5 different matrices, which is expected to improve diversity and potentially accuracy of our idea.

Acknowledgements. This work has been partially supported by the Greek General Secretariat for Research and Technology, via act *Supporting Groups of Small and Medium-Sized Enterprises for Resarch and Technological Development Activities*, project 22SMEs2010, *Intelligent System in Supply Chain Monitoring and Optimization*.

References

1. Kužnar, D., Možina, M., Bratko, I.: Curve prediction with kernel regression. In: Proceedings of the 1st Workshop on Learning from Multi-Label Data, pp. 61–68 (2009)
2. Kocev, D., Džeroski, S., White, M.D., Newell, G.R., Griffioen, P.: Using single- and multi-target regression trees and ensembles to model a compound index of vegetation condition. Ecological Modelling 220(8), 1159–1168 (2009)
3. Dzeroski, S., Demsar, D., Grbovic, J.: Predicting chemical parameters of river water quality from bioindicator data. Appl. Intell. 13(1), 7–17 (2000)
4. Dzeroski, S., Kobler, A., Gjorgjioski, V., Panov, P.: Using decision trees to predict forest stand height and canopy cover from landsat and lidar data. In: Proc. 20th Int. Conf. on Informatics for Environmental Protection - Managing Environmental Knowledge - ENVIROINFO (2006)
5. Tsoumakas, G., Katakis, I., Vlahavas, I.: Mining multi-label data. In: Maimon, O., Rokach, L. (eds.) Data Mining and Knowledge Discovery Handbook, 2nd edn., pp. 667–685. Springer, Heidelberg (2010)
6. Zhang, M.L., Zhou, Z.H.: A review on multi-label learning algorithms. IEEE Transactions on Knowledge and Data Engineering 99 (PrePrints), 1 (2013)
7. Spyromitros-Xioufis, E., Tsoumakas, G., Groves, W., Vlahavas, I.: Multi-label classification methods for multi-target regression. arXiv preprint arXiv:1211.6581 [cs.LG] (2014)
8. Tsoumakas, G., Katakis, I., Vlahavas, I.: Random k-labelsets for multi-label classification. IEEE Transactions on Knowledge and Data Engineering 23, 1079–1089 (2011)
9. Friedman, J.H.: Greedy function approximation: A gradient boosting machine. The Annals of Statistics 29(5), 1189–1232 (2001)

10. Kocev, D., Vens, C., Struyf, J., Džeroski, S.: Ensembles of multi-objective decision trees. In: Kok, J.N., Koronacki, J., Lopez de Mantaras, R., Matwin, S., Mladenič, D., Skowron, A. (eds.) ECML 2007. LNCS (LNAI), vol. 4701, pp. 624–631. Springer, Heidelberg (2007)
11. Izenman, A.J.: Reduced-rank regression for the multivariate linear model. Journal of Multivariate Analysis 5(2), 248–264 (1975)
12. Breiman, L., Friedman, J.H.: Predicting multivariate responses in multiple linear regression. Journal of the Royal Statistical Society: Series B (Statistical Methodology) 59(1), 3–54 (1997)
13. Blockeel, H., Raedt, L.D., Ramong, J.: Top-down induction of clustering trees. In: Proceedings of the 15th International Conference on Machine Learning, pp. 55–63. Morgan Kaufmann (1998)
14. Appice, A., Džeroski, S.: Stepwise induction of multi-target model trees. In: Kok, J.N., Koronacki, J., Lopez de Mantaras, R., Matwin, S., Mladenič, D., Skowron, A. (eds.) ECML 2007. LNCS (LNAI), vol. 4701, pp. 502–509. Springer, Heidelberg (2007)
15. Piccart, B., Struyf, J., Blockeel, H.: Empirical asymmetric selective transfer in multi-objective decision trees. In: Boulicaut, J.-F., Berthold, M.R., Horváth, T. (eds.) DS 2008. LNCS (LNAI), vol. 5255, pp. 64–75. Springer, Heidelberg (2008)
16. Jalali, A., Ravikumar, P., Sanghavi, S., Ruan, C.: A dirty model for multi-task learning. In: Proc. of the Conference on Advances in Neural Information Processing Systems (NIPS), pp. 964–972 (2010)
17. Blockeel, H., Džeroski, S., Grbović, J.: Simultaneous prediction of multiple chemical parameters of river water quality with TILDE. In: Żytkow, J.M., Rauch, J. (eds.) PKDD 1999. LNCS (LNAI), vol. 1704, pp. 32–40. Springer, Heidelberg (1999)
18. Ženko, B., Džeroski, S.: Learning classification rules for multiple target attributes. In: Washio, T., Suzuki, E., Ting, K.M., Inokuchi, A. (eds.) PAKDD 2008. LNCS (LNAI), vol. 5012, pp. 454–465. Springer, Heidelberg (2008)
19. Aho, T., Ženko, B., Džeroski, S.: Rule ensembles for multi-target regression. In: Proc. of the 9th IEEE International Conference on Data Mining, pp. 21–30. IEEE Computer Society (2009)
20. Aho, T., Ženko, B., Džeroski, S., Elomaa, T.: Multi-target regression with rule ensembles. Journal of Machine Learning Research 1, 1–48 (2012)
21. Hsu, D., Kakade, S., Langford, J., Zhang, T.: Multi-label prediction via compressed sensing. In: NIPS, pp. 772–780. Curran Associates, Inc. (2009)
22. Zhang, Y., Schneider, J.G.: Multi-label output codes using canonical correlation analysis. In: AISTATS 2011 (2011)
23. Zhang, Y., Schneider, J.G.: Maximum margin output coding. In: ICML. icml.cc / Omnipress (2012)
24. Tai, F., Lin, H.T.: Multilabel classification with principal label space transformation. Neural Comput. 24(9), 2508–2542 (2012)
25. Dietterich, T.G.: Ensemble Methods in Machine Learning. In: Kittler, J., Roli, F. (eds.) MCS 2000. LNCS, vol. 1857, pp. 1–15. Springer, Heidelberg (2000)
26. Breiman, L.: Random forests. Machine Learning 45(1), 5–32 (2001)
27. Hinton, G.E., Srivastava, N., Krizhevsky, A., Sutskever, I., Salakhutdinov, R.: Improving neural networks by preventing co-adaptation of feature detectors. CoRR abs/1207.0580 (2012)
28. Tsoumakas, G., Spyromitros-Xioufis, E., Vilcek, J., Vlahavas, I.: Mulan: A java library for multi-label learning. Journal of Machine Learning Research (JMLR) 12, 2411–2414 (2011)

29. Hall, M., Frank, E., Holmes, G., Pfahringer, B., Reutemann, P., Witten, I.H.: The weka data mining software: An update. SIGKDD Explorations 11 (2009)
30. Karalič, A., Bratko, I.: First order regression. Mach. Learn. 26(2-3), 147–176 (1997)
31. Asuncion, A., Newman, D.: UCI machine learning repository (2007)
32. Groves, W., Gini, M.: Improving prediction in TAC SCM by integrating multivariate and temporal aspects via PLS regression. In: David, E., Robu, V., Shehory, O., Stein, S., Symeonidis, A.L. (eds.) AMEC/TADA. LNBIP, vol. 119, pp. 28–43. Springer, Heidelberg (1981)
33. Pardoe, D., Stone, P.: The 2007 TAC SCM prediction challenge. In: Ketter, W., La Poutré, H., Sadeh, N., Shehory, O., Walsh, W. (eds.) AMEC 2008. LNBIP, vol. 44, pp. 175–189. Springer, Heidelberg (2010)
34. Demsar, J.: Statistical comparisons of classifiers over multiple data sets. Journal of Machine Learning Research 7, 1–30 (2006)
35. Dembczynski, K., Waegeman, W., Cheng, W., Hüllermeier, E.: On label dependence in multi-label classification. In: International Conference on Machine Learning (ICML)-2nd International Workshop on Learning from Multi-Label Data (MLD 2010), pp. 5–12 (2010)

Ratio-Based Multiple Kernel Clustering

Grigorios Tzortzis and Aristidis Likas

Department of Computer Science & Engineering, University of Ioannina,
GR 45110, Ioannina, Greece
{gtzortzi,arly}@cs.uoi.gr

Abstract. Maximum margin clustering (MMC) approaches extend the large margin principle of SVM to unsupervised learning with considerable success. In this work, we utilize the ratio between the margin and the intra-cluster variance, to explicitly consider both the separation and the compactness of the clusters in the objective. Moreover, we employ multiple kernel learning (MKL) to jointly learn the kernel and a partitioning of the instances, thus overcoming the kernel selection problem of MMC. Importantly, the margin alone cannot reliably reflect the quality of the learned kernel, as it can be enlarged by a simple scaling of the kernel. In contrast, our ratio-based objective is scale invariant and also invariant to the type of norm constraints on the kernel parameters. Optimization of the objective is performed using an iterative gradient-based algorithm. Comparative clustering experiments on various datasets demonstrate the effectiveness of the proposed formulation.

Keywords: maximum margin clustering, unsupervised multiple kernel learning, kernel k-means.

1 Introduction

The success of large margin techniques in supervised learning, particularly that of support vector machines (SVM), has generated great interest in extending such techniques to the unsupervised setting, leading to the, so called, maximum margin clustering (MMC) problem [21]. Given a dataset $\mathcal{X} = \{\mathbf{x}_i\}_{i=1}^{N}$, $\mathbf{x}_i \in \Re^d$, MMC approaches attempt to find a labeling (clustering) $\mathbf{y} = [y_1, \ldots, y_N]^\top$, $y_i \in \{\pm 1\}$, of the instances, such that a subsequent training of a standard SVM would result in a margin that is maximal over all possible labellings. MMC is formulated as:

$$\min_{\mathbf{y}} \min_{\mathbf{w}, b, \boldsymbol{\xi}} \frac{1}{2} \|\mathbf{w}\|^2 + C \sum_{i=1}^{N} \xi_i, \tag{1}$$

$$s.t. \quad -\ell \le \sum_{i=1}^{N} y_i \le \ell, \ \mathbf{y} \in \{\pm 1\}^N, \ y_i \left(\mathbf{w}^\top \phi(\mathbf{x}_i) + b \right) \ge 1 - \xi_i, \ \xi_i \ge 0,$$

where $\mathbf{w}$, b are the coefficients of the SVM hyperplane ($\|\mathbf{w}\|$ is the reciprocal of the margin), $\boldsymbol{\xi}$ the slack variables capturing the misclassification error and

T. Calders et al. (Eds.): ECML PKDD 2014, Part III, LNCS 8726, pp. 241–257, 2014.

$C > 0$ the regularizer. Instances are implicitly mapped through transformation ϕ to a higher dimensional feature space using the kernel trick ($\mathcal{K}(\mathbf{x}_i, \mathbf{x}_j) = \phi(\mathbf{x}_i)^\top \phi(\mathbf{x}_j)$). Moreover, to prevent the trivially "optimal" solution of assigning all instances to the same cluster and thus obtaining an infinite margin ($\|\mathbf{w}\| = 0$), a cluster balance constraint ($-\ell \leq \sum_{i=1}^N y_i \leq \ell$) was introduced by Xu et al. [21], where $\ell \geq 0$ is a constant controlling the imbalance of the clusters. The MMC problem is non-convex with integer parameters $\mathbf{y}$, making the optimization much trickier than that of (convex) supervised SVM. To solve (1), some approaches employ semidefinite programing (SDP) [18, 21, 22], others exploit the cutting plane method [20, 25] and others rely on alternating between the outer and the inner minimization [24].

It is well-known that the performance of kernel-based approaches, like MMC, heavily depends on the choice of the kernel. However, it is often unclear which is the best kernel for a particular task. Multiple kernel learning (MKL) [9], which has been mainly studied under the SVM paradigm, attempts to simultaneously locate the hyperplane with the largest margin and also learn a suitable kernel. The kernel, $\widetilde{\mathcal{K}}(\mathbf{x}_i, \mathbf{x}_j) = \widetilde{\phi}(\mathbf{x}_i)^\top \widetilde{\phi}(\mathbf{x}_j)$, is usually parametrized by a vector $\boldsymbol{\theta} = [\theta_1, \ldots, \theta_V]^\top$ of parameters. Most existing MKL approaches focus on supervised learning and in principle derive from the following optimization (subject to some slight modifications) (e.g. [11, 12, 14, 23]):

$$\min_{\boldsymbol{\theta}, \mathbf{w}, b, \boldsymbol{\xi}} \frac{1}{2}\|\mathbf{w}\|^2 + C\sum_{i=1}^N \xi_i, \qquad (2)$$

$$s.t. \ \theta_v \geq 0, \ \|\boldsymbol{\theta}\|_p^p \leq 1, \ y_i\left(\mathbf{w}^\top \widetilde{\phi}(\mathbf{x}_i) + b\right) \geq 1 - \xi_i, \ \xi_i \geq 0.$$

Kernel parameters θ_v are limited to nonnegative values to ensure the learned kernel is positive semidefinite and the p-norm constraint is employed to avoid overfitting. Usually the kernel is parametrized as a linear combination of some given basis kernels and either the 1-norm that promotes sparsity [14, 16, 26], or a more general p-norm, $p \geq 1$, [11, 12, 23], is chosen. There also exist a few studies that consider nonlinear combinations of basis kernels [3, 8], or even general types of parametric kernels [7, 19]. The optimization problem in (2) is non-convex due to $\boldsymbol{\theta}$. Depending on the form of the kernel parametrization and the choice of p-norm, various optimization strategies have been proposed, several of which alternate between updating $\boldsymbol{\theta}$ and solving a standard SVM to obtain $\mathbf{w}$, b and $\boldsymbol{\xi}$. For example, semi-infinite linear programming [11, 16, 26], gradient-based methods [7, 8, 14, 19] and closed-form methods [12, 23].

Extending MKL to the clustering domain, and in particular to MMC problems, is an interesting research direction, however, existing work is rather limited. The methods of [18, 25] seek to find a linear mixture of the basis kernels together with the cluster assignments, such that the margin is maximized, in essence combining (1) and (2). In this paper, we follow a similar path, but propose a novel objective that considers the ratio between the margin (a notion of cluster separability) and the intra-cluster variance criterion of kernel k-means [5] (a notion of cluster coherence). Hence, both the separation and the compactness of the clusters are explicitly taken into account, which can possibly improve on the solutions returned by approaches utilizing either of the two. Importantly, the margin has been shown to suffer from a major deficiency when applied to supervised MKL [7]. It can become arbitrarily large by a simple scaling of the kernel, thus it is inappropriate for assessing the quality of the learned kernel. The same can be demonstrated to hold for unsupervised MKL and we prove that our ratio-based objective is invariant to kernel scaling, thus overcoming this deficiency. Moreover, its global optimum solution is invariant to the type of p-norm constraint on the kernel parameters $\boldsymbol{\theta}$ (when a linear combination of basis kernels is employed), making the selection of a suitable norm less crucial.

A simple gradient-based optimization procedure that alternates between updating the kernel parameters $\boldsymbol{\theta}$ and the cluster assignments $\mathbf{y}$ is devised, avoiding the invocation of complex optimizers, such as the SDP solvers [18] and the cutting plane method [25]. Experiments on several datasets, including two collections of handwritten numerals and two image collections, reveal the superiority of the proposed method over approaches that rely solely on the margin or the intra-cluster variance.

The rest of this paper is organized as follows. Section 2 introduces our ratio-based formulation and presents its invariance properties and optimization details. Experiments follow in Section 3, before the concluding remarks of Section 4.

2   The RMKC Algorithm

2.1   Problem Formulation

Consider a dataset $\mathcal{X} = \{\mathbf{x}_i\}_{i=1}^{N}$, $\mathbf{x}_i \in \Re^d$, for which we want to simultaneously infer the cluster labels and also perform kernel learning under the large margin framework. While presenting our method we shall restrict ourselves on a linear combination of basis kernels, which is the most common technique of parametrizing kernels for MKL [12, 14, 23]. Later we will show that our model can accommodate more general parametric forms of kernels.

Assume that V basis kernels, $\mathcal{K}^{(v)} : \mathcal{X} \times \mathcal{X} \to \Re$, are available, each implicitly inducing a transformation $\phi^{(v)} : \mathcal{X} \to \mathcal{H}^{(v)}$ on the instances to a feature space $\mathcal{H}^{(v)}$ through $\mathcal{K}^{(v)}(\mathbf{x}_i, \mathbf{x}_j) = \phi^{(v)}(\mathbf{x}_i)^\top \phi^{(v)}(\mathbf{x}_j)$. A linear mixture of kernels gives rise to a composite kernel $\widetilde{\mathcal{K}}$:

$$\widetilde{\mathcal{K}}(\mathbf{x}_i, \mathbf{x}_j) = \sum_{v=1}^{V} \theta_v \mathcal{K}^{(v)}(\mathbf{x}_i, \mathbf{x}_j), \ \theta_v \geq 0, \tag{3}$$

that is parametrized by $\boldsymbol{\theta} = [\theta_1, \ldots, \theta_V]^\top$. Since $\widetilde{\mathcal{K}}$ is a valid kernel it holds that $\widetilde{\mathcal{K}}(\mathbf{x}_i, \mathbf{x}_j) = \widetilde{\phi}(\mathbf{x}_i)^\top \widetilde{\phi}(\mathbf{x}_j)$, $\widetilde{\phi} : \mathcal{X} \to \widetilde{\mathcal{H}}$, and actually $\widetilde{\phi}(\mathbf{x}_i) = \left[\sqrt{\theta_1}\phi^{(1)}(\mathbf{x}_i)^\top, \ldots, \sqrt{\theta_V}\phi^{(V)}(\mathbf{x}_i)^\top\right]^\top$ due to the linear combination.

We propose a new formulation that does not depend only on the margin, like most existing MMC and MKL studies, but utilizes the ratio between the margin and the intra-cluster variance objective of kernel k-means [5] in feature space $\widetilde{\mathcal{H}}$. Minimizing such a ratio can lead to superior partitionings as both compact and well-separated clusters are sought. Moreover, as it will be proved, it makes our formulation invariant to kernel scaling, an important property when kernel learning is involved [7]. Denoting by $\mathbf{y} = [y_1, \ldots, y_N]^\top$, $y_i \in \{\pm 1\}$, the vector of the instances' cluster labels, we consider the following optimization problem:

$$\min_{\boldsymbol{\theta},\mathbf{y}} \mathcal{J}(\boldsymbol{\theta},\mathbf{y}), \; s.t. \; \theta_v \geq 0, \; \|\boldsymbol{\theta}\|_p^p = 1, \; -\ell \leq \sum_{i=1}^N y_i \leq \ell, \; \mathbf{y} \in \{\pm 1\}^N, \tag{4}$$

$$\mathcal{J}(\boldsymbol{\theta},\mathbf{y}) = \min_{\mathbf{w},b,\boldsymbol{\xi}} \frac{1}{2}\mathcal{E}(\boldsymbol{\theta},\mathbf{y})\|\mathbf{w}\|^2 + C\sum_{i=1}^N \xi_i, \tag{5}$$

$$s.t. \; y_i\left(\mathbf{w}^\top\widetilde{\phi}(\mathbf{x}_i) + b\right) \geq 1 - \xi_i, \; \xi_i \geq 0.$$

Here $\mathcal{E}(\boldsymbol{\theta},\mathbf{y})$ is the kernel k-means criterion (6) describing the intra-cluster variance[1], where $\widetilde{\mathbf{m}}_k$ is the k-th cluster center and δ_{ik} is a cluster indicator variable with $\delta_{i1} = 1$ if $y_i = -1$ and $\delta_{i2} = 1$ if $y_i = 1$. Note that due to the SVM-like formulation we are limited to two-cluster solutions, i.e. $k \in \{1,2\}$, which is the typical case for MMC methods.

$$\mathcal{E}(\boldsymbol{\theta},\mathbf{y}) = \frac{1}{N}\sum_{i=1}^N\sum_{k=1}^2 \delta_{ik}\|\widetilde{\phi}(\mathbf{x}_i) - \widetilde{\mathbf{m}}_k\|^2, \tag{6}$$

$$\delta_{ik} = \begin{cases} 1, \; y_i = 2k - 3 \\ 0, \; \text{otherwise} \end{cases}, \; \widetilde{\mathbf{m}}_k = \frac{\sum_{i=1}^N \delta_{ik}\widetilde{\phi}(\mathbf{x}_i)}{\sum_{i=1}^N \delta_{ik}}$$

Note that the squared Euclidean distances in $\mathcal{E}(\boldsymbol{\theta},\mathbf{y})$ can be posed solely in terms of the entries of the kernel matrix $\widetilde{K} \in \Re^{N\times N}$ corresponding to $\widetilde{\mathcal{K}}$, i.e. $\widetilde{K}_{ij} = \widetilde{\mathcal{K}}(\mathbf{x}_i,\mathbf{x}_j)$ [5]. Additionally, by using (3), this composite kernel matrix can be written as the sum of the basis kernel matrices $K^{(v)} \in \Re^{N\times N}$, i.e. $\widetilde{K} = \sum_{v=1}^V \theta_v K^{(v)}$, thus getting (7).

$$\mathcal{E}(\boldsymbol{\theta},\mathbf{y}) = \frac{1}{N}\sum_{v=1}^V \theta_v \sum_{i=1}^N\sum_{k=1}^2 \delta_{ik}\left(K_{ii}^{(v)} - \frac{2\sum_{j=1}^N \delta_{jk}K_{ij}^{(v)}}{\sum_{j=1}^N \delta_{jk}} + \frac{\sum_{j=1}^N\sum_{l=1}^N \delta_{jk}\delta_{lk}K_{jl}^{(v)}}{\sum_{j=1}^N\sum_{l=1}^N \delta_{jk}\delta_{lk}}\right) \tag{7}$$

For the above optimization problem (4), it is easy to verify that its objective function $\mathcal{J}(\boldsymbol{\theta},\mathbf{y})$ at a given $\{\boldsymbol{\theta}, \mathbf{y}\}$ is defined as the optimal objective value of a problem (5) that closely resembles the standard SVM. The only difference is that the variance to margin ratio is employed in place of the margin. Similar to MMC methods [21,24], a cluster balance constraint ($-\ell \leq \sum_{i=1}^N y_i \leq \ell$) must be

---

[1] For simplicity, on the following, we shall refer to the intra-cluster variance as the variance of the clusters.

imposed to prevent meaningless solutions from arising. Finally, the composite kernel coefficients θ_v are required to be nonnegative so that $\widetilde{K}$ is a valid kernel and a p-norm constraint is introduced to avoid overfitting, as in (2).

Hence, the optimization in (4) searches for a pair of $\{\theta, \mathbf{y}\}$ values that yields a small variance to margin ratio $(\mathcal{E}(\theta, \mathbf{y})\|\mathbf{w}\|^2)$ regularized by the misclassification error (captured by the slack variables ξ). We shall call this approach Ratio-based Multiple Kernel Clustering, abbreviated as RMKC.

It should be clarified that the actual problem we are trying to solve is (s.t. the constraints in (4)-(5)):

$$\min_{\theta, \mathbf{y}, \mathbf{w}, b, \xi} \frac{1}{2}\mathcal{E}(\theta, \mathbf{y})\|\mathbf{w}\|^2 + C\sum_{i=1}^{N} \xi_i, \tag{8}$$

which is rather difficult to directly optimize, since it constitutes a non-convex problem with integer parameters $\mathbf{y}$. Reformulating it as in (4), analogously to Rakotomamonjy et al. [14], will enable us to devise an alternating optimization strategy, that benefits from differentiability w.r.t. θ and does not demand the use of complex solvers.

2.2   Properties of RMKC

In this section, two properties of RMKC are presented, which highlight some important advantages of combining the margin with the variance of the clusters.

Suppose the composite kernel $\widetilde{K}$ (3) is scaled by $\alpha > 0$, i.e. $\widetilde{K}' = \alpha\widetilde{K}$. Then the corresponding transformation becomes $\widetilde{\phi}' = \sqrt{\alpha}\widetilde{\phi}$. Moreover, as $\widetilde{K}$ is a linear combination of basis kernels, its scaling can be equivalently posed as a scaling on its parameters, i.e. $\theta' = \alpha\theta$.

Proposition 1. (Scale Invariance) *If a kernel $\widetilde{K}$ of the form defined in (3) is scaled by a scalar $\alpha > 0$, then $\mathcal{J}(\alpha\theta, \mathbf{y}) = \mathcal{J}(\theta, \mathbf{y})$.*

Proof. From (7) it is evident that $\mathcal{E}(\alpha\theta, \mathbf{y}) = \alpha\mathcal{E}(\theta, \mathbf{y})$, hence:

$$\mathcal{J}(\alpha\theta, \mathbf{y}) = \min_{\mathbf{w}, b, \xi} \frac{1}{2}\alpha\mathcal{E}(\theta, \mathbf{y})\|\mathbf{w}\|^2 + C\sum_{i=1}^{N} \xi_i,$$

$$s.t. \ y_i\left(\mathbf{w}^\top\left(\sqrt{\alpha}\widetilde{\phi}(\mathbf{x}_i)\right) + b\right) \geq 1 - \xi_i, \ \xi_i \geq 0.$$

Setting $\mathbf{w} = \mathbf{w}'/\sqrt{\alpha}$ and substituting in the above equation completes the proof, as (5) is recovered. $\qquad\square$

Our quest for an objective that satisfies Proposition 1 was inspired by Gai et al. [7], where it was illustrated that relying solely on the margin is not sufficient to perform kernel learning in the supervised case. Analogously, if $\mathcal{J}(\theta, \mathbf{y})$ in (4) is replaced with the more conventional margin-based objective:

$$\mathcal{J}'(\theta, \mathbf{y}) = \min_{\mathbf{w}, b, \xi} \frac{1}{2}\|\mathbf{w}\|^2 + C\sum_{i=1}^{N} \xi_i, \ s.t. \ y_i\left(\mathbf{w}^\top\widetilde{\phi}(\mathbf{x}_i) + b\right) \geq 1 - \xi_i, \ \xi_i \geq 0,$$

$$\tag{9}$$

it can be shown that an arbitrarily small $\mathcal{J}'(\boldsymbol{\theta}, \mathbf{y})$ value can be achieved by scaling the composite kernel, thus constituting the margin criterion unsuitable for evaluating the true quality of the kernel while learning $\{\boldsymbol{\theta}, \mathbf{y}\}$. Note that in the linear combination case (3), where scaling the composite kernel is equivalent to scaling its parameters, the scaling issue can be handled through the p-norm constraint on $\boldsymbol{\theta}$. However, this is not possible for nonlinear mixtures of basis kernels. On the contrary, our ratio-based objective (5) is scale invariant for arbitrary forms of composite kernels (the proof is analogous to Proposition 1) and also allows for norm invariance.

Proposition 2. (Norm Invariance) *Consider a kernel $\widetilde{\mathcal{K}}$ of the form defined in (3) as well as a) the optimization problem described by (4) without the p-norm constraint on $\boldsymbol{\theta}$ (p1) and b) the same problem (4), but with the slightly more general p-norm constraint $\|\boldsymbol{\theta}\|_p^p = c$, $c > 0$, in place of $\|\boldsymbol{\theta}\|_p^p = 1$ (p2). If $\{\boldsymbol{\theta}_a^*, \mathbf{y}_a^*\}$ is a global optimal solution of p1 then $\left\{ \frac{c^{1/p}}{\|\boldsymbol{\theta}_a^*\|_p} \boldsymbol{\theta}_a^*, \mathbf{y}_a^* \right\}$ is a global optimal solution of p2. Also, if $\{\boldsymbol{\theta}_b^*, \mathbf{y}_b^*\}$ is a global optimal solution of p2 then $\{\boldsymbol{\theta}_b^*, \mathbf{y}_b^*\}$ is a global optimal solution of p1.*

Proof. From the scale invariance property and since $\{\boldsymbol{\theta}_a^*, \mathbf{y}_a^*\}$ is a global optimum of p1 we get $\mathcal{J}\left(\frac{c^{1/p}}{\|\boldsymbol{\theta}_a^*\|_p} \boldsymbol{\theta}_a^*, \mathbf{y}_a^* \right) = \mathcal{J}(\boldsymbol{\theta}_a^*, \mathbf{y}_a^*) \leq \mathcal{J}(\boldsymbol{\theta}, \mathbf{y})$ for any $\{\boldsymbol{\theta}, \mathbf{y}\}$ satisfying the constraints of p1. Note that the admissible $\boldsymbol{\theta}$ values for problem p2 are a subset of those allowed in p1, hence the above inequality also holds for every $\{\boldsymbol{\theta}, \mathbf{y}\}$ adhering to the constraints of p2 (the constraints for $\mathbf{y}$ are identical in p1 and p2)). Together with the fact that $\left\| \frac{c^{1/p}}{\|\boldsymbol{\theta}_a^*\|_p} \boldsymbol{\theta}_a^* \right\|_p^p = c$ the first part of the proof is completed.

For any $\{\boldsymbol{\theta}, \mathbf{y}\}$ complying to the constraints of p1 it holds that $\left\{ \frac{c^{1/p}}{\|\boldsymbol{\theta}\|_p} \boldsymbol{\theta}, \mathbf{y} \right\}$ is admissible for p2, since $\left\| \frac{c^{1/p}}{\|\boldsymbol{\theta}\|_p} \boldsymbol{\theta} \right\|_p^p = c$. The scale invariance property and the global optimality of $\{\boldsymbol{\theta}_b^*, \mathbf{y}_b^*\}$ w.r.t. p2 yields $\mathcal{J}(\boldsymbol{\theta}_b^*, \mathbf{y}_b^*) \leq \mathcal{J}\left(\frac{c^{1/p}}{\|\boldsymbol{\theta}\|_p} \boldsymbol{\theta}, \mathbf{y} \right) = \mathcal{J}(\boldsymbol{\theta}, \mathbf{y})$, thus completing the second part of the proof. $\square$

Proposition 2 implies that the global optimal solution of the proposed formulation (4) is insensitive to the selected type of p-norm constraint, up to a scaling on the composite kernel parameters. The norm constraint can be even dropped from (4) without affecting its optimal solution. Of course, a solver that locates local optima of the ratio-based objective may produce different solutions when different p-norms are employed for the same problem, but at least the overall best will be the same, making the choice of the p-norm less crucial.

2.3   Optimizing the RMKC Objective

An iterative algorithm that alternates between updating the cluster labels $\mathbf{y}$ and reestimating the composite kernel coefficients $\boldsymbol{\theta}$, starting from some initial $\{\boldsymbol{\theta}, \mathbf{y}\}$ value, is presented and its main steps are summarized in Algorithms 1-2.

Evaluating the Objective Function. To compute the value of the objective function $\mathcal{J}(\boldsymbol{\theta}, \mathbf{y})$ for some fixed $\{\boldsymbol{\theta}, \mathbf{y}\}$, we need to solve the convex SVM-like optimization problem in (5). This can be facilitated by turning to its dual, which can be obtained by incorporating the constraints into the primal via Lagrange multipliers and setting the derivatives w.r.t. $\mathbf{w}$, b, and $\boldsymbol{\xi}$ to zero. After some manipulation the following dual emerges:

$$\max_{\boldsymbol{\alpha}} \sum_{i=1}^{N} \alpha_i - \frac{1}{2\mathcal{E}(\boldsymbol{\theta}, \mathbf{y})} \sum_{i=1}^{N} \sum_{j=1}^{N} \alpha_i \alpha_j y_i y_j \widetilde{K}_{ij}, \ s.t. \ 0 \leq \alpha_i \leq C, \ \sum_{i=1}^{N} \alpha_i y_i = 0.$$
(10)

Since the cluster variance $\mathcal{E}(\boldsymbol{\theta}, \mathbf{y})$ is a constant for given $\{\boldsymbol{\theta}, \mathbf{y}\}$, it can be included in the kernel matrix and, thus, (10) actually coincides with the dual of the standard SVM, with $\frac{1}{\mathcal{E}(\boldsymbol{\theta}, \mathbf{y})} \widetilde{K}$ as the kernel matrix. Hence, the optimal solution for (10), denoted by $\boldsymbol{\alpha}^*$, can be located using any of the existing SVM solvers (the optimal values for $\mathbf{w}$, b, and $\boldsymbol{\xi}$ in (5) are calculated based on the solution of the dual). Moreover, due to strong duality, the value of $\mathcal{J}(\boldsymbol{\theta}, \mathbf{y})$ can be directly acquired from the dual:

$$\mathcal{J}(\boldsymbol{\theta}, \mathbf{y}) = \sum_{i=1}^{N} \alpha_i^* - \frac{1}{2\mathcal{E}(\boldsymbol{\theta}, \mathbf{y})} \sum_{i=1}^{N} \sum_{j=1}^{N} \alpha_i^* \alpha_j^* y_i y_j \widetilde{K}_{ij}.$$
(11)

Updating the Kernel Parameters. Changing the composite kernel coefficients so that the ratio-based objective $\mathcal{J}(\boldsymbol{\theta}, \mathbf{y})$ is reduced, while keeping the cluster labels $\mathbf{y}$ fixed, can be effectively performed by means of gradient descent. Due to strong duality between (5) and (10) (Section 2.3), we can exploit (11) to compute the gradient of $\mathcal{J}(\boldsymbol{\theta}, \mathbf{y})$ w.r.t. $\boldsymbol{\theta}$.

Proof for the differentiability of $\mathcal{J}(\boldsymbol{\theta}, \mathbf{y})$ comes from Danskin's theorem [4], similar to [14, 19]. To apply this theorem to our problem, two conditions must be satisfied. First, the optimal solution $\boldsymbol{\alpha}^*$ of (10) must be unique. This can be ensured by demanding the composite kernel matrix $\widetilde{K}$ to be strictly positive definite for every admissible $\boldsymbol{\theta}$. Second, the objective function optimized in the dual (10) must be continuously differentiable w.r.t. $\boldsymbol{\theta}$, which can be ensured by demanding $\widetilde{K}$ to be continuously differentiable w.r.t. $\boldsymbol{\theta}$. As $\widetilde{K}$ is a linear mixture of basis kernel matrices $K^{(v)}$, both requirements are fulfilled as long as every $K^{(v)}$ is strictly positive definite. The theorem also states that $\mathcal{J}(\boldsymbol{\theta}, \mathbf{y})$ can be differentiated as if $\boldsymbol{\alpha}^*$ does not depend on $\boldsymbol{\theta}$. Therefore, the derivatives can be obtained from (11) as:

$$\frac{\partial \mathcal{J}(\boldsymbol{\theta}, \mathbf{y})}{\partial \theta_v} = \frac{1}{2\mathcal{E}(\boldsymbol{\theta}, \mathbf{y})^2} \sum_{i=1}^{N} \sum_{j=1}^{N} \alpha_i^* \alpha_j^* y_i y_j \widetilde{K}_{ij} \frac{\partial \mathcal{E}(\boldsymbol{\theta}, \mathbf{y})}{\partial \theta_v}$$

$$- \frac{1}{2\mathcal{E}(\boldsymbol{\theta}, \mathbf{y})} \sum_{i=1}^{N} \sum_{j=1}^{N} \alpha_i^* \alpha_j^* y_i y_j \frac{\partial \widetilde{K}_{ij}}{\partial \theta_v},$$
(12)

where $\frac{\partial \widetilde{K}_{ij}}{\partial \theta_v} = K_{ij}^{(v)}$ and $\frac{\partial \mathcal{E}(\theta, \mathbf{y})}{\partial \theta_v}$ follows directly from (7). Note that in order to calculate the derivatives, we must first obtain $\boldsymbol{\alpha}^*$ by solving (10) for the current $\{\theta, \mathbf{y}\}$ values.

The procedure for updating θ for given $\mathbf{y}$, begins by executing a standard gradient descent update on θ, using (12). Afterwards, θ is projected back to its feasible set, so that the positivity and p-norm constraints (4) are enforced. In this work, we consider the values $p = 1, 2$ and execute the projections as shown in [6, 15]. Note that the gradient descent step size, η, is adjusted according to the Armijo rule, which may require additional optimizations of the dual.

Updating the Cluster Labels. Finding a new set of cluster assignments $\mathbf{y}'$ that will further decrease $\mathcal{J}(\theta, \mathbf{y})$ (keeping the kernel parameters θ fixed) is not straightforward, since the underlying optimization is a non-convex integer problem. Some single kernel MMC approaches relax $\mathbf{y}$ on the continuous domain to ease the optimization (e.g. [18, 21]), however, in the end the relaxed solution should be mapped back to the discrete space. Here, on the contrary, our aim is to work directly on the discrete cluster labels without any relaxations.

We have developed a practical search framework, where an improved cluster labeling $\mathbf{y}'$ is obtained by moving instances between the two clusters. One possible direction would be to change the cluster label of a single instance only and then proceed with reestimating θ. However, we have empirically found that such a minor modification on $\mathbf{y}$ results in premature convergence as the algorithm overcommits to the initial assignments. A better strategy is to change the labels of multiple instances before reestimating θ. The strategy we follow is motivated by several graph partitioning heuristics that have been applied to clustering, prominently the Kernighan-Lin algorithm [10]: an initial split of the graph is revamped by exchanging several nodes (specified in an incremental fashion) between partitions and selecting the best subset of these nodes. Based on this idea, we build a sequence of L candidate cluster label vectors, $\mathbf{y}^{(1)}, \ldots, \mathbf{y}^{(L)}$, ($L$ is user-defined) and select the one generating the greatest improvement on $\mathcal{J}(\theta, \mathbf{y})$ in order to update $\mathbf{y}$. These L candidate label vectors are constructed incrementally (one after the other), such that compared to the previous candidate label vector, the next contains one more instance whose label has been changed (i.e. they differ in one element). Given $\mathbf{y}^{(l)}$, the $(l + 1)$-th instance to change clusters is selected to be the one that is expected to produce the smallest objective value when added to the current l changes, thus constructing $\mathbf{y}^{(l+1)}$.

A meaningful approach for picking the $(l + 1)$-th instance is to rank the contending instances based on the confidence about their labeling according to the current (after l cluster moves) separating hyperplane and select the one with the smallest $y_i(\mathbf{w}^\top \widetilde{\phi}(\mathbf{x}_i) + b)$ value. This way misclassified instances (if any exist) have a higher priority to change clusters, since $y_i(\mathbf{w}^\top \widetilde{\phi}(\mathbf{x}_i) + b) < 0$, followed by those falling inside the margin (if any exist), since $0 \leq y_i(\mathbf{w}^\top \widetilde{\phi}(\mathbf{x}_i) + b) < 1$, and finally those away from the margin, since $y_i(\mathbf{w}^\top \widetilde{\phi}(\mathbf{x}_i) + b) \geq 1$.

More specifically, let $\mathbf{y}^{(0)}$ to be the vector of the cluster labels before commencing the update process. Assume that $\mathbf{y}^{(l)}$ has already been generated, thus at this point l instances have already changed clusters w.r.t $\mathbf{y}^{(0)}$. As mentioned, the $(l+1)$-th instance is selected to be the one we are the less confident about its labeling according to the separating hyperplane. However, when the labels change so does the hyperplane. Therefore, we must solve the dual (10) for the current assignments $\mathbf{y}^{(l)}$ to obtain the corresponding optimal hyperplane parameters $\mathbf{w}^{(l)*}$ and $b^{(l)*}$. Then, the index of the $(l+1)$-th instance is given by:

$$i^* = \underset{i:y_i^{(l)}=y_i^{(0)}}{\operatorname{argmin}}\ y_i^{(l)}\left(\mathbf{w}^{(l)*\top}\widetilde{\phi}(\mathbf{x}_i) + b^{(l)*}\right), \tag{13}$$

and the $(l+1)$-th candidate label vector is defined as:

$$y_i^{(l+1)} = \begin{cases} y_i^{(l)}, & i \neq i^* \\ -y_i^{(l)}, & i = i^* \end{cases}. \tag{14}$$

From (13), it is obvious, that an instance $\mathbf{x}_i$ whose label has already changed is not considered again as a contender, since $y_i^{(l)} \neq y_i^{(0)}$, and the selected one flips its label (14). Moreover, observe that the label changes of all previous steps are retained when constructing $\mathbf{y}^{(l+1)}$, leading to an incremental reassignment of the instances. The above is repeated for $l = 0, 1, \ldots, L-1$.

The returned cluster assignments that are used to update $\mathbf{y}$ correspond to the cluster label vector $\mathbf{y}^{(l^*)}$ attaining the smallest objective value (i.e. $\mathbf{y}' = \mathbf{y}^{(l^*)}$):

$$l^* = \underset{0 \leq l \leq L}{\operatorname{argmin}}\ \mathcal{J}(\boldsymbol{\theta}, \mathbf{y}^{(l)}). \tag{15}$$

Note that if none of the candidate label vectors $\mathbf{y}^{(l)}$ reduces the objective, then $l^* = 0$ from (15), and no label change is accepted. This ensures that the ratio-based objective never increases after updating $\mathbf{y}$.

The procedure for modifying $\mathbf{y}$, as described up to this point, selects L instances belonging to either of the two clusters and flips their label to construct the candidate label vectors. Some trial experiments indicated that a better approach is to restrict all L instances that change clusters to originate from the same (i.e. a single) cluster. For this reason, our final procedure is divided into two phases. In the first phase the candidate vectors are formed by moving L instances from the cluster associated with the $+1$ label to the cluster associated with the -1 label, while in the second phase the opposite movement direction is considered. The two phases are independent from each other, both starting from $\mathbf{y}^{(0)}$. Hence, one phase does not take into account the cluster changes of the other. At the end, the best of the $2L$ candidate vectors is selected to update the cluster labels. To implement the above idea, in (13) we must, additionally to $y_i^{(l)} = y_i^{(0)}$, require that $y_i^{(l)} = +1$ ($y_i^{(l)} = -1$) for the first (second) phase contending instances. Our complete, two phase, framework is shown in Algorithm 2.

An issue we have yet to touch on is how to impose the cluster balance constraint (4). Fortunately, this is rather straightforward under our framework, since we can define an upper bound on the number L of candidate label vectors in each phase and, therefore, on the number of instances allowed to change

---

Algorithm 1. RMKC

Input: Basis kernel matrices $\{K^{(v)}\}_{v=1}^V$, Initial composite kernel coefficients $\theta^{(0)}$ and cluster assignments $\mathbf{y}^{(0)}$

Output: Final kernel coefficients θ and cluster assignments $\mathbf{y}$

1: Set $t = 0$
2: Set parameters L, ℓ and C
3: Set $\widetilde{K}^{(0)} = \sum_{v=1}^V \theta_v^{(0)} K^{(v)}$
4: **repeat**
5:     Solve the dual (10) for $\widetilde{K}^{(t)}$ (i.e. $\theta^{(t)}$) and $\mathbf{y}^{(t)}$ to obtain $\boldsymbol{\alpha}^{(t)*}$
6:     **for** $v = 1$ to V **do** // Update θ.
7:         $\theta_v^{(t+1)} = \theta_v^{(t)} - \eta^{(t)} \left. \frac{\partial \mathcal{J}(\theta,\mathbf{y})}{\partial \theta_v} \right|_{\theta=\theta^{(t)},\mathbf{y}=\mathbf{y}^{(t)},\boldsymbol{\alpha}^*=\boldsymbol{\alpha}^{(t)*}}$
8:     **end for**
9:     Project $\theta^{(t+1)}$ to satisfy the constraints in (4)
10:     $\widetilde{K}^{(t+1)} = \sum_{v=1}^V \theta_v^{(t+1)} K^{(v)}$
11:     $\mathbf{y}^{(t+1)} = \text{CLUSTER\_UPD}(\widetilde{K}^{(t+1)}, \mathbf{y}^{(t)})$ // Update $\mathbf{y}$.
12:     $t = t + 1$
13: **until** converged
14: **return** $\theta = \theta^{(t)}, \mathbf{y} = \mathbf{y}^{(t)}$

---

clusters, to guarantee that the constraint is never violated. For the first phase $L \leq (\ell + \sum_{i=1}^N y_i^{(0)})/2$, while for the second $L \leq (\ell - \sum_{i=1}^N y_i^{(0)})/2$. Note that $\sum_{i=1}^N y_i^{(0)}$ describes the initial imbalance before moving any instances (which, of course, satisfies the constraint) and $\ell \geq 0$ the maximum admissible imbalance.

2.4   Discussion

This section examines some additional aspects of the proposed RMKC method, starting with the convergence of the iterative algorithm used to optimize (4). In each iteration, the gradient descent update on θ reduces the ratio-based objective value. Moreover, the subsequent update on $\mathbf{y}$ selects a candidate cluster label vector that further decreases the objective. Hence, the overall process is guaranteed to monotonically converge. The final solution, though, depends on the initial $\{\theta, \mathbf{y}\}$ values, thus a local, and not the global, minimum of $\mathcal{J}(\theta, \mathbf{y})$ is located. The solution also depends on the user-specified constants C, ℓ and L, as well as, on the selected p-norm for the composite kernel coefficients constraint.

An important advantage of RMKC is that it can be readily extended to learning general forms of parametric composite kernels $\widetilde{\mathcal{K}}$, such as a nonlinear mixture of basis kernels, without being restricted to just the linear combination case (3). The formulation itself remains unchanged (e.g. (4), (5), (6), (10), (11)) and the iterative algorithm is applicable out of the box, if the gradient of the ratio-based objective can be computed. This is possible when the composite kernel matrix is strictly positive definite and continuously differentiable w.r.t. its parameters θ (see Section 2.3). Of course, $\frac{\partial \widetilde{K}_{ij}}{\partial \theta_v}$ and $\frac{\partial \mathcal{E}(\theta,\mathbf{y})}{\partial \theta_v}$ in (12) depend on the specific form of the composite kernel. Moreover, the scale invariance of our objective

---

Algorithm 2. RMKC - cluster update

Input: Current composite kernel matrix $\widetilde{K}$ and cluster assignments $\mathbf{y}$
Output: Updated cluster assignments $\mathbf{y}'$

1: **function** CLUSTER_UPD($\widetilde{K}, \mathbf{y}$)
    // First phase.
2:    Set $\mathbf{y}^{(0)} = \mathbf{y}$
3:    **for** $l = 0$ to $L - 1$ **do**
4:        Solve the dual (10) for $\widetilde{K}$ and $\mathbf{y}^{(l)}$ to obtain $\mathbf{w}^{(l)^*}$ and $b^{(l)^*}$
5:        Calculate $\mathbf{y}^{(l+1)}$ (14) with the added constraint $y_i^{(l)} = +1$ in (13)
6:    **end for**
    // Second phase. This phase ignores the cluster moves of the first.
7:    Set $\mathbf{y}^{(L+1)} = \mathbf{y}$
8:    **for** $l = L + 1$ to $2L$ **do**
9:        Solve the dual (10) for $\widetilde{K}$ and $\mathbf{y}^{(l)}$ to obtain $\mathbf{w}^{(l)^*}$ and $b^{(l)^*}$
10:        Calculate $\mathbf{y}^{(l+1)}$ (14) with the added constraint $y_i^{(l)} = -1$ in (13)
11:    **end for**
12:    $l^* = \operatorname{argmin}_{0 \leq l \leq 2L+1} \mathcal{J}(\boldsymbol{\theta}, \mathbf{y}^{(l)})$
13:    **return** $\mathbf{y}' = \mathbf{y}^{(l^*)}$
14: **end function**

(i.e. scaling $\widetilde{\mathcal{K}}$ by a scalar $\alpha > 0$) also holds in the general case (the proof is analogous to that in Proposition 1), but the same is not true for the norm invariance. Note that scaling $\widetilde{\mathcal{K}}$ is no more equivalent to scaling the parameters $\boldsymbol{\theta}$. The ability to accommodate general kernel forms broadness the applicability of RMKC and constitutes an advantage over existing MKL approaches that are usually limited to a particular type of composite kernel.

3   Empirical Evaluation

To investigate the potential of combining the margin with the variance in the clustering objective and perform kernel learning, the presented RMKC framework is compared to: a) kernel k-means, which serves as our baseline method, b) iterSVR [24], an iterative margin-based MMC approach that follows formulation (1), and c) two iterative variance-based MKL approaches that optimize (6), namely multi-view kernel k-means (MVKKM) and multi-view spectral clustering (MVSpec) [17]. The evaluation is made on various diverse datasets from the UCI repository[2] (Ionosphere, Letter, Satellite, Multiple Features and Optdigits), as well as on the COIL-20 image library of objects [13] and a subset of the Corel image collection[3]. Apart from Ionosphere, all other datasets contain instances of more than two categories. For this reason, we conduct experiments using pairs of the included categories. For Letter and Satellite we simply focus on the first two classes, i.e. A-B and C1(red soil)-C2(cotton crop), respectively, as in [24]. For

---

[2] http://archive.ics.uci.edu/ml
[3] http://www.cs.virginia.edu/~xj3a/research/CBIR/Download.htm

Fig. 1. The COIL-20 objects considered in the experiments

Fig. 2. Indicative images of the Corel categories considered in the experiments

the two databases of handwritten digits (i.e. Multiple Features and Optdigits) we try several pairs of the contained numerals (0-9), while for the two image collections we consider pairs of the classes depicted in Figures 1-2. The tested pairs are shown in Tables 3-4. Since ground truth information is available for every dataset, we employ the clustering accuracy metric to measure performance[4].

Multiple Features and Corel are multi-view datasets, hence, for the same instance multiple sets of attributes are available. Each attribute set naturally defines a basis kernel and the linear kernel is employed here to represent each view. For the other, single view, datasets, we follow [18,20] and construct 10 basis RBF kernels, where the kernel width σ varies from 10% to 100% of the range of distance between any two instances. Kernels are multiplicatively normalized [12].

Throughout the experiments, our algorithm is configured as follows: we fix the number of candidate label vectors in each phase to $L = 30$, the cluster imbalance parameter to $\ell = 0.5N$ (for the Corel images only, $\ell = 0.2N$) and conduct a grid search on the set $\{10^{-2}, 10^{-1}, \ldots, 10^{2}\}$ to locate the best performing value for the C regularizer in each dataset. The basis kernels are linearly combined (3) and their coefficients are uniformly initialized, i.e. $\theta_v = \frac{1}{V^{1/p}}$. To initialize the cluster assignments $\mathbf{y}$, we extract several pairs of instances (usually $0.25N$ pairs) using a k-means++-like procedure [1], where the first instance is chosen randomly and the second is picked with a probability that is proportional to its distance from the first. For each such pair, the remaining $N - 2$ instances are assigned to the closest of the two instances in the pair, thus producing a partitioning of the data. The partitioning $\mathbf{y}$ with the minimum $\mathcal{J}(\boldsymbol{\theta}, \mathbf{y})$ value is used to initialize a run of RMKC. Since the procedure for choosing the initial $\mathbf{y}$ is nondeterministic, the RMKC performance is averaged over 30 runs for each tried set of parameters $(L, \ell, C, p\text{-norm})$. Finally, the LIBSVM toolbox [2] is utilized for solving (10).

3.1   Norm Invariance in Practice

In Proposition 2, it was proved that the global optimal solution of our formulation (4) is invariant to the p-norm applied on the composite kernel coefficients $\boldsymbol{\theta}$,

---

[4] To evaluate performance, we make the typical assumption that clusters correspond to classes and set their number equal to the number of classes (e.g. [18, 20, 22, 25]).

Table 1. RMKC clustering accuracy (%) (averaged over all pairs of categories considered in each dataset) for different p-norm constraints

Dataset	No-norm	1-norm	2-norm
Ionosphere	71.51 ± 0.00	71.51 ± 0.00	71.51 ± 0.00
Letter	94.47 ± 0.00	94.47 ± 0.00	94.47 ± 0.00
Satellite	96.17 ± 0.50	96.19 ± 0.52	96.16 ± 0.51
COIL-20	98.75 ± 2.60	98.61 ± 2.65	98.43 ± 2.73
Corel	94.55 ± 1.62	94.64 ± 1.58	94.69 ± 1.62
Multiple Features	99.58 ± 0.22	99.53 ± 0.37	99.59 ± 0.23
Optdigits	97.77 ± 2.45	97.65 ± 2.71	97.75 ± 2.50

Table 2. Clustering accuracy (%) of the compared methods on three popular UCI datasets

Dataset	RMKC (1-norm)	MVKKM	MVSpec	Kernel k-means	IterSVR (best)	IterSVR (average)
Ionosphere	71.51 ± 0.00	71.23	70.66	73.22 ± 2.90	$\mathbf{74.83 \pm 1.65}$	71.83 ± 1.99
Letter (A-B)	94.47 ± 0.00	93.50	88.68	93.63 ± 0.00	$\mathbf{94.51 \pm 1.70}$	92.29 ± 1.97
Satellite (C1-C2)	96.19 ± 0.52	94.19	96.24	94.15 ± 0.03	$\mathbf{96.42 \pm 0.00}$	91.53 ± 5.58

Table 3. Clustering accuracy (%) of the compared methods on image clustering

Dataset	RMKC (1-norm)	MVKKM	MVSpec	Kernel k-means	IterSVR (best)	IterSVR (average)
COIL-20						
3-19	100.00 ± 0.00	100.00	100.00	94.05 ± 10.27	100.00 ± 0.00	100.00 ± 0.00
4-11	100.00 ± 0.00	77.78	100.00	96.30 ± 10.41	98.47 ± 8.37	98.34 ± 8.34
15-18	100.00 ± 0.00	90.28	95.83	97.57 ± 3.74	99.72 ± 0.35	99.21 ± 0.21
15-19	$\mathbf{94.44 \pm 10.59}$	68.06	86.11	86.57 ± 14.84	93.43 ± 14.30	91.86 ± 14.52
Corel						
700-4990	$\mathbf{97.62 \pm 0.65}$	95.00	95.00	85.98 ± 9.58	96.43 ± 0.25	83.19 ± 1.85
700-5530	92.60 ± 1.42	$\mathbf{94.00}$	$\mathbf{94.00}$	85.50 ± 0.00	88.63 ± 6.40	68.03 ± 3.49
770-840	$\mathbf{97.55 \pm 0.91}$	94.50	90.00	90.47 ± 0.37	94.20 ± 3.04	87.85 ± 0.58
770-1350	$\mathbf{94.03 \pm 1.72}$	93.50	92.00	88.72 ± 0.96	92.67 ± 1.27	84.10 ± 1.89
1340-1350	$\mathbf{95.50 \pm 0.00}$	95.00	95.00	91.00 ± 0.00	92.50 ± 0.00	83.71 ± 0.00
2890-4990	$\mathbf{90.57 \pm 4.79}$	87.00	86.00	85.00 ± 0.00	90.00 ± 0.00	73.04 ± 5.68

if $\widetilde{\mathcal{K}}$ is a linear mixture of basis kernels (3). However, the RMKC method locates local optima of the ratio-based objective. Hence, it is of particular interest to explore how these local optima vary for different choices of p-norm constraints.

To demonstrate this, RMKC is executed (according to the above configuration) for $p = 1, 2$ and also for the case where no norm constraint is imposed on $\boldsymbol{\theta}$ and the results are illustrated in Table 1. It can be observed that the solutions obtained across the different norms are very similar, therefore, in practice, the uncovered local optima are not significantly influenced by the choice of p-norm, although this cannot be theoretically guaranteed. On the following, we shall focus on the 1-norm, when presenting the results of our approach.

Table 4. Clustering accuracy (%) of the compared methods on the task of handwritten digits recognition

Dataset	RMKC (1-norm)	MVKKM	MVSpec	Kernel k-means	IterSVR (best)	IterSVR (average)
Mult. Feat.						
1-7	99.62 ± 0.78	98.75	98.75	98.00 ± 0.00	$\mathbf{99.75 \pm 0.00}$	96.85 ± 0.00
2-7	$\mathbf{100.00 \pm 0.00}$	99.00	99.75	97.92 ± 0.24	99.75 ± 0.00	97.61 ± 1.73
2-3	$\mathbf{99.70 \pm 0.23}$	99.25	99.00	99.50 ± 0.00	99.50 ± 0.00	94.13 ± 7.16
3-8	99.28 ± 0.38	99.50	99.50	97.50 ± 0.00	$\mathbf{99.75 \pm 0.00}$	98.78 ± 0.04
5-6	$\mathbf{99.42 \pm 0.48}$	98.50	98.50	98.29 ± 0.09	98.75 ± 0.00	95.68 ± 2.37
6-8	$\mathbf{99.15 \pm 0.33}$	97.25	98.50	97.33 ± 0.16	99.00 ± 0.00	94.94 ± 6.47
Optdigits						
1-7	99.56 ± 1.41	100.00	100.00	89.38 ± 16.06	96.93 ± 9.83	94.26 ± 13.14
2-7	98.03 ± 1.31	96.35	92.42	95.03 ± 8.40	$\mathbf{99.32 \pm 0.16}$	98.88 ± 0.84
2-3	96.29 ± 5.44	90.56	88.89	89.92 ± 9.10	$\mathbf{96.50 \pm 0.82}$	95.59 ± 2.70
3-8	92.43 ± 8.00	94.12	93.28	92.56 ± 7.80	$\mathbf{96.20 \pm 0.16}$	95.01 ± 4.08
5-6	$\mathbf{99.72 \pm 0.00}$	99.45	99.45	99.57 ± 0.14	$\mathbf{99.72 \pm 0.00}$	99.33 ± 0.01
6-8	$\mathbf{99.89 \pm 0.14}$	99.15	98.87	99.32 ± 0.26	99.72 ± 0.00	99.45 ± 0.06

3.2 Comparative Results

We have conducted a comprehensive evaluation of RMKC, kernel k-means, iterSVR, MVKKM and MVSpec on all datasets. RMKC is set up as previously described. Kernel k-means is restarted 30 times, from randomly picked initial centers. For iterSVR we employ a similar setup to [24], i.e. the cluster imbalance parameter is fixed to $\ell = 0.03N$ for balanced and to $\ell = 0.3N$ for unbalanced datasets, while the initial cluster labels are obtained from the kernel k-means solution (iterSVR is, thus, repeated 30 times). For the C regularizer, the same grid search as for RMKC is implemented. Finally, the sparsity controlling parameter p for MVKKM and MVSpec is selected by a grid search on the values $\{1, 1.5, \ldots, 5\}$.

Performance is measured in terms of average clustering accuracy (and its deviation) over the 30 restarts (MVKKM and MVSpec are deterministically initialized [17], thus we have no restarts). Let us stress, that both kernel k-means and iterSVR are single kernel methods that do not implement kernel learning. For this reason, these algorithms are independently executed for each of the individual basis kernels in each data collection and the kernel attaining the highest accuracy is reported. Moreover, for iterSVR the average performance over all basis kernels is also shown. It is important to make clear that it is not possible to know a priori which is the best basis kernel for a given dataset.

In Table 2 we observe that iterSVR with the optimal basis kernel achieves the best accuracy, being closely matched by RMKC. Only for Ionosphere the difference is large, where, surprisingly, all three MKL approaches (RMKC, MVKKM and MVSpec) are even inferior to kernel k-means. However, this is a difficult dataset to cluster and all methods yield rather poor outcomes (accuracy does not exceed 75%).

Turning our attention to image clustering (Table 3), it is evident that our ratio-based objective constantly outperforms the other methods. For the

COIL-20 objects, whose images are taken from different angles in a neutral background, hence are easy to distinguish, our approach manages to find the correct clusters for 3/4 of subsets and iterSVR appears to be its closest competitor. Clustering the Corel images is a more difficult task, due to variations in the composition of the depicted scene within each class. Here the differences of RMKC to iterSVR are more distinct and its closest competitor is MVKKM, which clearly displays the benefits of combining information from multiple views under MKL.

For the task of handwritten digits recognition (Table 4) the best performance is equally shared between RMKC and iterSVR across the two datasets. Note that for Multiple Features, which, like Corel, is a multi-view dataset, RMKC is superior. MVKKM and MVSpec achieve the highest accuracy on a single case (Optdigits for the pair 1-7) and are superior to RMKC for only 3/12 of subsets.

Overall, the proposed RMKC algorithm obtains a higher clustering accuracy for the majority of the tested category pairs. The margin-based iterSVR approach seems to be close, or even better, for some cases, provided the optimal basis kernel is used (iterSVR(best)). However, in practice, the best kernel for a particular dataset is not a priori known. By looking at the Tables' last column, one can notice that iterSVR results degrade significantly if an inappropriate basis kernel is chosen. On the contrary, RMKC is able to automatically infer a meaningful kernel by combining the basis kernels.

4 Conclusions

We have proposed a novel MKL formulation that considers the ratio between the margin and the intra-cluster variance. Its objective is optimized by an iterative, gradient-based algorithm to get both the cluster assignments and the composite kernel parameters. Moreover, it is characterized by two important properties: it is invariant to scalings of the learned kernel and, when basis kernels are linearly mixed, is also invariant (on its global optimum) to the type of p-norm constraint on the composite kernel parameters. Our framework compares favorably to existing approaches that rely either on the margin or the intra-cluster variance.

Although multiple cluster problems can be tackled by iteratively solving a sequence of two-cluster problems, an interesting research direction would be to extend our formulation to directly handle multiple clusters, following the ideas in [22, 25, 26]. Moreover, evaluating different parametric forms for the composite kernel is in our plans.

References

1. Arthur, D., Vassilvitskii, S.: k-means++: the advantages of careful seeding. In: ACM-SIAM Symposium on Discrete Algorithms (SODA), pp. 1027–1035 (2007)
2. Chang, C.C., Lin, C.J.: LIBSVM: A library for support vector machines. ACM Transactions on Intelligent Systems and Technology 2(3), 1–27 (2011), http://www.csie.ntu.edu.tw/~cjlin/libsvm

3. Cortes, C., Mohri, M., Rostamizadeh, A.: Learning non-linear combinations of kernels. In: Advances in Neural Information Processing Systems (NIPS), pp. 396–404 (2009)
4. Danskin, J.M.: The theory of max-min, with applications. SIAM Journal on Applied Mathematics 14(4), 641–664 (1966)
5. Dhillon, I.S., Guan, Y., Kulis, B.: Weighted graph cuts without eigenvectors a multilevel approach. IEEE Transactions on Pattern Analysis and Machine Intelligence 29(11), 1944–1957 (2007)
6. Duchi, J.C., Shalev-Shwartz, S., Singer, Y., Chandra, T.: Efficient projections onto the l_1-ball for learning in high dimensions. In: International Conference on Machine Learning (ICML), pp. 272–279 (2008)
7. Gai, K., Chen, G., Zhang, C.: Learning kernels with radiuses of minimum enclosing balls. In: Advances in Neural Information Processing Systems (NIPS), pp. 649–657 (2010)
8. Gönen, M., Alpaydın, E.: Localized multiple kernel learning. In: International Conference on Machine Learning (ICML), pp. 352–359 (2008)
9. Gönen, M., Alpaydın, E.: Multiple kernel learning algorithms. Journal of Machine Learning Research 12, 2211–2268 (2011)
10. Kernighan, B.W., Lin, S.: An efficient heuristic procedure for partitioning graphs. The Bell System Technical Journal 49(2), 291–308 (1970)
11. Kloft, M., Brefeld, U., Sonnenburg, S., Laskov, P., Müller, K.R., Zien, A.: Efficient and accurate l_p-norm multiple kernel learning. In: Advances in Neural Information Processing Systems (NIPS), pp. 997–1005 (2009)
12. Kloft, M., Brefeld, U., Sonnenburg, S., Zien, A.: lp-norm multiple kernel learning. Journal of Machine Learning Research 12, 953–997 (2011)
13. Nene, S.A., Nayar, S.K., Murase, H.: Columbia Object Image Library (COIL-20). Tech. Rep. CUCS-005-96, Department of Computer Science, Columbia University (1996), http://www.cs.columbia.edu/CAVE/software/softlib/coil-20.php
14. Rakotomamonjy, A., Bach, F.R., Canu, S., Grandvalet, Y.: SimpleMKL. Journal of Machine Learning Research 9, 2491–2521 (2008)
15. Songsiri, J.: Projection onto an l_1-norm ball with application to identification of sparse autoregressive models. In: Asean Symposium on Automatic Control, ASAC (2011)
16. Sonnenburg, S., Rätsch, G., Schäfer, C., Schölkopf, B.: Large scale multiple kernel learning. Journal of Machine Learning Research 7, 1531–1565 (2006)
17. Tzortzis, G., Likas, A.: Kernel-based weighted multi-view clustering. In: International Conference on Data Mining (ICDM), pp. 675–684 (2012)
18. Valizadegan, H., Jin, R.: Generalized maximum margin clustering and unsupervised kernel learning. In: Advances in Neural Information Processing Systems (NIPS), pp. 1417–1424 (2006)
19. Varma, M., Babu, B.R.: More generality in efficient multiple kernel learning. In: International Conference on Machine Learning (ICML), pp. 1065–1072 (2009)
20. Wang, F., Zhao, B., Zhang, C.: Linear time maximum margin clustering. IEEE Transactions on Neural Networks 21(2), 319–332 (2010)
21. Xu, L., Neufeld, J., Larson, B., Schuurmans, D.: Maximum margin clustering. In: Advances in Neural Information Processing Systems (NIPS), pp. 1537–1544 (2004)
22. Xu, L., Schuurmans, D.: Unsupervised and semi-supervised multi-class support vector machines. In: AAAI Conference on Artificial Intelligence (AAAI). pp. 904–910 (2005)

23. Xu, Z., Jin, R., Yang, H., King, I., Lyu, M.R.: Simple and efficient multiple kernel learning by group lasso. In: International Conference on Machine Learning (ICML), pp. 1175–1182 (2010)
24. Zhang, K., Tsang, I.W., Kwok, J.T.: Maximum margin clustering made practical. In: International Conference on Machine Learning (ICML), pp. 1119–1126 (2007)
25. Zhao, B., Kwok, J.T., Zhang, C.: Multiple kernel clustering. In: SIAM International Conference on Data Mining (SDM), pp. 638–649 (2009)
26. Zien, A., Ong, C.S.: Multiclass multiple kernel learning. In: International Conference on Machine Learning (ICML), pp. 1191–1198 (2007)

Evidence-Based Clustering for Scalable Inference in Markov Logic

Deepak Venugopal and Vibhav Gogate

Computer Science Department,
The University of Texas at Dallas,
Richardson, USA
{dxv021000,vibhav.gogate}@utdallas.edu

Abstract. Markov Logic is a powerful representation that unifies first-order logic and probabilistic graphical models. However, scaling-up inference in Markov Logic Networks (MLNs) is extremely challenging. Standard graphical model inference algorithms operate on the propositional Markov network obtained by grounding the MLN and do not scale well as the number of objects in the real-world domain increases. On the other hand, algorithms which perform inference directly at the first-order level, namely *lifted inference algorithms*, although more scalable than propositional algorithms, require the MLN to have specific symmetric structure. Worse still, evidence breaks symmetries, and the performance of lifted inference is the same as propositional inference (or sometimes worse, due to overhead). In this paper, we propose a general method for solving this "evidence" problem. The main idea in our method is to approximate the given MLN having, say, n objects by an MLN having k objects such that $k << n$ and the results obtained by running potentially much faster inference on the smaller MLN are as close as possible to the ones obtained by running inference on the larger MLN. We achieve this by finding clusters of "similar" groundings using standard clustering algorithms (e.g., K-means), and replacing all groundings in the cluster by their cluster center. To this end, we develop a novel distance (or similarity) function for measuring the similarity between two groundings, based on the evidence presented to the MLN. We evaluated our approach on many different benchmark MLNs utilizing various clustering and inference algorithms. Our experiments clearly show the generality and scalability of our approach.

1 Introduction

Markov Logic Networks (MLNs) [18,4] unify first-order logic and probabilistic models and are arguably the most popular representation for statistical relational learning. They have been used in a wide variety of application domains including natural language understanding [17], computer vision [22] and planning [21]. Just as in conventional probabilistic models such as Bayesian networks and Markov networks, the key challenge in MLNs is to develop scalable inference algorithms. However, this challenge is more pronounced in MLNs because MLNs are template models, compactly specified using a first-order logic representation and as a result even a seemingly simple MLN can yield an arbitrary large (propositional) probabilistic model as the number of objects in the real-world domain increases.

T. Calders et al. (Eds.): ECML PKDD 2014, Part III, LNCS 8726, pp. 258–273, 2014.

Existing MLN inference algorithms can be broadly classified into two categories, propositional algorithms, which operate on the Markov network obtained by grounding the MLN and lifted algorithms, which operate directly on the first-order representation, grounding only as necessary. Propositional algorithms such as Gibbs Sampling [5] and Belief Propagation [28] do not scale well as the number of objects gets large, because they perform inference over the Markov network obtained by grounding the MLN, which for large domain-sizes can be huge. On the other hand, lifted inference algorithms [15,3,6,26,20,10,7,2,27,13] either directly operate on the first-order structure or exploit symmetries in the propositional model and can therefore, in principle, scale significantly better than propositional inference algorithms.

Lifted inference algorithms typically suffer from two problems. First, they require MLNs to have a specific symmetric structure [3,9,23], which is not always the case in real-world applications. For example, to apply certain inference operations, the MLN needs to be composed of purely singleton atoms [9]. Second, a far more serious problem is that, in the presence of evidence most MLNs are not liftable because evidence breaks symmetries. As a concrete example, the symmetrical marginal probabilities in Fig. 1 (a) are broken with evidence (b). Therefore, a lifted algorithm that could potentially exploit the symmetry in (a) can no longer do so in (c). Thus, in the presence of evidence, lifted inference algorithms often resort to grounding the MLN. This is problematic because most interesting inference problems are almost always of the form $P(Q|E)$, i.e., computing the probability of a query given evidence. Therefore, there is a pressing need for inference algorithms that work without restrictions on the MLN structure or evidence. The main contribution of this paper is presenting one such method.

Our main idea is to reduce the number of objects in the domain of the MLN, thereby approximating it by a much smaller MLN such that the results obtained by performing inference on the smaller MLN are as close as possible to the ones obtained by running an expensive inference algorithm on the original MLN. To achieve this domain-reduction, we pre-process the MLN utilizing standard clustering algorithms such as K-means to merge together objects that are in some sense "similar" to each other from an inference perspective. Importantly, this pre-processing step allows us to plug-in existing grounded/lifted inference algorithms where the sampling-space (for sampling-based inference) or search-space (for search-based inference) can be controlled, which makes inference feasible even when the original MLN's domain is extremely large.

In order to obtain an accurate domain-reduced approximation of the original MLN, we specify a novel distance function that measures similarity based on the evidence presented to the MLN. This distance function helps cluster together objects having similar evidence-structure. The inherent symmetry in MLN representation makes it more likely that similar evidence structure translates to approximately similar marginal probabilities. Thus, we compute the marginal probability for a single element of the cluster and project the same results to all elements in the cluster, thereby drastically reducing the complexity of inference.

We evaluated our approach on several benchmark MLNs available on the Alchemy website [11]. Also, in our experiments, we leverage a number of clustering algorithms from data-mining/machine learning literature implemented in Weka [8] to scale-up inference to very large domain-sizes. We experimented with two inference algorithms,

Wins(A,A)	0.56
Wins(A,B)	0.56
Wins(A,C)	0.56
Wins(B,A)	0.56
Wins(B,B)	0.56
Wins(B,C)	0.56
Wins(C,A)	0.56
Wins(C,B)	0.56
Wins(C,C)	0.56

Strong(C)
Wins(A,C)
Wins(B,B)
Wins(B,C)
Wins(C,A)

Wins(A,A)	0.6
Wins(A,B)	0.6
Wins(B,A)	0.63
Wins(C,B)	0.85
Wins(C,C)	0.85

(a) Original Marginals   (b) Evidence   (c) New Marginals

Fig. 1. Effect of evidence on an MLN with one formula, 1.75 Strong(x) $\Rightarrow$ Wins(x,y). The marginal probabilities which were equal in (a) become unequal in (c) due to evidence (b).

a propositional sampling-based algorithm, Gibbs sampling [5] and a lifted message-passing algorithm, Lifted Belief Propagation [20] to show the generality of our approach. Our results clearly illustrate that, using a fraction of the true groundings, we are able to approximate the marginal probabilities quite consistently on a wide variety of MLN structures with arbitrary evidence.

2 Preliminaries

First-order logic (FOL) consists of predicates (e.g., Friends) that represent relations between objects, logical connectives (e.g., $\vee$, $\neg$, etc.) and quantifiers ($\forall$, $\exists$). Each predicate has a parenthesized list of arguments which can be substituted by a term which can either be a logical variable (x), a constant (X) or a function. A formula in first order logic is a predicate (atom), or any complex sentence that can be constructed from atoms using logical connectives and quantifiers. For example, the formula $\forall x$ Smokes(x) $\Rightarrow$ Asthma(x) states that all persons who smoke have asthma. A *ground* atom corresponding to a predicate is one where each term is substituted by a constant symbol.

We use a strict subset of FOL. Specifically, we make the following assumptions. First, we assume that there is a one-to-one mapping between the constant symbols and objects (Herbrand semantics). This means that any possible *world* is simply an assignment of True or False to every distinct *ground* atom. Second, we assume a function-free language where each variable is typed and the number of constant symbols is finite. Therefore, for any variable x, we can define a finite set Δ_x (domain of x) which consists of all the constant symbols that can be substituted for x. We refer to the constants corresponding to a domain as the domain's groundings. A ground formula is a formula obtained by substituting all of its variables with a constant. A ground KB is a KB containing all possible groundings of all of its formulas. For example, the grounding of a KB containing one formula, Smokes(x) $\Rightarrow$ Asthma(x) where $\Delta_x = \{Ana, Bob\}$, is a KB containing two ground formulas: Smokes(Ana) $\Rightarrow$ Asthma(Ana) and Smokes(Bob) $\Rightarrow$ Asthma(Bob).

Markov logic [4] extends FOL by softening the hard constraints expressed by the formulas. A soft formula or a weighted formula is a pair (f, w) where f is a formula in FOL and w is a real-number. A MLN denoted by $\mathcal{M}$, is a set of weighted formulas (f_i, w_i). Given a set of constants that represent objects in the domain, a Markov logic network defines a Markov network or a log-linear model. The Markov network is obtained by grounding the weighted first-order knowledge base and represents the following probability distribution.

$$P_{\mathcal{M}}(\omega) = \frac{1}{Z(\mathcal{M})} \exp \left(\sum_i w_i N(f_i, \omega) \right) \tag{1}$$

where ω is a world, $N(f_i, \omega)$ is the number of groundings of f_i that evaluate to True in the world ω and $Z(\mathcal{M})$ is a normalization constant or the partition function.

In this paper, we assume that the input MLN to our algorithm is in normal form [9,12]. A *normal* MLN [9] is an MLN that satisfies the following two properties: (1) There are no constants in any formula, and (2) If two distinct atoms with the same predicate symbol have variables x and y in the same position then $\Delta_x = \Delta_y$. An important distinction here is that, unlike in previous work on lifted inference that use normal forms [9,6] which require the MLN along with the associated evidence to be normalized, here we only require the MLN in normal form.

The two main inference problems in MLNs are computing the partition function and the marginal probabilities of query atoms given evidence. In this paper, we focus on the latter.

3 Domain Clustering

3.1 Problem Formulation

Let $\mathcal{M}$ denote an MLN with M predicates $R_1, R_2, \ldots, R_M$, and N weighted formulas $f_1, f_2, \ldots, f_N$. Let $G_{\mathcal{M}}$ denote the propositional Markov network obtained by grounding all the formulas in $\mathcal{M}$. Let $\mathbf{E} = \{E_k\}_{k=1}^S$ be the set of *evidences*. Each $E_k \in \mathbf{E}$ represents a single ground atom that is known to be either True or False. Let $\mathbf{I}$ be a set of indices of the form (i, j) such that $1 \leq i \leq M, 1 \leq j \leq A_i$, where A_i is the arity of the i-th predicate in $\mathcal{M}$. In other words, (i, j) is an index to the j-th argument of the i-th predicate in $\mathcal{M}$.

Let R be a binary relation on $\mathbf{I}$ such that $(i, j) \, R \, (a, b)$ iff there exists a formula $f \in \mathcal{M}$ such that: (1) f contains atoms having predicate symbols indexed by i and a, and (2) a logical variable x of f appears as the j-th argument and as the b-th argument of atoms having predicate symbols indexed by i and a respectively. Clearly, R is symmetric and reflexive. Let R^+ be the transitive closure of R on $\mathbf{I}$. R^+ is an equivalence relation on $\mathbf{I}$. Let $\mathcal{I} = \{\mathcal{I}_1 \mathcal{I}_2 \ldots \mathcal{I}_P\}$ denote the set of equivalence classes of $\mathbf{I}$ due to the equivalence relation R^+. Let $\Delta_{\mathcal{I}_k}$ denote the domain (possible groundings) of an element of $\mathcal{I}_k$. Note that since we assume that the MLN is in normal form, all elements of $\Delta_{\mathcal{I}_k}$ have the same domain.

Example 1. Let $\mathcal{M}$ contain exactly one formula $R_1(x,y) \wedge R_2(y,z) \Rightarrow R_3(z,x)$. Let $\Delta_x = \Delta_y = \Delta_z = \{A,B\}$. $\mathcal{I} = \{\{(1,1), (3,2)\}, \{(1,2), (2,1)\}, \{(2,2), (3,1)\}\}$. $\Delta_{\mathcal{I}_1} = \{A,B\}$ and grounding $\mathcal{I}_1$ with A, yields the partially ground formula, $R_1(A,y) \wedge R_2(y,z) \Rightarrow R_3(z,A)$.

To reduce the total number of formulas in $G_{\mathcal{M}}$, we reduce the number of groundings in each $\mathcal{I}_k \in \mathcal{I}$ independently. Specifically, for each $\Delta_{\mathcal{I}_k}$, we learn a new domain, $\hat{\Delta}_{\mathcal{I}_k}$ and a surjective mapping $\zeta : \Delta_{\mathcal{I}_k} \to \hat{\Delta}_{\mathcal{I}_k}$, i.e., $\forall \mu \in \hat{\Delta}_{\mathcal{I}_k}$, $\exists C \in \Delta_{\mathcal{I}_k}$ such that $\zeta(C) = \mu$. We formulate this domain-reduction problem ($|\hat{\Delta}_{\mathcal{I}_k}| << |\Delta_{\mathcal{I}_k}|$) as a standard clustering problem below.

Definition 1. *Given a distance measure d between any two groundings of $\mathcal{I}_k \in \mathcal{I}$ and the number of clusters for $\mathcal{I}_k$ equal to r_k, we define the clustering problem as,*

$$\min_{\mathbf{C}_1 \ldots \mathbf{C}_P} \sum_{k=1}^{P} \sum_{j=1}^{r_k} \sum_{C_{kj} \in \mathbf{C}_{kj}} d(C_{kj}, \mu_{kj}) \tag{2}$$

where $\mathbf{C}_{kj}$ corresponds to all groundings of $\mathcal{I}_k$ that are placed in cluster j, μ_{kj} is the cluster-center of $\mathbf{C}_{kj}$, i.e., it represents the "average grounding" for that cluster, $\zeta^{-1}(\mu_{kj}) = \mathbf{C}_{kj}$.

Each cluster-center in some sense "compresses" the original domain, and we generate a new MLN $\hat{\mathcal{M}}$ from $\mathcal{M}$ by replacing each $\Delta_{\mathcal{I}_k}$ with $\hat{\Delta}_{\mathcal{I}_k} = \{\mu_{kj}\}_{j=1}^{r_k}$. Importantly, the formulation in Eq. (2) allows us control the inference-complexity in $\hat{\mathcal{M}}$ even when $G_{\mathcal{M}}$ is extremely large. This is important because, for arbitrary MLN structures or for inference with evidence, even state-of-the-art inference techniques end up working on a model whose size is comparable to $G_{\mathcal{M}}$ and in most cases, $G_{\mathcal{M}}$ grows rapidly with domain-size. For example, consider the MLN, $\text{R}(x, y) \wedge \text{S}(y, z) \Rightarrow \text{T}(z, x)$, even for $\Delta_x = \Delta_y = \Delta_z = \Delta_u = 100$, the number of formulas in $G_{\mathcal{M}}$ is already one million. Further, the search space (for search-based algorithms) or the sampling space (for sampling-based algorithms) is massive, i.e., exponential in the total number of ground atoms in the MLN. By clustering, we are essentially compressing this large space and now any existing inference algorithm can be used to solve large problems as they implicitly work in this reduced space. The key advantage is that this space complexity can now be controlled based on the cluster-size. Specifically,

Proposition 1. *The number of ground atoms in $\hat{\mathcal{M}}$ is $O(Mr^A)$, where M is the number of predicates in $\mathcal{M}$, $r = \max_k r_k$ and A is the maximum arity of a predicate in $\mathcal{M}$.*

Clearly, the ground atoms in $\hat{\mathcal{M}}$ are different from those in $\mathcal{M}$. Specifically, an atom in $\hat{\mathcal{M}}$ is ground with cluster-centers rather than concrete objects of the original MLN. Thus, one ground atom in $\hat{\mathcal{M}}$ implicitly corresponds to multiple ground atoms in $\mathcal{M}$. This also means that in $\hat{\mathcal{M}}$, the original evidence $\mathbf{E}$ needs to be modified because it is specified on the ground atoms of $\mathcal{M}$. Therefore, we approximate $\mathbf{E}$ with $\hat{\mathbf{E}}$ which specifies the evidence on atoms ground with cluster-centers instead of the original objects in $\mathcal{M}$. To specify this, we define the *expansion* of a ground atom in $\hat{\mathcal{M}}$ as the set of all groundings that it represents in $\mathcal{M}$. Formally,

Definition 2. *The expansion of the j-th ground atom corresponding to the i-th predicate $(\text{R}_i(\mu_{i_1 j_1}, \ldots \mu_{i_{A_i} j_{A_i}}))$ in $\hat{\mathcal{M}}$ is denoted by π_{ij} and consists of all distinct ground atoms of the form $\text{R}_i(C_1, \ldots, C_{A_i})$ where $C_k \in \zeta^{-1}(\mu_{i_k j_k})$.*

Clearly, if we assert in $\hat{\mathbf{E}}$ that a ground atom in $\hat{\mathcal{M}}$ is True (or False), this implicitly asserts that every grounding in its expansion is True (or False). Given a clustering of the domains, in order to best approximate $\mathbf{E}$ for this clustering, we minimize the approximation error as follows.

$$\min_{\hat{\mathbf{E}}} |\mathbf{E} \triangle \bar{\pi}(\hat{\mathbf{E}})| \tag{3}$$

where $\hat{\mathbf{E}}$ is a subset of the ground atoms in $\hat{\mathcal{M}}$ and each grounding is assigned a sign (positive/True or negative/False), $\bar{\pi}(\hat{\mathbf{E}})$ expands every grounding in $\hat{\mathbf{E}}$ and assigns each grounding in the expansion the same sign as its corresponding grounding in $\hat{\mathbf{E}}$. The $\triangle$ operator computes the symmetric difference between $\mathbf{E}$ and $\bar{\pi}(\hat{\mathbf{E}})$. (Note that a grounding with different signs is treated as distinct elements for our purpose). $\hat{\mathbf{E}}$ can be optimally chosen using the following proposition.

Proposition 2. *Let* π_{ij} *be the expansion of one grounding* $(\hat{E})$ *in* $\hat{\mathbf{E}}$. *Let* n_+ *be the count of positive-sign elements and* n_-, *the count of negative-sign elements in* $\pi_{ij} \cap \mathbf{E}$. *Eq. (3) is optimized by assigning* $\hat{E}$ *as positive (negative) if* $n_+ \geq \frac{|\pi_{ij}|}{2}$ $\left(n_- \geq \frac{|\pi_{ij}|}{2} \right)$.

Algorithm 1 shows a schematic illustration of our algorithm to compute the marginal probabilities in an MLN given evidence. Algorithm 1 needs three other algorithms to be specified namely, the distance function, clustering algorithm and the inference algorithm. The amount of reduction applied to each domain is specified as the cluster-bound α. The algorithm starts by computing the partition $\mathcal{I}$ from the term dependencies in $\mathcal{M}$. Next, to each $\mathcal{I}_k \in \mathcal{I}$, the clustering algorithm $\mathcal{L}$ is applied which outputs the clustered domain $\Delta_{\mathcal{I}_k}$ as well as the mapping function ζ. $\Delta_{\mathcal{I}_k}$ is now replaced by its approximation in the new MLN $\hat{\mathcal{M}}$. Once all the domains are suitably reduced, the next step is to approximate the evidence based on the reduced domains. Using Proposition 2, for every grounding of every atom in $\hat{\mathcal{M}}$, we make a decision as to whether it is to be considered positive evidence, negative evidence or treated as a grounding whose truth value is unknown. This yields the approximate evidence set $\hat{\mathbf{E}}$. We then invoke the inference algorithm $\mathcal{F}$ to compute the marginals in $\hat{\mathcal{M}}$. Finally, we project the results obtained on $\hat{\mathcal{M}}$ back to the original domains. Specifically, if a grounding in $\hat{\mathcal{M}}$ has a marginal probability p, then each grounding in its expansion is assigned the same probability.

3.2   Distance Function

The distance function is a key parameter that affects the quality of the generated clusters in Eq. (2) and in turn the inference results computed in Algorithm 1. The advantage of our formulation is that it is quite easy to plug-in a new distance function and generate "new" inference algorithms targeted towards specific applications or datasets. Here, we develop a generic distance measure using the evidence specified on the MLN.

Example 2. Consider the MLN with one formula $R(x) \Rightarrow S(x,y)$ with weight 1.75 and domain $\Delta_x = \{A, B, C\}$. Let the evidence $\mathbf{E} = \{R(A), R(B)\}$. The task is to compute the marginal probabilities of all groundings of $S(x,y)$ which we refer to as the query. The exact marginal probabilities for the query are, $S(A,y) = S(B,y) = 0.5$, $S(C,y) =$

Algorithm 1. Compute-Marginals

Input: MLN $\mathcal{M}$, Evidence $\mathbf{E}$, set of query predicates $\mathbf{Q}$, Distance function d, Clustering function $\mathcal{L}$, Inference algorithm $\mathcal{F}$, cluster-bound α
Output: Marginal probabilities $\mathcal{P}$ for each ground atom corresponding to a predicate in $\mathbf{Q}$
1 Compute the partition $\mathcal{I}$ from $\mathcal{M}$
2 $\hat{\mathcal{M}} = \mathcal{M}$
3 **for** $\mathcal{I}_k \in \mathcal{I}$ **do**
4      $numclusters = \alpha \times \Delta_{\mathcal{I}_k}$
5      $(\hat{\Delta}_{\mathcal{I}_k}, \zeta) = \mathcal{L}(numclusters, d)$
6   Replace $\Delta_{\mathcal{I}_k}$ with $\hat{\Delta}_{\mathcal{T}_k}$ in $\hat{\mathcal{M}}$

7 Construct $\hat{\mathbf{E}}$ based on Proposition 2
8 $\hat{\mathcal{P}} = \mathcal{F}(\hat{\mathcal{M}}, \hat{\mathbf{E}}, \mathbf{Q})$
9 **for** *Each* $R_k \in \mathbf{Q}$ **do**
10   **for** *Each j, where j indexes the possible groundings of R_k in $\hat{\mathcal{M}}$* **do**
11   **for** *Each t, where t indexes the possible groundings of R_k in the expansion π_{kj}*
           do
12             $\mathcal{P}(R_k, t) = \hat{\mathcal{P}}(R_k, j)$

13 return $\mathcal{P}$

0.56. Thus, an ideal distance function should give us a clustering of Δ_x where A and B are placed in the same cluster as they have the same marginals w.r.t the query variable. To do this, we observe that the evidence on $R(A) \Rightarrow S(A,y)$ and $R(B) \Rightarrow S(B,y)$ are "symmetrical", i.e., they satisfy the same number of groundings and consequently the number of groundings that are left unsatisfied in both the formulas is the same. In other words, when $x = A$, the relevant evidence yields MLN $\mathcal{M}'$ and when $x = B$, its yields $\mathcal{M}''$ and if $\mathcal{M}'$ is sufficiently close to $\mathcal{M}''$, we would want all the groundings where $x = A$ clustered together with the groundings where $x = B$ because they are likely to have the same marginal probabilities. We formalize this intuitive idea below.

Let $\mathcal{M}_{C_{kj}}$ represent the MLN obtained after grounding $\mathcal{I}_k$ with the j-th constant in $\Delta_{\mathcal{I}_k}$. Clearly, in the general case, for any two distinct $j_1, j_2, \mathcal{M}_{C_{kj_1}}$ and $\mathcal{M}_{C_{kj_2}}$ are not necessarily independent MLNs as there may be atoms in $\mathcal{M}_{C_{kj_1}}$ that are also present in $\mathcal{M}_{C_{kj_2}}$. However, in our distance function, we relax the constraints/dependencies between $\mathcal{M}_{C_{kj_1}}, \mathcal{M}_{C_{kj_2}}$ and assume these to be independent MLNs and compute the distance between these two MLNs. Specifically, we define a feature vector $\mathbf{U}_{C_{kj}} = c_{f_1},$ $\ldots c_{f_N}$, where c_{f_k} is the number of groundings in formula f_k of MLN $\mathcal{M}_{C_{kj}}$ satisfied due to the evidence $\mathbf{E}$. The distance is computed as $d(C_{kj_1}, C_{kj_2}) = \|\mathbf{U}_{C_{kj_1}} - \mathbf{U}_{C_{kj_2}}\|$.

Even though the above distance function seems like an intuitive and reasonable heuristic, it turns out that computing the distance function efficiently is infeasible in the general case because computing the counts in $\mathbf{U}_{C_{kj}}$ is a hard problem when $\mathbf{E}$ is large. Formally, the following result has been shown in [4],

Theorem 1. *Computing the number of satisfied groundings of a first-order clause in a database is #P-complete in the length of the clause.*

Using database terminology, computing the counts in $\mathbf{U}_{C_{kj}}$ requires computing joins over an arbitrary number of relations (or tables). Therefore, we further relax the constraints/dependencies within the atoms in a formula to guarantee feasibility of computing $\mathbf{U}_{C_{kj}}$ when the size of the evidence-set is very large. In order to formalize this clearly, we specify the ground atoms using a relational database. Further, we also assume that each formula is reduced to a clausal form. The i-th predicate R_i is stored as a relational database table R_i with $A_i + 1$ columns (A_i is the arity), namely, $id_1,, id_2 \ldots id_{A_i}$ and val. The first A_i columns correspond to a specific grounding and the val column specifies whether that ground atom is True ($val = 1$), False ($val = 0$) or unknown ($val = -1$). Given such a database, computing the feature vector involves counting the number of groundings of the formulas in $\mathcal{M}_{C_{kj}}$ that are satisfied by the evidence, which according to Theorem 1 is a hard problem in the general case. Though Theorem 1 is not an issue when the number of evidence atoms is small, to scale-up inference to arbitrarily large evidence-sets, we adopt the following approach. Instead of computing the exact number of groundings for a formula satisfied by the evidence, which involves an arbitrary number of joins over the relations in the formula, we approximate this with a vector of counts, where each count is computed on a subset of relations and the computation involves a bounded number of joins over these relations.

Example 3. Let $\mathcal{M}$ contain one formula, $\neg R(x, y) \vee \neg S(y,z) \vee T(z,x)$, where $\Delta_x = \{A, B, C\}$. To compute the count of satisfied groundings for $x = A$, we compute its inverse, i.e., the number of unsatisfied groundings for $x = A$. The satisfied count is simply the difference between the total number of groundings and the number of unsatisfied groundings. Since the total number of groundings $\Delta_y \times \Delta_z$ is a constant for all groundings of x, it does not affect the clustering and we simply ignore it. The unsatisfied groundings for $x = A$ is given by the following relational algebra expression

$$\sigma_{R.val=1 \wedge S.val=1 \wedge T.val=0} \big((\sigma_{R.id_1=A}(R) \bowtie_{R.id_2=S.id_1} S) $$
$$\bowtie_{S.id_2=T.id_1 \wedge R.id_1=T.id_2} T \big) \qquad (4)$$

where σ is the selection operator and $\bowtie$ is the join operator. Clearly, the above expression has two joins. However, if we impose a constraint that no joins are allowed during the computation of the feature vector, we approximate Eq. (4) by implicitly assuming that each predicate in the formula is independent i.e. we ignore the joins to obtain a vector of counts by counting the tuples returned by 3 separate queries, $\sigma_{R.val=1 \wedge R.id_1=A}(R)$, $\sigma_{S.val=1}(S)$ and $\sigma_{T.val=0 \wedge T.id_2=A}(T)$. An alternate distance function can be obtained if we only allow exactly one join in a query. In this case, we can get a better approximation of Eq. (4) by considering two queries,

$$\sigma_{R.id_1=A \wedge R.val=1}(R) \bowtie_{R.id_2=S.id_1} \sigma_{S.val=1}(S)$$

$$\sigma_{S.val=1}(S) \bowtie_{S.id_2=T.id_1} \sigma_{T.val=0 \wedge T.id_2=A}(T)$$

The algorithm illustrated in Fig. 3 generalizes the idea in the above example and computes the feature vectors for a specific $\mathcal{I}_k \in \mathcal{I}$. The algorithm generates multiple queries corresponding to each grounding of $\mathcal{I}_k$ such that the number of joins in each query is lesser than J. For this, we go over each formula f_t, and first check if f_t is

---

Algorithm 1: Build-Query

    Input: Clausal formula f_t
    Output: Relational-Algebra expression $\mathcal{Q}$
1 $\mathcal{Q} = \emptyset$
2 **for** $R_i \in f_t$ **do**
3 | $Rvalue = 1$
4 | **if** R_i *is positive* **then**
5 | $\lfloor$ $Rvalue = 0$
6 | **if** $\mathcal{Q} = \emptyset$ **then**
7 | $\lfloor$ $\mathcal{Q} = \mathcal{Q} + \sigma_{R_i.val=Rvalue}(R_i)$
8 | **else**
9 | $\lfloor$ $\mathcal{Q} = \mathcal{Q} \bowtie_\theta \sigma_{R_i.val=Rvalue}(R_i)$

Fig. 2. Building a query for the feature vector

relevant to $\mathcal{I}_k$, i.e., if f_t contains at least one atom corresponding to R_i such that for some p, $(i, p) \in \mathcal{I}_k$, then f_t is a relevant formula for clustering $\mathcal{I}_k$, otherwise, we ignore f_t. This is because, the features from f_t which are not relevant to $\mathcal{I}_k$ remains identical for every grounding of x and therefore never affects the clustering. For every relevant f_t, we first build the complete query which is a sequence of θ-joins on the tables corresponding to every atom in f_t. The query selects the the groundings of f_t that are not satisfied by the evidence. The θ in the join specifies variables shared among atoms in f_t. For e.g. In a formula $\neg R(x) \vee S(x)$, the θ-join is specified as $\sigma_{R.val=1}(R) \bowtie_{R.id_1=S.id_1} \sigma_{S.val=0}(S)$. Once we build the full query, we simply walk through the query executing no more than J joins at a time. For each atom which has a variable that corresponds to some element of $\mathcal{I}_k$, we ground the variable by enforcing the select condition in line 11 of the algorithm. We execute the partial query $\mathcal{Q}'$ with a maximum of J joins and store the result (count) in the feature vector. Next, we remove $\mathcal{Q}'$ from $\mathcal{Q}$ and relax the next θ- join condition as follows. Among all the tables mentioned in $\mathcal{Q}'$, we select one table R_s, that participates in the next join operation in $\mathcal{Q} - \mathcal{Q}'$. We only retain the join conditions related to R_s in the next join in $\mathcal{Q} - \mathcal{Q}'$ and remove the rest of the conditions. We continue until we empty the original query $\mathcal{Q}$. Finally, we return the vector of counts accumulated across all queries for each grounding of $\mathcal{I}_k$.

4   Related Work

Several previous approaches have been suggested for improving the scalability of inference in MLNs. Most of these approaches can be termed as lifted inference algorithms since they either use rules that can be directly applied on the first-order structure or identify symmetries in the ground representation to perform efficient inference. Both exact [3,6,26,1] as well as approximate [20,10,7,13,27,2] lifted algorithms have been developed that can greatly improve scalability. However, all these algorithms are efficient only when given the right MLN structure/evidence. Specifically, [1,25] show that efficient inference is possible when presented with specific evidence-structure. More

---

Algorithm 1: Compute-Features

Input: $\mathcal{M}$ and its associated relational DB, join-bound J, $\mathcal{I}_k \in \mathcal{I}$

Output: Feature vector set $\{\mathbf{U}_{C_{kj}}\}_{j=1}^{\Delta_{\mathcal{I}_k}}$

1   $\mathbf{U} = \emptyset$
2   **for** $C_{kj} \in \Delta_{\mathcal{I}_k}$ **do**
3     $\mathbf{U}_{C_{kj}} = \emptyset$
4   **for** $f_t \in \mathbf{F}$ **do**
5   **if** f_t *is not relevant to* $\mathcal{I}_k$ **then**
6   continue
7       $\mathcal{Q} = $ Build-Query(f_t)
8   **while** $\mathcal{Q}$ *not empty* **do**
9         $\mathcal{Q}' = $ Select a sub-query containing up to the first J joins in $\mathcal{Q}$
10   **for** $R_i \in \mathcal{Q}'$ **do**
11   **if** $\exists\, p$ *such that* $(i, p) \in \mathcal{I}_k$ **then**
12   Wrap a select $(\sigma_{R_i.id_p = C_{kj}})$ around R_i
13        $\mathbf{U}_{C_{kj}}$.append(Count($\mathcal{Q}'$))
14   Let R_s be a table in $\mathcal{Q}'$ whose attribute participates in the θ-join after $\mathcal{Q}'$
15   **if** $R_s = \emptyset$ **then**
16          $\mathcal{Q} = (\mathcal{Q} - \mathcal{Q}')$
17   **else**
18   Relax the θ-join and include only those constraints involving R_s
19          $\mathcal{Q} = R_s \bowtie_\theta (\mathcal{Q} - \mathcal{Q}')$
20     $\mathbf{U}$.append($\mathbf{U}_{C_{kj}}$)
21 **return** $\mathbf{U}$

Fig. 3. Algorithm to compute the feature vectors

recently, [24] have proposed to counter the evidence-problem by adding more symmetries that make the MLN liftable. Specifically, they compute a low-rank boolean matrix factorization of the evidence matrix which implicitly induces a clustering whereas we explicitly cast it as a clustering problem thereby allowing us the flexibility to use a range of clustering algorithms and also better control of the inference-complexity. Further, [24] handles only binary evidence while our approach is much more general. Finally, our approach of pre-processing the MLN is related to [19] which develops a systematic grounding procedure that can reduce the ground MLN size in many cases, and our approach of leveraging databases for MLN inference is related to [14].

5 Experiments

5.1 Setup

We evaluate our approach on 4 benchmark MLNs available in Alchemy [11], namely Entity Resolution (ER), Segmentation (Seg), Web Linkage analysis (WebKB) and Protein Interaction (Protein). Additionally, we added two new MLNs that have different structures called Student (Teaches(i, c) $\wedge$ Prereq(c, $c1$) $\Rightarrow$ Takes(s, $c1$)) and Relation (Related(i, j) $\wedge$ Friends(j, k) $\Rightarrow$ Loves(k, i)). For our experiments, we implemented our system using MySQL. To speed up query processing, we created n indexes for a table corresponding to a n-ary predicate, where the column corresponding to each argument of a predicate is indexed separately. For the distance function, we limit the number of joins (J) to 1. In the inference subroutine, we used two algorithms, a lifted algorithm based on message passing, Lifted-BP [20] and a propositional algorithm based on sampling, Gibbs sampling [5]. We used the implementation of both these algorithms from Alchemy. For the clustering subroutine, we experimented with four different algorithms available in Weka [8] namely, KMeans++ (KM), Expectation-Maximization (EM), Hierarchical clustering (HC) and XMeans (XM). We ran all our experiments on a quad-core Ubuntu machine with 6 GB RAM.

5.2   Approximation Results on Benchmarks

Fig. 4 illustrates the approximation error on each benchmark for all combinations of the two inference and four clustering algorithms. The *x-axis* plots the inverse compression ratio $ICR = \frac{N_C}{N_G}$, where N_C is the total number of ground formulas in the approximated MLN after clustering and N_G is the total number of ground formulas in the original MLN. The *y-axis* shows the approximation error calculated as follows. $Err = \frac{\sum_{g \in G} D_{KL}(P_g||P'_g)}{|G|}$, where D_{KL} is the standard KL-Divergence distance measure, G refers to all groundings of a query predicate, P_g is the marginal distribution of g computed from the original MLN and P'_g is computed from the approximate MLN using clustered domains. For fairness, both marginals are computed using the same inference algorithm. We set 50% of arbitrary groundings as evidence, where 25% are True and 25% are False.

Fig. 4 illustrates the trade-off between accuracy and complexity. As ICR increases, the complexity increases, however, the approximation error reduces because we map the original domains to a larger set thereby reducing the difference between the original MLN and the approximate MLN. The structure of the MLN also plays an important role in determining the accuracy of the approximation. For some cases such as Student in Fig. 4 (a), (g) the error goes down quite rapidly initially and stays consistently low afterwards. In some other cases such as ER, Fig. 4 (f), (l), the change is more gradual. This is because ER contains complex formulas with multiple self-joins such as the transitive relation which make it harder to approximate. In almost all cases for Lifted-BP (except ER), the approximation error was below 0.2 for even small compression ratios. For Gibbs sampling though, in general, it took slightly larger compression ratios before the approximation was close to ground inference such as in the benchmarks Protein (k) and ER (l). One of the reasons for this could be that deterministic dependencies or hard constraints tend to be problematic for Gibbs sampling [16], i.e., if the probabilities lie at the extremes then the Gibbs sampler mixes very slowly. Therefore the approximations given by Gibbs sampling are not very accurate even for the fully ground model. Among the clustering algorithms, KM and HC clearly outperformed XM and EM. In almost all cases, KM and HC produced clusterings that produced more stable and consistent results compared to EM and XM. For example, in (a), (d) and (h) the EM algorithm gave poor results while in (i), XM gave poor results.

5.3   Effect of Evidence

Fig. 5 illustrates the error for different values of cluster-bounds (α) and varying amount of evidence. The results shown Fig. 5 use K-Means++ for clustering and Lifted-BP for

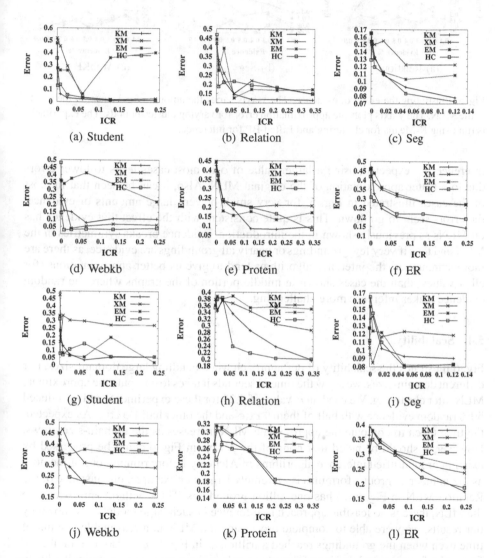

Fig. 4. Approximation-error vs ICR. The y-axis shows the average KL-Divergence of the marginals computed on the clustered MLN from the marginals computed on the original MLN (smaller is better). (a) - (f) show the results using Lifted-BP, (g) - (l) using Gibbs sampling.

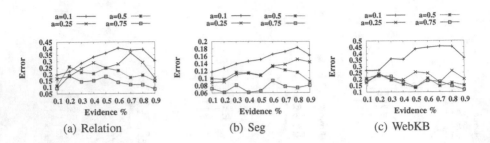

Fig. 5. Illustrating the effect of evidence. The x-axis varies the amount of evidence on the atoms in the MLN. The y-axis plots the approximation error for varying cluster-bounds. The experiment is run using K-Means for clustering and Lifted-BP for inference.

inference. As expected, using a larger value of α in most cases leads to lower errors due to a better approximation of the original MLN. Also, it can be seen that in most of the cases illustrated in Fig. 5, for very small or very large amounts of evidence, the errors seem to go down. This is quite consistent with the effect that evidence has on MLNs as previously shown in the introduction. Evidence breaks symmetries in the MLN and thus if very few groundings or nearly all groundings are evidence, as there are more symmetries, the inference algorithms tend to give us better approximations (for all α values) than the cases shown in middle portion of the graphs where the random evidence makes inference more challenging.

5.4   Scalability

Fig. 6 illustrates the scalability of our approach when handling large domain-sizes. For different domain-sizes, we show the time in seconds it takes to compute the approximate MLN after clustering. We used an α value of 0.25 for these experiments and introduced 50% random evidence with half of them `True` and the other half `False`. As expected, the time taken to compute the approximate MLN increases as the domain-size grows. However, it should be noted that none of the MLNs in Fig. 6 could be processed by existing ground/lifted inference algorithms in Alchemy before running out of memory as the number of ground formulas is extremely large. For example, one instance of the Relation MLN in Fig. 6 (a) has one billion groundings. Thus, without approximating the MLN, there is no feasible approach to inference in such large models. As shown by our results, we were able to complete processing the MLN in a reasonable amount of time even when the groundings reached a trillion as in Fig. 6 (e). Also, the number of first-order formulas and their structure play a role in determining the complexity due to the distance function computation. Recall that we compute a vector for every formula in the MLN. Therefore, a larger number of formulas mean more computations on the database. For instance, Fig. 6 (e) has just one formula while (f) has 8 formulas which have more complex structure. Therefore, even though the number of ground formulas in (e) is a trillion while in (f) it is a billion, we took more time to process (f). Further, it can be seen that for each of the benchmarks, the third instance (the largest MLN) takes a visibly longer time when compared to the first two instances. This is expected because,

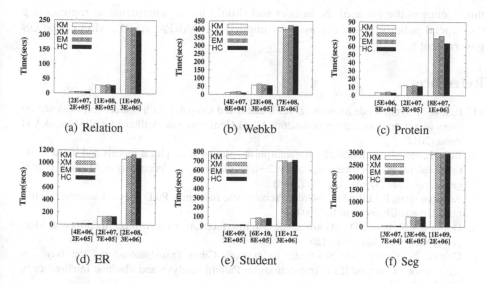

Fig. 6. Scalability experiments. The y-axis shows the time taken to form the approximate MLN and the x-axis shows $[N_f, N_a]$, where N_f is the number of ground formulas and N_a is the number of ground atoms.

when the size of the database grows really large as is the case for very large domain-sizes, it typically requires many more hard disk accesses for query processing which causes it to slow down. Finally, as seen in the results, the type of clustering has minimal impact on the time taken to process the MLN, i.e., nearly all clustering methods took approximately the same amount of time.

6   Conclusion

In this paper, we presented an approach for scaling up inference in MLNs. Existing approaches either ground the MLN which makes it too large to process or use rules to identify symmetries and perform lifted inference. However, lifting rules are applicable only in certain specific, symmetric cases and more importantly, in the presence of ev-idence, these symmetries are broken, rendering lifted inference powerless. To achieve scalable inference for such hard cases in which we can have arbitrary MLN structures with arbitrary evidence, we proposed to compress the original MLN. Specifically, we defined a novel distance function that is sensitive to the evidence presented to the MLN and used it to replace groups of similar objects in the MLN by their cluster centers. Our experimental results on several benchmark MLNs clearly illustrated the high accuracy and scalability of our approach.

Acknowledgements. This research was partly funded by ARO MURI grant W911NF-08-1-0242, by the AFRL under contract number FA8750-14-C-0021 and by the DARPA Probabilistic Programming for Advanced Machine Learning Program under AFRL prime contract number FA8750-14-C-0005. The views and conclusions contained in this document are those of the authors and should not be interpreted as representing the official policies, either expressed or implied, of DARPA, AFRL, ARO or the US government.

References

1. Bui, H., Huynh, T., de Salvo Braz, R.: Exact lifted inference with distinct soft evidence on every object. In: Proceedings of the 26th AAAI Conference on Artificial Intelligence. AAAI Press (2012)
2. Bui, H., Huynh, T., Riedel, S.: Automorphism groups of graphical models and lifted variational inference. In: Proceedings of the 29th Conference on Uncertainty in Artificial Intelligence, pp. 132–141. AUAI Press (2013)
3. de Salvo Braz, R.: Lifted First-Order Probabilistic Inference. Ph.D. thesis, University of Illinois, Urbana-Champaign, IL (2007)
4. Domingos, P., Lowd, D.: Markov Logic: An Interface Layer for Artificial Intelligence. Morgan & Claypool, San Rafael (2009)
5. Geman, S., Geman, D.: Stochastic Relaxation, Gibbs Distributions, and the Bayesian Restoration of Images. IEEE Transactions on Pattern Analysis and Machine Intelligence 6, 721–741 (1984)
6. Gogate, V., Domingos, P.: Probabilistic Theorem Proving. In: Proceedings of the Twenty-Seventh Conference on Uncertainty in Artificial Intelligence. pp. 256–265. AUAI Press (2011)
7. Gogate, V., Jha, A., Venugopal, D.: Advances in Lifted Importance Sampling. In: Proceedings of the 26th AAAI Conference on Artificial Intelligence. AAAI Press (2012)
8. Hall, M., Frank, E., Holmes, G., Pfahringer, B., Reutemann, P., Witten, I.H.: The weka data mining software: An update. SIGKDD Explor. Newsl. 11(1), 10–18 (2009)
9. Jha, A., Gogate, V., Meliou, A., Suciu, D.: Lifted Inference from the Other Side: The tractable Features. In: Proceedings of the 24th Annual Conference on Neural Information Processing Systems (NIPS), pp. 973–981 (2010)
10. Kersting, K., Ahmadi, B., Natarajan, S.: Counting Belief Propagation. In: Proceedings of the 25th Conference on Uncertainty in Artificial Intelligence. pp. 277–284. AUAI Press (2009)
11. Kok, S., Sumner, M., Richardson, M., Singla, P., Poon, H., Lowd, D., Wang, J., Domingos, P.: The Alchemy System for Statistical Relational AI. Tech. rep., Department of Computer Science and Engineering, University of Washington, Seattle, WA (2008), http://alchemy.cs.washington.edu
12. Milch, B., Zettlemoyer, L.S., Kersting, K., Haimes, M., Kaelbling, L.P.: Lifted Probabilistic Inference with Counting Formulas. In: Proceedings of the Twenty-Third AAAI Conference on Artificial Intelligence, pp. 1062–1068 (2008)
13. Niepert, M.: Markov chains on orbits of permutation groups. In: Proceedings of the 28th Conference on Uncertainty in Artificial Intelligence. pp. 624–633. AUAI Press (2012)
14. Niu, F., Ré, C., Doan, A., Shavlik, J.W.: Tuffy: Scaling up statistical inference in markov logic networks using an rdbms. PVLDB 4(6), 373–384 (2011)
15. Poole, D.: First-Order Probabilistic Inference. In: Proceedings of the 18th International Joint Conference on Artificial Intelligence, pp. 985–991 (2003)
16. Poon, H., Domingos, P.: Sound and Efficient Inference with Probabilistic and Deterministic Dependencies. In: Proceedings of the 21st National Conference on Artificial Intelligence. pp. 458–463. AAAI Press (2006)
17. Poon, H., Domingos, P.: Joint Unsupervised Coreference Resolution with Markov Logic. In: Proceedings of the 2008 Conference on Empirical Methods in Natural Language Processing. pp. 649–658. ACL (2008)

18. Richardson, M., Domingos, P.: Markov Logic Networks. Machine Learning 62, 107–136 (2006)
19. Shavlik, J.W., Natarajan, S.: Speeding up inference in markov logic networks by preprocessing to reduce the size of the resulting grounded network. In: Proceedings of the 21st International Joint Conference on Artificial Intelligence, pp. 1951–1956 (2009)
20. Singla, P., Domingos, P.: Lifted First-Order Belief Propagation. In: Proceedings of the Twenty-Third AAAI Conference on Artificial Intelligence. pp. 1094–1099. AAAI Press (2008)
21. Singla, P., Mooney, R.J.: Abductive Markov Logic for Plan Recognition. In: Proceedings of the 25th AAAI Conference on Artificial Intelligence, pp. 1069–1075. AAAI Press (2011)
22. Tran, S.D., Davis, L.S.: Event modeling and recognition using markov logic networks. In: Forsyth, D., Torr, P., Zisserman, A. (eds.) ECCV 2008, Part II. LNCS, vol. 5303, pp. 610–623. Springer, Heidelberg (2008)
23. van den Broeck, G.: On the completeness of first-order knowledge compilation for lifted probabilistic inference. In: Proceedings of the 25th Annual Conference on Neural Information Processing Systems (NIPS), pp. 1386–1394 (2011)
24. van den Broeck, G., Darwiche, A.: On the complexity and approximation of binary evidence in lifted inference. In: Proceedings of the 27th Annual Conference on Neural Information Processing Systems (NIPS), pp. 2868–2876 (2013)
25. van den Broeck, G., Davis, J.: Conditioning in first-order knowledge compilation and lifted probabilistic inference. In: Proceedings of the 26th AAAI Conference on Artificial Intelligence. AAAI Press (2012)
26. van den Broeck, G., Taghipour, N., Meert, W., Davis, J., De Raedt, L.: Lifted Probabilistic Inference by First-Order Knowledge Compilation. In: Proceedings of the 22nd International Joint Conference on Artificial Intelligence, pp. 2178–2185 (2011)
27. Venugopal, D., Gogate, V.: On lifting the gibbs sampling algorithm. In: Proceedings of the 26th Annual Conference on Neural Information Processing Systems (NIPS), pp. 1664–1672 (2012)
28. Yedidia, J.S., Freeman, W.T., Weiss, Y.: Generalized Belief Propagation. In: Proceedings of the 14th Annual Conference on Neural Information Processing Systems (NIPS), pp. 689–695 (2000)

On Learning Matrices with Orthogonal Columns or Disjoint Supports

Kevin Vervier[1,2,3], Pierre Mahé[1], Alexandre D'Aspremont[4],
Jean-Baptiste Veyrieras[1], and Jean-Philippe Vert[2,3]

[1] Data and Knowledge Lab, Biomerieux, 69280 Marcy l'Etoile, France
[2] Centre for Computational Biology, Mines ParisTech, 77300 Fontainebleau, France
[3] Institut Curie, INSERM U900, 75005 Paris, France
[4] CNRS and D.I. UMR 8548, École normale supérieure, 75005 Paris, France

Abstract. We investigate new matrix penalties to jointly learn linear models with orthogonality constraints, generalizing the work of Xiao et al. [24] who proposed a strictly convex matrix norm for orthogonal transfer. We show that this norm converges to a particular atomic norm when its convexity parameter decreases, leading to new algorithmic solutions to minimize it. We also investigate concave formulations of this norm, corresponding to more aggressive strategies to induce orthogonality, and show how these penalties can also be used to learn sparse models with disjoint supports.

1   Introduction

Learning several models simultaneously instead of separately, a framework often referred to as multitask or transfer learning, is a powerful setting to leverage information across related but different problems [10,22,4,2,12]. In particular it has been empirically shown that when different tasks share some similarity, such as learning binding models for similar proteins [14], predicting exams score for students of different schools [2,12] or learning models for semantically related concepts in a hierarchy [16,8], jointly learning the different models with a multitask strategy leads to better performance. In all aforementioned examples (and many others), the underlying assumption is that different tasks share some similarity, and the different multitask strategies exploit this assumption by, e.g., imposing shared parameters estimated jointly across the tasks, or penalizing differences between the models learned in different tasks.

Alternatively, in some situations we would like to solve different tasks under the opposite assumption, namely, that the models are *different*, e.g., that they use different features or should be orthogonal to each other. This is the case for example when we want to learn unrelated tasks, such as recognizing the identity and the emotion of a person on a picture, where we know from literature that these two recognition problems depend on different and uncorrelated features of the same image [9,19]. In structured learning such as classification in a hierarchical taxonomy, it has been proposed to learn local models at each node of the hierarchy and to encourage the classifier at each node to be different from

T. Calders et al. (Eds.): ECML PKDD 2014, Part III, LNCS 8726, pp. 274–289, 2014.
© Springer-Verlag Berlin Heidelberg 2014

the classifiers at its ancestors, in order to better reflect the natural coarse-to-fine nature of the classifiers at different levels of the hierarchy [24,13]. Several approaches have been proposed recently to learn such different models. [24] proposed to penalize a weighted ℓ_1 norm of the off-diagonal entries of the covariance matrix between the tasks, in order to promote sparsity of inner products hence orthogonality between tasks; however some extra ridge term must be added in order to make the penalty convex and amenable to efficient optimization, leading to potentially unwanted over-regularization. [19] proposed also a convex penalty to learn two groups of tasks based on orthogonal subspaces; again, due to the non-convex nature of the norm applied to inner products between vectors, an extra ridge term is needed to make the penalty convex. Finally, [13] proposed a method to learn a tree of metrics, enforcing disjoint sparsity between the different metrics. The convex penalty of [13], though, only promotes sparsity for nonnegative vectors, such as the diagonals of metric matrices, and can not easily be extended to enforce disjoint sparsity on general vectors.

In this work, we extend the work of [24] in two directions. First, we investigate generalization of the penalty proposed by [24] when we decrease its convexity, in order to make it more "aggressive" in promoting orthogonality. Our main findings can be visualized in Figure 1, which shows the level sets of penalties we consider. Starting from the strictly convex penalty of [24], corresponding to a strictly convex unit ball with singularities at matrices with orthogonal columns (left), we show that by reducing its convexity it converges to a convex atomic norm [11], whose unit ball is the convex hull of the singularities of the first ball. This shows that for particular choices of parameters the penalty of [24] is "optimal" to learn matrices with pairwise orthogonal columns, in the sense that it is the tightest convex function which is equal to the Frobenius norm on the subset of matrices that we are interested in. This observation has also algorithmic consequences: while [24] propose an optimization scheme that only works when the penalty is strictly convex, we show that the dual norm in the limit case of the atomic norm can be estimated efficiently by solving a small semidefinite program (SDP), leading to new algorithmic solutions to use this norm as regularizer in a learning problem. We also propose and investigate empirically more concave extensions of this norm in order to increase the propensity to learn matrices with orthogonal columns (right). Our second extension is to show how these penalties can be modified to learn sparse models with disjoint supports, a particular case of orthogonal models which is relevant when different tasks are know to involve different features.

2   An Atomic Norm to Learn Matrices with Orthogonal Columns

We consider the problem of learning a $d \times T$ matrix $W = (w_1, \ldots, w_T)$, where each column w_i is a d-dimensional vector corresponding to a task such as a linear classification model at a node of a taxonomy. We call such a matrix *scaled orthogonal* if $W^\top W$ is diagonal, i.e., if all columns of W are orthogonal to each

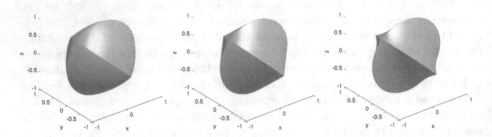

Fig. 1. Level sets of the penalty Ω_K defined in (2) for 2-by-2 symmetric matrices parametrized as $\begin{pmatrix} x & y \\ y & z \end{pmatrix}$, when $K = \begin{pmatrix} \gamma & 1 \\ 1 & \gamma \end{pmatrix}$ and we vary γ from $\gamma = 2$ (left), which corresponds to a strictly convex penalty proposed by [24], to $\gamma = 1$ (center), which is a limit case where the penalty is convex but not strictly convex and turns out to be an atomic norm (Theorem 1), and to $\gamma = 1/2$ (right), which corresponds to a non convex penalty.

other, and denote by $\mathcal{O}$ the set of $d \times T$ scaled orthogonal matrices. Note that this should not be confused with the stronger concept of orthogonal matrix often used in mathematics, which means that W is square and $W^\top W$ is the identity, i.e., that the columns form an orthonormal basis.

A general approach to estimate W from observations is to formulate the inference as an optimization problem:

$$\min_W f(W) + \frac{\lambda}{2}\, \Omega(W)^2 \,, \tag{1}$$

where $f(W)$ is an empirical risk which measures the fit to data (e.g., variance captured in the case of dimensionality reduction, empirical error on the training set in the case of regression and classification), $\Omega(W)$ is a penalty that enforces some constraints on the solution such as sparseness or low-rankness, and $\lambda > 0$ is a parameter adjusting the tradeoff between these two objectives. When $f(W)$ and $\Omega(W)$ are convex functions, then (1) is a convex optimization problem that can often be solved efficiently and lead to a unique solution. Classical examples of penalties $\Omega(W)$ include the ℓ_1 norm to promote sparsity in W [23], the nuclear norm to learn low-rank matrices [21], and the ℓ_1/ℓ_2 norm to perform joint feature selection across tasks [17].

Suppose we know that some or all of the columns of W should be orthogonal to each other. [24] proposed an orthogonal regularizer of the form $\sum_{i,j} K_{i,j}|w_i^\top w_j|$, where $K_{i,j}$ is a nonnegative weight to enforce more or less the orthogonality between w_i and w_j. This is however not a convex function of W, and [24] propose to define a convex penalty by adding ridge terms to this regularizer, namely:

$$\Omega_K(W)^2 = \sum_{i=1}^{T} K_{ii}||w_i||^2 + \sum_{i \neq j} K_{ij}|w_i^\top w_j|, \tag{2}$$

where K is an hyperparameter matrix representing structure among different models. [24] give a sufficient condition on K to ensure that (2) is convex, but there remains a lot of freedom in the choice of K.

Let us consider the case where we choose $K_{ii} = 1$ and $K_{ij} > 0$ in (2). Then we see that for scaled orthogonal matrices $W \in \mathcal{O}$ the penalty (2) boils down to the Frobenius norm:

$$\forall W \in \mathcal{O}, \quad \Omega_K(W)^2 = \sum_{i=1}^{T} ||w_i||^2 = ||W||_F^2.$$

The extra terms $K_{ij}|w_i^\top w_j|$ in (2) ensure that, in addition, the penalty is not differentiable at scaled orthogonal matrices, allowing under some conditions the recovery of such matrices when (2) is plugged into (1) [1,11].

There are however many penalties, including (2), that are convex, singular on $\mathcal{O}$ and which equal the Frobenius norm in $\mathcal{O}$. Among them, we propose to consider the *tightest* one, namely, the atomic norm in the sense of [11] induced by the set of atoms $\mathcal{A} = \{W \in \mathcal{O} : ||W||_F = 1\}$. This norm, which we denote below by $\Omega_\mathcal{O}(X)$ for any $d \times T$ matrix X, can be expressed as

$$\Omega_\mathcal{O}(X) = \inf \left\{ \sum_{Y \in \mathcal{A}} \lambda_Y : X = \sum_{Y \in \mathcal{A}} \lambda_Y Y, \lambda_Y \geq 0 \right\}. \tag{3}$$

In other words, this last expression writes $\Omega_\mathcal{O}(X)$ as the ℓ_1 norm of the vector of coefficients λ in a decomposition of X into atoms, namely, scaled orthogonal matrices of unit Frobenius norms. Plugging (3) into (1) provides a convex problem to infer an atom, or a sparse combination of atoms. Note that, contrary to Ω_K (2), $\Omega_\mathcal{O}$ is always convex without technical conditions. In addition, since both norms are equal on the atoms $\mathcal{A}$, the tangent cone of $\Omega_\mathcal{O}$ at any scaled orthogonal matrix $W \in \mathcal{O}$ is contained in the tangent cone of Ω_K at the same point, suggesting that the recovery and inference of a scaled orthogonal matrix through the convex procedure (1) is easier with $\Omega_\mathcal{O}$ than with Ω_K [11].

The following result shows that, surprisingly, the norms Ω_K with adequate weights and $\Omega_\mathcal{O}$ coincide on matrices with two columns. This theorem is illustrated in Figure 1, where we show the unit ball of Ω_K when we change K. The ball at the center corresponds to a limit situation where Ω_K is still convex, but not strictly convex anymore. We see in this picture that the ball can equivalently be defined as the convex hull of two circles, which correspond precisely to the set of matrices with orthogonal columns and unit Frobenius norm; i.e., that Ω_K in this case is precisely the atomic norm induced by these atoms.

Theorem 1. *For any $d \geq 1$ and any $d \times 2$ matrix $W = (w_1, w_2)$, it holds that:*

$$\Omega_{\mathcal{O}}(W) = \Omega_K(W),\tag{4}$$

with

$$K = \begin{pmatrix} 1 & 1 \\ 1 & 1 \end{pmatrix}.\tag{5}$$

Proof. Since K in (5) is entry-wise nonnegative, and since the companion matrix

$$\bar{K} = \begin{pmatrix} 1 & -1 \\ -1 & 1 \end{pmatrix}$$

is positive semidefinite, we know from [24, Theorem 1] that Ω_K^2 is convex in this case. Since (4) obviously holds for $W \in \mathcal{O}$, and since $\Omega_{\mathcal{O}}$ is the tightest convex function such that (4) holds on $\mathcal{O}$, we directly get that $\Omega_{\mathcal{O}}(W) \leq \Omega_K(W)$ for any $W \in \mathbb{R}^{d \times 2}$. To prove the converse inequality, it suffices to find, for any $W \in \mathbb{R}^{d \times 2}$, a decomposition of the form $W = \lambda U + (1 - \lambda)V$, with $U, V \in \mathcal{O}$, $\lambda \in [0, 1]$, such that $\Omega_K(U) = \Omega_K(V) = \Omega_K(W)$. Geometrically, this would mean that any point on the unit ball of Ω_K lies on a straight segment that connects two atoms on this ball, meaning that the unit ball of Ω_K is precisely the convex hull of the unit ball restricted to the atoms. The following lemma, which can be proved by direct calculation, shows that this is indeed possible by explicitly providing such a decomposition.

Lemma 1. *For any $W = (w_1, w_2) \in \mathbb{R}^{d \times 2}$, let:*

- *if $w_1^\top w_2 \geq 0$, $U = (w_1 + w_2, 0)$ and $V = \left(w_1 - \frac{w_1^\top w_2}{\| w_2 \|^2} w_2, \left(1 + \frac{w_1^\top w_2}{\| w_2 \|^2} \right) w_2 \right)$,*
- *if $w_1^\top w_2 < 0$, $U = (w_1 - w_2, 0)$ and $V = \left(w_1 - \frac{w_1^\top w_2}{\| w_2 \|^2} w_2, \left(1 - \frac{w_1^\top w_2}{\| w_2 \|^2} \right) w_2 \right)$,*

and let $\lambda = \frac{| w_1^\top w_2 |}{| w_1^\top w_2 | + \| w_2 \|^2}$. Then it holds that:

- *$U, V \in \mathcal{O}$,*
- *$\lambda \in [0, 1]$ and $W = \lambda U + (1 - \lambda)V$,*
- *$\Omega_K(W) = \Omega_K(U) = \Omega_K(V)$.*

∎

Theorem 1 can be easily generalized (with a different set of atoms) when K is any 2-by-2 symmetric, positive semidefinite matrix with non-negative entries and with 0 as eigenvalue, corresponding to the limit case where Ω_K is convex but not strictly convex: it is then always an atomic norm. The extension of Theorem 1 to more than 2 columns, however, is not true. Atoms of $\Omega_{\mathcal{O}}$ are matrices with *all* columns orthogonal to each other, so using $\Omega_{\mathcal{O}}$ as a penalty on matrices with $T > 2$ columns may either lead to such an atom, or to a sparse linear combination of atoms, which would in general have no pair of column orthogonal to each other. The following theorem, which is a simple consequence of Theorem 1, shows that for some choices of K in the $T > 2$ case, the penalty Ω_K can be written as a sum of $\Omega_{\mathcal{O}}$ that penalizes pairs of columns.

Theorem 2. *For any $T \geq 2$, let K be a symmetric T-by-T matrix with non-negative entries and such that, for any $i = 1,\ldots,T$,*

$$\forall i = 1,\ldots,T \quad K_{ii} = \sum_{j \neq i} K_{ij}.$$

Then, for any $d \geq 1$ and any $d \times T$ matrix $W = (w_1,\ldots,w_T)$, it holds that:

$$\Omega_K(W) = \sum_{i<j} K_{ij}\Omega_\mathcal{O}((w_i, w_j)),$$

where $(w_i, w_j) \in \mathbb{R}^{d \times 2}$ is the matrix with columns w_i and w_j.

Proof. Let $A = \begin{pmatrix} 1 & 1 \\ 1 & 1 \end{pmatrix}$. By Theorem 1, we know that $\Omega_A((w_i, w_j)) = \Omega_\mathcal{O}((w_i, w_j))$ for al $i \neq j$, therefore:

$$\sum_{i<j} K_{ij}\Omega_\mathcal{O}((w_i, w_j)) = \sum_{i<j} K_{ij}\Omega_A((w_i, w_j))$$

$$= \sum_{i<j} K_{ij}\left(\|w_i\|^2 + \|w_j\|^2 + 2|w_i^\top w_j|\right)$$

$$= \sum_{i=1}^{T}\left(\sum_{j \neq i} K_{ij}\right)\|w_i\|^2 + \sum_{i \neq j}|w_i^\top w_j|$$

$$= \Omega_K(W).$$

∎

3   The Dual of the Atomic Norm

In this section we consider the atomic norm $\Omega_\mathcal{O}$ for matrices with 2 columns, and show that we can efficiently compute its dual and a subgradient of its dual by solving a 6-dimensional SDP. This can be useful to provide simple duality gaps and stopping criteria to learn with convex but not strictly convex penalties Ω_K, which are in particular not amenable to optimization with the method of [24].

Remember that the dual of a norm $\Omega(X)$ is

$$\Omega^*(X) = \sup_{Y \,:\, \Omega(Y) \leq 1} \mathbf{Tr}(X^\top Y).$$

Since $\Omega_\mathcal{O}$ is an atomic norm induced by the atom set $\mathcal{A}$, its dual satisfies [11]:

$$\Omega_\mathcal{O}^*(X) = \sup_{Y \in \mathcal{A}} \mathbf{Tr}(X^\top Y), \tag{6}$$

and in addition any atom $Y \in \mathcal{A}$ which achieves the maximum in (6) is a subgradient of $\Omega_\mathcal{O}^*$ at X. We now show that computing $\Omega_\mathcal{O}^*(X)$ and a subgradient can be done efficiently:

Theorem 3. *For any $d \geq 1$ and $X \in \mathbb{R}^{d \times 2}$, a solution to*

$$\Omega_{\mathcal{O}}^*(X) = \sup_{Y \in \mathcal{A}} \mathbf{Tr}(X^{\top} Y) \qquad (7)$$

can be obtained from the solution of a SDP over matrices of size 6×6.

Proof. From the definition of $\mathcal{A}$ we can reformulate (7) as:

$$\begin{aligned}
\Omega_{\mathcal{O}}^*(X) = \text{maximize} \quad & \mathbf{Tr}(Y^{\top} X) \\
\text{subject to} \quad & Y^{\top} Y \text{ diagonal} \\
& \|Y\|_F = 1,
\end{aligned}$$

in the variable $Y \in \mathbb{R}^{d \times 2}$. Because $-Y$ is a feasible point whenever Y is, this problem is equivalent to

$$\begin{aligned}
\Omega_{\mathcal{O}}^*(X)^2 = \text{maximize} \quad & \mathbf{Tr}(Y^{\top} X)^2 \\
\text{subject to} \quad & Y^{\top} Y \text{ diagonal} \qquad (8) \\
& \|Y\|_F = 1,
\end{aligned}$$

which is a *non-convex* quadratic program in Y. We first reformulate this problem in "vector" terms and write $z = \mathbf{vec}(Y) \in \mathbb{R}^{2d}$, so that $z^{\top} = (z_1^{\top}, z_2^{\top})$ with $z_1 = Y_1$ and $z_2 = Y_2$. Problem (8) becomes

$$\begin{aligned}
\text{maximize} \quad & (X_1^{\top} z_1 + X_2^{\top} z_2)^2 \\
\text{subject to} \quad & z_1^{\top} z_2 = 0 \\
& \|z_1\|_2^2 + \|z_2\|_2^2 = 1,
\end{aligned}$$

which is again

$$\begin{aligned}
\text{maximize} \quad & (\mathbf{vec}(X)^{\top} z)^2 \\
\text{subject to} \quad & z^{\top} \begin{pmatrix} 0 & I \\ I & 0 \end{pmatrix} z = 0 \\
& z^{\top} z = 1.
\end{aligned}$$

Following the classical lifting technique derived by [20,15], we can produce a semidefinite relaxation of this last problem by changing variables, setting $Z = zz^{\top}$, and dropping the implicit rank constraint on Z, to get

$$\begin{aligned}
\text{maximize} \quad & \mathbf{Tr}\left(\mathbf{vec}(X)\,\mathbf{vec}(X)^{\top} Z\right) \\
\text{subject to} \quad & \mathbf{Tr}\left(\begin{pmatrix} 0 & I \\ I & 0 \end{pmatrix} Z\right) = 0 \qquad (9) \\
& \mathbf{Tr}(Z) = 1, \ Z \succeq 0,
\end{aligned}$$

which is a SDP in the matrix variable $Z \in \mathbf{S}_{2d}$. The quadratic convexity results of [7] (see also [3], §II.14), also known as the $\mathcal{S}$-procedure or Brickman's theorem, tells us that the optimal value of the semidefinite program (9) is equal to the optimal value of the non-convex QP in (8), and a solution Y to (8) can be constructed from an optimal solution Z of (9) (see, e.g.,, [6] App. B.3 for an explicit recursive procedure).

Problem (9) is an SDP over $2d \times 2d$ matrices, which can be prohibitive in practice as soon as d gets large. Let us now show that a simple decomposition allows to reformulate the problem as a SDP of fixed dimension 6. We can compute the QR decomposition of X written $X = QR_2$ where $Q \in \mathbb{R}^{d \times d}$ is an orthogonal matrix and $R_2 \in \mathbb{R}^{d \times 2}$ with $R_2 = (R^\top, 0)^\top$ where $R \in \mathbb{R}^{2 \times 2}$ is an upper triangular matrix. This means that without loss of generality, the original problem of computing $\Omega^*(X)$ can be rewritten

$$
\begin{aligned}
&\text{maximize } \mathbf{Tr}(Y^T Q R_2) \\
&\text{subject to } Y^T Q Q^T Y \text{ diagonal} \\
&\qquad\qquad \|Q^T Y\|_F = 1 ,
\end{aligned}
\tag{10}
$$

which is equivalent to

$$
\begin{aligned}
&\text{maximize } \mathbf{Tr}(Y^T R_2) \\
&\text{subject to } Y^T Y \text{ diagonal} \\
&\qquad\qquad \|Y\|_F = 1 ,
\end{aligned}
$$

in the variable $Y \in \mathbb{R}^{d \times 2}$. This means that we can always assume that X is block upper diagonal with lower block equal to zero. This program can be rewritten

$$
\begin{aligned}
&\text{maximize } (\mathbf{vec}(R_2)^T z)^2 \\
&\text{subject to } z^T \begin{pmatrix} 0 & \mathbf{I}_d \\ \mathbf{I}_d & 0 \end{pmatrix} z = 0 \\
&\qquad\qquad z^T z = 1 ,
\end{aligned}
$$

in the variable $z = \mathbf{vec}(Y) \in \mathbb{R}^{2d}$. Now notice that

$$
\begin{pmatrix} 0 & \mathbf{I}_d \\ \mathbf{I}_d & 0 \end{pmatrix} = \begin{pmatrix} 0 & 1 \\ 1 & 0 \end{pmatrix} \otimes \mathbf{I}_d = (P^T \, \mathbf{diag}(-1,1) P) \otimes \mathbf{I}_d ,
$$

where $P = \frac{1}{\sqrt{2}} \begin{pmatrix} -1 & 1 \\ 1 & 1 \end{pmatrix}$ is an orthogonal matrix. Let us write $S = P \otimes \mathbf{I}_d$ (also an orthogonal matrix), $w = Sz$ and $b = S \, \mathbf{vec}(R_2)$, we can rewrite the QP above as

$$
\begin{aligned}
&\text{maximize } (\mathbf{vec}(R_2)^T S^T w)^2 \\
&\text{subject to } w^T \begin{pmatrix} -\mathbf{I}_d & 0 \\ 0 & \mathbf{I}_d \end{pmatrix} w = 0 \\
&\qquad\qquad w^T w = 1 ,
\end{aligned}
$$

in the variable $w \in \mathbb{R}^{2d}$. Now $b = S \, \mathbf{vec}(R_2)$ means

$$
b = (P \otimes \mathbf{I}_d) \, \mathbf{vec}(R_2) = \mathbf{vec}(R_2 P) ,
$$

so if $R_2 = (R^T, 0)^T$ where $R \in \mathbb{R}^{T \times T}$ as above, then $b = \mathbf{vec}((P^T R^T, 0)^T)$ hence the b has only four nonzero coefficients at indices $J = \{1, 2, d+1, d+2\}$. This means that the QP can be reformatted as

$$
\begin{aligned}
&\text{maximize } w_J^T (b_J b_J^T) w_J \\
&\text{subject to } w_J^T \begin{pmatrix} -\mathbf{I}_2 & 0 \\ 0 & \mathbf{I}_2 \end{pmatrix} w_J = y_1^T y_1 - y_2^T y_2 \\
&\qquad\qquad w_J^T w_J + y_1^T y_1 + y_2^T y_2 = 1 ,
\end{aligned}
$$

in the variables $w_J \in \mathbb{R}^4$ and $y_1, y_2 \in \mathbb{R}^{d-2}$, where we have defined $z_1^T = (w_3, \ldots, w_d)$ and $z_1^T = (w_{d+3}, \ldots, w_{2d})$. By symmetry, we can assume w.l.o.g. that the coefficients of the vectors y_1 and y_2 are uniformly equal to scalars $y_1, y_2 \in \mathbb{R}$, so the last problem is equivalent to

$$\text{maximize } w_J^T (b_J b_J^T) w_J$$
$$\text{subject to } w_J^T \begin{pmatrix} -\mathbf{I}_2 & 0 \\ 0 & \mathbf{I}_2 \end{pmatrix} w_J = (d-2)y_1^2 - (d-2)y_2^2$$
$$w_J^T w_J + (d-2)y_1^2 + (d-2)y_2^2 = 1,$$

which is now a QP of dimension 6 in the variables $w_J \in \mathbb{R}^4$ and $y_1, y_2 \in \mathbb{R}$. This last problem can then be lifted as above, to become

$$\text{maximize } \mathbf{Tr}\, W \begin{pmatrix} b_J b_J^T & 0 \\ 0 & 0 \end{pmatrix}$$
$$\text{subject to } \mathbf{Tr}\, W \begin{pmatrix} -\mathbf{I}_2 & 0 & 0 \\ 0 & \mathbf{I}_2 & 0 \\ 0 & 0 & \mathbf{diag}(-(d-2),(d-2)) \end{pmatrix} = 0 \qquad (11)$$
$$\mathbf{Tr}\, W \begin{pmatrix} \mathbf{I}_4 & 0 \\ 0 & (d-2)\mathbf{I}_2 \end{pmatrix} = 1, \; W \succeq 0,$$

which is a semidefinite program in the variable $W \in \mathbf{S}_6$. The optimal values of programs (10) and (11) are equal and a solution to (10) can be constructed from an optimal solution to (11). Because (11) is a semidefinite program of fixed dimension 6, it can be solved efficiently *independently of the dimension d*. All we need is the QR decomposition of X which can be formed with cost $O(d)$ when $X \in \mathbb{R}^{d \times 2}$. ∎

4   Algorithms

In order to learn with the penalty Ω_K we need to solve problems of the form

$$\min_W f(W) + \frac{\lambda}{2} \Omega_K(W)^2. \qquad (12)$$

When Ω_K is strictly convex, [24] propose a regularized dual averaging (RDA) method based on subgradient descent, and show that a subgradient of $\Omega_K(W)$ in that case is given by $G = (g_1, \ldots, g_t)$ where

$$g_i = K_{ii}w_i + \sum_{j \neq i} \text{sign}\left(w_i^T w_j\right) K_{ij}w_j, \qquad (13)$$

with the convention $\text{sign}(0) = 0$. When Ω_K is not strictly convex, e.g., when it is a sum of atomic norms as in Theorem 2 or when it is not even convex (as on the right-hand plot of Figure 1), the RDA methods can not be used anymore. In that case, we propose to use a classical subgradient descent scheme using the subgradient (13), and a step size decreasing with $t^{-1/2}$ where t is the iteration.

Note that [24] only prove that (13) is a valid subgradient when Ω_K is convex; we keep the same formula in the general case since Ω_K is differentiable almost everywhere. In the non-convex case, subgradient descent will converge to a stationary point, so one may run it several times with random initializations before taking the best solution. In the experiments below, we always run subgradient descent starting from the null matrix, and observed empirically that it often leads to a good solution compared to random initialization.

Let us now discuss another possible optimization scheme when K satisfies the conditions of Theorem 2, i.e., when the penalty is a linear combination of nuclear norms over pairs of columns. In that case, by Theorem 2 the optimization problem has the form:

$$\min_W f(W) + \frac{\lambda}{2} \sum_{i<j} K_{ij} \Omega_{\mathcal{O}}((w_i, w_j))^2 . \tag{14}$$

We can then write an equivalent dual problem amenable to optimization. Let us first consider the simple case of $T = 2$ columns, in which case (14) boils down to

$$\min_W \left\{ f(W) + \frac{\lambda}{2} \Omega_{\mathcal{O}}^2(W) \right\} \tag{15}$$

in the variable $W \in \mathbb{R}^{d \times 2}$. Remember that for any norm, if $h(x) = \|x\|^2/2$ then the Fenchel dual of h is $h^*(y) = \|y\|_*^2/2$ [6, §3.3.1]). Then [5, Th. 3.3.5] shows that the dual of (15) is written

$$\sup_Z \left\{ -f^*(Z) - \frac{1}{2\lambda} (\Omega_{\mathcal{O}}^*(Z))^2 \right\} \tag{16}$$

in the variable $Z \in \mathbb{R}^{d \times 2}$. Under mild technical conditions, the optimal values of both problems are equal. Back to the general case (14), note that the conjugate of the function $\Omega_{\mathcal{O}}((w_i, w_j))$, which we write $\tilde{\Omega}_{ij}^*(W)$, is given by

$$\tilde{\Omega}_{ij}^*(W) = \begin{cases} \Omega_{\mathcal{O}}^*((W_i, W_j)) & \text{if } W_l = 0 \text{ for } l \neq i, j \\ +\infty & \text{otherwise.} \end{cases}$$

Then, using the following inf-convolution result [18, Th. 16.4]:

$$(f_1 + \ldots + f_s)^*(y) = \inf_{y_1, \ldots, y_s} \{ f_1^*(y_1) + \ldots + f_s^*(y_s) : y_1 + \ldots + y_s = y \},$$

we obtain that the Fenchel dual of problem (14) is written

$$\sup_Z \left\{ -f \left(\sum_{i<j} Z_{ij} \right) - \sum_{i<j} \frac{1}{2\lambda K_{ij}} \tilde{\Omega}_{ij}^*(Z_k)^2 \right\} \tag{17}$$

in the variables $(Z_{ij})_{i<j} \in \mathbb{R}^{d \times T}$. Note that the definitions of $\tilde{\Omega}_{ij}^*$ mean that each Z_{ij} only has two nonzero columns at positions i and j. Now, note that by Theorem 3, the function to be optimized in (17) can be efficiently estimated and a subgradient can be computed. Any value of (17) provides a lower bound to (14), thus giving a duality gap that can be used to monitor convergence of the subgradient descent method.

5   Learning Disjoint Supports

An interesting particular case of learning orthogonal vectors is the situation where we seek sparse vectors with disjoint supports. In this section we briefly discuss how Ω_K can help in this situation, too. For simplicity we only discuss the case of $T = 2$ vectors, an extension to the general case being straightforward. The matrix $W \in \mathbb{R}^{d \times 2}$ has columns with complementary supports if, for $i = 1, \ldots, d$,

$$W_{1,i} \neq 0 \implies W_{2,i} = 0 \text{ and } W_{2,i} \neq 0 \implies W_{1,i} = 0,$$

or in other words $W_1 \circ W_2 = 0$ where $\circ$ denotes the Hadamard (entrywise) product of matrices. If we denote by $|W|$ the matrix whose entries are the absolute values of the entries of W, then we further observe that $|W_1 \circ W_2| = |W_1| \circ |W_2|$, so $W_1 \circ W_2 = 0$ if and only if $|W_1| \circ |W_2| = 0$. Interestingly, if $V \in \mathbb{R}^{d \times 2}$ is a matrix with non-negative entries, then $V_1 \circ V_2 = 0$ is equivalent to $V_1^T V_2 = 0$; this shows that W has columns with complementary supports if and only if $|W_1|$ and $|W_2|$ are orthogonal.

This suggest a general way to learn a matrix with disjoint supports, by solving a problem of the form:

$$\min_W f(W) + \frac{\lambda}{2} \Omega_K(|W|)^2, \tag{18}$$

where Ω_K is a penalty that induces orthogonality among columns. To solve (18), we introduce a non-negative matrix V such that $-V \leq W \leq V$ (where $\leq$ refers to elementwise comparisons), and solve the following problem:

$$\min_{-V \leq W \leq V} f(W) + \frac{\lambda}{2} \Omega_K(V)^2. \tag{19}$$

At the optimum of (19), we have $V = |W|$ which shows that (19) is indeed equivalent to (18). Since a subgradient of (19) in (V, W) can easily be computed, we propose to solve a(19) by a projected subgradient scheme, where at each iteration we update V and W with a move along a subgradient, and then project the new point to the constraint set $-V \leq W \leq V$ and $V \geq 0$.

6   Experiments

In this section, we present numerical experiments on two simulated datasets. We benchmark the following methods:

- Xiao: this is the method described in [24] where we solve (1) with the penalty (2). We consider both with convex and non-convex versions, by changing the matrix K in (2).
- Disjoint Supports: this is the approach where we solve (18), with non-convex and convex versions.
- Ridge Regression: this standard method corresponds to learning the tasks independently by ridge regression.

- LASSO: this is the classical approach inducing sparsity over all tasks, without sharing information across the tasks.

In all experiments involving Ω_K, we consider a symmetric matrix K parametrized by its diagonal value γ,

$$K = \begin{pmatrix} \gamma & & 1 \\ & \ddots & \\ 1 & & \gamma \end{pmatrix}. \tag{20}$$

Based on the conditions for the convexity of Ω_K studied by [24], we control the convexity of Ω_K used in the Xiao and Disjoint Supports approaches with the following rule on γ:

- $\gamma > T - 1$ leads to strictly convex Ω_K function, as described in [24],
- $\gamma = T-1$ is the the limit case where Ω_K satisfies the conditions of Theorem 2, i.e., where it is a sum of atomic norms over pairs of columns:

$$\Omega_K(W) = \sum_{i<j} \Omega_\mathcal{O}((w_i, w_j)), \tag{21}$$

- $\gamma < T - 1$ corresponds to the case where Ω_K is not convex.

We test the different methods on regression problem where, given a matrix of covariates $X \in \mathbb{R}^{n \times d}$ and a matrix of T response variables $Y \in \mathbb{R}^{n \times T}$, we seek to minimize the squared error $f(W) = \|Y - XW\|^2$.

6.1 The Effect of Convexity

We use simulated data to test whether theoretical differences between $\Omega_K, \Omega_\mathcal{O}$ and concave formulations have an impact on analytical performances. In particular, by playing with γ in (20), we investigate to what extent the convexity constraint imposed by [24] is restrictive in terms of performance.

For that purpose, we randomly generate models W consisting of $T = 10$ tasks in $d = 10$ dimensions, such that all tasks are orthogonal to each other. The training set X_{train} is composed of $n = 50$ instances, each element of X_{train} being sampled from a normal distribution $\mathcal{N}(0, 1)$. We simulate the response variable $Y_{train} \in \mathbb{R}^{n \times T}$ according to $Y_{train} = X_{train}W + \epsilon$, where ϵ is a noise matrix of i.i.d. centered Gaussian variables with variance σ^2. We estimate the performance of each model on a test set of 1000 samples generated similarly. We also measure how orthogonal the models are, by the mean absolute difference between the angle between two columns of W and $\pi/2$. For each value of γ we estimate the Xiao model with different regularization parameters λ over a grid of 21 values regularly spaced after log transform; the grid was set to ensure that it covered good parameters for all methods. For each γ, we report the performance of the best λ in terms of test MSE. We repeat the full procedure 100 times and report the average results over the 100 repeats.

Figure 2 shows the performance of the methods in terms of test error (top), and in terms of how far the models learned are from orthogonal models (bottom). On each plot, the horizontal axis is the γ parameter on the diagonal of K defined in (20), and the vertical dotted line corresponds to the atomic norm (21) and is the transition from convex (to its right) to non-convex (to its left). From left to right, we show results corresponding to increasing noise in the response variable, with the variance of ϵ set respectively to $1, 2.5$ and 4. We see that in the small noise regime (left), non-convex formulations perform better while with high noise (right), the convex formulations are more adapted. Inbetween (middle), the best performance is reached for slightly non-convex penalties. In all cases, the models learned are similar in terms of how non-orthogonal they are; we see that non-convex formulations lead to significantly more orthogonal models than convex formulations. Overall, these results suggest that restricting ourselves to strictly convex penalties may be restrictive and sub-optimal in some cases; they show that non-convex penalties can allow to learn more orthogonal models with better performance.

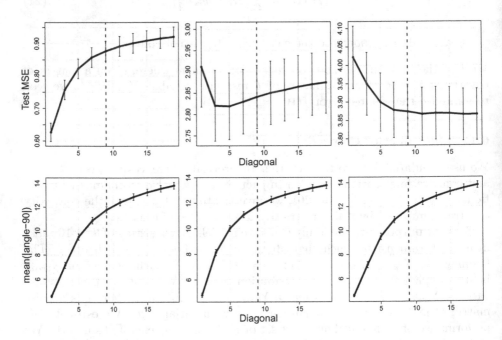

Fig. 2. Test MSE (top) and deviation from pairwise orthogonality (bottom) as a function of the convexity parameter γ, from low to high noise regimes (from left to right: $\sigma^2 \in \{1, 2.5, 4\}$). On each plot, the horizontal axis is the γ parameter on the diagonal of K defined in (20). The vertical dotted line corresponds to the atomic norm (21).

6.2 Regression with Disjoint Supports

As a second proof of concept, we check the relevance of the formulation presented in Section 5 to jointly learn linear models with disjoint support. For that purpose, we simulate data as in Section 6.1, with the additional constraint that the columns of W are orthogonal and have disjoint supports. Since $d = T = 10$, this means that W is simply diagonal. We fix the noise level at $\sigma^2 = 1$, and simulate training sets of increasing size between 10 and 50 samples, repeating the full procedure 100 times. We compare four methods: (i) the Xiao model with varying parameter γ according to (20), leading to orthogonal but non-sparse vectors, (ii) our new method (18) again with convex and non-convex formulations by varying γ in (20), (iii) a baseline ridge regression model and (iv) a LASSO regression model leading to sparse but not necessarily orthogonal vectors. For each model, a 5-fold cross-validation is performed on the training set to select an optimal regularization parameter, which is then used to train the model on the full training set before doing a prediction on an independent test set. We assess the performance of each method on the test set in terms of accuracy (measured by the MSE), and in terms of disjoint support recovery, measured as the proportion of features which are correctly selected in a single column of W.

The results are shown in Figure 3, where for sake of clarity we only report the results of Xiao and Disjoint Supports for the optimal diagonal value γ, which in both cases is equal to 0.1, corresponding to a very non-convex penalty. In terms of performance, we see that Xiao is a bit better than Ridge regression for $n = 50$ training point, which is coherent with the observation made in Section 6.1 in the small-noise regime, although for less than 30 samples Ridge regression is better.

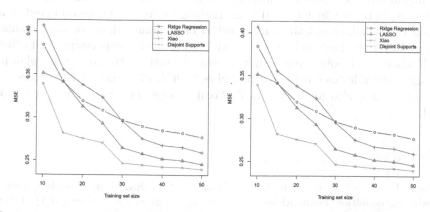

Fig. 3. Sparse regression with disjoint supports. Test MSE for training set of increasing size (left), and proportion of correctly affected features (right). Ridge regression and Xiao are not shown on the right plot because they are not sparse.

Both methods are outperformed by LASSO, which in this case benefits from the very sparse structure of W. Interestingly, the new Disjoint Support model significantly outperforms all other methods for all training set sizes (P-value $< 10^{-3}$). As for the ability of different methods to correctly recover the disjoint supports, we see that Disjoint Supports shows increasing support recovery score for large training set size, and outperforms LASSO which induces global sparsity but is not able to affect features to an unique column. Ridge Regression and Xiao are not shown because they do not achieve any sparsity in the model they learn. In summary, this simulation shows that the Disjoint Supports model has the potential to outperform other methods when the model to learn is sparse with disjoint supports.

7   Conclusion

We have extended the work of [24] in two directions: on the one hand, we have investigated the possibility to work with non strictly convex or non convex formulations, leading to more agressive control of model orthogonality, and on the other hand we have shown how models to learn orthogonal columns can be extended to learn sparse models with disjoint supports. In the two-columns case, we have proved that the penalty of [24] is an atomic norm derived from the set of scaled orthogonal matrices, and for the general case $T > 2$ we have shown that for suitable choices of parameters it can be written as a linear combination of atomic norms applied to pairs of columns. In terms of algorithms, the RDA algorithm proposed by [25] is only suitable to solve the problem (12) in the strictly convex case, and we have shown that in the limit case where Ω_K is convex but not strictly convex we can solve iteratively with a series of 6-dimensional SDP. Our simulations show that considering non-convex versions of the penalty can be relevant, in particular for small noise regime. Interestingly, we observed that non-convex formulations lead to more orthogonal models than convex formulations, and that the Disjoint Support model significantly outperformed all other models when the disjoint support hypothesis was met. In the future, we plan to investigate the relevance of this approaches with more structured matrices K, such as the ones used for hierarchical classifications [24] or learning groups of models [19].

References

1. Bach, F., Jenatton, R., Mairal, J., Obozinski, G.: Optimization with sparsity-inducing penalties. Foundations and Trends® in Machine Learning 4(1), 1–106 (2011)
2. Bakker, B., Heskes, T.: Task clustering and gating for bayesian multitask learning. J. Mach. Learn. Res. 4, 83–99 (2003)
3. Barvinok, A. A Course in Convexity. American Mathematical Society (2002)
4. Baxter, J.: A model of inductive bias learning. Journal of Artificial Intelligence Research 12, 149–198 (2000)

5. Borwein, J.M., Lewis, A.S.: Convex Analysis and Nonlinear Optimization: Theory and Examples. Cms Books in Mathematics Series. Springer (2000)
6. Boyd, S., Vandenberghe, L.: Convex Optimization. Cambridge University Press (2004)
7. Brickman, L.: On the field of values of a matrix. Proceedings of the American Mathematical Society, 61–66 (1961)
8. Cai, L., Hofmann, T.: Hierarchical document categorization with support vector machines. In: Proceedings of the Thirteenth ACM International Conference on Information and Knowledge Management, pp. 78–87. ACM, New York (2004)
9. Calder, A.J., Burton, A.M., Miller, P., Young, A.W., Akamatsu, S.: A principal component analysis of facial expressions. Vision Res. 41(9), 1179–1208 (2001)
10. Caruana, R.: Multitask learning. Mach. Learn. 28(1), 41–75 (1997)
11. Chandrasekaran, V., Recht, B., Parrilo, P.A., Willsky, A.S.: The convex geometry of linear inverse problems. Found. Comput. Math. 12(6), 805–849 (2012)
12. Evgeniou, T., Micchelli, C., Pontil, M.: Learning multiple tasks with kernel methods. J. Mach. Learn. Res. 6, 615–637 (2005)
13. Hwang, S.J.J., Grauman, K., Sha, F.: Learning a tree of metrics with disjoint visual features. In: Shawe-Taylor, J., Zemel, R.S., Bartlett, P., Pereira, F.C.N., Weinberger, K.Q. (eds.) Adv. Neural. Inform. Process Syst. 24, pp. 621–629 (2011)
14. Jacob, L., Vert, J.-P.: Protein-ligand interaction prediction: an improved chemogenomics approach. Bioinformatics 24(19), 2149–2156 (2008)
15. Lovász, L., Schrijver, A.: Cones of matrices and set-functions and 0-1 optimization. SIAM Journal on Optimization 1(2), 166–190 (1991)
16. McCallum, A., Rosenfeld, R., Mitchell, T.M., Ng, A.Y.: Improving text classification by shrinkage in a hierarchy of classes. In: Proceedings of the Fifteenth International Conference on Machine Learning, pp. 359–367. Morgan Kaufmann Publishers Inc., San Francisco (1998)
17. Obozinski, G., Taskar, B., Jordan, M.I.: Joint covariate selection and joint subspace selection for multiple classification problems. Statistics and Computing 20(2), 231–252 (2010)
18. Rockafellar, R.T.: Convex Analysis. Princeton University Press, Princeton (1970)
19. Romera-Paredes, B., Argyriou, A., Berthouze, N., Pontil, M.: Exploiting unrelated tasks in multi-task learning. J. Mach. Learn. Res. - Proceedings Track 22, 951–959 (2012)
20. Shor, N.Z.: Quadratic optimization problems. Soviet Journal of Computer and Systems Sciences 25, 1–11 (1987)
21. Srebro, N., Rennie, J.D.M., Jaakkola, T.S.: Maximum-margin matrix factorization. In: Saul, L.K., Weiss, Y., Bottou, L. (eds.) Adv. Neural. Inform. Process Syst. 17, pp. 1329–1336. MIT Press, Cambridge (2005)
22. Thrun, S., Pratt, L. (eds.): Learning to learn. Kluwer Academic Publishers, Norwell (1998)
23. Tibshirani, R.: Regression shrinkage and selection via the lasso. J. R. Stat. Soc. Ser. B 58(1), 267–288 (1996)
24. Xiao, L., Zhou, D., Wu, M.: Hierarchical classification via orthogonal transfer. In: Getoor, L., Scheffer, T. (eds.) Proceedings of the 28th International Conference on Machine Learning, ICML 2011, Bellevue, Washington, USA, June 28-July 2, pp. 801–808. Omnipress (2011)
25. Xiao, L.: Dual averaging methods for regularized stochastic learning and online optimization. J. Mach. Learn. Res. 9999, 2543–2596 (2010)

Scalable Moment-Based Inference for Latent Dirichlet Allocation

Chi Wang, Xueqing Liu, Yanglei Song, and Jiawei Han

Computer Science Department, University of Illinois at Urbana-Champaign,
Urbana, IL 61801, USA
{chiwang1,xliu93,ysong44,hanj}@illinois.edu

Abstract. Topic models such as Latent Dirichlet Allocation have been useful text analysis methods of wide interest. Recently, moment-based inference with provable performance has been proposed for topic models. Compared with inference algorithms that approximate the maximum likelihood objective, moment-based inference has theoretical guarantee in recovering model parameters. One such inference method is tensor orthogonal decomposition, which requires only mild assumptions for exact recovery of topics. However, it suffers from scalability issue due to creation of dense, high-dimensional tensors. In this work, we propose a speedup technique by leveraging the special structure of the tensors. It is efficient in both time and space, and only requires scanning the corpus twice. It improves over the state-of-the-art inference algorithm by one to three orders of magnitude, while preserving equal inference ability.

1 Introduction

Statistical topic modeling techniques are powerful tools for exploring large data sets such as text and social networks. They are frequently used for text summarization, dimensionality reduction and community detection. One important model is latent Dirichlet allocation (LDA) [6], which has widespread use and variations in data mining and machine learning. It models documents as mixtures of multiple topics, while every topic is modeled as a multinomial distribution over a vocabulary. We consider the unsupervised inference problem for LDA: estimating the unknown word distribution of every topic so as to fit the observed word occurrences in the documents.

The inference can be performed under different principles. Maximum likelihood is the most commonly employed principle, but exact inference based on this objective is proved to be intractable [5]. Recently, researchers have found that a new inference principle, *method of moments*, enables tractable computations to recover the topics with theoretical bound [2, 3]. The intuition is to relate model parameters to *population moments*, which are expected frequencies of co-occurred word pairs, triples *etc.*, and infer parameters from empirical estimation of the population moments. Under mild assumptions, a *tensor orthogonal decomposition* algorithm in [3] can perform error-bounded topic recovery, with

T. Calders et al. (Eds.): ECML PKDD 2014, Part III, LNCS 8726, pp. 290–305, 2014.
© Springer-Verlag Berlin Heidelberg 2014

best known sample complexity and numerical stability. However, it has a severe scalability issue, which is the the problem we tackle in this paper.

The tensor orthogonal decomposition algorithm has two steps. In the first step, it collects empirical moments (*i.e.*, expectation of word co-occurrence frequencies) from the data, and constructs two large and dense tensors (*i.e.*, hypermatrices). In the second step, it uses the two large tensors to compute a small tensor, and performs orthogonal decomposition for the small tensor based on an iterative tensor power method. With their scale being V^2 and V^3 respectively, where V is the vocabulary size, the large and dense tensors are prohibitive to construct. [3] suggests an alternative by computing the tensor power iterations on the fly scanning through the original data, without creating any tensor in memory. However, it requires one scan of the data *per iteration*. The efficiency is not satisfactory for large scale corpora.

In this work, we propose a novel strategy to scale up the tensor orthogonal decomposition algorithm. By careful analysis of the problem, we advocate to avoid explicit creation of large and dense tensors, but still construct the small tensor and store it in memory. With this strategy, we bypass the scale bottleneck, yet are still able to perform efficient tensor power iterations. To directly construct the small tensor without creating the large and dense tensors, we leverage the special structures of the moments: *sparse*, *low rank* and *decomposable*. We design an efficient algorithm that only requires two scans of data in total while consuming much smaller space. With experiments on both synthetic and real data, we demonstrate that our method can be 20-3000 times faster than the state-of-the-art inference method, while preserving robust topic recovery capability.

2   Related Work

In the last decade, statistical topic modeling techniques have gained popularity. Two important methods are probabilistic latent semantic analysis (PLSA) [12] and its Bayesian extension latent Dirichlet allocation (LDA) [6]. They model the generative process of each word from each document in a corpus. The model parameters can be partitioned into corpus-level (the unknown word distribution of every topic) and document-level (the unknown topic distribution of every document). The goal of inference is to find parameters that best explain the observed data, *i.e.*, word occurrences in the documents. Yet there are different principles to quantify what it means by 'best explain the observed data'.

Most of existing topic model inference methods are based on the *maximum likelihood* (ML) principle (including its Bayesian version *maximum a posterior*). For example, PLSA [12] uses an Expectation-Maximization algorithm to approximately optimize the data likelihood. For LDA, two most popular approximate inference methods have been variational Bayesian inference [6] and Markov Chain Monte Carlo (especially Gibbs sampling) [9]. In spite of the vast body of followup work, the computational complexity of ML inference is not studied until 2011. Sontag and Roy [19] show that the document-level inference is not always well defined, and Arora, Ge and Moitra [5] prove the NP-hardness of exact corpus-level ML inference.

In accordance with their theoretical hardness, the above inference methods tend to suffer from slow convergence and long runtime. As a result, there has been a substantial amount of work targeting on accelerating the above methods. e.g., by leveraging sparsity [11, 18, 20] and parallelization [16, 21], or online learning mechanism [1, 8, 10]. However, none of them have theoretical guarantee of convergence within a bounded number of iterations, and are nondeterministic either due to sampling or the random initialization.

Recently, an alternative inference of topic models has been proposed based on the *method of moments* [2], and improved in [3]. Compared with ML inference, it has the following two advantages: i) the distance between inferred corpus-level parameters and the true parameters has a theoretical upper bound that is inversely related to sample size; ii) the convergence is guaranteed with a bounded number of iterations. Another related study [5] assumes the existence of *anchor word* that only exists in one topic, and uses that assumption to bound the recovery error. Its efficiency is improved in [4]. This method requires stronger assumptions than [2] and the error bound is weaker.

3   Preliminaries

We first introduce the notations:

- The input to the inference problem is a corpus of D documents with vocabulary size V. The i-th document d_i has l_i word tokens, and the whole corpus has L tokens in total. We use a k-dimensional vector θ_i $(i \in [D])$ to denote the document-level topic distribution for d_i, and $\alpha = (\alpha_1, \ldots, \alpha_k)$ to denote the Dirichlet prior where θ_i's are drawn from. Larger α_t posits stronger prior at $\theta_{i,t}$. Define $\alpha_0 = \sum_{t=1}^{k} \alpha_t$. We use a V-dimensional vector ϕ_t $(t \in [k])$ to denote the corpus-level word distribution for topic t. To generate a token $w_{i,j} \in [V]$ in position j of document d_i, one first samples a topic $z_{i,j}$ according to θ_i, and then samples a word from $\phi_{z_{i,j}}$;
- A tensor is a hypermatrix that can contain more than two degrees. The outer product $\otimes$ of any p-degree tensor $A \in \mathbb{R}^{s_1 \times \cdots \times s_p}$ and any q-degree tensor $B \in \mathbb{R}^{s_{p+1} \times \cdots \times s_{p+q}}$ is a $(p+q)$-degree tensor $A \otimes B \in \mathbb{R}^{s_1 \times \cdots \times s_{p+q}}$: $A \otimes B[i_1, \ldots, i_{p+q}] = A[i_1, \ldots, i_p]B[i_{p+1}, \ldots, i_{p+q}]$;
- For any tensor $A \in \mathbb{R}^{s \times s \times s}$, matrix $B \in \mathbb{R}^{s \times s_1}, C \in \mathbb{R}^{s \times s_2}, D \in \mathbb{R}^{s \times s_3}$, $A(B, C, D)$ is a tensor in $\mathbb{R}^{s_1 \times s_2 \times s_3}$, $A(B, C, D)[i_1, i_2, i_3] = \sum_{j_1, j_2, j_3 \in [s]} A[j_1, j_2, j_3]B[j_1, i_1]C[j_2, i_2]D[j_3, i_3]$;
- W^+ denotes the Moore-Penrose pseudoinverse of W;
- $\Omega(A, a, b, c)$ permutes the modes of tensor A, s.t. $\Omega(A, a, b, c)[i_1, i_2, i_3] = A[i_a, i_b, i_c]$.

We focus on the corpus-level inference problem in this paper. The document-level inference problem can be solved using the method in [19] after we infer the corpus-level parameters. Our goal is to recover the unknown ϕ_t's based on the observed corpus.

Anandkumar et al. [3] propose a tractable exact inference method based on the *method of moments*. In statistics, the ξ-th order *population moment* of a random variable is the expectation of its ξ-th power. The method of moments derives equations that relate the population moments to the model parameters. Then, it collects empirical moments from observed samples, and solves the equations using the empirical moments in place of the population moments.

In our case, the random variable is a token $w_{i,j}$ in a document d_i. The value of $w_{i,j}$ is a word in $[V]$. The ξ-th population moment is the expected co-occurrence of words in ξ token positions. We can collect empirical moments from the corpus, and estimate ϕ_t's by fitting the empirical moments with population moments. For example, the 2nd order moment is a matrix $E_2 \in \mathbb{R}^{V \times V}$. The element $E_2[x_1, x_2]$ in x_1-th row and x_2-th column of E_2 is equal to the probability $w_{i,1}$ being x_1 and $w_{i,2}$ being x_2 given α.

$$E_2[x_1, x_2] = p(w_{i,1} = x_1, w_{i,2} = x_2 | \alpha)$$

$$= \int_{\theta_i} p(\theta_i | \alpha) \sum_{t_1=1}^{k} \sum_{t_2=1}^{k} p(z_{i,1} = t_1 | \theta_i) p(z_{i,2} = t_2 | \theta_i) p(w_{i,1} = x_1 | z_{i,1} = t_1)$$

$$\cdot p(w_{i,2} = x_2 | z_{i,2} = t_2) d\theta_i = \sum_{t_1 \neq t_2} \frac{\alpha_{t_1} \alpha_{t_2}}{\alpha_0 (\alpha_0 + 1)} \phi_{t_1,x_1} \phi_{t_2,x_2} + \sum_{t=1}^{k} \frac{\alpha_t (\alpha_t + 1)}{\alpha_0 (\alpha_0 + 1)} \phi_{t,x_1} \phi_{t,x_2} \tag{1}$$

Likewise, we can derive the 3rd order moment as a tensor $E_3 \in \mathbb{R}^{V \times V \times V}$. The element $E_3[x_1, x_2, x_3]$ is equal to the probability $w_{i,1}$ being x_1, $w_{i,2}$ being x_2 and $w_{i,3}$ being x_3 given α. The equation below follows a similar derivation as Eq. (1), written in a more concise form:

$$E_3 = \sum_{t_1 \neq t_2 \neq t_3 \neq t_1} \frac{\alpha_{t_1} \alpha_{t_2} \alpha_{t_3}}{\alpha_0 (\alpha_0 + 1)(\alpha_0 + 2)} \phi_{t_1} \otimes \phi_{t_2} \otimes \phi_{t_3}$$

$$+ \sum_{t_1 \neq t_2} \frac{\alpha_{t_1} \alpha_{t_2} (\alpha_{t_1} + 1)}{\alpha_0 (\alpha_0 + 1)(\alpha_0 + 2)} (\phi_{t_1} \otimes \phi_{t_1} \otimes \phi_{t_2} + \phi_{t_1} \otimes \phi_{t_2} \otimes \phi_{t_1} +$$

$$+ \phi_{t_2} \otimes \phi_{t_1} \otimes \phi_{t_1}) + \sum_{t=1}^{k} \frac{\alpha_t (\alpha_t + 1)(\alpha_t + 2)}{\alpha_0 (\alpha_0 + 1)(\alpha_0 + 2)} \phi_t \otimes \phi_t \otimes \phi_t \tag{2}$$

Moment-based inference methods set the left hand sides to empirical estimation of moments, and solve these equations to estimate the parameters ϕ_t's.

In general, one can compute moments of an arbitrary order, but the cost can be high for computing high-order moments. Fortunately, Anandkumar et al. [3] find that we only need up to 3rd order moments to infer an LDA model, under some mild non-degeneracy conditions. We restate their algorithm in Section 4, and propose a more scalable algorithm in Section 5.

4   Tensor Orthogonal Decomposition

The tensor orthogonal decomposition algorithm for LDA relies on the following theorem (revised statement of Theorem 4.3 in [3]).

Theorem 1. *Assume that*

$$M_2 = \sum_{t=1}^{k} \lambda_t v_t \otimes v_t, \quad M_3 = \sum_{t=1}^{k} \lambda_t v_t \otimes v_t \otimes v_t \tag{3}$$

where $\lambda_t > 0, v_t \in \mathbb{R}^V, t \in [k]$ are linearly independent and $\|v_t\| = 1$. Given $M_2 \in \mathbb{R}^{V \times V}$ and $M_3 \in \mathbb{R}^{V \times V \times V}$ as input, the equations can be uniquely solved for unknown variables λ_t and v_t in polynomial time.

Proof sketch. Let $M_2 = M\Sigma M^T$ be the spectral decomposition of M_2, define $W = M\Sigma^{-\frac{1}{2}}$ as the *whitening matrix* of M_2, then $\tilde{v}_t = \sqrt{\lambda_t} W^T v_t$ are orthonormal since $\sum_{t=1}^{k} \tilde{v}_t \otimes \tilde{v}_t = I$. It follows that $\sum_{t=1}^{k} \frac{1}{\sqrt{\lambda_t}} \tilde{v}_t \otimes \tilde{v}_t \otimes \tilde{v}_t = \sum_{t=1}^{k} \lambda_t (W^T v_t) \otimes (W^T v_t) \otimes (W^T v_t) = M_3(W, W, W) \equiv \tilde{T}$. Due to the uniqueness of tensor's orthogonal decomposition [3], $\tilde{v}_t$ and $\frac{1}{\sqrt{\lambda_t}}$ are uniquely determined from $\tilde{T}$ in polynomial time. So each $v_t = \frac{1}{\sqrt{\lambda_t}} (W^T)^+ \tilde{v}_t$ and λ_t are uniquely determined. ∎

Note that the sole equation $M_2 = \sum_{t=1}^{k} \lambda_t v_t \otimes v_t$ is not sufficient to determine v_t uniquely, because v_t's are not constrained to be orthogonal. By defining M_1 and M_2 in the following way, M_2 fits into Eq. (3):

$$M_1 = \sum_{t=1}^{k} \frac{\alpha_t}{\alpha_0} \phi_t, \quad M_2 = (\alpha_0 + 1)E_2 - \alpha_0 M_1 \otimes M_1 = \sum_{t=1}^{k} \frac{\alpha_t}{\alpha_0} \phi_t \otimes \phi_t \tag{4}$$

And the following definition of M_3 fits into Eq. (3):

$$U_1 = E_2 \otimes M_1, \quad U_2 = \Omega(U_1, 1, 3, 2), \quad U_3 = \Omega(U_1, 2, 3, 1) \tag{5}$$

$$M_3 = \frac{(\alpha_0 + 1)(\alpha_0 + 2)}{2} E_3 + \alpha_0^2 M_1 \otimes M_1 \otimes M_1$$
$$- \frac{\alpha_0(\alpha_0 + 1)}{2} [U_1 + U_2 + U_3] = \sum_{t=1}^{k} \frac{\alpha_t}{\alpha_0} \phi_t \otimes \phi_t \otimes \phi_t \tag{6}$$

It is clear now that the corpus-level parameters ϕ_t's can be uniquely determined by up to 3rd order moments. Algorithm 1 outlines the tensor orthogonal decomposition method for recovering the components, given the summation α_0 of Dirichlet prior α as input. It includes two main parts:

1. Lines 1.1 to 1.5 to compute the $k \times k \times k$ tensor $\tilde{T}$;
2. Lines 1.6 to 1.16 to perform orthogonal decomposition of $\tilde{T}$ via a robust *power method*, and recover the unique λ_t's and v_t's (Line 1.13).

Lemma 5.1 in [3] ensures that the power iteration loop Line 1.10 with iteration # n converges in a quadratic rate when the tensor $\tilde{T}$ is accurate. The outer loop with iteration # N ensures the convergence when $\tilde{T}$ is perturbed.

Algorithm 1. Tensor Orthogonal Decomposition (TOD)

Input: Corpus with L tokens and vocabulary size V, number of components k, number
    of outer and inner iterations N and n, α_0
Output: The model parameters $(\alpha_t, \phi_t), t = 1, \ldots, k$

1.1 Compute $M_2 \in \mathbb{R}^{V \times V}$ and $M_3 \in \mathbb{R}^{V \times V \times V}$;

1.2 Compute k orthonormal eigenpairs (σ_t, μ_t) of M_2;

1.3 Compute the whitening matrix $W = M\Sigma^{-\frac{1}{2}}$;

1.4 Compute $(W^T)^+ = M\Sigma^{\frac{1}{2}}$;

1.5 Compute a $k \times k \times k$ tensor $\widetilde{T} = M_3(W, W, W)$;

1.6 **for** $t = 1..k$ **do**

1.7 $\lambda^* \leftarrow 0$;   // the largest eigenvalue so far

1.8   **for** $outIter = 1..N$ **do**

1.9 $v \leftarrow$ a random unit-form vector in $\mathbb{R}^k$;

1.10   **for** $innerIter = 1..n$ **do** $v \leftarrow \frac{\widetilde{T}(I,v,v)}{||\widetilde{T}(I,v,v)||}$;

1.11   **if** $\widetilde{T}(v,v,v) > \lambda^*$ **then** $(\lambda^*, v^*) \leftarrow (\widetilde{T}(v,v,v), v)$;

1.12   **end**

1.13 $\lambda_t = \frac{1}{(\lambda^*)^2}, v_t = \lambda_t (W^T)^+ v^*$;

1.14 $\widetilde{T} \leftarrow \widetilde{T} - \lambda^* v^* \otimes v^* \otimes v^*$;   // deflation

1.15 $\alpha_t = \alpha_0 \lambda_t, \phi_t = v_t$;

1.16 **end**

1.17 **return** $(\alpha_t, \phi_t), t = 1, \ldots, k$

---

5   Scalable Tensor Orthogonal Decomposition

5.1   Scalability Analysis of TOD

Although Algorithm 1 theoretically guarantees robust convergence, it is not scalable. In general, when we directly deal with large and dense 2nd or 3rd order tensors, computation cost is huge in both time and space, and this hinders the scalability of part 1 described in Section 4, where explicit computations for a matrix of size V^2 and a tensor of size V^3 are involved. In contrast, part 2 ($k \times k \times k$ tensor orthogonal decomposition) can be efficient in practice, because in most cases of LDA inference, only a small number of topics are desired. In total, the space complexity of Algorithm 1 is $\mathcal{O}(V^3)$ and the time complexity is $\mathcal{O}(V^3 k + L\hat{l}^2 + Nnk^4)$, where $\hat{l}$ is the maximum document length.

Anandkumar et al. [3] discusses a plausible way to reduce the memory cost. It suggests no explicit creation of the tensors M_3 and $\widetilde{T}$, but going through the document-word occurrence data for computing the power iteration update Line 1.10. This mitigates the space challenge of part 1, but gives away the efficiency of part 2 of Algorithm 1. One obvious disadvantage is that it needs to scan the whole corpus for Nnk times to execute Line 1.10. The space complexity is $\mathcal{O}(V^2)$ and the time complexity is $\mathcal{O}(V^2 k + LNnk)$.

We make key contributions to solving the challenge in a different approach. We avoid explicit creation of both tensor M_3 and M_2, but we do explicitly create $\widetilde{T}$ since it is memory efficient. It reduces the cost of part 1, and retains efficient power iteration updates as in part 2 of Algorithm 1. Utilizing the special structure of the tensors in our problem, we show that $\widetilde{T}$ can be created by scanning the corpus only twice, without incurring creations of any dense V^2 or

V^3 tensors. One scan is needed for computing the whitening matrix W, and the other for computing the tensor product $M_3(W, W, W)$, as discussed in the following two sections.

5.2   Scalable Computation of Whitening Matrix

To compute the whitening matrix, the straightforward approach by spectral decomposition of the dense matrix M_2 requires $\mathcal{O}(V^2 k)$ time and $\mathcal{O}(V^2)$ space. However, by observation of Eq. (4), we can decouple M_2 into matrix E_2 and $M_1 \otimes M_1$. Taking advantage of the *low rank* and *sparsity* of E_2, we can compute the spectral decomposition of M_2 in an efficient alternative procedure.

Low Rank. We notice that M_1, E_2 and M_2 are in the same *column space* $\mathcal{S}$ spanned by k linearly independent vectors $\phi_t, t \in [k]$. Thus E_2 has a low rank.

Sparsity. Let vector $c_i \in \mathbb{R}^V$ be the counts of word 1 to V in document d_i. An empirical estimation of M_1 and E_2 is:

$$M_1 = \sum_{i=1}^{D} \frac{1}{l_i} c_i, E_2 = \sum_{i=1}^{D} \frac{1}{l_i(l_i - 1)} [c_i \otimes c_i - diag(c_i)] \tag{7}$$

where $l_i = \sum_{x=1}^{V} c_{i,x}$ is the length of document d_i. The estimated M_1 and E_2 can be computed by one scan of the data. E_2 is sparse because many word pairs do not co-occur in the real documents.

Our alternative procedure performs two spectral decompositions, one on the sparse and low rank matrix E_2 and the other on a small size matrix.

1. Let $E_2 = U \Sigma_1 U^T$ be its spectral decomposition, where $U \in \mathbb{R}^{V \times k}$ is the matrix of k eigenvectors, and $\Sigma_1 \in \mathbb{R}^{k \times k}$ is the diagonal eigenvalue matrix. The k column vectors of U form an orthonormal basis of $\mathcal{S}$. M_1's representation in this basis is $M_1' = U^T M_1$. Now, M_2 can be written as:

$$M_2 = U[(\alpha_0 + 1)\Sigma_1 - \alpha_0 M_1' \otimes M_1'] U^T = U M_2' U^T$$

2. A second spectral decomposition can be performed on $M_2' \in \mathbb{R}^{k \times k}$. Let the decomposition be $M_2' = U' \Sigma U'^T$. It follows that:

$$M_2 = U M_2' U^T = (UU') \Sigma (UU')^T$$

Let $M = UU'$. Now we effectively obtain the spectral decomposition of $M_2 = M \Sigma M^T$ without explicitly creating M_2. With this new procedure, we only need to store a sparse matrix E_2 with $m \ll V^2$ nonzero elements, and the time complexity is reduced to $\mathcal{O}(km + k^3) = \mathcal{O}(km)$.

5.3   Scalable Product of M_3 and W

The straightforward computation of $\widetilde{T} = M_3(W, W, W)$ using explicit M_3 and W requires $\mathcal{O}(V^3)$ space and $\mathcal{O}(V^3 k + L\hat{l}^2)$ time, where $\hat{l}$ is the maximal document length. To solve this challenge, we utilize two *decomposing laws*:

i) $(v \otimes v \otimes v)(W, W, W) = (W^T v) \otimes (W^T v) \otimes (W^T v) = (W^T v)^{\otimes 3}$; and
ii) $(v \otimes B)(W, W, W) = (W^T v) \otimes B(W, W) = (W^T v) \otimes (W^T B W)$
where v is a vector and B is a matrix.

We break down M_3 as a summation of multiple tensors, such that the product between each tensor and W has a **decomposable** form as in the left hand side of one decomposing law. According to Eq. (2), M_3 is a linear combination of tensors E_3, U_1, U_2, U_3 and $M_1^{\otimes 3}$. We discuss how to efficiently compute the product between each part and W.

Compute $E_3(W, W, W)$. E_3 can be estimated by averaging the frequency of all the 3-word triples in each document. Using the word count vector c_i we defined before, we have:

$$E_3 = \frac{1}{D}[A_1 - A_2 - \Omega(A_2, 2, 1, 3) - \Omega(A_2, 2, 3, 1) + 2A_3]$$

$$A_1 = \sum_{i=1}^{D} \rho_i c_i \otimes c_i \otimes c_i, \quad A_2 = \sum_{i=1}^{D} \rho_i c_i \otimes diag(c_i), \quad A_3 = \sum_{i=1}^{D} \rho_i tridiag(c_i) \tag{8}$$

where $\rho_i = \frac{1}{l_i(l_i-1)(l_i-2)}$, $tridiag(v)$ is a tensor with vector v on its diagonal: $tridiag(v)_{i,i,i} = v_i$.

According to decomposing law i) and ii):

$$A_1(W, W, W) = \sum_{i=1}^{D} \rho_i (W^T c_i)^{\otimes 3} \tag{9}$$

$$A_2(W, W, W) = \sum_{i=1}^{D} \rho_i (W^T c_i) \otimes W^T diag(c_i) W \tag{10}$$

Let W_x^T be the x-th column of W^T. We have:

$$A_3(W, W, W) = \sum_{x=1}^{V} \sum_{i=1}^{D} \rho_i c_{i,x} (W_x^T)^{\otimes 3} \tag{11}$$

Using Eq. (9)-(11), we can compute $E_3(W, W, W)$ without explicit creation of E_3. The time complexity is $\mathcal{O}(Lk^2)$.

Compute $M_1^{\otimes 3}(W, W, W), U_i(W, W, W), i = 1, 2, 3$. Using the two decomposing laws, we can obtain:

$$(M_1 \otimes M_1 \otimes M_1)(W, W, W) = (W^T M_1)^{\otimes 3} \tag{12}$$

$$U_1(W, W, W) = W^T E_2 W \otimes W^T M_1 \tag{13}$$

Eq. (13) requires $O(k^2 m)$ time to compute, where m is the number of nonzero elements in E_2. We can further speed it up. By definition we have $W^T M_2 W = I$. Substituting M_2 with Eq. (4), we have:

$$W^T[(\alpha_0 + 1)E_2 - \alpha_0 M_1 \otimes M_1]W = I \tag{14}$$

$$\Rightarrow W^T E_2 W = \frac{1}{(\alpha_0 + 1)}[I + \alpha_0 (W^T M_1)^{\otimes 2}] \tag{15}$$

Plugging Eq. (15) into (13) further reduces the complexity of computing $U_1(W, W, W)$ to $\mathcal{O}(Vk + k^3)$. $U_2(W, W, W)$ and $U_3(W, W, W)$ can be obtained by permuting $U_1(W, W, W)$'s modes, in $\mathcal{O}(k^3)$ time.

Putting these together based on the distributive law, we can compute $\widetilde{T} = M_3(W, W, W)$ by one scan of the data:

$$
\widetilde{T} = M_3(W, W, W) = \frac{(\alpha_0 + 1)(\alpha_0 + 2)}{2} E_3(W, W, W)
$$
$$
- \frac{\alpha_0(\alpha_0 + 1)}{2}[(U_1 + U_2 + U_3)(W, W, W)] + \alpha_0^2 (W^T M_1)^{\otimes 3}
$$

(16)

which requires $O(Lk^2 + Vk + k^3) = O(Lk^2)$ time.

Algorithm 2. Scalable Tensor Orthogonal Decomposition (STOD)

Input: Corpus with L tokens and vocabulary size V, number of topics k,
        number of outer and inner iterations N, n, α_0
Output: The model parameters $(\alpha_t, \phi_t), t = 1, \ldots, k$

2.1 First scan of data: Compute M_1 and E_2 according to Eq. (7);
2.2 Find k largest orthonormal eigenpairs (σ_t, μ_t) of E_2;
2.3 $M_1' = U M_1$;   // $U = [\mu_1, \ldots, \mu_k], \Sigma_1 = diag(\sigma_1, \ldots, \sigma_k)$
2.4 Compute spectral decomposition for
    $M_2' = (\alpha_0 + 1)\Sigma_1 - \alpha_0 M_1' \otimes M_1' = U'\Sigma U'^T$;
2.5 $M = UU', W = M\Sigma^{-\frac{1}{2}}, (W^T)^+ = M\Sigma^{\frac{1}{2}}$;
2.6 Second scan of data: Compute $\widetilde{T} = M_3(W, W, W)$ according to Eq. (16);
2.7 Perform power method Line 1.6 to 1.16 in Algorithm 1;
2.8 **return** $(\alpha_t, \phi_t), t = 1, \ldots, k$

5.4   Our Final Algorithm

Algorithm 2 outlines our scalable tensor orthogonal decomposition algorithm. Line 2.1 scans the data once to collect E_2, and Line 2.2–2.5 are asymptotically equivalent to Line 1.2–1.4. Line 2.6 uses a second scan of the data to compute $\widetilde{T}$, which is equivalent to Line 1.5 but much more efficient. The power method on $\widetilde{T}$ remains the same as in Algorithm 1.

In most applications, $V^3 \gg L, V^2 \gg m, V \gg \hat{l} > k$. We reduce the time complexity for constructing the small tensor $\widetilde{T} \in \mathbb{R}^{k \times k \times k}$ to $\mathcal{O}(Lk^2 + km)$, and the space complexity to $\mathcal{O}(m)$. The total time complexity for STOD is $\mathcal{O}(Lk^2 + km + Nnk^4)$. Comparing with TOD, STOD is superior in both space and time. The practical speedup is significant with orders of magnitude, as we will demonstrate in experiments.

6   Experiments

In this section we first introduce the methods used for comparison, then present evaluation on synthetic and real datasets respectively.

Methods for Comparison. Our main contribution is the STOD algorithm. It accelerates the TOD algorithm, which has bounded error for LDA inference. We also compare STOD with one of the most popular LDA inference methods collapsed Gibbs sampling, although its error is not theoretically bounded. We do not include other maximum-likelihood based inference methods for LDA, e.g., collapsed variational Bayesian inference, because they mostly have a similar performance with collapsed Gibbs sampling, and no theoretical error bound either. We list the implementation details below:

- STOD – our Algorithm 2. Outer iteration $\# N$ and inner iteration $\# n$ are both set to 30. They are sufficient for our experiments. In fact, in most cases, the algorithm converges with $N = n = 10$. α_0 is fixed to be 1.
- TOD – A faster implementation of Algorithm 1 as we discussed in Section 5.1. It skips the tensor construction and computes the power iteration on the fly. Outer iteration $\# N$ and inner iteration $\# n$ are both set to 30. $\alpha_0 = 1$.
- Collapsed Gibbs sampling. We use a fast implementation by Griffiths and Steyvers [9]. The iteration $\#$ is set to 1500, following the common practice. From now on, we use 'Gibbs' or 'Gibbs sampling' for short.

For fair comparison, we do not use distributed computation for any of the methods. We conduct all the experiments on a single Linux server running MATLAB 2013a with Inter Xeon CPU E5-2680 2.80GHz and 256GB RAM.

6.1   Synthetic Data

We use synthetic data to conduct controlled experiments with known topics and other parameters. With synthetic data we are able to evaluate the error of each method in recovering the known topics. We compare each method's: i) topic recovery error; and ii) runtime.

The generative process of synthetic data simply follows LDA [6]. The length of each document is generated from a Poisson distribution, where the Poisson parameter λ, or the expected document length, is set to 100. The Dirichlet prior α of each document-level topic distribution θ_i is uniform: $\alpha_t = \frac{1}{k}, t \in [k]$; the Dirichlet prior β of each corpus-level topic-word distribution ϕ_t is also uniform: $\beta_x = \frac{200}{V}, x \in [V]$. The same Dirichlet prior is used for Gibbs sampling inference. We creat three controlled sets of pseudo corpora by varying the following parameters:

1. D, ranging from 5,000 to 500,000, with fixed $V = 10000, k = 50$.
2. V, ranging from 3,000 to 100,000, with fixed $D = 100,000, k = 50$.
3. k, ranging from 10 to 100, with fixed $D = 100,000, V = 10,000$.

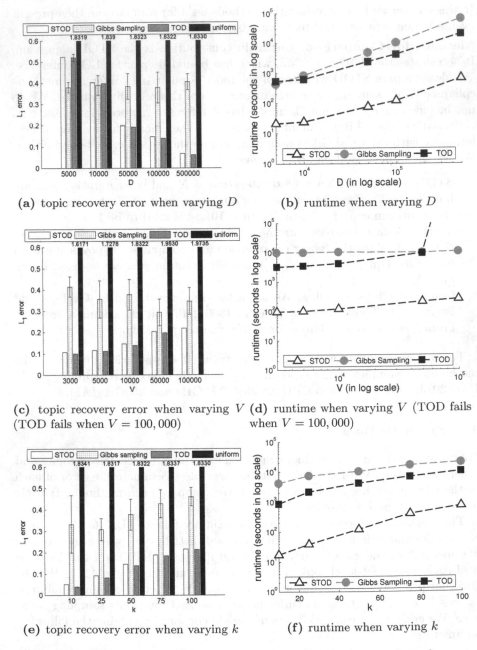

(a) topic recovery error when varying D

(b) runtime when varying D

(c) topic recovery error when varying V (TOD fails when $V = 100,000$)

(d) runtime when varying V (TOD fails when $V = 100,000$)

(e) topic recovery error when varying k

(f) runtime when varying k

Fig. 1. Performance study on synthetic data (lower values are better)

Topic Recovery Error. We measure the topic recovery error in the following way. For each run of each algorithm, let $\tilde{\phi}_t, t \in [k]$ denote the corpus-level multinomial distributions inferred by the algorithm; and $\phi_t^*, t \in [k]$ the ground-truth generated from Dirichlet allocation. We compute all the k^2 L1 distances: $\|\tilde{\phi}_{t_1} - \phi_{t_2}^*\|_1, t_1, t_2 \in [k]$, and build a bipartite graph with negative L1 distances as edge weights. Then we use the Hungarian algorithm to compute a maximum matching between the inferred topics and the ground truth topics. Finally, we average the k L1 distances between matched pairs as the error for this run.

Figure 1a, 1c and 1e show the recovery errors of different methods. For each fixed triple of (D, V, k), we run each algorithm for 5 times, then plot the mean and standard deviation of the 5 recovery errors using an error bar. To put the numbers in context, we include a baseline 'uniform', which reflects the average distance of a uniform distribution over the vocabulary to every topic.

We observe that TOD and STOD have almost zero variance across multiple runs, due to the robustness of tensor decomposition. Gibbs sampling produces large variance comparing with the other two, which is a drawback most existing maximum likelihood based LDA inference algorithms suffer.

In general, the errors of STOD and TOD decrease when D increases or V, k decrease, *i.e.*, the sample size increases or the model complexity decreases. This is because the error of tensor orthogonal decomposition is bounded by the distance of empirical moments from theoretical moments. For Gibbs sampling, this trend is not as clear as the moment-based methods. It has no error bound of topic recovery.

In all these datasets, the moment-based methods TOD and STOD have almost equal errors. When the corpus size is sufficiently large ($D \geq 50,000$ in these datasets), the error is 37–85% lower than Gibbs sampling. This verifies that TOD has the state-of-the-art capability of topic recovery accuracy, and that our STOD algorithm preserves that capability.

Runtime. From Figure 1b,1d and 1f, we see a clear superiority in efficiency of our STOD algorithm in all the datasets. STOD is faster than TOD and Gibbs sampling by 1 to 3 orders of magnitude. While TOD is generally faster than Gibbs sampling, it consumes much larger memory, and fails to terminate when $V = 100,000$.

The runtime of STOD grows more slowly with respect to D than Gibbs sampling, because STOD only scans the corpus twice while Gibbs sampling iteratively passes through the corpus for thousands of times. The runtime of STOD grows more tenderly with respect to V than TOD, because the former does not even need to construct the dense tensor of size V^2. The runtime of STOD grows more rapidly with respect to k than Gibbs sampling and TOD, because it constructs a tensor of size k^3 explicitly. Therefore, the advantage of STOD is most prominent when the corpus size and vocabulary size are large, while the number of topics is small.

6.2   Real-World Data

We use two real-world datasets to evaluate the performance of our algorithm in practice (we perform stemming to the favor of baseline methods, and remove stopwords in the corpus):

- TREC AP news: A TREC news dataset (1998). It contains 106K full articles, 170K unique words, and 19M tokens. After preprocessing, the size of vocabulary is 45,105.
- CS abstract: A dataset of computer science paper abstracts from Arnetminer[1]. The set has 529K papers, 186K unique words, and 39M tokens. After preprocessing, the size of vocabulary is 51,069.

Runtime. Table 1 shows the overall runtime in these datasets, in two scenarios. One scenario is that the data can be all loaded into memory, and the other scenario is that the data are too large to be loaded into memory. STOD is one to two orders of magnitude faster than the other methods in the first scenario, and two to three orders of magnitude faster in the second scenario. On the largest dataset it reduces the runtime of TOD/Gibbs sampling from 3 weeks/1.2 days to 9.6 minutes.

Table 2 shows the decomposed runtime for STOD and TOD. In both datasets, the most time consuming part for STOD is the spectral decomposition (Line 2.1–2.5) and tensor construction (Line 2.6). The news dataset has longer documents but fewer tokens than the CS dataset. As a result, the spectral decomposition in news dataset bears a larger fraction though the total runtime is shorter than in CS. The practical implementation of TOD does not create tensors but goes through the corpus many times to compute the power iteration on the fly, and that part accounts for the slow execution.

Table 1. Total runtime (in seconds) on real-world datasets (K=50)

dataset \ method	loaded into memory			not loaded into memory		
	STOD	TOD	Gibbs sampling	STOD	TOD	Gibbs sampling
news	**293**	6877	21641	**310**	768110	48999
CS	**541**	14439	47293	**577**	1661101	102136

Table 2. Decomposed runtime (%) on real-world datasets for STOD and TOD

dataset \ method	STOD			TOD	
	spectral decomp	construct tensor	power iter	spectral decomp	power iter
news	38.0	47.8	14.2	1.2	98.8
CS	11.1	80.7	8.3	1.2	98.8

Quality of Inferred Topics. Lack of gold standard is a well known challenge for unsupervised topic modeling methods. As such, people have proposed evaluation

---

[1] http://www.arnetminer.org

metrics without relying on labels. The conventional evaluation using the held-out perplexity of test data has been challenged [7, 17], and found to have negative correlation with human interpretability. According to the most recent work in topic model quality evaluation [14], there are two major approaches to measure the human interpretability: indirect approach with *word intrusion*, and direct approach with *observed coherence*. In this study we take the direct approach, and use the automated evaluation measure OC-Auto-NPMI in [14], which was reported to have above 0.9 Pearson correlation with human judgment.

The OC-Auto-NPMI measure for one topic is defined to be the average of normalized pointwise mutual information between every pair of the top-X words:

$$\text{OC-Auto-PMI}(t) = \frac{2}{X(X-1)} \sum_{j=2}^{X} \sum_{i=1}^{j-1} \log \frac{p(w_j, w_i)}{p(w_i)p(w_j)} \qquad (17)$$

where $w_1, \ldots, w_X$ are the top-X words in topic t. Then the mean of the OC-Auto-NPMI measure for all the topics can be used to measure the quality of the inferred topics by an inference algorithm (the higher the better).

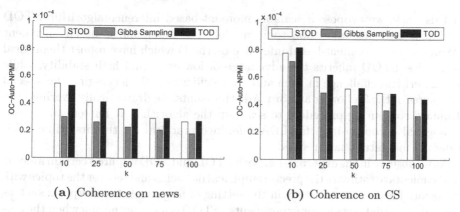

(a) Coherence on news   (b) Coherence on CS

Fig. 2. Quality of inferred topics on real-world data (the higher the better)

As shown in Figure 2, STOD and TOD again have close performance, and both outperform Gibbs sampling[2], by as much as 80% in the news dataset, and as much as 40% in the CS dataset. The moment-based methods not only have theoretical low error, but also good practical performance with real-world data.

Table 3 visualizes several example topics with top-ranked words from the TREC AP news dataset. Since STOD and TOD have identical results in this experiment, we only keep STOD in the table. We can see when k is set to 25, these four topics are interpretable in both methods, although a few top ranked words are less intuitive to interpret in Gibbs. For example, in topic 4 of Gibbs sampling, word 'oil' cannot be recognized as a part of *weather* topic, while most of the other words have a strong correlation with *weather*.

---

[2] We experimented with hyperparameter optimization for Gibbs sampling as well, and it does not affect the conclusion

Table 3. Example topics from a 25-topic run of STOD & Gibbs on news

topic 1: finance		topic 2: politics		topic 3: law		topic 4: weather	
STOD	*Gibbs*	*STOD*	*Gibbs*	*STOD*	*Gibbs*	*STOD*	*Gibbs*
dollar	cents	vote	city	court	court	fair	oil
yen	market	house	black	case	state	cloudy	fair
prices	stock	democratic	state	state	law	city	coast
late	trade	senate	democratic	judge	ruling	northern	state
trade	prices	republican	mayor	abortion	union	part	rain
close	dollar	bill	white	ruling	abortion	rain	texas
london	exchange	committee	campaign	law	strike	central	national
gold	higher	election	election	appeals	judge	north	northen
rate	futurers	members	republican	federal	case	coast	part
bid	lower	party	year	supreme	federal	south	north

7 Discussion

In this work, we propose a scalable moment-based inference algorithm STOD for latent Dirichlet allocation topic model. The algorithm is based on recent advancement of moment-based inference methods which have robust theoretical properties. STOD inherits the advantage of low error and high stability, while solving critical challenge in time and space efficiency. By leveraging the special structures of the 2nd order and 3rd order moments, we dramatically overhaul the standard computing procedure to scale up the algorithm. It renders the tensor orthogonal decomposition for LDA inference practical for the first time, with orders of magnitude faster speed.

As we observe in the experiments, both STOD and TOD require a certain amount of documents to estimate the precise empirical moments and recover the topics with low error. This is easy to satisfy in the setting of large-scale text corpora, such as the real-world datasets in our experiments. STOD is most promising when the corpus size is large, and when the number of topics is small. This makes it a desirable method to summarize a large corpus' topics in a hierarchical structure where every topic has a few number of subtopics, which is one of our ongoing study.

Although we do not compare with distributed or online inference mechanism for MCMC or variational Bayesian inference, we would like to point out that: i) STOD can be easily parallelized by employing distributed spectral decomposition method such as [13, 15], with theoretically guaranteed performance; and ii) STOD scans the data only twice, which is similar to online inference methods requiring only one pass of data, but STOD does not trade in inference accuracy. Besides parallelization, the advantage of STOD can be further fulfilled by adaptation to dynamic text collections, or more advanced spectral decomposition methods.

Acknowledgments. This work was supported in part by the U.S. Army Research Laboratory under Cooperative Agreement No. W911NF0920053 (NS-CTA), the Army Research Office under Cooperative Agreement No. W911NF-13-1-0193, National Science Foundation CNS-1027965, IIS-1017362, IIS-1320617, and IIS-1354329, DTRA, and MIAS, a DHS-IDS Center for Multimodal Information Access and Synthesis at UIUC.

References

1. Ahmed, A., Ho, Q., Teo, C.H., Eisenstein, J., Xing, E.P., Smola, A.J.: Online inference for the infinite topic-cluster model: Storylines from streaming text. In: AISTATS (2011)
2. Anandkumar, A., Foster, D.P., Hsu, D., Kakade, S., Liu, Y.-K.: A spectral algorithm for latent dirichlet allocation. In: NIPS (2012)
3. Anandkumar, A., Ge, R., Hsu, D., Kakade, S.M., Telgarsky, M.: Tensor decompositions for learning latent variable models. arXiv preprint arXiv:1210.7559 (2012)
4. Arora, S., Ge, R., Halpern, Y., Mimno, D., Moitra, A., Sontag, D., Wu, Y., Zhu, M.: A practical algorithm for topic modeling with provable guarantees. In: ICML (2013)
5. Arora, S., Ge, R., Moitra, A.: Learning topic models–going beyond svd. In: FOCS (2012)
6. Blei, D.M., Ng, A.Y., Jordan, M.I.: Latent dirichlet allocation. Journal of Machine Learning Research 3, 993–1022 (2003)
7. Chang, J., Boyd-Graber, J., Wang, C., Gerrish, S., Blei, D.M.: Reading tea leaves: How humans interpret topic models. In: NIPS (2009)
8. Foulds, J., Boyles, L., DuBois, C., Smyth, P., Welling, M.: Stochastic collapsed variational bayesian inference for latent dirichlet allocation. In: KDD (2013)
9. Griffiths, T.L., Steyvers, M.: Finding scientific topics. Proc. of the National Academy of Sciences of USA 101(suppl. 1), 5228–5235 (2004)
10. Hoffman, M., Blei, D., Wang, C., Paisley, J.: Stochastic variational inference. Journal of Machine Learning Research 14, 1303–1347 (2013)
11. Hoffman, M., Blei, D.M., Mimno, D.M.: Sparse stochastic inference for latent dirichlet allocation. In: ICML (2012)
12. Hofmann, T.: Unsupervised learning by probabilistic latent semantic analysis. Machine Learning 42(1-2), 177–196 (2001)
13. Kempe, D., McSherry, F.: A decentralized algorithm for spectral analysis. In: STOC (2004)
14. Lau, J.H., Newman, D., Baldwin, T.: Machine reading tea leaves: Automatically evaluating topic coherence and topic model quality. In: EACL (2014)
15. Maschhoff, K.J., Sorensen, D.: P_ARPACK: An efficient portable large scale eigenvalue package for distributed memory parallel architectures. In: Madsen, K., Olesen, D., Waśniewski, J., Dongarra, J. (eds.) PARA 1996. LNCS, vol. 1184, Springer, Heidelberg (1996)
16. Newman, D., Asuncion, A., Smyth, P., Welling, M.: Distributed algorithms for topic models. Journal of Machine Learning Research 10, 1801–1828 (2009)
17. Newman, D., Lau, J.H., Grieser, K., Baldwin, T.: Automatic evaluation of topic coherence. In: NAACL-HLT (2010)
18. Porteous, I., Newman, D., Ihler, A., Asuncion, A., Smyth, P., Welling, M.: Fast collapsed gibbs sampling for latent dirichlet allocation. In: KDD (2008)
19. Sontag, D., Roy, D.: Complexity of inference in latent dirichlet allocation. In: NIPS (2011)
20. Yao, L., Mimno, D., McCallum, A.: Efficient methods for topic model inference on streaming document collections. In: KDD (2009)
21. Zhai, K., Boyd-Graber, J., Asadi, N., Alkhouja, M.L.: Mr. lda: A flexible large scale topic modeling package using variational inference in mapreduce. In: WWW (2012)

Unsupervised Feature Selection via Unified Trace Ratio Formulation and K-means Clustering (TRACK)

De Wang, Feiping Nie, and Heng Huang*

Department of Computer Science and Engineering, University of Texas at Arlington,
Arlington, TX 76019, USA
{wangdelp,feipingnie}@gmail.com, heng@uta.edu

Abstract. Feature selection plays a crucial role in scientific research and practical applications. In the real world applications, labeling data is time and labor consuming. Thus, unsupervised feature selection methods are desired for many practical applications. Linear discriminant analysis (LDA) with trace ratio criterion is a supervised dimensionality reduction method that has shown good performance to improve classifications. In this paper, we first propose a unified objective to seamlessly accommodate trace ratio formulation and K-means clustering procedure, such that the trace ratio criterion is extended to unsupervised model. After that, we propose a novel unsupervised feature selection method by integrating unsupervised trace ratio formulation and structured sparsity-inducing norms regularization. The proposed method can harness the discriminant power of trace ratio criterion, thus it tends to select discriminative features. Meanwhile, we also provide two important theorems to guarantee the unsupervised feature selection process. Empirical results on four benchmark data sets show that the proposed method outperforms other sate-of-the-art unsupervised feature selection algorithms in all three clustering evaluation metrics.

1 Introduction

Feature selection is to select relevant and informative features from the high-dimensional feature space. Because it can improve the mode generalization capability, prevent model over-fitting, identify useful features, and hugely reduce the computational time, feature selection has been playing a crucial role in many scientific and practical applications, such as text mining [7], bioinformatics [5,23,3], medical image analysis [22,24], computer vision [4,12], *etc.*

There are three types of feature selection methods: filter method [20,13,19,5], wrapper method [11], and embedded method [26]. The filter methods compute a score to each feature, so the computational cost is relatively low, but the selected features often cannot achieve good classification performance. Wrapper methods treat the classifier as a black box, and use classification results to evaluate potential feature subset, thus the features selected by wrapper methods usually have good performance. However, their computational cost is very high since it need to use the classifier all the way through the

---

* This work was partially supported by NSF IIS-1117965, IIS-1302675, IIS-1344152, DBI-1356628.

T. Calders et al. (Eds.): ECML PKDD 2014, Part III, LNCS 8726, pp. 306–321, 2014.

process of feature selection. The embedded methods treat classifier as a white box, and incorporate feature selection and classification model into a single optimization problem. Thus, the classification performance is good, and the computational cost is much lower than wrapper method.

From another point of view, feature selection techniques can be categorized into supervised method (using label information) and unsupervised method (without using label information). Supervised feature selection methods determine the importance of a feature by evaluating the feature's correlation with label. The higher correlation indicates a more important feature. Unsupervised feature selection approaches select features with maximum representative and discriminant power. In the real world data mining applications, labeling data is time and labor consuming. Thus, the unsupervised feature selection methods are crucial for practical applications.

Many unsupervised feature selection methods have been proposed. Among them, maximum-variance is the simplest one, which just selects top ranked features with maximum variance. Although selected features are representative for data variance, they are not guaranteed to be discriminant for classification [9]. Laplacian Score [9] selects features that can preserve the local manifold structure of data, and such features are supposed to be discriminative. SPEC [27] selects features that are most consistent with the graph structure of data. MCFS [2] first performs regression using the eigenvector of graph Laplacian, and then selects features with maximum spectral regression coefficients.

In this work, we focus on the unsupervised feature selection model design. Most existing unsupervised feature selection methods are similar to filter methods in supervised learning, and define different score systems to select features. Considering the advantages of embedded feature selection methods in supervised learning, we hope to use the embedded feature selection mechanism in an unsupervised way. In this paper, we address this problem using the unsupervised trace ratio formulation, and rigorously prove that our unsupervised trace ratio formulation is the unified and unique objective of both trace ratio linear discriminant analysis (LDA) and K-means clustering. After that, we propose an unsupervised feature selection method using unsupervised trace ratio formulation and $\ell_{1,2}$-norm regularization. The proposed method can harness the discriminant power of trace ratio formulation, thus it tends to select discriminative features. The optimization algorithm is derived with rigorous convergence analysis. Moreover, we provide important theoretical analysis to guarantee the unsupervised feature selection process. Empirical results on four benchmark data sets show that the proposed method outperforms other sate-of-the-art unsupervised feature selection methods on all three standard evaluation metrics.

2 Notations and Definitions

In this paper, matrices are written as uppercase letters and vectors are written as bold lowercase letters. Given a matrix $W = \{w_{ij}\}$, its i-th row, j-th column are denoted as $\mathbf{w}^i$, $\mathbf{w}_j$, respectively. The $\ell_{1,2}$-norm of matrix W is defined as $||W||_{1,2} = \sum_{i=1}^{d} ||\mathbf{w}^i||_2$. $Tr(W)$ means the trace operation for matrix W.

Given data matrix $X = [\mathbf{x}_1, \cdots, \mathbf{x}_n] \in \Re^{d \times n}$, d is the number of features and n is the number of data samples. In the classic Linear Discriminant Analysis (LDA), the total scatter matrix S_t, within-class scatter matrix S_w, and between-class scatter matrix S_b are defined as following:

$$S_t = \sum_{i=1}^{n} (\mathbf{x}_i - \bar{\mathbf{x}})(\mathbf{x}_i - \bar{\mathbf{x}})^T,$$

$$S_w = \sum_{k=1}^{c} \sum_{\mathbf{x}_i \in l_k} (\mathbf{x}_i - \mathbf{m}_k)(\mathbf{x}_i - \mathbf{m}_k)^T,$$

$$S_b = \sum_{k=1}^{c} n_k (\mathbf{m}_k - \bar{\mathbf{x}})(\mathbf{m}_k - \bar{\mathbf{x}})^T,$$

where $\mathbf{x}_i \in \Re^{d \times 1}$ is the i-th data sample, c is the number of clusters, n_k is the number of data points belong to class l_k, $\mathbf{m}_k$ is the center of the k-th cluster, $i.e.$ $\mathbf{m}_k = \frac{1}{n_k} \sum_{\mathbf{x}_i \in l_k} \mathbf{x}_i$, $\bar{\mathbf{x}}$ is the center of all data, $i.e.$ $\bar{\mathbf{x}} = \frac{1}{n} \sum_{i=1}^{n} \mathbf{x}_i$. It is well known that $S_t = S_b + S_w$.

Suppose $X \in \Re^{d \times n}$ is the data matrix after centralization, $i.e.$ $\bar{x} = 0$, the formulations of total scatter matrix S_t and between-class scatter matrix S_b can be thus reduced to:

$$S_t = \sum_{i=1}^{n} \mathbf{x}_i \mathbf{x}_i^T = XX^T, \quad S_b = \sum_{k=1}^{c} n_k \mathbf{m}_k \mathbf{m}_k^T . \tag{1}$$

Denote $G \in \Re^{n \times c}$ as the class indicator matrix, where $G_{ij} = 1$ if x_i belongs to the j-th class and $G_{ij} = 0$ otherwise. We define a cluster centroid matrix M to include the centroid vector of each cluster as $M = [\mathbf{m}_1, \mathbf{m}_2, \cdots, \mathbf{m}_c]$. Using the class indicator matrix G, we can represent the cluster centroid matrix M as:

$$M = XG(G^T G)^{-1} . \tag{2}$$

Using matrices G and M, we can re-write the scatter matrices into more compact manner as:

$$S_b = MG^T GM^T \tag{3}$$

$$S_w = (X - MG^T)(X - MG^T)^T \tag{4}$$

3 Trace Ratio Linear Discriminant Analysis Review

In recent research, Linear Discriminant Analysis (LDA) with trace ratio criterion has shown better performance than the traditional LDA with ratio trace criterion [18,10]. Thus, the trace ratio LDA has attracted more and more attention and has been well studied. The problem of trace ratio LDA is as follows:

$$\max_{W^T W = I} \frac{Tr(W^T S_b W)}{Tr(W^T S_w W)}, \tag{5}$$

where $W \in \Re^{d \times m}$ is the projection matrix, which is constrained to be orthonormal, and m is the reduced dimension.

Using the optimal solution W of the problem (5), the data points are projected to a lower dimensional subspace such that the Euclidean distances of data pairs within the same class are minimized while the Euclidean distances of data pairs between different classes are maximized. That is to say, the data points are easy to be classified after the dimensionality reduction with W.

Because of $S_t = S_b + S_w$, problem (5) is equivalent to the following problem:

$$\max_{W^T W = I} \frac{Tr(W^T S_b W)}{Tr(W^T S_t W)}. \tag{6}$$

4 Discriminative Unsupervised Feature Selection

Because the LDA can enhance the classification tasks, several recent research works have used this criterion for supervised feature selection and shown promising results [15,21]. However, the LDA strategy cannot be applied to unsupervised feature selection, because the unsupervised learning models don't provide the data labels which are required to compute the within-class and between-class scatters. In previous work [6], the authors utilized the clustering results to calculate S_b and S_w and iteratively do LDA and K-means clustering, such that the LDA criterion can be applied to improve clustering results. However, the authors only presented a heuristic algorithm and didn't have a unified objective for two different processes, *i.e.* the LDA and K-means clustering minimize different objectives. Thus, the optimality and convergence of their algorithm are NOT guaranteed.

In this work, we are interested in designing a powerful unsupervised feature selection method. To address the above problems, we will derive a new formulation and rigorously prove it unifies both trace ratio LDA and K-means clustering, such that the trace ratio LDA criterion can be applied to unsupervised model seamlessly. Combining with the structured sparsity-inducing norms, we will propose a novel unsupervised feature selection method.

4.1 Unsupervised Dimensionality Reduction Using Trace Ratio Criterion

Trace ratio LDA is a supervised dimensionality reduction method. Plugging Eq. (3) into Eq. (6), the trace ratio LDA objective can be written as:

$$\max_{W^T W = I} \frac{Tr(W^T X G (G^T G)^{-1} G^T X^T W)}{Tr(W^T S_t W)}, \tag{7}$$

where $G \in \Re^{n \times c}$ is the class indicator matrix defined in Section 2.

In unsupervised circumstance where there is no label information, we don't know both projection matrix W and class indicator matrix G. If we apply the trace ratio strategy to unsupervised dimensionality reduction, we need solve the following problem:

$$\max_{W^T W = I, G \in Ind} \frac{Tr(W^T X G (G^T G)^{-1} G^T X^T W)}{Tr(W^T S_t W)}, \tag{8}$$

where Ind is the set of clustering indicator matrices and $G \in Ind$ means G is constrained to be a clustering indicator matrix. This is not LDA anymore. How does problem (8) reduce the data dimensionality to an unsupervised way? Our following theorem will rigorously show that the problem (8) is a unified and unique objective of both trace ratio LDA and K-means clustering.

Solving problem (8) is exactly equivalent to iteratively solving trace ratio LDA and doing K-means clustering. When G is fixed, obviously solving problem (8) is to solve the trace ratio LDA *w.r.t.* W, *i.e.* solving problem (7).

When W is fixed, $Tr(W^T S_t W)$ is irrelevant to G. Thus, we need to solve the following problem:

$$\max_{G \in Ind} Tr(W^T X G (G^T G)^{-1} G^T X^T W). \qquad (9)$$

Although the problem (9) only has one variable, it is difficult to solve due to the intractable constraint. Because $Tr(W^T S_t W)$ is a constant now (W is fixed), maximizing between-class distance in problem (9) is equivalent to minimizing within-class distance. Problem (9) is equivalent to the following problem:

$$\min_{G \in Ind, M} Tr(W^T S_w W), \qquad (10)$$

where $S_w = (X - MG^T)(X - MG^T)^T$ as shown in Eq. (4). Thus, we need to optimize:

$$\min_{G \in Ind, M} Tr(W^T(X - MG^T)(X - MG^T)^T W)$$

$$\implies \min_{G \in Ind, M} \left\| W^T X - W^T MG^T \right\|_F^2$$

$$\implies \min_{G \in Ind, F} \left\| W^T X - FG^T \right\|_F^2, \qquad (11)$$

where $F = W^T M$.

Problem (11) can be easily solved by alternating optimization, *i.e.*, iteratively optimizing F when G is fixed and optimizing G when F is fixed. Interestingly, this iterative procedure is exactly the procedure of traditional K-means clustering algorithm on the projected data $W^T X$: that is, when G is fixed, the optimal solution of F is the centers of the clusters in the projected subspace; when F is fixed, the optimal solution of G can be computed by assigning the data points to their closest centers. Thus, the objective function in (9) is equivalent to K-means clustering objective.

Therefore, solving problem (8) is equivalent to iteratively solving trace ratio LDA (fix G to solve W) and doing K-means clustering (fix W to solve G).

Therefore, the objective in (8) is a good trace ratio formulation to reduce the dimensionality in an unsupervised way. The K-means clustering indicators can be used as labels to calculate scatter matrices, such that the projection matrix is discriminative to separate different data groups.

Please notice that our method is significantly different from the method in [6], where the traditional ratio trace LDA and K-means clustering algorithms are heuristically combined *without* any optimality and convergence guarantee. Our new Theorem 1 rigorously proves that the trace ratio formulation in (8) is the *unified and unique* objective when we iteratively solve trace ratio LDA and K-means clustering. Thus, this procedure is guaranteed to converge.

Based on our above derivations, the unsupervised trace ratio formulation in (8) is equivalent to the following problem:

$$\min_{W^T W=I, G \in Ind, F} \frac{\left\| W^T X - F G^T \right\|_F^2}{Tr(W^T S_t W)} \tag{12}$$

4.2 Unsupervised Feature Selection Using Structured Sparse Trace Ratio Formulation

Both supervised and unsupervised trace ratio LDA are dimensionality reduction methods, where the projected feature is a linear combination of all original features. However, in many applications (*e.g.* bioinformatics and document mining), we are more interested in the feature selection model, *i.e.*, selecting a few relevant features. To address this problem, we integrate the structured sparsity-inducing norms with the above unsupervised trace ratio formulation, such that we can select informative features in an unsupervised way.

We hope to learn a row sparse projection matrix W in which only a few rows of W are non-zeros. With this row sparse projection matrix W, only a few important features are involved in the projection. This goal can be achieved by minimizing $\|W\|_{1,2}$. Therefore, problem (12) can be changed to the following objective for unsupervised feature selection:

$$\min_{W^T W=I, G \in Ind, F} \frac{\left\| W^T X - F G^T \right\|_F^2}{Tr(W^T S_t W)} + \gamma \|W\|_{1,2}, \tag{13}$$

where γ is a regularization parameter which controls the row sparsity of the projection matrix W. The greater the γ is, the more sparse rows the projection matrix W has.

The optimal solution of the problem (13) can harness the discriminative power of the unsupervised trace ratio model, thus it tends to select discriminative features. Only those discriminative features would have non-zero weights in W, and thus each new projected feature is a linear combination of only these discriminative features. In this way, only discriminative information are retained.

5 Optimization Algorithm

We use the alternating optimization method to solve the problem (13). When W is fixed, the problem becomes problem (11), which can be solved by alternating optimization. Specifically, when G is fixed, the optimal F is:

$$F = W^T X G (G^T G)^{-1}; \tag{14}$$

when F is fixed, the optimal G is:

$$G_{ij} = \begin{cases} 1, \ j = \arg\min_k \left\| W^T x_i - f_k \right\|_2^2 \\ 0, \qquad\qquad other \end{cases} \tag{15}$$

As mentioned before, this update of F and G is exactly the K-means procedure.

When G and F are fixed, we substitute Eq. (14) into the problem (13) and thus the problem (13) becomes

$$\min_{W^T W = I} \frac{Tr(W^T S_w W)}{Tr(W^T S_t W)} + \gamma \|W\|_{1,2}, \tag{16}$$

where

$$S_w = (X - XG(G^T G)^{-1} G^T)(X - XG(G^T G)^{-1} G^T)^T. \tag{17}$$

Due to the trace ratio formulation, the above objective is difficult to optimize. The standard proximal gradient, Augmented Lagrange Multiplier, fixed point, proximal methods cannot efficiently optimize it. We will use the iterative re-weighted optimization strategy to solve this objective. Solving the above objective is equivalent to solve:

$$\min_{W^T W = I} \frac{Tr(W^T S_w W)}{Tr(W^T S_t W)} + \gamma Tr(W^T D W), \tag{18}$$

where D is a diagonal matrix with the i-th diagonal element $d_i = \frac{1}{2\|\mathbf{w}^i\|_2}$. When $\|\mathbf{w}^i\|_2 = 0$, the original objective is not differentiable. Following [8], we can introduce a small perturbation to regularize the i-th diagonal element of D as $\frac{1}{2\sqrt{\|\mathbf{w}^i\|_2^2 + \varsigma}}$. Then it can be verified that the algorithm minimizes the following problem: $\frac{Tr(W^T S_w W)}{Tr(W^T S_t W)} + \gamma \sum_{i=1}^{d} \sqrt{\|\mathbf{w}^i\|_2^2 + \varsigma}$ is apparently reduced to problem Eq. (16) when $\varsigma \to 0$.

In the following, we derive an iterative algorithm to solve the problem (18) with a similar trick used in [17]. The Lagrangian function of the problem (18) is:

$$\mathcal{L}(W, \Lambda) = \frac{Tr(W^T S_w W)}{Tr(W^T S_t W)} + \gamma Tr(W^T D W) \\ - Tr(\Lambda(W^T W - I)). \tag{19}$$

By taking the derivative *w.r.t.* W to zero, we have

$$\left(S_w - \frac{Tr(W^T S_w W)}{Tr(W^T S_t W)} S_t + \gamma Tr(W^T S_t W) D \right) W = W \Lambda. \tag{20}$$

Thus, the optimal solution of W is the m smallest eigenvectors of the matrix:

$$S_w - \frac{Tr(W^T S_w W)}{Tr(W^T S_t W)} S_t + \gamma Tr(W^T S_t W) D. \tag{21}$$

We can iteratively update the D and the W such that the Eq. (20), *i.e.* KKT condition, is satisfied. Please notice that D is not a variable to optimize. In the iterative steps, we optimize Eq. (21) to get W, and then re-calculate Eq. (21), where D is only an intermediate value to help calculation.

In summary, the algorithm to solve the discriminative unsupervised feature selection problem (13) is outlined in Algorithm 1. Since our formulation is based on TRACe ratio and K-means formulations, we call this algorithm as TRACK for short.

Algorithm 1. Algorithm to solve the objective of our TRACK method in (13).

Initialize D as an identity matrix.

repeat

1. Iteratively update F by Eq. (14) and update G by Eq. (15) till to converge.
2. Iteratively update the diagonal matrix D with the i-th diagonal element as $d_i = \frac{1}{2\|\mathbf{w}^i\|_2}$, and update W by the m eigenvectors corresponding to the m smallest eigenvalues of

$$S_w - \frac{Tr(W^T S_w W)}{Tr(W^T S_t W)} S_t + \gamma Tr(W^T S_t W) D,$$

till converge.

until Converges

5.1 Convergence Analysis

In Algorithm 1, the Step 1 is the K-means clustering procedure and converges to local optimal solution. Step 2 is the iterative re-weighted algorithm to solve problem (16), *i.e.* problem (18). In each iteration within Step 2, the objective value of problem (18) is decreased until the algorithm converges. The proof is similar to [1,16], and thus we omit it due to limited space. When the Step 2 converges, Eq. (20) is satisfied. Note that Eq. (20) is the KKT condition of problem (18), therefore the converged solution satisfies the KKT condition of problem (18), and thus is a local optimal solution to the problem (18).

It deserves to be mentioned that, based on our unified and unique objective for both steps, Step 1 and Step 2 in Algorithm 1 are guarantied to mutually benefit each other. On the one hand, the better clustering results in Step 1 will result in better scatter matrices, and thus results in more discriminative projection matrix in Step 2; On the other hand, the more discriminative projection matrix in Step 2 will make the data more separable, thus lead to better clustering results in Step 1.

5.2 Theoretical Analysis for Feature Selection

To guarantee the unsupervised feature selection process, we provide the following important theorems on the problem (13). First, we will show that our method guarantees to have m features for selection, *i.e.* the sparsity shrinkage won't over suppress the non-zero rows in W. Second, we will prove that using the $\ell_{1,2}$-norm regularization in our TRACK objective is equivalent to using the $\ell_{0,2}$-norm regularization, which is the ideal feature selection formulation.

Theorem 1. *The number of non-zero rows of the optimal solution to the problem (13) will not be less than m.*

Proof: Because $W \in \Re^{d \times m}$ and $W^T W = I$, the rank of W is m. Thus, the number of non-zero rows of any feasible solution to the problem (13) will not smaller than m, otherwise the rank of W is smaller than m. □

Theorem 1 indicates the selected feature number is at least m by solving the problem (13) with even a very large γ. This is important, because the sparse learning based feature selection methods could over suppress the non-zero rows such that there are no enough features for selection.

Moreover, we have the following theorem, which indicates that minimizing the $\ell_{1,2}$-norm of W in our TRACK objective is equivalent to minimizing the $\ell_{0,2}$-norm of W under the constraint of $W^T W = I$.

Theorem 2. *Let* $W \in \Re^{d \times m}$. *The optimal solutions to the problem* $\min\limits_{W^T W = I} \|W\|_{1,2}$ *and the optimal solutions to the problem* $\min\limits_{W^T W = I} \|W\|_{0,2}$ *are the same.*

Proof: Obviously, the optimal solution W^* to the problem $\min\limits_{W^T W = I} \|W\|_{0,2}$ is any matrix with only m non-zero rows, and the matrix with the m non-zero rows is an orthonormal matrix. Without loss of generality, suppose $W^* = \begin{bmatrix} W_1 \\ 0 \end{bmatrix}$, where $W_1 \in \Re^{m \times m}$ is an orthonormal matrix, then we have $\|W^*\|_{1,2} = m$. For any matrix $W \in \Re^{d \times m}$ with the constraint $W^T W = I$, we can construct an orthonormal matrix $\hat{W} = [W, W^\perp] \in \Re^{d \times d}$, then the i-th row of $\hat{W}$ has $\|\hat{\mathbf{w}}_i\|_2 = 1$, and then the i-th row of W has $\|\mathbf{w}_i\|_2 \leq 1$. So we have $\|\mathbf{w}_i\|_2 \geq \|\mathbf{w}_i\|_2^2$, and then:

$$\|W\|_{1,2} \geq \|W\|_F^2 = m = \|W^*\|_{1,2} . \tag{22}$$

Therefore, W^* is the optimal solution to the problem $\min\limits_{W^T W = I} \|W\|_{1,2}$.   □

Therefore, in our TRACK method, features selected by the $\ell_{1,2}$-norm regularization are the same as using the ideal $\ell_{0,2}$-norm regularization.

6   Experimental Results

In this section, we compare the proposed TRACK feature selection algorithm with other state-of-the-art unsupervised feature selection algorithms: Maximum-Variance (Max-Var), Laplacian Score (LS) [9], SPEC [27] and MCFS [2], and ldaKm [6].

6.1   Brief Descriptions of Comparison Methods

We briefly describe the comparison methods in this section. MaxVar is the simplest unsupervised feature selection algorithm, which just select top ranked features with maximum variance. Although selected features are representative for data variance, they are not guaranteed to be discriminant for classification [9].

Laplacian Score selects features that can preserve the local manifold structure of data, and such features are supposed to be discriminative. It computes the score for each feature as $S_i = \frac{\hat{f}_i^T \mathcal{L} \hat{f}_i}{\hat{f}_i^T D \hat{f}_i}$, where $\mathcal{L}$ is the graph Laplacian, and $\hat{f}_i = f_i - \frac{\hat{f}_i^T D 1}{1^T D 1} 1$.

SPEC algorithm selects features that are most consistent with the graph structure of data. It computes the score for each feature as $S_i = \hat{f}_i^T \mathcal{L} \hat{f}_i$, where $\hat{f}_i = \frac{D^{\frac{1}{2}} f_i}{\|f_i\|}$.

MCFS algorithm first performs regression using the eigenvector of graph Laplacian, and then selects features with maximum spectral regression coefficients. The regression problem is formulated as $\min_{a_k} \left\| y_k - X^T a_k \right\|_F^2$, where y_k is the k_{th} eigenvector of the graph Laplacian matrix, a_k is the spectral regression coefficients. The score for the i_{th} feature is defined as $S_i = \max_k |a_{k,i}|$.

LdaKm is an adaptive dimensionality reduction method that integrates K-means clustering and LDA. The ldaKm alternatively performs the following two steps: (1) perform K-means clustering on projected space; (2) perform traditional ratio trace LDA to get the projection matrix. Following our approach, $\ell_{1,2}$-norm regularization is used to select features for ldaKm method.

6.2   Data Sets and Evaluation Metrics

Four real world data sets are used to validate the effectiveness of our TRACK feature selection algorithm: MSRC-V1, ORL, JAFFE, and XM2VTS.

MSRC-V1 database is from Microsoft Research in Cambridge. This data set contains coarse pixel-wise labeled images, and it is commonly used for full scene segmentation.

ORL database contains a set of face images taken between April 1992 and April 1994 at the ATT lab. Ten different images are taken for each of the 40 distinct subjects. For some subjects, the images were taken at different times, with different light condition, facial expressions (*i.e.*: smiling or not smiling, open or closed eyes). All the images were taken against a dark homogeneous background with the subjects in an upright, frontal position.

JAFFE (Japanese Female Facial Expression) database contains 213 images of 7 facial expressions (6 basic facial expressions + 1 neutral) posed by 10 Japanese female models, which were taken at the Psychology Department in Kyushu University. Each image has been rated on 6 emotion adjectives by 60 Japanese subjects.

XM2VTS (Extended Multi Modal Verification for Teleservices and Security applications) database is a large multi-modal database which was captured onto high quality digital video. It contains four recordings of 295 subjects taken over a period of four months. Sets of data taken from this database are available including high quality color images, 32 KHz 16-bit sound files, video sequences and a 3d Model.

Important statistics of the data sets are summarized in Table 1.

Table 1. Data set descriptions

	sample #	feature #	class #
MSRC-V1	210	1302	7
ORL	400	644	40
JAFFE	213	1024	10
XM2VTS	1180	1024	295

Three measures are used to evaluate the clustering performance of all methods: accuracy, normalized mutual information (NMI) and purity.

Accuracy is the percentage of correct predicted label. Because the real label of each cluster is unknown, the Hungarian algorithm [14] is used to get the best map to the real label. Let C denotes the ground truth label, C' denotes the label obtained from a clustering algorithm, the mutual information (MI) is defined as:

$$MI(C, C') = \Sigma_{c_i \in C, c'_j \in C'} p(c_i, c'_j) log \frac{p(c_i, c'_j)}{p(c_i)p(c'_j)} \qquad (23)$$

where $p(c_i), p(c'_j)$ are the probability of a arbitrarily selected sample belongs to cluster c_i, c'_j, respectively. $p(c_i, c'_j)$ is the probability of a arbitrarily selected sample belongs to both cluster c_i and c'_j.

NMI is the normalized MI as following:

$$NMI(C, C') = \frac{MI(C, C')}{\max(H(C), H(C'))} \qquad (24)$$

where $H(C) and H(C')$ are the entropies of C and C', respectively.

Purity is computed by assigning the label of a cluster to the most frequent class. More formally, it is defined as:

$$purity(C, C') = \frac{1}{N} \Sigma_j \max_i (c'_j \cap c_i) \qquad (25)$$

6.3 Demonstration of Discriminant Power of Selected Features

In this section, we show the discriminant power of selected features by various algorithms. We use different unsupervised feature selection algorithms to select top 30 features on the MSRC-V1 data set. Then selected features are used to perform principle component analysis (PCA), and data samples are projected onto the first 2 principle components (PC), as shown in Figure 1 (PCA performed using top 30 features). For the baseline method, all features are used to perform PCA.

From Figure 1, we can see that: The TRACK algorithm separates data much better than other feature selection algorithms. The MCFS and ldaKm algorithms perform slightly better than the remaining algorithms. Data are much more entangled with each other using the MaxVar and SPEC algorithm. This shows that: the TRACK algorithm can harness the discriminant power of trace ratio formulation, therefore, features selected by the TRACK algorithm are much more discriminant than those selected by other algorithms, and using those discriminant features can separate data from different classes well.

6.4 Clustering Performance Comparison

We select top 10 till to top 100 features using different methods, and perform K-means using the selected features to evaluate the clustering performance. Since K-means clustering is sensitive to initialization, we perform 20 trials and record the average clustering metric. The result of using all features is also reported as a baseline. The

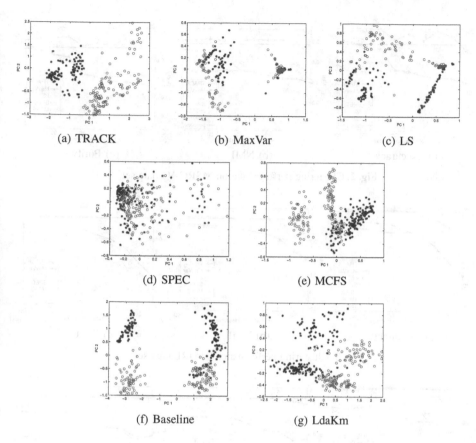

(a) TRACK   (b) MaxVar   (c) LS

(d) SPEC   (e) MCFS

(f) Baseline   (g) LdaKm

Fig. 1. Projection on first two principle components (PC) using top 30 features selected by various feature selection algorithms on the MSRC-V1 data set. The horizontal axis is the score of the first principle component, and the vertical axis is the score of the second principle component. Different shape or color mark samples from different classes.

regularization parameter is tuned from $\{10^{-4}, 10^{-3}, 10^{-2}, 10^{-1}, 1, 10, 10^2, 10^3, 10^4\}$ for both the TRACK algorithm and the ldaKm algorithm. The reduced dimension m in our method is set as: $m = c - 1$ if $d <= n$, and $m = c - 1 + d - n$ if $d > n$, as suggested in the paper [25]. Clustering accuracy, NMI, purity on the four data sets are reported in Figures 2- 5.

From those figures, we can conclude that:

(1) On all the four data sets, our method can outperform other state-of-the-art unsupervised feature selection algorithms on all evaluation metrics. The TRACK algorithm can outperform the baseline (using all features) using just 20 to 50 features, which justifies that the TRACK algorithm is able to select the most discriminant features.

(2) Generally, clustering performance becomes better when more features are selected.

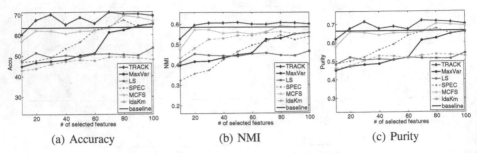

(a) Accuracy   (b) NMI   (c) Purity

Fig. 2. Clustering performance on MSRC-V1 data set

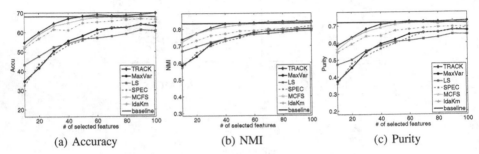

(a) Accuracy   (b) NMI   (c) Purity

Fig. 3. Clustering performance on ORL data set

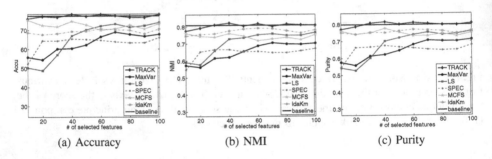

(a) Accuracy   (b) NMI   (c) Purity

Fig. 4. Clustering performance on JAFFE data set

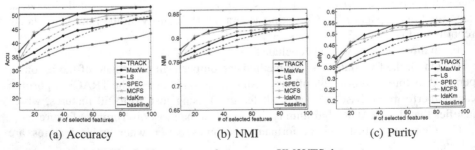

(a) Accuracy   (b) NMI   (c) Purity

Fig. 5. Clustering performance on XM2VTS data set

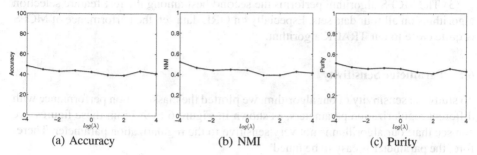

Fig. 6. Clustering performance versus the regularization parameter on MSRC-V1 data set

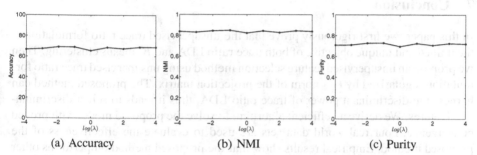

Fig. 7. Clustering performance versus the regularization parameter on ORL data set

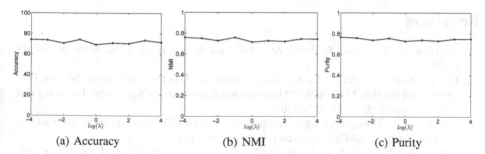

Fig. 8. Clustering performance versus the regularization parameter on JAFFE data set

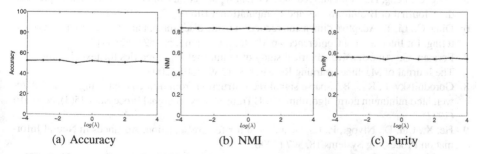

Fig. 9. Clustering performance versus the regularization parameter on XM2VTS data set

(3) The MCFS algorithm performs the second best among the rest feature selection algorithms on all four data sets. Especially on ORL data set, the performance of MCFS is quite close to our TRACK algorithm.

6.5 Parameter Sensitivity

To study the sensitivity of our algorithm, we plotted the classification performance with different regularization parameters, as shown in Figure 6 to 9. From these figures, we can see that: our algorithm is not very sensitive to the regularization parameter. Therefore, the parameter is easy to be tuned.

7 Conclusion

In this paper, we first rigorously prove that the unsupervised trace ratio formulation is the unified and unique objective of both trace ratio LDA and K-means clustering. Then we propose an unsupervised feature selection method using unsupervised trace ratio formulation regularized by $\ell_{1,2}$-norm of the projection matrix. The proposed method can harness the discriminant power of trace ratio LDA, thus it tends to select discriminative features. We derive an efficient algorithm to solve the proposed model with proved convergence. Four real world data sets are used to evaluate the effectiveness of the proposed method. Empirical results show that the proposed method outperforms other sate-of-the-art unsupervised feature selection algorithms on all three valuation metrics.

References

1. Argyriou, A., Evgeniou, T., Pontil, M.: Multi-task feature learning. In: NIPS, pp. 41–48 (2007)
2. Cai, D., Zhang, C., He, X.: Unsupervised feature selection for multi-cluster data. In: Proceedings of the 16th ACM SIGKDD International Conference on Knowledge Discovery and Data Mining, pp. 333–342. ACM (2010)
3. Cai, X., Nie, F., Huang, H., Ding, C.: Feature selection via l2,1-norm support vector machine. In: IEEE International Conference on Data Mining (2011)
4. Chen, C.H., Pau, L.F., Wang, P.S.P.: Handbook of pattern recognition and computer vision. World Scientific (2010)
5. Ding, C., Peng, H.: Minimum redundancy feature selection from microarray gene expression data. Journal of Bioinformatics and Computational Biology 3(02), 185–205 (2005)
6. Ding, C., Li, T.: Adaptive dimension reduction using discriminant analysis and k-means clustering. In: International Conference on Machine Learning, pp. 521–528 (2007)
7. Forman, G.: An extensive empirical study of feature selection metrics for text classification. The Journal of Machine Learning Research 3, 1289–1305 (2003)
8. Gorodnitsky, I., Rao, B.: Sparse signal reconstruction from limited data using focuss: A reweighted minimum norm algorithm. IEEE Transactions on Signal Processing 45(3), 600–616 (1997)
9. He, X., Cai, D., Niyogi, P.: Laplacian score for feature selection. Advances in Neural Information Processing Systems 18, 507 (2006)
10. Jia, Y., Nie, F., Zhang, C.: Trace ratio problem revisited. IEEE Transactions on Neural Networks 20(4), 729–735 (2009)

11. Kohavi, R., John, G.H.: Wrappers for feature subset selection. Artificial Intelligence 97(1-2), 273–324 (1997)
12. Kong, D., Ding, C., Huang, H., Zhao, H.: Multi-label relieff and f-statistic feature selections for image annotation. In: The 25th IEEE Conference on Computer Vision and Pattern Recognition (CVPR), pp. 2352–2359 (2012)
13. Kononenko, I.: Estimating attributes: analysis and extensions of relief. In: Bergadano, F., De Raedt, L. (eds.) ECML 1994. LNCS, vol. 784, pp. 171–182. Springer, Heidelberg (1994)
14. Kuhn, H.W.: The hungarian method for the assignment problem. Naval Research Logistics Quarterly 2(1-2), 83–97 (1955)
15. Masaeli, M., Fung, G., Dy, J.G.: From transformation-based dimensionality reduction to feature selection. In: ICML, pp. 751–758 (2010)
16. Nie, F., Huang, H., Cai, X., Ding, C.: Efficient and robust feature selection via joint l2,1-norms minimization. Advances in Neural Information Processing Systems 23, 1813–1821 (2010)
17. Nie, F., Xiang, S., Jia, Y., Zhang, C.: Semi-supervised orthogonal discriminant analysis via label propagation. Pattern Recognition 42(11), 2615–2627 (2009)
18. Nie, F., Xiang, S., Jia, Y., Zhang, C., Yan, S.: Trace ratio criterion for feature selection. In: AAAI, pp. 671–676 (2008)
19. Peng, H., Long, F., Ding, C.: Feature selection based on mutual information criteria of max-dependency, max-relevance, and min-redundancy. IEEE Transactions on Pattern Analysis and Machine Intelligence 27(8), 1226–1238 (2005)
20. Raileanu, L.E., Stoffel, K.: Theoretical comparison between the gini index and information gain criteria. Ann. Math. Artif. Intell. 41(1), 77–93 (2004)
21. Wang, C., Caob, L., Miao, B.: Optimal feature selection for sparse linear discriminant analysis and its applications in gene expression data. Computational Statistics and Data Analysis 66, 140–149 (2013)
22. Wang, D., Nie, F., Huang, H., Yan, J., Risacher, S.L., Saykin, A.J., Shen, L.: Structural brain network constrained neuroimaging marker identification for predicting cognitive functions. In: Gee, J.C., Joshi, S., Pohl, K.M., Wells, W.M., Zöllei, L. (eds.) IPMI 2013. LNCS, vol. 7917, pp. 536–547. Springer, Heidelberg (2013)
23. Wang, H., Nie, F., Huang, H., Kim, S., Nho, K., Risacher, S.L., Saykin, A.J., Shen, L.: Identifying quantitative trait loci via group-sparse multitask regression and feature selection: an imaging genetics study of the adni cohort. Bioinformatics 28(2), 229–237 (2012)
24. Wang, H., Nie, F., Huang, H., Risacher, S., Ding, C., Saykin, A.J., Shen, L.: ADNI: Sparse multi-task regression and feature selection to identify brain imaging predictors for memory performance. In: IEEE Conference on Computer Vision (2011)
25. Xiang, S., Nie, F., Zhang, C.: Learning a mahalanobis distance metric for data clustering and classification. Pattern Recognition 41(12), 3600–3612 (2008)
26. Yuan, M., Lin, Y.: Model selection and estimation in regression with grouped variables. Journal of The Royal Statistical Society Series B 68(1), 49–67 (2006)
27. Zhao, Z., Liu, H.: Spectral feature selection for supervised and unsupervised learning. In: Proceedings of the 24th International Conference on Machine Learning, pp. 1151–1157. ACM (2007)

On the Equivalence between Deep NADE and Generative Stochastic Networks

Li Yao, Sherjil Ozair, Kyunghyun Cho, and Yoshua Bengio*

Département d'Informatique et de Recherche Opérationelle
Université de Montréal, Canada

Abstract. Neural Autoregressive Distribution Estimators (NADEs) have recently been shown as successful alternatives for modeling high dimensional multimodal distributions. One issue associated with NADEs is that they rely on a particular order of factorization for $P(\mathbf{x})$. This issue has been recently addressed by a variant of NADE called Orderless NADEs and its deeper version, Deep Orderless NADE. Orderless NADEs are trained based on a criterion that stochastically maximizes $P(\mathbf{x})$ with all possible orders of factorizations. Unfortunately, ancestral sampling from deep NADE is very expensive, corresponding to running through a neural net separately predicting each of the visible variables given some others. This work makes a connection between this criterion and the training criterion for Generative Stochastic Networks (GSNs). It shows that training NADEs in this way also trains a GSN, which defines a Markov chain associated with the NADE model. Based on this connection, we show an alternative way to sample from a trained Orderless NADE that allows to trade-off computing time and quality of the samples: a 3 to 10-fold speedup (taking into account the waste due to correlations between consecutive samples of the chain) can be obtained without noticeably reducing the quality of the samples. This is achieved using a novel sampling procedure for GSNs called annealed GSN sampling, similar to tempering methods that combines fast mixing (obtained thanks to steps at high noise levels) with accurate samples (obtained thanks to steps at low noise levels).

1 Introduction

Unsupervised representation learning and deep learning have progressed rapidly in recent years [5]. On one hand, supervised deep learning algorithms have achieved great success. The authors of [15], for instance, claimed the state-of-the-art recognition performance in a challenging object recognition task using a deep convolutional neural network. Despite the promise given by supervised deep learning, its unsupervised counterpart is still facing several challenges [3]. A large proportion of popular unsupervised deep learning models are based on either directed or undirected graphical models with latent variables [12, 13, 20]. One problem of these unsupervised models is that it is often intractable to compute the likelihood of a model exactly.

---

* CIFAR Fellow.

T. Calders et al. (Eds.): ECML PKDD 2014, Part III, LNCS 8726, pp. 322–336, 2014.

The Neural Autoregressive Distribution Estimator (NADE) was proposed in [16] to avoid this problem of computational intractability. It was inspired by the early work in [4], which like NADE modeled a binary distribution by decomposing it into a product of multiple conditional distributions of which each is implemented by a neural network, with parameters, representations and computations shared across all these networks. These kinds of models therefore implement a fully connected directed graphical model, in which ancestral sampling of the joint distribution is simple (but not necessarily efficient when the number of variables, e.g., pixel images, is large). Consequently, unlike many other latent variable models, it is possible with such directed graphical models to compute the exact probability of an observation tractably. NADEs have since been extended to model distributions of continuous variables in [23], called a real-valued NADE (RNADE) which replaces a Bernoulli distribution with a mixture of Gaussian distributions for each conditional probability (see ,e.g., [9]). The authors of [22] proposes yet another variant of NADE, called a Deep NADE, that uses a *deep* neural network to compute the conditional probability of each variable. In order to make learning tractable, they proposed a modified training procedure that effectively trains an ensemble of multiple NADEs.

Another thread of unsupervised deep learning is based on the family of autoencoders (see, e.g., [25]). The autoencoder has recently begun to be understood as a density estimator [1, 7]. These works suggest that an autoencoder trained with some arbitrary noise in the input is able to learn the distribution of either continuous or discrete random variables. This perspective on autoencoders has been further extended to a generative stochastic network (GSN) proposed in [6].

Unlike a more conventional approach of directly estimating the probability distribution of data, a GSN aims to learn a transition probability of a Markov Chain Monte Carlo (MCMC) sampler whose stationary distribution estimates the data generating distribution. The authors of [6] were able to show that it is possible to learn the distribution of data with a GSN having a network structure inspired by a deep Boltzmann machine (DBM) [21] using this approach. Furthermore, a recently proposed multi-prediction DBM (MP-DBM) [11], which models the joint distribution of data instance and its label, can be considered a special case of a GSN and achieves state-of-the-art classification performance on several datasets.

In this paper, we find a close relationship between the deep NADE and the GSN. We show that training a deep NADE with the order-agnostic (OA) training procedure [22] can be cast as GSN training. This equivalence allows us to have an alternative theoretical explanation of the OA training procedure. Also, this allows an alternative sampling procedure for a deep NADE based on a MCMC method, rather than ancestral sampling.

In Sec. 2.1 and Sec. 3, we describe both NADE and GSN in detail. Based on these descriptions we establish the connection between the order-agnostic training procedure for NADE and the training criterion of GSN in Sec. 4 and propose a novel sampling algorithm for deep NADE. In Sec. 5, we introduce a novel sampling strategy for GSN called annealed GSN sampling, which is inspired by tempering methods and does a good trade-off between computing time and accuracy. We empirically investigate the effect of the proposed GSN sampling procedure for deep NADE models in Sec. 6.

2   Deep NADE and Order-Agnostic Training

In this section we describe the deep NADE and its training criterion, closely following [22].

2.1   NADE

NADE [16] models a joint distribution $p(\mathbf{x})$ where $\mathbf{x} \in \mathbb{R}^D$. D is the dimensionality of $\mathbf{x}$. NADE factorizes $p(\mathbf{x})$ into

$$p(\mathbf{x}) = \prod_{d=1}^{D} p(x_{o_d} | \mathbf{x}_{o_{<d}}) \tag{1}$$

where o is a predefined ordering of D indices. $o_{<d}$ denotes the first $d - 1$ indices of the ordering o.

The NADE then models each factor in Eq. (1) with a neural network having a single hidden layer H. That is,

$$p(x_{o_d} = 1 | \mathbf{x}_{o_{<d}}) = \sigma(\mathbf{V}_{.,o_d} \mathbf{h}_d + b_{o_d}),$$

where

$$\mathbf{h}_d = \phi(\mathbf{W}_{.,o_{<d}} + \mathbf{c}).$$

$\mathbf{V} \in \mathbb{R}^{H \times D}$, $b \in \mathbb{R}^D$, $\mathbf{W} \in \mathbb{R}^{H \times D}$ and $\mathbf{c} \in \mathcal{R}^H$ are the output weights, the output biases, the input weights and the hidden biases, respectively. σ is a logistic sigmoid function, and ϕ can be any nonlinear activation function.

To train such a model, one maximizes the log-likelihood function of the training set

$$\theta^* = \arg\max_{\theta} \mathcal{L}_o(\theta) = \arg\max_{\theta} \frac{1}{N} \sum_{n=1}^{N} \sum_{d=1}^{D} \log p(x_{o_d}^n | \mathbf{x}_{o_{<d}}^n, o), \tag{2}$$

where θ denotes all the parameters of the model.

2.2   Deep NADE

One issue with the original formulation of the NADE is that the ordering of variables needs to be predefined and fixed. Potentially, this limits the inference capability of a trained model such that when the model is asked to infer the conditional probability which is not one of the factors in the predefined factorization (See Eq. (1)). For instance, a NADE trained with D visible variables with an ordering $(1, 2, \ldots, D)$, one cannot easily infer $x_2 \| x_1, x_D$ except by expensive (and intractable) marginalization over all the other variables.

Another issue is that it is not possible to build a deeper architecture for NADE with the original formulation without losing a lot in efficiency. When there is only a single hidden layer with H units in the neural network modeling each conditional probability of a NADE, it is possible to share the parameters (the input weights and the hidden biases) to keep the computational complexity linear with respect to the number of parameters, i.e., $O(DH)$. However, if there are more than one hidden layers, it is not possible to re-use computations in the same way. In this case, the computational complexity is $O(DH + DH^2L)$ where L is the number of hidden layers. Notice the extra D factor in front, compared to the number of parameters which is $O(DH + H^2L)$. This comes about because we cannot re-use the computations performed after the first hidden layer for predicting the i-th variable, when predicting the following ones. In the one-layer case, this sharing is possible because the hidden units weighted sums needed when predicting the $i+1$-th variable are the same as the weighted sums needed when predicting the i-th variable, plus the scalar contributions w_{ki} associated with the k-th hidden unit and the extra input x_i that is now available when predicting x_{i+1} but was not available when predicting x_i.

To resolve those two issues, the authors of [22] proposed the order-agnostic (OA) training procedure that trains a factorial number of NADEs with shared parameters. In this case, the following objective function is maximized, instead of $\mathcal{L}_o$ in Eq. (2):

$$\mathcal{L}(\theta) = \mathbb{E}_{\mathbf{x}^n} \sum_{d=1}^{D} \mathbb{E}_{o<d} \mathbb{E}_{o_d} \log p(x_{o_d}^n | \mathbf{x}_{o<d}^n, \theta, o). \tag{3}$$

This objective function is, however, intractable, since it involves the factorial number of summations. Instead, in practice, when training, we use a stochastic approximation $\widehat{\mathcal{L}}$ by sampling an ordering o, the index of predicted variable d and a training sample $\mathbf{x}^n$ at each time:

$$\widehat{\mathcal{L}}(\theta) = \frac{D}{D - d + 1} \sum_{i \notin o<d} \log p(x_i^n | \mathbf{x}_{o<d}^n, \theta, o). \tag{4}$$

Computing $\widehat{\mathcal{L}}$ is identical to a forward computation in a regular feedforward neural network except for two differences. Firstly, according to the sampled ordering o, the input variables of indices $o_{>d}$ are set to 0, and the identity of the zeroed indices is provided as extra inputs (through a binary vector of length D). Secondly, the conditional probabilities of only those variables of indices $o_{>d}$ are used to compute the objective function $\widehat{\mathcal{L}}$.

This order-agnostic procedure solves the previously raised issues of the original NADE. Since the model is optimized for all possible orderings, it does not suffer from being inefficient at inferring any conditional probability. Furthermore, the lack of predefined ordering makes it possible to use a single set of parameters for modeling all conditional distributions. Thus, the computational cost of training a deep NADE with the OA procedure is linear with respect to the number of parameters, regardless of the depth of each neural network.

From here on, we call a NADE trained with the OA procedure simply a deep NADE to distinguish it from a NADE trained with a usual training algorithm other than the OA procedure.

3   Generative Stochastic Networks

In [6, 7] a new family of models called generative stochastic networks (GSN) was proposed, which tackles the problem of modeling a data distribution, $p(\mathbf{x})$, although without providing a tractable expression for it.

The underlying idea is to learn a transition operator of a Markov Chain Monte Carlo (MCMC) sampler that samples from the distribution $p(\mathbf{x})$, instead of learning the whole distribution directly. If we let $p(\mathbf{x}' \mid \mathbf{x})$ be the transition operator, then we may rewrite it by introducing a latent variable h into

$$p(\mathbf{x}' \mid \mathbf{x}) = \sum_{\forall \mathbf{h}} p(\mathbf{x}' \mid \mathbf{h})p(\mathbf{h} \mid \mathbf{x}). \tag{5}$$

In other words, two separate conditional distributions $p(\mathbf{x}' \mid \mathbf{h})$ and $p(\mathbf{h} \mid \mathbf{x})$ jointly define the transition operator. In [6, 7] it is argued that it is easier to learn these simpler conditional distributions because they have less modes (they only consider small changes from the previous state), meaning that the associated normalization constants can be estimated more easily (either by an approximate parametrization, e.g., a single or few component mixture, or by MCMC on a more powerful parametrization, which will have less variance if the number of modes is small).

A special form of GSN also found with denoising auto-encoders predefines $p(\mathbf{h} \mid \mathbf{x})$ such that it does not require learning from data. Then, we only learn $p(\mathbf{x}' \mid \mathbf{h})$. This is the case in [7], where they proposed to use a user-defined corruption process, such as randomly masking out some variables with a fixed probability, for $p(\mathbf{h} \mid \mathbf{x})$. They, then, estimated $p(\mathbf{x}' \mid \mathbf{h})$ as a denoising autoencoder f_θ, parameterized with θ, that reverses the corruption process $p(\mathbf{h} \mid \mathbf{x})$ [24].

It was shown in [7] that if the denoising process f_θ is a consistent estimator of $p(\mathbf{x}' \mid \mathbf{h})$, this leads to consistency of the Markov chain's stationary distribution as an estimator of the data generating distribution. This is under some conditions ensuring the irreducibility, ergodicity and aperiodicity of the Markov chain, i.e., that it mixes. In other words, training f_θ to match $p(\mathbf{x}' \mid \mathbf{h})$ is enough to learn implicitly the whole distribution $p(\mathbf{x})$, albeit indirectly, i.e., through the definition of a Markov chain transition operator. The result from [6] further suggests that it is possible to also parameterize the corruption process $p(\mathbf{h} \mid \mathbf{x})$ and learn both $p(\mathbf{h} \mid \mathbf{x})$ and $p(\mathbf{x}' \mid \mathbf{h})$ together.

From the qualitative observation on some of the learned transition operators of GSNs (see, e.g., [6]), it is clear that the learned transition operator quickly finds a plausible mode in the whole distribution, even when the Markov chain was started from a random configuration of $\mathbf{x}$. This is because the GSN reconstruction criterion encourages the learner to quickly move from low probability configurations to high-probability ones, i.e., to burn-in quickly. This is in contrast to using a Gibbs sampler to generate samples from other generative models that explicitly model the whole distribution $p(\mathbf{x})$, which requires often many more *burn-in* steps before the Markov chain finds a plausible mode of the distribution.

4   Equivalence between Deep NADE and GSN

Having described both deep NADE and GSN, we now establish the relationship, or even equivalence, between them. In particular, we show in this section that the order-agnostic (OA) training procedure for NADE is one special case of GSN learning.

We start from the stochastic approximation to the objective function of the OA training procedure for deep NADE in Eq. (4). We notice that the sampled ordering o in the objective function $\hat{\mathcal{L}}$ can be replaced with another random variable $\mathbf{m} \in \{0,1\}^D$, where D is the dimensionality of an observation x. The binary mask $\mathbf{m}$ is constructed such that

$$m_i = \begin{cases} 1, \text{ if } i \in o_{<d} \\ 0, \text{ otherwise} \end{cases}$$

Then, we rewrite Eq. (4) by

$$\hat{\mathcal{L}}(\theta) \propto \sum_{i=1}^{D} (1 - m_i) \log p(x_i^n \mid \mathbf{m} \odot \mathbf{x}^n, \theta, \mathbf{m})$$

$$= \sum_{i=1}^{D} \log \left(m_i + (1 - m_i) p(x_i^n \mid \mathbf{m} \odot \mathbf{x}^n, \theta, \mathbf{m}) \right)$$

$$= \sum_{i=1}^{D} \log \left(m_i + (1 - m_i) p(x_i^n \mid \mathbf{h}^{(n)}, \theta) \right) \tag{6}$$

where m_i is the i-th element of the binary mask $\mathbf{m}$, and $\odot$ is an element-wise multiplication. We introduced a new variable $\mathbf{h} = [\mathbf{m}, \mathbf{m} \odot \mathbf{x}^n] \in \mathbb{R}^{2D}$ which is a concatenation of the corrupted copy (some variables masked out) of $\mathbf{x}^n$ and the sampled mask $\mathbf{m}$.

It is now easy to see the connection between the objective function of the OA training procedure in Eq. (6) to a GSN training criterion using a user-defined (not learned) corruption process which we described in the earlier section.

In this case, the corruption process $p(\mathbf{h} \mid \mathbf{x})$ (Eq. (5)) is

$$p(\mathbf{h} \mid \mathbf{x}) = p([\mathbf{m}, \mathbf{m} \odot \mathbf{x}] \mid \mathbf{x}) = \prod_{i=1}^{D} k \prod_{j=1}^{D} \delta_{m_j x_j}(h_{j+D}), \tag{7}$$

where k is a random number sampled uniformly between 0 and 1, and $\delta_\mu(a)$ is a shifted Dirac delta function which is 1 only when $a = \mu$ and 0 otherwise. This means that sampling is done by first generating an uniformly random binary mask $\mathbf{m}$ and then taking $\mathbf{m} \odot \mathbf{x}$ as the corrupted version of $\mathbf{x}$.

The conditional probability of $\mathbf{x}'$ given $\mathbf{h}$ is

$$p(\mathbf{x}' \mid \mathbf{h}) = \prod_{i=1}^{D} \left[m_i \delta_{x_i}(x_i') + (1 - m_i) p(x_i' \mid r_i(\mathbf{x} \odot \mathbf{m} \mid \theta)) \right], \tag{8}$$

where r_i is a parametric function (neural network) that models the conditional probability.

If we view the estimation of $p(\mathbf{x}' \mid \mathbf{h})$ in Eq. (8) as a denoising autoencoder, one effectively ignores each variable x_i' with its mask m_i set to 1, since the sample of x_i' from Eq. (8) is always x_i due to $\delta_{x_i}(x_i')$. A high-capacity auto-encoder could learn that when $m_i = 1$, it can just copy the i-th input to the i-th output. On the other hand, when m_i is 0, training this denoising autoencoder would maximize $\log p(x_i' \mid r_i(\mathbf{x} \odot \mathbf{m} \mid \theta))$, making it assign high probability to the original x_i given the non-missing inputs. Therefore, maximizing the logarithm of $p(\mathbf{x}' \mid \mathbf{h})$ in Eq. (8) is equivalent to maximizing $\widehat{L}$ in Eqs. (6) and (4) up to a constant.

In essence, maximizing $\widehat{\mathcal{L}}$ in Eq. (4) is equivalent to training a GSN with the conditional distributions defined in Eqs. (7)–(8). Furthermore, the chain defined in this way is ergodic as every state $\mathbf{x}$ has a non-zero probability at each step $(\mathbf{x} \to \mathbf{x}')$, making this GSN chain a valid MCMC sampler.

4.1   Alternative Sampling Method for NADE

Although the training procedure of the deep NADE introduced in [22] is order-agnostic, sampling from the deep NADE is not.

The authors of [22] proposed an ancestral sampling method for a deep NADE. Firstly, one randomly selects an ordering uniformly from all possible orderings. One generates a sample of each variable from its conditional distribution following the selected ordering. When D, H and L are respectively the dimensionality of the observation variable, the number of hidden units in each hidden layer and the number of layers, the time complexity of sampling a single sample using this ancestral approach is $O(DLH^2)$.

We propose here an alternative sampling strategy based on our observation of the equivalence between the deep NADE and GSN. The new strategy is simply to alternating between sampling from $p(\mathbf{h} \mid \mathbf{x})$ in Eq. (7) and $p(\mathbf{x}' \mid \mathbf{h})$ in Eq. (8), which corresponds to performing Markov Chain Monte Carlo (MCMC) sampling on $p(\mathbf{x})$. The computational complexity of a single step $(\mathbf{x} \to \mathbf{h} \to \mathbf{x}')$ in this case is $O(DH + LH^2)$.

Unlike the original ancestral strategy, the proposed approach does not generate an exact sample in a single step. Instead, one often needs to run the chain K steps until the exact, independent sample from the stationary distribution of the chain is collected, which we call *burn-in*. In other words, the new approach requires $O(KDH + KLH^2)$ to collect a single sample in the worst case.[1]

If we assume that H is not too larger than D ($H = O(D)$ or $H = \Theta(D)$), which is an usual practice, the time complexity of the ancestral approach is $O(D^3)$, and that of the proposed GSN approach is $O(KD^2)$, where we further assume that L is a small constant. Effectively, if the MCMC method used in the latter strategy requires only a small, controllable number K of steps to generate a single exact, independent sample such that $K \ll D$, the new approach is more efficient in collecting samples from a trained deep NADE. Importantly, as we have already mentioned earlier, a GSN has been shown to learn a transition operator of an MCMC method that requires only a small number of burn-in steps.

In the experiments, we investigate empirically whether this new sampling strategy is computationally more efficient than the original ancestral approach in a realistic setting.

4.2   The GSN Chain Averages an Ensemble of Density Estimators

As discussed in [22], maximizing $\widehat{\mathcal{L}}$ in Eq. (3) can be considered as training a factorial number of different NADEs with shared parameters. Each NADE differs from each other by the choice of the ordering of variables and may assign a different probability to the same observation.[2] Based on this observation, it is suggested in [22] to use the average of the assigned probabilities by all these NADE, or a small randomly chosen subset of them, as the actual probability.

This interpretation of seeing the deep NADE as an ensemble of multiple NADEs and our earlier argument showing that the deep NADE training is special case of GSN training naturally leads to a question: *does the GSN Markov chain average an ensemble of density/distribution estimators?*.

We claim that the answer is yes. From the equivalence we showed in this paper, it is clear that a GSN trained with a criterion such as the NADE criterion *learns* an ensemble of density/distribution estimators (in this case, masking noise with the reconstruction conditional distribution in Eq. (8)). Furthermore, when one samples from the associated GSN Markov chain, one is averaging the contributions associated with different orders. So, although each of these conditionals (predicting a subset given another subset) may not be consistent with a single joint distribution, the associated GSN Markov chain which combines them randomly does define a clear joint distribution: the stationary distribution of the Markov chain. Clearly, this stationary distribution is an ensemble average over all the possible orderings.

---

[1] Since it is a usual practice to collect every t-th samples from the same chain, where $t << K$, we often do not need KN steps to collect N samples.

[2] The fact that all ensembles share the exact same parameters makes it similar to the recently proposed technique of dropout [14].

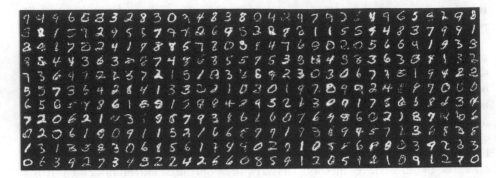

Fig. 1. Independent samples generated by the ancestral sampling procedure from the deep NADE

5 Annealed GSN Sampling

With the above proposal for GSN-style sampling of a Deep NADE model, one can view the average fraction p of input variables that are resampled at each step as a kind of noise level, or the probability of resampling any particular visible variable x_i. With uniform sampling of subsets, we obtain $p = 0.5$, but both higher and lower values are possible. When $p = 1$, all variables are re-sampled independently and the resulting samples are coming from the marginal distributions of each variable, which would be a very poor rendering of the Deep NADE distribution, but would mix very well. With p as small as possible (or more precisely, resampling only one randomly chosen variable given the others), we obtain a *Gibbs sampler* associated with the Deep NADE distribution, which we know has the same stationary distribution as Deep NADE itself. However, this would mix very slowly and would not bring any computational gain over ancestral sampling in the Deep NADE model (in fact it would be considerably worse because the correlation between consecutive samples would reduce the usefulness of the Markov chain samples, compared to ancestral sampling that provides i.i.d. samples). With intermediate values of p, we obtain a compromise between the fast computation and the quality of samples.

However, an even better trade-off can be reached by adopting a form of annealed sampling for GSNs, a general recipe for improving the compromise between accuracy of the sampling distribution and mixing for GSNs. For this purpose we talk about a generic noise level, although in this paper we refer to p, the probability of resampling any particular visible variable.

The idea is inspired by annealing and tempering methods that have been useful for undirected graphical models [18, 19]: *before sampling from the low-noise regime, we run the high-noise version of the transition operator and gradually reduce the noise level over a sequence of steps.* The steps taken at high noise allow to mix quickly while the steps taken at low noise allow to burn-in near high probability samples. Therefore we consider an approximation of the GSN transition operator which consists of the successive application of a sequence of

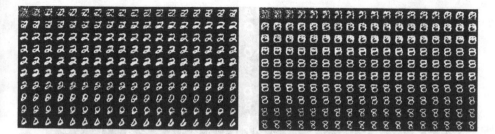

Fig. 2. The consecutive samples from the two independent GSN sampling chains without any annealing strategy. Both chains started from uniformly random configurations. Note the few spurious samples which can be avoided with the annealing strategy (see Figure 3).

instances of the operator associated with gradually reduced noise levels, ending at the target noise level. Conceptually, it is as if the overall Markov chain was composed, for each of its steps, by a short chain of steps with gradually decreasing noise levels. By making the annealing schedule have several steps at or near the target low noise level, and by controlling the lengths of these annealing runs, we can trade-off between accuracy of the samples (improved by a longer annealing run length) and speed of computation.

In the experiments, we used the following annealing schedule:

$$p_t = \max(p_{\min}, p_{\max} - (t - 1) * (p_{\max} - p_{\min})/(\alpha * (T - 1)))$$

where $p_{\max}$ is the high noise level, $p_{\min}$ is the low (target) noise level, T is the length of the annealing run, and $\alpha \geq 1$ controls which fraction of the run is spent in annealing vs doing burn-in at the low noise level.

6   Experiments

6.1   Settings: Dataset and Model

We run experiments using the handwritten digits dataset (MNIST, [17]) which has 60,000 training samples and 10,000 test samples. Each sample has 784 dimensions, and we binarized each variable by thresholding at 0.5. The training set is split into two so that the first set of 50,000 samples is used to train a model and the other set of 10,000 samples is used for validation.

Using MNIST we trained deep NADE with various architectures and sets of hyperparameters using the order-agnostic (OA) training procedure (see Sec. 2.2). The best deep NADE model according to the validation performance has two hidden layers with size 2000 and was trained with a linearly decaying learning rate schedule (from 0.001 to 0) for 1000 epochs. We use this model to evaluate the two sampling strategies described and proposed earlier in this paper.

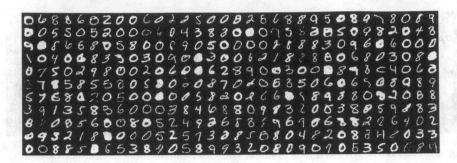

Fig. 3. Samples generated by the annealed GSN sampling procedure for the same deep NADE model. Visually the quality is comparable to the ancestral samples, and mixing is very fast. This is obtained with $p_{max} = 0.9$, $p_{min} = 0.1$, $\alpha = 0.7$ and $T = 20$.

6.2   Qualitative Analysis

Fig. 1 shows a subset of 10,000 samples collected from the deep NADE using the conventional ancestral sampling. The average log-probability of the samples is -70.36 according to the deep NADE. As each sample by the ancestral sampling is exact and independent from others, we use these samples and their log-probability as a baseline for assessing the proposed GSN sampling procedure.

We first generate samples from the deep NADE using the GSN sampling procedure without any annealing strategy. A sampling chain is initialized with a uniformly random configuration, and a sample is collected at each step. The purpose of this sampling is to empirically confirm that the GSN sampling does not require many steps for burn-in. We ran two independent chains and visualize the initial 240 samples from each of them in Fig. 2, which clearly demonstrates that the chain rapidly finds a plausible mode in only a few steps.

Although this visualization suggests a faster burn-in, one weakness is clearly visible from these figures (Fig. 2. The chain generates many consecutive samples of a single digit before it starts generating samples of another digit. That is, the samples are highly correlated temporally, suggesting potentially slow convergence to the stationary distribution.

We then tried sampling from the deep NADE using the novel annealed GSN sampling proposed in Sec. 5. Fig. 3 visualizes the collected, samples over the *consecutive* annealing runs. Compared to the samples generated using the ordinary GSN sampling method, the chain clearly mixes well. One can hardly notice a case where a successive sample is a realization of the same digit from the previous sample. Furthermore, the samples are qualitatively comparable to those exact samples collected with the ancestral sampling (see Fig. 1).

In the following section, we further investigate the proposed annealed GSN sampling in a more quantitative way, in comparison to the ancestral sampling.

6.3 Quantitative Results

We first evaluate the effect of using a user-defined noise level in $p(\mathbf{h} \mid \mathbf{x})$ (Eq. (7)). We generated 1000 samples from GSN chains with five different noise levels; 0.1, 0.3, 0.4, 0.5 and 0.6. For each noise level, we ran 100 independent chains and collected every 200-th sample from each chain. As a comparison, we also generated 1000 samples from a chain with the proposed annealed GSN sampling with $p_{max} = 0.9$, $p_{min} = 0.1$ and $\alpha = 0.7$.

We computed the log-probability of the set of samples collected from each chain with the deep NADE to evaluate the quality of the samples. Tab. 1 lists the log-probabilities of the sets of samples, which clearly shows that as the noise level increases the quality of the samples degrades. Importantly, none of the chains were able to generate samples from the model that are close to those generated by the ancestral sampling. However, the annealed GSN sampling was able to generate samples that are quantitatively as good as those from the ancestral sampling.

Noise	Log-Probability
0.1	-77.1
0.3	-78.93
0.4	-77.9
0.5	-81.1
0.6	-88.1
Annealed	-69.72
Ancestral	-70.36

Table 1. Log-probability of 1000 samples when anealing is not used. To collect samples, 100 parallel chains are run and 10 samples are taken from each chain and combined together. The noise level is fixed at a particular level during the sampling. We also report the best log-probability of samples generated with an annealed GSN sampling.

We also perform quantitative analysis to measure the computational gain when using the GSN sampling procedure to generate samples. The speedup by using annealed GSN sampling instead of ancestral sampling is shown in Figure 4. To compute the speedup factor, we timed both the ancestral NADE sampling and GSN sampling on the same machine running single process. NADE sampling takes 3.32 seconds per sample and GSN sampling takes 0.009 seconds. That means the time to get one sample in ancestral sampling can get 369 samples in GSN sampling. Although the the direct speedup factor is 369, it must be discounted because of the autocorrelation of successive samples in the GSN chain. Then we perform different GSN sampling runs with different settings of α. Figure 4 shows the results with different α. For each α, a GSN sampling starting at random is run and we collect one out of every K samples till 1000 samples are collected. The effective sample size [10] is then estimated based on the sum of the autocorrelations in the autocorrelation factor. The speedup factor is discounted accordingly.

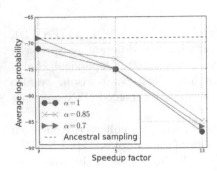

Fig. 4. The annealed GSN sampling procedure is compared against NADE ancestral sampling, trading off the computational cost (computational wrt to ancestral sampling on x-axis) against log-likelihood of the generated samples (y-axis). The computational cost discards the effect of the Markov chain autocorrelation by estimating the effective number of samples and increasing the computational cost accordingly.

7 Conclusions

This paper introduced a new view of the orderless NADE training procedure as a GSN training procedure, which yields several interesting conclusions:

- The orderless NADE training procedure also trains a GSN model, where the transition operator randomly selects a subset of input variables to be resampled given the others.
- Whereas orderless NADE models really represent an ensemble of conditionals that are not all compatible, the GSN interpretation provides a coherent interpretation of the estimated distribution through the stationary distribution of the associated Markov chain.
- Whereas ancestral sampling in NADE is exact, it is very expensive for deep NADE models, multiplying computing cost (of running once through the neural network to make a prediction) by the number of visible variables. On the other hand, each step of the associated GSN Markov chain only costs running once through the predictor, but because each prediction is made in parallel for all the resampled variables, each such step is also less accurate, unless very few variables are resampled. This introduces a trade-off between accuracy and computation time that can be controlled. This was validated experimentally.
- A novel sampling procedure for GSNs was introduced, called annealed GSN sampling, which permits a better trade-off by combining high-noise steps with a sequence of gradually lower noise steps, as shown experimentally.

Acknowledgments. We would like to thank the developers of Theano [2, 8]. We would also like to thank CIFAR, and Canada Research Chairs for funding, and Compute Canada, and Calcul Québec for providing computational resources.

References

[1] Alain, G., Bengio, Y.: What regularized auto-encoders learn from the data generating distribution. In: International Conference on Learning Representations, ICLR 2013 (2013)

[2] Bastien, F., Lamblin, P., Pascanu, R., Bergstra, J., Goodfellow, I.J., Bergeron, A., Bouchard, N., Bengio, Y.: Theano: new features and speed improvements. In: Deep Learning and Unsupervised Feature Learning NIPS 2012 Workshop (2012)

[3] Bengio, Y.: Deep learning of representations: Looking forward. In: Dediu, A.-H., Martín-Vide, C., Mitkov, R., Truthe, B. (eds.) SLSP 2013. LNCS, vol. 7978, pp. 1–37. Springer, Heidelberg (2013)

[4] Bengio, Y., Bengio, S.: Modeling high-dimensional discrete data with multi-layer neural networks. In: NIPS 1999, pp. 400–406. MIT Press (2000)

[5] Bengio, Y., Courville, A., Vincent, P.: Unsupervised feature learning and deep learning: A review and new perspectives. IEEE Trans. Pattern Analysis and Machine Intelligence, PAMI (2013)

[6] Bengio, Y., Thibodeau-Laufer, E., Alain, G., Yosinski, J.: Deep generative stochastic networks trainable by backprop. Technical Report arXiv:1306.1091 (2014)

[7] Bengio, Y., Yao, L., Alain, G., Vincent, P.: Generalized denoising auto-encoders as generative models. In: Advances in Neural Information Processing Systems 26, NIPS 2013 (2013)

[8] Bergstra, J., Breuleux, O., Bastien, F., Lamblin, P., Pascanu, R., Desjardins, G., Turian, J., Warde-Farley, D., Bengio, Y.: Theano: a CPU and GPU math expression compiler. In: Proceedings of the Python for Scientific Computing Conference (SciPy). Oral Presentation (June 2010)

[9] Bishop, C.M.: Mixture density networks (1994)

[10] Geyer, C.J.: Practical markov chain monte carlo. Statistical Science, 473–483 (1992)

[11] Goodfellow, I., Miraz, M., Courville, A., Bengio, Y.: Multi-prediction deep Boltzmann machines. In: Burges, C., Bottou, L., Welling, M., Ghahramani, Z., Weinberger, K. (eds.) Advances in Neural Information Processing Systems 26, pp. 548–556 (December 2013)

[12] Hinton, G.E., Osindero, S., Teh, Y.: A fast learning algorithm for deep belief nets. Neural Computation 18, 1527–1554 (2006)

[13] Hinton, G.E., Salakhutdinov, R.: Reducing the dimensionality of data with neural networks. Science 313(5786), 504–507 (2006)

[14] Hinton, G.E., Srivastava, N., Krizhevsky, A., Sutskever, I., Salakhutdinov, R.: Improving neural networks by preventing co-adaptation of feature detectors. Technical report, arXiv:1207.0580 (2012)

[15] Krizhevsky, A., Sutskever, I., Hinton, G.: ImageNet classification with deep convolutional neural networks. In: Advances in Neural Information Processing Systems 25, NIPS 2012 (2012)

[16] Larochelle, H., Murray, I.: The Neural Autoregressive Distribution Estimator. In: Proceedings of the Fourteenth International Conference on Artificial Intelligence and Statistics (AISTATS 2011). JMLR: W&CP, vol. 15 (2011)

[17] LeCun, Y., Bottou, L., Bengio, Y., Haffner, P.: Gradient-based learning applied to document recognition. Proceedings of the IEEE 86(11), 2278–2324 (1998)

[18] Neal, R.M.: Sampling from multimodal distributions using tempered transitions. Technical Report 9421, Dept. of Statistics, University of Toronto (1994)

[19] Neal, R.M.: Annealed importance sampling. Statistics and Computing 11(2), 125–139 (2001)

[20] Salakhutdinov, R., Hinton, G.: Deep Boltzmann machines. In: Proceedings of the Twelfth International Conference on Artificial Intelligence and Statistics (AISTATS 2009), vol. 8 (2009)

[21] Salakhutdinov, R., Hinton, G.: Deep Boltzmann machines. In: Proceedings of the International Conference on Artificial Intelligence and Statistics, vol. 5, pp. 448–455 (2009)
[22] Uria, B., Murray, I., Larochelle, H.: A deep and tractable density estimator. Technical Report arXiv:1310.1757 (2013)
[23] Uria, B., Murray, I., Larochelle, H.: Rnade: The real-valued neural autoregressive density-estimator. In: NIPS 2013 (2013)
[24] Vincent, P., Larochelle, H., Bengio, Y., Manzagol, P.-A.: Extracting and composing robust features with denoising autoencoders. In: ICML 2008 (2008)
[25] Vincent, P., Larochelle, H., Lajoie, I., Bengio, Y., Manzagol, P.-A.: Stacked denoising autoencoders: Learning useful representations in a deep network with a local denoising criterion. J. Machine Learning Res. 11 (2010)

Scalable Nonnegative Matrix Factorization with Block-wise Updates

Jiangtao Yin[1], Lixin Gao[1], and Zhongfei (Mark) Zhang[2]

[1] University of Massachusetts Amherst, Amherst, MA 01003, USA
[2] Binghamton University, Binghamton, NY 13902, USA
{jyin,lgao}@ecs.umass.edu, zhongfei@cs.binghamton.edu

Abstract. Nonnegative Matrix Factorization (NMF) has been applied with great success to many applications. As NMF is applied to massive datasets such as web-scale dyadic data, it is desirable to leverage a cluster of machines to speed up the factorization. However, it is challenging to efficiently implement NMF in a distributed environment. In this paper, we show that by leveraging a new form of update functions, we can perform local aggregation and fully explore parallelism. Moreover, under the new form of update functions, we can perform frequent updates, which aim to use the most recently updated data whenever possible. As a result, frequent updates are more efficient than their traditional concurrent counterparts. Through a series of experiments on a local cluster as well as the Amazon EC2 cloud, we demonstrate that our implementation with frequent updates is up to two orders of magnitude faster than the existing implementation with the traditional form of update functions.

1 Introduction

Nonnegative matrix factorization (NMF) [8] is a popular dimension reduction and factor analysis method that has attracted a lot of attention recently. It arises from a wide range of applications, including genome data analysis [3], text mining [15], recommendation systems [7], and social network analysis [13, 20]. NMF factorizes an original matrix into two low-rank factor matrices by minimizing a loss function that measures the discrepancy between the original matrix and the product of the two factor matrices. NMF algorithms typically use update functions to iteratively and alternately refine factor matrices.

Many practitioners have to deal with NMF on massive datasets. For example, recommendation systems in web services such as Netflix have been dealing with NMF on web-scale dyadic datasets, which involve millions of users, millions of movies, and billions of ratings. For such web-scale matrices, it is desirable to leverage a cluster of machines to speed up the factorization. MapReduce [4] has emerged as a popular distributed framework for data intensive computation. It provides a simple programming model where a user can focus on the computation logic without worrying about the complexity of parallel computation. Prior approaches (e.g., [12]) of handling NMF on MapReduce usually select an existing NMF algorithm and then focus on implementing matrix operations.

T. Calders et al. (Eds.): ECML PKDD 2014, Part III, LNCS 8726, pp. 337–352, 2014.

In this paper, we present a new form of factor matrix update functions. This new form operates on blocks of matrices. In order to support the new form, we partition the factor matrices into blocks along the short dimension and split the original matrix into corresponding blocks. The new form of update functions allows us to update distinct blocks independently and simultaneously when updating a factor matrix. It also facilitates a distributed implementation. Different blocks of one factor matrix can be updated in parallel. Additionally, the blocks can be distributed in memories of all the machines in a cluster, thus avoiding overflowing the memory of one single machine. Storing factor matrices in memory allows random access and local aggregation. As a result, the new form of update functions leads to an efficient MapReduce implementation.

Moreover, under the new form of update functions, we can update only a subset of its blocks when we update a factor matrix, and the number of blocks in the subset can be adjusted. The only requirement is that when one factor matrix is being updated, the other one has to be fixed. For instance, we can update one block of a factor matrix and then immediately update all blocks of the other factor matrix. We refer to this kind of updates as *frequent block-wise updates*. Frequent block-wise updates aim to utilize the most recently updated data whenever possible. As a result, frequent block-wise updates are more efficient than their traditional concurrent counterparts, *concurrent block-wise updates*, which update all blocks of either factor matrix alternately. Additionally, frequent block-wise updates maintain the convergence property in theory.

We implement concurrent block-wise updates on MapReduce and implement both concurrent and frequent block-wise updates on an extended version of MapReduce, iMapReduce [25], which supports iterative computations more efficiently. We evaluate these implementations on a local cluster as well as the Amazon EC2 cloud. With both synthetic and real-world datasets, the evaluation results show that our MapReduce implementation for concurrent block-wise updates is 19x - 57x faster than the existing MapReduce implementation [12] (with the traditional form of update functions) and that our iMapReduce implementation further achieves up to 2x speedup over our MapReduce implementation. Furthermore, the iMapReduce implementation with frequent block-wise updates is up to 2.7x faster than that with concurrent block-wise updates. Accordingly, our iMapReduce implementation with frequent block-wise updates is up to two orders of magnitude faster than the existing MapReduce implementation.

2   Background

NMF aims to factorize an original matrix A into two low-rank factor matrices W and H. Matrix A's elements must be nonnegative by assumption. The achieved factorization has the property of $A \simeq WH$, and the factor matrices W and H are also nonnegative. A loss function is used to measure the discrepancy between A and WH. The NMF problem can be formulated as follows.

Given $A \in \mathbb{R}_+^{m \times n}$ and a positive integer $k \ll \min\{m, n\}$, find $W \in \mathbb{R}_+^{m \times k}$ and $H \in \mathbb{R}_+^{k \times n}$ such that a loss function $L(A, WH)$ is minimized.

Loss function $L(A, WH)$ is typically not convex in both W and H together. Hence, it is unrealistic to have an approach that finds the global minimum. Fortunately, there are many techniques for finding local minima.

A general approach is to adopt the block coordinate descent rules [11]:
- Initialize W, H with nonnegative W^0, H^0, $t \leftarrow 0$.
- Repeat until a convergence criterion is satisfied:
  Find H^{t+1}: $L(A, W^t H^{t+1}) \leq L(A, W^t H^t)$;
  Find W^{t+1}: $L(A, W^{t+1} H^{t+1}) \leq L(A, W^t H^{t+1})$.

When the loss function is the square of the Euclidean distance, i.e.,

$$L(A, WH) = ||A - WH||_F^2, \tag{1}$$

where $|| \cdot ||_F$ is the Frobenius norm, one of the most well-known algorithms for implementing the above rules is Lee and Seung's multiplicative update approach [9]. It updates W and H as follows:

$$H = H * \frac{W^T A}{W^T W H}, \qquad W = W * \frac{A H^T}{W H H^T}. \tag{2}$$

3 Distributed NMF

In this section, we present how to efficiently apply the block coordinate descent rules to NMF in a distributed environment.

3.1 Decomposition

The loss function is usually decomposable [17]. That is, it can be represented as the sum of losses for each element in the matrix. For example, the widely adopted loss function, the square of the Euclidean distance, is decomposable. We list several popular decomposable loss functions in Table 1. We focus on NMF with decomposable loss functions in this paper.

Table 1. Decomposable loss functions

Square of Euclidean distance	$\sum_{(i,j)} (A_{ij} - [WH]_{ij})^2$
KL-divergence	$\sum_{(i,j)} A_{ij} \log \frac{A_{ij}}{[WH]_{ij}}$
Generalized I-divergence	$\sum_{(i,j)} (A_{ij} \log \frac{A_{ij}}{[WH]_{ij}} - (A_{ij} - [WH]_{ij}))$
Itakura-Saito distance	$\sum_{(i,j)} (\frac{A_{ij}}{[WH]_{ij}} - \log \frac{A_{ij}}{[WH]_{ij}} - 1)$

Distributed NMF needs to partition the matrices W, H, and A across computing nodes. To this end, we leverage a popular scheme in gradient descent algorithms [6,18] that partitions W and H into blocks along the short dimension to fully explore parallelism and splits the original matrix A into corresponding blocks. We use symbol $W^{(I)}$ to denote the Ith block of W, $H^{(J)}$ to denote the Jth

block of H, and $A^{(I,J)}$ to denote the corresponding block of A (i.e., the $(I, J)^{th}$ block). Under this partition scheme, $A^{(I,J)}$ is only related to $W^{(I)}$ and $H^{(J)}$, and it is independent of other blocks of W and H, in terms of the loss value (computed by the loss function). We refer to the partition scheme as *block-wise partition*. The view of the block-wise partition scheme is shown in Figure 1.

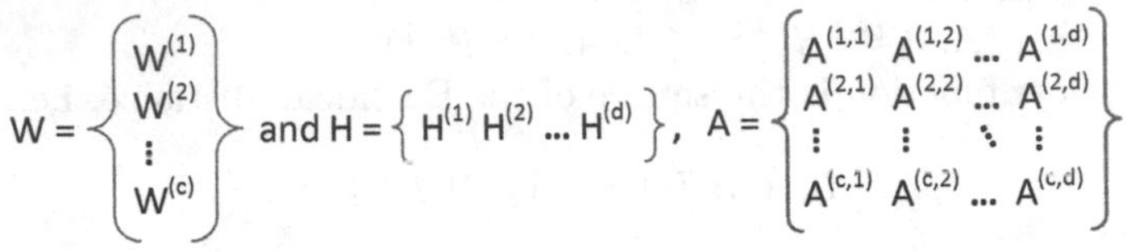

$$W = \begin{Bmatrix} W^{(1)} \\ W^{(2)} \\ \vdots \\ W^{(c)} \end{Bmatrix} \text{ and } H = \left\{ H^{(1)} \, H^{(2)} \, ... \, H^{(d)} \right\}, \quad A = \begin{Bmatrix} A^{(1,1)} \, A^{(1,2)} \, ... \, A^{(1,d)} \\ A^{(2,1)} \, A^{(2,2)} \, ... \, A^{(2,d)} \\ \vdots \quad \vdots \quad \ddots \quad \vdots \\ A^{(c,1)} \, A^{(c,2)} \, ... \, A^{(c,d)} \end{Bmatrix}$$

Fig. 1. The block-wise partition scheme for distributed NMF

Due to its decomposability, loss function $L(A, WH)$ can be expressed as

$$L(A, WH) = \sum_I \sum_J L(A^{(I,J)}, W^{(I)} H^{(J)}). \tag{3}$$

Let $F_I = \sum_J L(A^{(I,J)}, W^{(I)} H^{(J)})$ and $G_J = \sum_I L(A^{(I,J)}, W^{(I)} H^{(J)})$, then

$$L(A, WH) = \sum_I F_I = \sum_J G_J. \tag{4}$$

F_I and G_J can be considered as local loss functions. The overall loss function $L(A, WH)$ is the sum of the local loss functions. By fixing H, F_I is independent of one another. Therefore, F_I can be minimized independently and simultaneously by fixing H. Similarly, G_J can be minimized independently and simultaneously by fixing W.

3.2   Block-wise Updates

In this paper, we use the square of the Euclidean distance as an example of decomposable loss functions. Nevertheless, the techniques derived in this section can be applied to any decomposable loss function.

The block-wise partition allows us to update its blocks independently when updating a factor matrix (by fixing the other factor matrix). In other words, each block can be treated as one update unit. We refer to this kind of updates as *block-wise updates*. In the following, we illustrate how to update one block of W (by minimizing F_I) and that of H (by minimizing G_J).

Here we first show how to update one block of H (i.e., $H^{(J)}$). When W is fixed, minimizing G_J can be expressed as follows:

$$\min_{H^{(J)}} G_J = \min_{H^{(J)}} \sum_I ||A^{(I,J)} - W^{(I)} H^{(J)}||_F^2. \tag{5}$$

We leverage gradient descent to update $H^{(J)}$:

$$H_{\alpha\mu}^{(J)} = H_{\alpha\mu}^{(J)} - \eta_{\alpha\mu}[\frac{\partial G_J}{\partial H^{(J)}}]_{\alpha\mu}, \tag{6}$$

where $H_{\alpha\mu}^{(J)}$ denotes the element at the αth row and the μth column of $H^{(J)}$, $\eta_{\alpha\mu}$ is an individual step size for the corresponding gradient element, and

$$\frac{\partial G_J}{\partial H^{(J)}} = \sum_I [(W^{(I)})^T W^{(I)} H^{(J)} - (W^{(I)})^T A^{(I,J)}]. \tag{7}$$

If all step sizes are set to a sufficiently small positive number, the update should reduce G_J. However, if the number is too small, the decrease speed can be very slow. To obtain a good speed and to guarantee convergence, we derive step sizes by following Lee and Seung's multiplicative update approach [9]:

$$\eta_{\alpha\mu} = \frac{H_{\alpha\mu}^{(J)}}{[\sum_I (W^{(I)})^T W^{(I)} H^{(J)}]_{\alpha\mu}}. \tag{8}$$

Then, substituting Eq. (7) and Eq. (8) into Eq. (6), we have:

$$H_{\alpha\mu}^{(J)} = H_{\alpha\mu}^{(J)} * \frac{[\sum_I (W^{(I)})^T A^{(I,J)}]_{\alpha\mu}}{[\sum_I (W^{(I)})^T W^{(I)} H^{(J)}]_{\alpha\mu}}. \tag{9}$$

Similarly, we can derive the update formula for $W^{(I)}$ as follows:

$$W_{\alpha\mu}^{(I)} = W_{\alpha\mu}^{(I)} * \frac{[\sum_J A^{(I,J)} (H^{(J)})^T]_{\alpha\mu}}{[\sum_J W^{(I)} H^{(J)} (H^{(J)})^T]_{\alpha\mu}}. \tag{10}$$

Block-wise updates can update each block of one factor matrix independently. This flexibility allows us to have different ways of updating the blocks. We can simultaneously update all the blocks of one factor matrix and then update all the blocks of the other factor matrix. Also, we can update a subset of blocks of one factor matrix and then update a subset of blocks of the other one, where the number of blocks in the subsets can be adjusted. Additionally, block-wise updates also facilitate a distributed implementation. Different blocks of one factor matrix can be updated in parallel. Furthermore, the blocks can be distributed in memories of all the machines in a cluster, thus avoiding overflowing the memory of one single machine (when there are large factor matrices). Storing factor matrices in memory allows random access and local aggregation, which are highly useful for updating them.

3.3 Concurrent Block-wise Updates

With block-wise updates, a straightforward way of fulfilling the block coordinate descent rules is to update all blocks of H and then update all blocks of W. Since this approach updates all blocks of H (or W) concurrently, we refer to it as *concurrent block-wise updates*.

From matrix operation perspective, we can show that concurrent block-wise updates (using Eq. (9) and Eq. (10)) are equivalent to the multiplicative update approach shown in Eq. (2). Without loss of generality, we assume that $H^{(J)}$ is a block of H from the J_0^{th} column to the J_b^{th} column. Let Y be a block of $W^T W H$ from the J_0^{th} column to the J_b^{th} column, then we have $Y = \sum_I (W^{(I)})^T W^{(I)} H^{(J)}$ since $W^T W = \sum_I (W^{(I)})^T W^{(I)}$. Assuming that X is a block of $W^T A$ from the J_0^{th} column to the J_b^{th} column, we can show that $X = \sum_I (W^{(I)})^T A^{(I,J)}$. As a result, for both the concurrent block-wise updates and the multiplicative update approach, the formula for updating $H^{(J)}$ is equivalent to $H^{(J)} = H^{(J)} * \frac{X}{Y}$. Therefore, Eq. (9) is equivalent to the formula for updating H in Eq. (2). Similarly, we can show that Eq. (10) is equivalent to the formula for updating W in Eq. (2).

3.4 Frequent Block-wise Updates

Since all blocks of one factor matrix can be updated independently when the other matrix is fixed, another (more general) way of fulfilling block coordinate descent rules is to update a subset of blocks of W and then update a subset of blocks of H. Since this approach updates the factor matrices more frequently (compared to concurrent block-wise updates), we refer to it as *frequent block-wise updates*. Frequent block-wise updates aim to utilize the most recently updated data whenever possible and thus can potentially accelerate convergence.

More formally, frequent block-wise updates start with an initial guess of W and H, and then seek to minimize the loss function by iteratively applying the following two steps:

Step I: Fix W, update a subset of blocks of H using Eq. (9).

Step II: Fix H, update a subset of blocks of W using Eq. (10).

In both steps, the size of the subset is a parameter, and we rotate the subset among all the blocks to guarantee that each block has an equal chance to be updated. The size of the subset controls the update frequency. For example, if we always set the subset to include all the blocks, frequent block-wise updates degrade to concurrent block-wise updates.

Frequent block-wise updates maintain the convergence property. Using techniques similar to that used in [9], we can prove that G_J and F_I are nonincreasing under formulae Eq. (9) and Eq. (10), respectively. Then, it is straightforward to prove that L is nonincreasing when frequent block-wise updates are applied and that L is constant if and only if W and H are at a stationary point of L.

Frequent block-wise updates provide a high flexibility in updating factor matrices. For simplicity, we update a subset of blocks of one factor matrix and then update all blocks of the other one in each iteration. Here, we assume that we update a subset of blocks of W and then update all the blocks of H. Intuitively, updating H frequently might incur additional overhead. However, we find that the formula for updating H can be incrementally computed. That is, the cost of updating H grows linearly with the number of W blocks that have been updated in the previous iteration. As a result, performing frequent updates on H does not necessarily introduce a large additional cost.

To incrementally update H when a subset of W blocks are updated, we introduce a few auxiliary matrices. Let $X^{(J)} = \sum_I (W^{(I)})^T A^{(I,J)}$, $X^{(I,J)} = (W^{(I)})^T A^{(I,J)}$, $S = \sum_I (W^{(I)})^T W^{(I)}$, and $S^{(I)} = (W^{(I)})^T W^{(I)}$. Then, $H_{\alpha\mu}^{(J)}$ can be updated by

$$H_{\alpha\mu}^{(J)} = H_{\alpha\mu}^{(J)} * \frac{X_{\alpha\mu}^J}{[SH^{(J)}]_{\alpha\mu}}. \tag{11}$$

We next show how to incrementally calculate $X^{(J)}$ and S by saving their values from the previous iteration. When a subset of $W^{(I)}$ ($I \in C$) have been updated, the new value of $X^{(J)}$ and S can be computed as follows:

$$X^{(J)} = X^{(J)} + \sum_{I \in C} [(W^{(I)new})^T A^{(I,J)} - X^{(I,J)}]; \tag{12}$$

$$S = S + \sum_{I \in C} [(W^{(I)new})^T W^{(I)new} - S^{(I)}]. \tag{13}$$

From Eq. (11), Eq. (12), and Eq. (13), we can see that the cost of incrementally updating $H^{(J)}$ depends on the number of W blocks that have been updated rather than the total number of blocks that W has.

4 Implementation on Distributed Frameworks

MapReduce [4] and its extensions (e.g, [21, 22, 25]) have emerged as distributed frameworks for data intensive computation. MapReduce expresses a computation task as a series of jobs. Each job typically has one map task (mapper) and one reduce task (reducer). In this section, we illustrate the implementation of concurrent block-wise updates on MapReduce. Also, we show how to implement frequent block-wise updates on an extended version of MapReduce, iMapReduce [25], which supports iterative computations more efficiently.

Block-wise updates enable efficient distributed implementation. With block-wise updates, the basic computation units in update functions (Eq. (9) and Eq. (10)) are blocks of factor matrices and of the original matrix. The size of a block can be adjusted. As a result, when performing essential matrix operations that involve two blocks of matrices (e.g., $(W^{(I)})^T$ and $A^{(I,J)}$), we can assume that at least the smaller block can be held in the memory of a single machine. Since W and H are low-rank factor matrices, they are usually much smaller than A, and thus the assumption that one of their blocks can be held in the memory of a single machine is reasonable. The result matrix of an essential matrix operation (e.g., $(W^{(I)})^T A^{(I,J)}$) is usually relatively small and can be held in the memory of a single machine as well. Storing a matrix (or a block of a matrix) in memory efficiently supports random and repeated access, which is commonly needed in a matrix operation such as multiplication. Maintaining the result matrix in memory supports local aggregation. Therefore, each single machine can complete an essential matrix operation locally and efficiently. Note that the other (larger) block (e.g., a block of A) is still in disk so as to scale to large NMF problems.

Accordingly, the MapReduce programming model fits block-wise updates well. An essential matrix operation with two blocks can be realized in one mapper, and the aggregation of the results of essential matrix operations can be realized in reducers. In contrast, the previous work [12], which implements the traditional form of update functions on MapReduce, has a poor performance. For example, to perform matrix multiplication (with two large matrices), a row (or column) of one matrix needs to join with each column (or row) of the other one. Since neither of these two large matrices can be held in memory, a huge amount of intermediate data has to be generated and shuffled.

4.1   Concurrent Block-wise Updates on MapReduce

We first show an efficient MapReduce implementation for concurrent block-wise updates. To realize matrix multiplication with two blocks of matrices in one mapper, we exploit the fact that a mapper can load data in memory before processing input key-value pairs and that a mapper can maintain state across the processing of multiple input key-value pairs and defer emission of intermediate key-value pairs until all input pairs have been processed.

The update formula for $H^{(J)}$ (Eq. (9)) can be split into three parts: $X^{(J)} = \sum_I (W^{(I)})^T A^{(I,J)}$, $Y^{(J)} = \sum_I (W^{(I)})^T W^{(I)} H^{(J)}$, and $H^{(J)} = H^{(J)} * \frac{X^{(J)}}{Y^{(J)}}$.

Computing $X^{(J)}$ can be done in one MapReduce job. The mapper calculates $(W^{(I)})^T A^{(I,J)}$, and the reducer performs summation. Let $X^{(I,J)} = (W^{(I)})^T A^{(I,J)}$. When holding $W^{(I)}$ in memory, a mapper can compute $X^{(I,J)}$ via continuously reading elements of $A^{(I,J)}$ from disk: $X^{(I,J)}_{\cdot j} = \sum_{i=1} A^{(I,J)}_{i,j} (W^{(I)}_{i \cdot})^T$, where $X^{(I,J)}_{\cdot j}$ is the jth column of $X^{(I,J)}$, and $W^{(I)}_{i \cdot}$ is the ith row of $W^{(I)}$. $X^{(I,J)}$ (which is usually small) stays in memory for local aggregation. Then, the aggregation $X^{(J)} = \sum_I X^{(I,J)}$ can be computed in a reducer. The operations of this job (Job-I) are illustrated as follows.

- **Map:** Load $W^{(I)}$ in memory first, then calculate $X^{(I,J)} = (W^{(I)})^T A^{(I,J)}$, and last emit $< J, X^{(I,J)} >$.
- **Reduce:** Take $< J, X^{(I,J)} >$ (for any I) and emit $< J, X^{(J)} >$, where $X^{(J)} = \sum_I X^{(I,J)}$.

Computing $Y^{(J)} = \sum_I (W^{(I)})^T W^{(I)} H^{(J)}$ naturally needs two MapReduce jobs. One job (Job-II) is used to compute $S = \sum_I (W^{(I)})^T W^{(I)}$, and the other one is used to calculate $Y^{(J)} = SH^{(J)}$. Let $S^{(I)} = (W^{(I)})^T W^{(I)}$. Calculating $S^{(I)}$ (a small $k \times k$ matrix) can be performed in one mapper. Then, all the mappers send $(W^{(I)})^T W^{(I)}$ to one particular reducer for a global summation. The MapReduce operations are stated as follows.

- **Map:** Load $W^{(I)}$ in memory first, then calculate $S^{(I)} = (W^{(I)})^T W^{(I)}$, and last emit $< 0, S^{(I)} >$ (sending to reducer 0).
- **Reduce:** Take $< 0, S^{(I)} >$ and emit $< 0, S >$, where $S = \sum_I S^{(I)}$.

After computing $S = \sum_I (W^{(I)})^T W^{(I)}$, calculating $Y^{(J)} = SH^{(J)}$ can be done in a MapReduce job (Job-III) with the map phase only, as follows.

– **Map:** Load S in memory. Emit tuples $< J, Y^{(J)} >$, where $Y^{(J)} = SH^{(J)}$.

Lastly, one MapReduce job (with the map phase only) can compute $H^{(J)} \leftarrow H^{(J)} * \frac{X^{(J)}}{Y^{(J)}}$. The operations of this job (Job-IV) are described as follows.

– **Map:** Read $< J, H^{(J)} >$, $< J, X^{(J)} >$, and $< J, Y^{(J)} >$ (column by column). Emit tuple $< J, H^{(J)new} >$, where $H^{(J)new} = H^{(J)} * \frac{X^{(J)}}{Y^{(J)}}$.

In the previous implementation, we try to minimize data shuffling by utilizing local aggregation. However, in each iteration it still needs four MapReduce jobs to update H. In addition, intermediate data (e.g., $X^{(J)}$) needs to be dumped into disk and be reloaded in latter jobs.

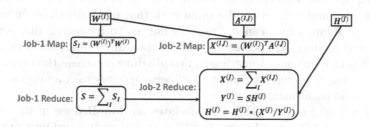

Fig. 2. Overview of the optimized implementation for updating $H^{(J)}$ on MapReduce

To avoid dumping and reloading intermediate data, such as $X^{(J)}$ and $Y^{(J)}$, and to minimize the number of jobs, we integrate Job-I, Job-III, and Job-IV into one job (Job-2). The integrated job has the same map phase as Job-I. However, in the reduce phase, besides computing $X^{(J)}$, it also computes $Y^{(J)}$ and finally calculates $H^{(J)new} = H^{(J)} * [X^{(J)}/Y^{(J)}]$. Job-II can be kept (as Job-1) for the simplicity of implementation since it only produces a small ($k \times k$) matrix and reloading its output does not take much time. The overview of our optimized implementation is presented in Figure 2, and the MapReduce operations in the integrated job (Job-2) are described as follows (the operations in Job-1 are skipped since they are the same with those in Job-II).

– **Map:** Load $W^{(I)}$ in memory first, then calculate $X^{(I,J)} = (W^{(I)})^T A^{(I,J)}$, and last emit $< I, X^{(I,J)} >$.

– **Reduce:** Take $< I, X^{(I,J)} >$, and first calculate $X^{(J)}$. Load S in memory. Then, read $H^{(J)}$ so as to compute $Y^{(J)}$. Last, calculate $H^{(J)new}$.

In the above, we describe the MapReduce operations used to complete the update of H for one iteration. Updating W can be performed in the same fashion. The formula for W (Eq. (10)) can be also treated as three parts: $U^{(I)} = \sum_J A^{(I,J)} (H^{(J)})^T$, $V^{(I)} = \sum_J W^{(I)} H^{(J)} (H^{(J)})^T$, and $W^{(I)} = W^{(I)} * \frac{U^{(I)}}{V^{(I)}}$. Due to space limitations, we omit the description of its MapReduce operations.

4.2   Frequent Block-wise Updates on iMapReduce

Although frequent block-wise updates have potential to speed up NMF, parallelizing frequent block-wise updates in a distributed environment is challenging.

Computations such as global summations need to be done in a centralized way. Synchronizing the global resources in a distributed environment may result in a considerable overhead, especially on MapReduce. MapReduce starts a new job for each computation errand. Each job needs to be initialized and to load its input data, even when the data is from the previous job. Frequent updates introduce more jobs. Consequently, the initialization overhead and the cost of repeatedly loading data may cause the benefit of frequent updates to vanish.

In this subsection, we propose an implementation of frequent block-wise updates on iMapReduce [25]. iMapReduce uses persistent mappers and reducers to avoid job initialization overhead. Each mapper is paired with one reducer. A pair of mapper and reducer can be seen as one logical worker. Data shuffling between mappers and reducers is the same with that of MapReduce. In addition, a reducer of iMapReduce can redirect its output to its paired mapper. Since mappers and reducers are persistent, state data can be maintained in memory across different iterations. Accordingly, iMapReduce decreases the overhead of a job. As a result, it provides frequent block-wise updates with an opportunity to achieve a good performance.

We implement frequent block-wise updates on iMapReduce in the following way. H is evenly split into r blocks, and W is evenly partitioned into $p*r$ blocks, where r is the number of workers and p is a parameter used to control update frequency. Each worker handles p blocks of W and one block of H. In each iteration a worker updates its H block and one selected W block. That is, there are r blocks of W in total to be updated in each iteration. Each worker rotates the selected W block among all its W blocks. The setting of p plays an important role on frequent block-wise updates. Setting p too large may incur considerable overhead for synchronization. Setting it too small may degrade the effect of the frequent updates. Nevertheless, we will show in our experiments (Section 5.4) that quite a wide range of p can enable frequent block-wise updates to have better performance than concurrent block-wise updates. Note that like the MapReduce implementation, our iMapReduce implementation still reads A from disk every time rather than holds it in memory so as to scale to large NMF problems. The operations in our iMapReduce implementation are illustrated as follows (Map-1x represents different stages of a mapper, and Reduce-1x represents different stages of a reducer).

- **Map-1a:** Load a subset (i.e., p) of W blocks (e.g., $(W^{(B)new})$) in memory (first iteration only) or receive one updated W block from the previous iteration. For all the loaded or received blocks, compute S_l via $S_l = \sum_B (W^{(B)new})^T W^{(B)new}$ (first iteration) or $S_l = S_l + ((W^{(B)new})^T W^{(B)new} - (W^{(B)})^T W^{(B)})$, and replace $W^{(B)}$ with $W^{(B)new}$. Broadcast $< 0, S_l >$ to all the reducers.
- **Reduce-1a:** Take $< 0, S_l >$, compute $S = \sum_l S_l$, and store S in memory.
- **Map-1b:** For each loaded or received W block in the previous stage (e.g., $(W^{(B)new})$), emit tuple $< J, X^{(B,J)} >$ where $X^J = (W^{(B)new})^T A^{(B,J)}$ (first iteration) or $< J, \Delta X^{(B,J)} >$ where $\Delta X^{(B,J)} = (W^{(B)new})^T A^{(B,J)} - X^{(B,J)}$.

- **Reduce-1b:** Take $< J, X^{(B,J)} >$ and calculate $X^{(J)} = \sum_B X^{(B,J)}$ (first iteration) or take $< J, \Delta X^{(B,J)} >$ and calculate $X^{(J)} = X^{(J)} + \sum_B \Delta X^{(B,J)}$. Then, load $H^{(J)}$ into memory (first iteration) and compute $Y^{(J)} = SH^{(J)}$. Last, calculate $H^{(J)new}$ by $(H^{(J)new} = H^{(J)} * \frac{X^{(J)}}{Y^{(J)}})$, store it in memory, and pass one copy to Map-1c in the form of $< J, H^{(J)new} >$.
- **Map-1c:** Receive (just updated) $H^{(J)}$ from Reduce-1b. Broadcast $< J, H^{(J)}(H^{(J)})^T >$ to all the reducers.
- **Reduce-1c:** Take $< J, H^{(J)}(H^{(J)})^T >$, compute $Z = \sum_J H^{(J)}(H^{(J)})^T$, and store Z in memory.
- **Map-1d:** For a W block that is selected in the current iteration (e.g., $(W^{(B)})$), emit tuples in the form of $< B, U^{(B,J)} >$, where $U^{(B,J)} = A^{(B,J)}(H^{(J)})^T$.
- **Reduce-1d:** Take $< B, U^{(B,J)} >$ and calculate $U^{(B)} = \sum_J U^{(B,J)}$. Then, compute $V^{(B)} = W^{(B)}Z$. Last, calculate $W^{(B)new} = W^{(B)} * \frac{U^{(B)}}{V^{(B)}}$, store it in memory, and pass one copy to Map-1a.

5   Evaluation

In this section, we evaluate the efficiency of block-wise updates. To show the performance improvement, we use the existing implementation [12] as a reference point. For MapReduce, we leverage its open source version, Hadoop [1].

5.1   Experiment Setup

We build both a local cluster and a large-scale cluster on Amazon EC2. The local cluster consists of 4 machines, and each one has a dual-core 2.66GHz CPU, 4GB of RAM, 1TB of disk. The Amazon cluster consists of 100 medium instances, and each instance has one core, 3.7GB of RAM, and 400GB of disk.

Table 2. Summary of datasets

Dataset	# of rows	# of columns	# of nonzero elements
Netflix	$480,189$	$17,770$	$100M$
NYTimes	$300,000$	$102,660$	$70M$
Syn-m-n	m	n	$0.1 * m * n$

Both synthetic and real-word datasets are used in our experiments. We use two Real-world datasets. One is a document-term matrix, NYTimes, from UCI Machine Learning Repository [2]. The other one is a user-movie matrix from the Netflix prize [7]. We also generate several matrices with different choices of m and n. The sparsity is set to 0.1, and each element is a random integer uniformly selected from the range $[1, 5]$. The datasets are summarized in Table 2.

Unless otherwise specified, we use rank $k = 10$, and use $p = 8$ for frequent block-wise updates (which means each worker updates $1/8$ of its local W blocks in each iteration).

5.2 Comparison with Existing Work

The first set of experiments focus on the advantage of our (optimized) implementation of concurrent block-wise updates on MapReduce. We compare it with the existing work of implementing the multiplicative update approach on MapReduce [12], which uses the traditional form of update functions. We also include the implementation of concurrent block-wise updates on iMapReduce in the comparison (by setting $p = 1$) to show iMapReduce's superiority over MapReduce. As described in Section 3.4, concurrent block-wise updates are equivalent to the multiplicative update approach, and thus we focus on the time taken in a single iteration to directly compare the performance. Figure 3 shows the time taken in one iteration for all the three implementations. Note that the y-axis is in log scale. Our implementation on MapReduce (denoted by "Block-wise on MR") is 19x - 57x faster than the existing MapReduce implementation (denoted by "Row/Column-wise on MR"). Moreover, the implementation on iMapReduce (denoted by "Block-wise on iMR") is up to 2x faster than that on MapReduce.

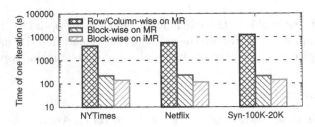

Fig. 3. Time taken in one iteration for different implementations on the local cluster

5.3 Effect of Frequent Block-wise updates

To evaluate the effect of frequent block-wise updates, we compare frequent block-wise updates with concurrent block-wise updates when both are implemented on iMapReduce. Both update approaches start with the same initial values when compared on the same dataset. Figure 4 plots the performance comparison. We can see that frequent block-wise updates ("Frequent") converge much faster than concurrent block-wise updates ("Concurrent") on all the three datasets. In other words, if we use a predefined loss value as the convergence criterion, frequent block-wise updates would have much shorter running time.

5.4 Tuning Update Frequency

As stated in Section 4.2, the update frequency affects the performance of frequent block-wise updates. In the experiments, we find that quite a wide range of p can allow frequent block-wise updates to have a better performance than their

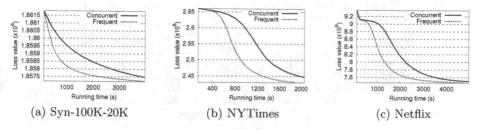

(a) Syn-100K-20K   (b) NYTimes   (c) Netflix

Fig. 4. Convergence speed on the local cluster

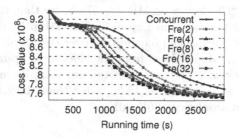

Fig. 5. Convergence speed vs. update frequency on dataset Netflix. The numbers associated with "Fre" represent different settings of p.

concurrent counterparts, and the best setting of p stays in the range from 4 to 32. This is also why we set $p = 8$ by default. For example, Figure 5 shows the convergence speed with different settings on dataset Netflix. Another interesting finding is that if a setting is better during a first few iterations, it will continue to be better. Hence, another way of obtaining a good setting of p is to test several candidate settings, each for a few iterations, and then choose the best one.

5.5 Different Data Sizes

We also measure how block-wise updates scale with the increasing size of matrix A. We generate synthetic datasets of different sizes by fixing the number of ($100k$) rows and increasing the number of columns. We use the loss value when concurrent block-wise updates run for 25 iterations as the convergence point. The time used to reach this convergence point is measured as the running time. This criterion also applies to the latter comparisons. As presented in Figure 6, the running times of both types of updates increase linearly with the size of the dataset. Moreover, frequent block-wise updates are up to 2.7x faster than concurrent block-wise updates.

To summarize, our iMapReduce implementation with frequent block-wise updates ("Frequent") is up to two orders of magnitude faster than the existing MapReduce implementation ("Row/Column-wise on MR"). Take dataset Syn-100K-20K for example. Our MapReduce implementation with concurrent

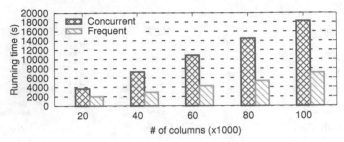

Fig. 6. Running time vs. dataset size on the local cluster

block-wise updates ("Block-wise on MR") is 57x faster than the existing MapReduce implementation (as shown in Figure 3). The iMapReduce implementation ("Block-wise on iMR") achieves 1.5x speedup over the MapReduce implementation. Furthermore, on iMapReduce, frequent block-wise updates are 1.8x faster than concurrent block-wise updates. In total, our iMapReduce implementation with frequent block-wise updates is 154x faster than the existing MapReduce implementation for dataset Syn-100K-20K.

5.6   Scaling Performance

To validate the scalability of our implementations, we evaluate them on the Amazon EC2 cloud. We use dataset Syn-1M-20K, which has 1 million rows, 20 thousand columns, and 2 billion nonzero elements. Figure 7a plots the time taken in a single iteration when all the three implementations are running on 100 nodes. Our implementation on MapReduce is 23x faster than the existing implementation. Moreover, the implementation on iMapReduce is 1.5x faster than that on MapReduce. Figure 7b shows the performance as the number of nodes being used increases from 20 to 100. We can see that the running times of both frequent block-wise updates and concurrent block-wise updates decrease smoothly as the number of nodes increases. In addition, frequent block-wise updates outperform concurrent block-wise updates with any number of nodes in the cluster.

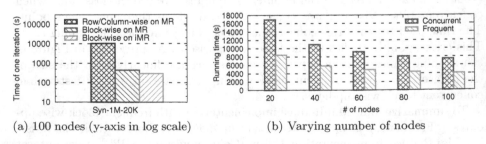

(a) 100 nodes (y-axis in log scale)   (b) Varying number of nodes

Fig. 7. Performance on the Amazon EC2 cloud

6   Related Work

Matrix factorization has been applied very widely [3,7,13,15,20]. Due to its popularity and increasingly larger datasets, many approaches for parallelizing it have been proposed. Zhou et al. [26] and Schelter et al. [16] show how to distribute the alternating least squares algorithm. Both approaches require each computing node to have a copy of one factor matrix when the other factor matrix is updated. This requirement limits their scalability. For large matrix factorization problems, it is important that factor matrices can be distributed. Several works handle matrix factorization using distributed gradient descent methods [6,10,18,24]. These approaches mainly focus on in-memory settings, in which both the original matrix and factor matrices are held in memory, and the forms of update functions used are different from the form we present. Additionally, our approach allows the original matrix to be in disk so as to scale to large NMF problems. A closely related work is from Liu et al. [12]. They propose a scheme of implementing the multiplicative update approach on MapReduce. Their scheme is based on the traditional form of update functions, which results in a poor performance.

It has been shown that frequent updates can accelerate expectation maximization (EM) algorithms [14,19,23]. Somewhat surprisingly, there has been no attempt to apply this method to NMF, even though there is equivalence between certain variations of NMF and some particular EM algorithms like K-means [5]. Our work demonstrates that frequent updates can also accelerate NMF.

7   Conclusion

In this paper, we find that by leveraging a new form of update functions, block-wise updates, we can perform local aggregation and thus have an efficient MapReduce implementation for NMF. Moreover, we propose frequent block-wise updates, which aim to use the most recently updated data whenever possible. As a result, frequent block-wise updates can further improve the performance, comparing with concurrent block-wise updates. We implement frequent block-wise updates on iMapReduce, an extended version of MapReduce. The evaluation results show that our iMapReduce implementation is up to two orders of magnitude faster than the existing MapReduce implementation.

Acknowledgments. We thank anonymous reviewers for their insightful comments. This work is partially supported by NSF grants CNS-1217284, CCF-1018114, and CCF-1017828. Any opinions, findings, conclusions or recommendations expressed in this paper are those of the authors and do not necessarily reflect the views of the sponsor.

References

1. Apache Hadoop, http://hadoop.apache.org/
2. UCI Machine Learning Repository, http://archive.ics.uci.edu/ml
3. Brunet, J.-P., Tamayo, P., Golub, T.R., Mesirov, J.P.: Metagenes and molecular pattern discovery using matrix factorization. PNAS 101(12), 4164–4169 (2004)

4. Dean, J., Ghemawat, S.: MapReduce: Simplified data processing on large clusters. In: OSDI 2004, pp. 107–113 (2004)
5. Ding, C., Li, T., Jordan, M.I.: Convex and semi-nonnegative matrix factorizations. TPAMI 32(1), 45–55 (2010)
6. Gemulla, R., Nijkamp, E., Haas, P.J., Sismanis, Y.: Large-scale matrix factorization with distributed stochastic gradient descent. In: KDD 2011, pp. 69–77 (2011)
7. Koren, Y., Bell, R., Volinsky, C.: Matrix factorization techniques for recommender systems. Computer 42(8), 30–37 (2009)
8. Lee, D.D., Seung, H.S.: Learning the parts of objects by non-negative matrix factorization. Nature 401, 788–791 (1999)
9. Lee, D.D., Seung, H.S.: Algorithms for non-negative matrix factorization. In: NIPS, pp. 556–562. MIT Press (2000)
10. Li, B., Tata, S., Sismanis, Y.: Sparkler: Supporting large-scale matrix factorization. In: EDBT 2013, pp. 625–636 (2013)
11. Lin, C.-J.: Projected gradient methods for nonnegative matrix factorization. Neural Comput. 19(10), 2756–2779 (2007)
12. Liu, C., Yang, H.-C., Fan, J., He, L.-W., Wang, Y.-M.: Distributed nonnegative matrix factorization for web-scale dyadic data analysis on mapreduce. In: WWW 2010, pp. 681–690 (2010)
13. Menon, A.K., Elkan, C.: Link prediction via matrix factorization. In: Gunopulos, D., Hofmann, T., Malerba, D., Vazirgiannis, M. (eds.) ECML PKDD 2011, Part II. LNCS, vol. 6912, pp. 437–452. Springer, Heidelberg (2011)
14. Neal, R., Hinton, G.E.: A view of the EM algorithm that justifies incremental, sparse, and other variants. In: Learning in Graphical Models, pp. 355–368 (1998)
15. Pauca, V.P., Shahnaz, F., Berry, M.W., Plemmons, R.J.: Text mining using nonnegative matrix factorizations. In: SDM, pp. 452–456 (2004)
16. Schelter, S., Boden, C., Schenck, M., Alexandrov, A., Markl, V.: Distributed matrix factorization with mapreduce using a series of broadcast-joins. In: RecSys 2013, pp. 281–284 (2013)
17. Singh, A.P., Gordon, G.J.: A unified view of matrix factorization models. In: Daelemans, W., Goethals, B., Morik, K. (eds.) ECML PKDD 2008, Part II. LNCS (LNAI), vol. 5212, pp. 358–373. Springer, Heidelberg (2008)
18. Teflioudi, C., Makari, F., Gemulla, R.: Distributed matrix completion. In: ICDM 2012, pp. 655–664 (2012)
19. Thiesson, B., Meek, C., Heckerman, D.: Accelerating EM for large databases. Mach. Learn. 45(3), 279–299 (2001)
20. Wang, F., Li, T., Wang, X., Zhu, S., Ding, C.: Community discovery using nonnegative matrix factorization. Data Min. Knowl. Discov. (May 2011)
21. Yin, J., Liao, Y., Baldi, M., Gao, L., Nucci, A.: Efficient analytics on ordered datasets using mapreduce. In: HPDC 2013, pp. 125–126 (2013)
22. Yin, J., Liao, Y., Baldi, M., Gao, L., Nucci, A.: A scalable distributed framework for efficient analytics on ordered datasets. In: UCC 2013, pp. 131–138 (2013)
23. Yin, J., Zhang, Y., Gao, L.: Accelerating expectation-maximization algorithms with frequent updates. In: CLUSTER 2012, pp. 275–283 (2012)
24. Yu, H.-F., Hsieh, C.-J., Si, S., Dhillon, I.: Scalable coordinate descent approaches to parallel matrix factorization for recommender systems. In: ICDM 2012 (2012)
25. Zhang, Y., Gao, Q., Gao, L., Wang, C.: iMapReduce: A distributed computing framework for iterative computation. In: DataCloud 2011, pp. 1112–1121 (2011)
26. Zhou, Y., Wilkinson, D., Schreiber, R., Pan, R.: Large-scale parallel collaborative filtering for the netflix prize. In: Fleischer, R., Xu, J. (eds.) AAIM 2008. LNCS, vol. 5034, pp. 337–348. Springer, Heidelberg (2008)

Convergence of Min-Sum-Min Message-Passing for Quadratic Optimization

Guoqiang Zhang* and Richard Heusdens

Department of Circuits and Systems,
Delft University of Technology,
Delft, The Netherlands
{g.zhang-1,r.heusdens}@tudelft.nl

Abstract. We propose a new message-passing algorithm for the quadratic optimization problem. As opposed to the min-sum algorithm, the new algorithm involves two minimizations and one summation at each iteration. The new min-sum-min algorithm exploits feedback from last iteration in generating new messages, resembling the Jacobi-relaxation algorithm. We show that if the feedback signal is large enough, the min-sum-min algorithm is guaranteed to converge to the optimal solution. Experimental results show that the min-sum-min algorithm outperforms two reference methods w.r.t. the convergence speed.

Keywords: quadratic optimization, Gaussian belief propagation, min-sum, min-sum-min.

1 Introduction

In this paper we consider solving a quadratic optimization problem in a distributed fashion, namely

$$\min_{x \in \mathbb{R}^n} f(x) \overset{\triangle}{=} \min_{x \in \mathbb{R}^n} \left(\frac{1}{2} x^\top J x - h^\top x \right), \tag{1}$$

where the quadratic matrix $J \in \mathbb{R}^{n \times n}$ is real symmetric positive definite and $h \in \mathbb{R}^n$. It is known that the optimal solution is given by $x^* = J^{-1}h$. We suppose that the quadratic matrix J is sparse and the dimensionality n is large. In this situation, the direct computation (without using the sparse structure of J) of the optimal solution may be expensive and unscalable. One natural question is how to exploit the sparse geometry to efficiently obtain the optimal solution.

A common approach that exploits the sparsity of J is to associate the function $f(x)$ with an undirected graph $G = (V, E)$. That is, the graph has a node for each variable x_i and an edge between node i and j only if the element J_{ij} is nonzero. By doing so, the sparsity of J is fully captured by the graph. As a consequence, the function can be decomposed with respect to $G = (V, E)$ as

$$f(x) = \sum_{i \in V} f_i(x_i) + \sum_{(i,j) \in E} f_{ij}(x_i, x_j), \tag{2}$$

---

* This work was supported by the COMMIT program, The Netherlands.

T. Calders et al. (Eds.): ECML PKDD 2014, Part III, LNCS 8726, pp. 353–368, 2014.
© Springer-Verlag Berlin Heidelberg 2014

where each edge-function $f_{ij}(x_i, x_j)$ characterizes the interaction of x_i and x_j as specified by J_{ij}. With the graphic model (2), distributed quadratic optimization (DQO) boils down to how to spread the global information of (J, h) in (1) over the graph efficiently by exchanging local information between neighboring nodes.

In the literature, the Jacobi algorithm is a natural approach for solving the problem over the associated graph [1]. At each iteration, the algorithm performs node-oriented minimizations over all the nodes in the graph, of which the messages are in a form of linear functions (see Table 1). It is known that when the matrix J is walk-summable[1], the Jacobi algorithm converges to the optimal solution [3,5]. To fix the convergence for a general matrix J, the Jacobi algorithm was under-relaxed by incorporating an estimate of x^* from last iteration in computing a new estimate (see Table 1). The Jacobi-relaxation algorithm possesses a guaranteed convergence if the relaxation parameter is properly chosen [1]. For the above two algorithms, once a node-estimate is updated, this estimate is broadcast to all its neighbors. Because the information transmitted is general, and not edge-specific, the two algorithms are known to converge slowly [1].

Table 1. Algorithm comparison. The min-sum-min algorithm is a new method that we will present in the paper.

J is walk-summable	J is general

Jacobi Alg.:
  * node-oriented minimization
  * linear message
	Jacobi-relaxation Alg.:
	  * introduce feedback in Jacobi Alg.

LiCD Alg.:
  * pairwise minimization
  * linear message
	GLiCD Alg.:
	  * introduce feedback in LiCD Alg.

min-sum Alg.:
  * pairwise minimization
  * quadratic message
	min-sum-min Alg.:
	  * introduce feedback in min-sum Alg.

To accelerate the convergence of the Jacobi algorithm, the linear coordinate descent (LiCD) algorithm was proposed in [9]. At each iteration, the LiCD algorithm performs pairwise minimizations over all the edges in the graph, of which the messages are in a form of linear functions (see Table 1). As shown in [9], if the quadratic matrix J is walk-summable, the LiCD algorithm converges to the optimal solution. To fix the convergence for a general matrix J, the LiCD algorithm was further extended in [10] by incorporating feedback from last iteration in computing new messages, which is referred to as the *generalized LiCD* (GLiCD) algorithm.

An alternative scheme for solving the quadratic problem is by using the framework of probability theory. The optimal solution x^* is viewed as the mean value of a random vector $x \in \mathbb{R}^n$ with Gaussian distribution

---

[1] See subsection 2.3 for the definition.

$$p(x) \propto \exp\left(-\frac{1}{2}x^\top Jx + h^\top x\right). \tag{3}$$

The min-sum (also known as max-product) algorithm is one popular approach to estimate both the mean value $x^* = J^{-1}h$ and individual variances [8,2]. At each iteration, the algorithm essentially performs pairwise minimizations over all the edges in the graph, of which the messages are in a form of quadratic functions (see Table 1). For a graph with a tree-structure, the min-sum algorithm converges to the optimal solution in finite steps [8]. The question of convergence for loopy graphic models has been proven difficult. In [3,5,6], it was shown when the matrix J is walk-summable, the min-sum algorithm converges to the optimal solution (see Table 1). In [4], a double-loop algorithm has been proposed to compute the optimal solution for a general matrix J, where the min-sum algorithm is used as a subroutine. We note that the double-loop algorithm is time-consuming. This motivates us to develop a single-loop min-sum based algorithm.

In this paper, we complete Table 1 by proposing a (single-loop) min-sum-min algorithm for a general quadratic optimization problem. Our primary motivation is to fix the convergence failure of the min-sum algorithm when the matrix J is general. Inspired by the GLiCD algorithm, the min-sum-min algorithm also incorporates feedback from last iteration in computing new messages. Compared to the min-sum algorithm, the min-sum-min algorithm involves one more minimization at each iteration. The additional minimization is performed to compute the estimate of x^*, which is used to construct the feedback signal in generating new messages.

We also study the convergence of the min-sum-min algorithm. We show that by setting the feedback signal large enough in computing new messages, the algorithm possesses a guaranteed convergence. Experimental results show that the min-sum-min algorithm converges faster than the Jacobi-relaxation and GLiCD algorithms.

2   Min-Sum-Min Message-Passing

In this section, we present the min-sum-min algorithm for the quadratic optimization problem. In particular, we describe how to construct feedback signal in update the messages.

2.1   Message-Passing Framework

Consider the quadratic optimization problem (1). Without loss of generality, we assume the quadratic matrix J is of unit diagonal (i.e., $J_{ii} = 1$, $i = 1, \ldots, n$). By using the sparsity of the matrix J, the quadratic function $f(x)$ can be decomposed w.r.t. a graph $G = (V, E)$

$$f(x) = \sum_{i \in V} f_i(x_i) + \sum_{(i,j) \in E} f_{ij}(x_i, x_j),$$

where the node and edge functions are given by

$$f_i(x_i) = \frac{1}{2}x_i^2 - h_i x_i \qquad i \in V \tag{4}$$

$$f_{ij}(x_i, x_j) = J_{ij}x_i x_j \qquad (i,j) \in E. \tag{5}$$

An edge exists between node i and j in the graph only if $J_{ij} \neq 0$. For each node $i \in V$, we denote the set of its neighbors as $N(i) \triangleq \{j \in V : (i,j) \in E\}$. For each edge $(i,j) \in E$, we use $[j,i]$ to denote the directed edge from node i to j. Correspondingly, we denote the set of all directed edges of the graph as $\overrightarrow{E}$.

The min-sum-min algorithm intends to minimize the quadratic function in an iterative, synchronous message-passing fashion. At time t, each node i keeps track of a message and an estimate of x_i^* from each neighbor $u \in N(i)$. We denote the message and the estimate from node u to i as $m_{ui}^{(t)}(x_i)$ and $\hat{x}_i^{u,(t)}$, respectively. Correspondingly, we use $\hat{x}_{edge}^{(t)}$ to denote the vector of all the estimates at time t. $\hat{x}_{edge}^{(t)}$ is of dimension $|\overrightarrow{E}|$, of which each component $\hat{x}_i^{j,(t+1)}$ corresponds to a directed edge $[i,j] \in \overrightarrow{E}$. Note that for each node $i \in V$, the estimates $\{\hat{x}_i^{u,(t)}, u \in N(i)\}$ reveal information about the optimal solution x_i^*. Thus, the estimates obtained at time t can be used as feedback in computing new messages and new estimates at time $t+1$.

Formally, we use the estimates at time t to construct $|E|$ penalty functions, one for each edge in the graph. In particular, we define the penalty function $p_{ij}^{(t)}(x_i, x_j)$ for $(i,j) \in E$ to be a quadratic function:

$$p_{ij}^{(t)}(x_i, x_j) = \frac{s}{2}\left(x_i - \hat{x}_i^{j,(t)}\right)^2 + \frac{s}{2}\left(x_j - \hat{x}_j^{i,(t)}\right)^2, \tag{6}$$

where the weighting factor $1 > s \geq 0$. Note that each penalty function only involves the estimates that are computed along the associated edge. The particular form of the penalty function enables the performance analysis of the algorithm (see Section 3.2).

With the penalty functions (6), we define new node and edge functions at time t as

$$g_i^{(t)}(x_i) = (1-s)f_i(x_i) + \sum_{u \in N(i)} m_{ui}^{(t)}(x_i) \quad i \in V \tag{7}$$

$$g_{ij}^{(t)}(x_i, x_j) = (1-s)f_{ij}(x_i, x_j) - m_{ji}^{(t)}(x_i) - m_{ij}^{(t)}(x_j)$$
$$+ p_{ij}^{(t)}(x_i, x_j) \quad (i,j) \in E. \tag{8}$$

As opposed to (4)-(5), the new edge and node functions (7)-(8) include both the current messages and the penalty functions.

In next subsection, we explain how to use (7)-(8) in computing new messages and estimates. Note that as the weighting factor s approaches to one, the original function $f(x)$ has less and less impact on the new local functions (7)-(8). While at the same time, the penalty function enlarges the impact of the estimates when computing new estimates and messages in next iteration.

Remark 1. We point out that when $s = 0$, the local-function formation (7)-(8) coincides with that of the min-sum algorithm [6]. It is the penalty functions that make the node and edge functions special.

2.2 Message-Updating Expressions

We have thus far presented the message-passing framework. In particular, we have defined the penalty functions (6). In this subsection, we derive the updating expressions for the messages and estimates. We then point out the difference between the min-sum-min and min-sum algorithms.

Suppose that the messages at time t take a quadratic form: (see [6] for a similar definition)

$$m_{ui}^{(t)}(x_i) = -\frac{1}{2}\gamma_{ui}^{(t)}(1-s)^2 J_{ui}^2 x_i^2 + z_{ui}^{(t)} x_i, \quad \forall\, [u,i] \in \overrightarrow{E}, \tag{9}$$

where $\{\gamma_{ui}\}$ and $\{z_{ui}\}$ are quadratic parameters and linear parameters, respectively. The weighting factor s is involved in (9) because of the penalty functions. We use $\gamma^{(t)}$ to denote the vector of all the quadratic parameters at time t. Similarly, we use $z^{(t)}$ to denote the vector of all the linear parameters. Both $\gamma^{(t)}$ and $z^{(t)}$ are of dimension $|\overrightarrow{E}|$.

We now compute the new estimates and messages for time $t+1$ given the information at time t. Without loss of generality, we focus on computing $\{m_{ij}^{(t+1)}, m_{ji}^{(t+1)}\}$ and $\{\hat{x}_i^{j,(t+1)}, \hat{x}_j^{i,(t+1)}\}$ that are associated with the edge $(i,j) \in E$. Note that the old messages $\{m_{ij}^{(t)}, m_{ji}^{(t)}\}$ and estimates $\{\hat{x}_i^{j,(t)}, \hat{x}_j^{i,(t)}\}$ are only involved in three local functions $\{g_i^{(t)}(x_i), g_j^{(t)}(x_j), g_{ij}^{(t)}(x_i, x_j)\}$. Thus, we use the three local functions in computing the corresponding new messages and estimates.

Formally, we define a function $L_{ij}^{(t)}(x_i, x_j)$ for $(i,j) \in E$ to be

$$L_{ij}^{(t)}(x_i, x_j) \triangleq g_i^{(t)}(x_i) + g_j^{(t)}(x_j) + g_{ij}^{(t)}(x_i, x_j). \tag{10}$$

The function $L_{ij}^{(t)}(x_i, x_j)$ is in a quadratic form. For the time being, we assume that $L_{ij}^{(t)}(x_i, x_j)$ is a strictly convex quadratic function. In other words, the 2×2 quadratic matrix in $L_{ij}^{(t)}(x_i, x_j)$ is assumed to be symmetric positive definite. In next subsection, we explain under what conditions the assumption holds. We compute the new estimates $\{\hat{x}_i^{j,(t+1)}, \hat{x}_j^{i,(t+1)}\}$ by minimizing the function $L_{ij}^{(t)}(\cdot, \cdot)$ over x_i and x_j:

$$\left(\hat{x}_i^{j,(t+1)}, \hat{x}_j^{i,(t+1)}\right) = \arg\min_{x_i, x_j} L_{ij}^{(t)}(x_i, x_j). \tag{11}$$

Since $L_{ij}^{(t)}(x_i, x_j)$ is a quadratic function, $\hat{x}_i^{j,(t+1)}$ and $\hat{x}_j^{i,(t+1)}$ have closed-form expressions.

Note that the information about $\hat{x}_i^{j,(t+1)}$ or $\hat{x}_j^{i,(t+1)}$ is embedded in both node i and j. We design the message $m_{ji}^{(t+1)}(x_i)$ with the purpose to bring all the information about $\hat{x}_i^{j,(t+1)}$ that is contained in node j to node i. In doing so, we reconsider the minimization of $L_{ij}^{(t)}(x_i, x_j)$:

$$
\begin{aligned}
\min_{x_i, x_j} L_{ij}^{(t)}(x_i, x_j) &= \min_{x_i} \left[g_i^{(t)}(x_i) + \min_{x_j} \left(g_j^{(t)}(x_j) + g_{ij}^{(t)}(x_i, x_j) \right) \right] \\
&= \min_{x_i} \left[g_i^{(t)}(x_i) + \frac{s}{2}(x_i - \hat{x}_i^{j,(t)})^2 - m_{ji}^{(t)}(x_i) \right. \\
&\qquad + \min_{x_j} \left((1-s)f_j(x_j) + (1-s)f_{ij}(x_i, x_j) \right. \\
&\qquad \left. \left. + \sum_{v \in N(j) \setminus i} m_{vj}^{(t)}(x_j) + \frac{s}{2}(x_j - \hat{x}_j^{i,(t)})^2 \right) \right].
\end{aligned}
\tag{12}
$$

By following (12), we define $m_{ji}^{(t+1)}(x_i)$ to be

$$
\begin{aligned}
m_{ji}^{(t+1)}(x_i) &\triangleq \min_{x_j} \left((1-s)f_j(x_j) + (1-s)f_{ij}(x_i, x_j) \right. \\
&\qquad \left. + \sum_{v \in N(j) \setminus i} m_{vj}^{(t)}(x_j) + \frac{s}{2}(x_j - \hat{x}_j^{i,(t)})^2 \right) + \kappa,
\end{aligned}
\tag{13}
$$

where κ represents an arbitrary offset term. The derivation of $m_{ij}^{(t+1)}(x_j)$ follows a similar procedure.

Based on the above computation guideline, we present the final expressions for the new messages and estimates. By combining (4)-(5), (9) and (13), we obtain the expressions for $\gamma_{ji}^{(t+1)}$ and $z_{ji}^{(t+1)}$ of $m_{ji}^{(t+1)}(x_i)$ as

$$
\gamma_{ji}^{(t+1)} = \frac{1}{1 - \sum_{v \in N(j) \setminus i} \gamma_{vj}^{(t)}(1-s)^2 J_{vj}^2},
\tag{14}
$$

$$
z_{ji}^{(t+1)} = (1-s)J_{ij}\gamma_{ji}^{(t+1)} \left((1-s)h_j + s\hat{x}_j^{i,(t)} - \sum_{v \in N(i) \setminus i} z_{vj}^{(t)} \right).
\tag{15}
$$

The parameters $\gamma_{ij}^{(t+1)}$ and $z_{ij}^{(t+1)}$ of $m_{ij}^{(t+1)}(x_i)$ can be computed similarly. By combining (11)-(15), we obtain the expressions for $\hat{x}_i^{j,(t+1)}$ and $\hat{x}_j^{i,(t+1)}$ as

$$
\begin{pmatrix} \hat{x}_i^{j,(t+1)} \\ \hat{x}_j^{i,(t+1)} \end{pmatrix} = \frac{1}{(1-s)J_{ij}} \begin{pmatrix} 1 & (1-s)J_{ij}\gamma_{ij}^{(t+1)} \\ (1-s)J_{ij}\gamma_{ji}^{(t+1)} & 1 \end{pmatrix}^{-1} \begin{pmatrix} z_{ij}^{(t+1)} \\ z_{ji}^{(t+1)} \end{pmatrix}.
\tag{16}
$$

The above expression fully characterizes the relationship between the estimates and the linear parameters. With (14)-(16) at hand, one can easily work out the updating-expressions of the messages and estimates associated with other edges in the graph.

Finally we reconsider the expression (12). Note that there are two minimizations and one summation involved in (12). As indicated in (13), the minimization over x_j and the summation of the incoming messages excluding $m_{ij}^{(t)}$ originate from the min-sum algorithm. The second minimization over x_i in (12) computes an estimate of x_i^*, which is used as feedback in generating new messages and estimates in next iteration. This is how the name *min-sum-min* message-passing comes up.

Remark 2. It is worth noting that when $s = 0$ in (14)-(15), we actually obtain the message-updating expressions for the min-sum algorithm. In other words, the min-sum-min algorithm includes the min-sum algorithm as a special case by setting $s = 0$.

2.3 Algorithm Implementation

In this subsection, we consider the algorithm implementation. We mainly study under what conditions the minimization problem (11) is well defined for $t \geq 0$ and for any $(i, j) \in E$.

Before formally presenting the algorithm implementation, we first provide the definition of the walk-summability of a positive definite matrix below. We emphasize that the min-sum algorithm converges to the optimal solution if the matrix J in (1) is walk-summable [3,5,6].

Definition 1. *[3,5] A symmetric positive definite matrix $J \in \mathbb{R}^{n \times n}$, with all ones on its diagonal, is walk-summable if the spectral radius of the matrix $\bar{R}$, where $R = I - J$ and $\bar{R} = [\|R_{ij}\|]_{i,j=1}^{n}$, is less than one (i.e., $\rho(\bar{R}) < 1$).*

To facilitate the analysis, we set the initial estimates and messages to be zero, i.e., $\hat{x}_{edge}^{(0)} = 0$, $\gamma^{(0)} = 0$ and $z^{(0)} = 0$. We note that $\hat{x}_{edge}^{(0)}$ and $z^{(0)}$ have to satisfy Equation (16). In order for the algorithm to evolve continuously by following (14)-(16), the minimization problem (11) should be correctly posed for any $t \geq 0$. By working on (11), a sufficient condition can be derived:

$$1 > (1 - s)^2 J_{ij}^2 \gamma_{ij}^{(t)} \gamma_{ji}^{(t)} \quad \forall (i, j) \in E, \tag{17}$$

$$\gamma^{(t)} > 0, \tag{18}$$

where $t = 1, 2, \ldots$. Note that the above two equations only involve the quadratic vector $\gamma^{(t)}$ and J.

Next we argue that if the parameter s is chosen such that the matrix

$$J_s = sI + (1 - s)J \tag{19}$$

is walk-summable, (17)-(18) hold for any $t \geq 1$. Note that J_s is again of unit-diagonal. It is not difficult to show that when $s \in ([1 - 1/\rho(\bar{R})]_+, 1)$, J_s is walk-summable. The operation $\lfloor w \rfloor_+ = \max(0, w)$ for $w \in \mathbb{R}$. From [6], it is known that if J_s is walk-summable, then $\gamma^{(t)}$ converges to a fixed point γ_s^* by following (14). Further,

$$\gamma_s^* \geq \gamma^{(t+1)} \geq \gamma^{(t)} \quad \forall t \geq 0. \tag{20}$$

Considering the fixed point γ_s^* in (17), we have

$$1 > (1-s)^2 J_{ij}^2 \gamma_{\{s,ij\}}^* \gamma_{\{s,ji\}}^* \quad \forall (i,j) \in E$$

$$\overset{(a)}{\Longleftrightarrow} 1 > \sum_{u \in N(i)} (1-s)^2 J_{ui}^2 \gamma_{\{s,ui\}}^* \quad \forall (i,j) \in E, \tag{21}$$

where step (a) follows from (14) and the fact that γ_s^* is stable. (21) holds when J_s is walk-summable [6,7]. By using (20)-(21) and the initialization $\gamma^{(0)} = 0$, it can be easily shown that (17)-(18) hold when J_s is walk-summable. We summarize the result in a lemma below:

Lemma 1. *if s is chosen from $(\lfloor 1 - 1/\rho(\bar{R})\rfloor_+, 1)$ such that J_s is walk-summable and $\gamma^{(0)} = 0$, then the minimization problem (11) is well defined for any $(i,j) \in E$, $t \geq 0$. The quadratic vector $\gamma^{(t)}$ monotonically converges to γ_s^*.*

Besides the quadratic vector $\gamma^{(0)}$, we also have to initialize $\hat{x}_{edge}^{(0)}$ and $z^{(0)}$. Due to the expression (16), we only need to initialize $\hat{x}_{edge}^{(0)}$, the linear vector $z^{(0)}$ can be computed accordingly. If the algorithm converges to the optimal solution as $t \to \infty$, we have

$$\hat{x}_i^{u,(\infty)} = x_i^* \quad \forall u \in N(i) \text{ and } i \in V.$$

For the estimation vector $\hat{x}_{edge}^{(t)}$, $t \geq 0$, we denote its corresponding optimal solution as x_{edge}^*. In practice, one can measure the difference of the estimates $\{\hat{x}_i^{u,(t)}, u \in N(i)\}$ for each variable x_i to terminate the iteration procedure.

To briefly summarize,the min-sum-min algorithm generalizes the min-sum algorithm by introducing the penalty functions. Our goal in this paper is to study whether the min-sum-min algorithm converges for an arbitrary positive definite matrix J by choosing the weighting factor s properly.

3   Convergence of Min-Sum-Min Algorithm

In this section, we study the convergence of the min-sum-min algorithm. We first reformulate the message updating-expressions into vector forms. We then present the convergence analysis for the min-sum-min algorithm.

3.1   Reformulation of the Message Updating-Expressions

In this subsection, we reformulate the two updating expressions (15)-(16) into vector forms. The vector forms provide a big picture of the evolution of the algorithm.

We first consider the evolution of the linear vector $z^{(t)}$. From (15), we have

$$z^{(t+1)} = (1-s)^2 BD^{(t)} y + s(1-s) BD^{(t)} \hat{x}_{edge}^{(t)}$$

$$- (1-s) BD^{(t)} C z^{(t)} \quad t \geq 0 \tag{22}$$

where the matrices $D^{(t)}, B, C \in \mathbb{R}^{|\vec{E}| \times |\vec{E}|}$, and the vector $y \in \mathbb{R}^{|\vec{E}|}$, are given by

$$D_{ij,uk}^{(t)} = \begin{cases} \gamma_{ij}^{(t)} & u = i, k = j \text{ and } [i,j] \in \vec{E} \\ 0 & \text{otherwise} \end{cases}$$

$$B_{ij,uk} = \begin{cases} J_{ij} & u = i, k = j \text{ and } [i,j] \in \vec{E} \\ 0 & \text{otherwise} \end{cases}$$

$$C_{ij,uk} = \begin{cases} 1 & u \neq j, k = i \text{ and } [i,j],[u,k] \in \vec{E} \\ 0 & \text{otherwise} \end{cases}$$

$$y_{ij} = h_i \qquad [i,j] \in \vec{E}.$$

$D^{(t)}$ and B are two diagonal matrices. In particular $\gamma^{(t)} = D^{(t)} \cdot \mathbf{1}$, where $\mathbf{1}$ is the all-one vector. As $\gamma^{(t)} \to \gamma_s^*$ over time, $D^{(t)}$ converges to D_s^*.

Next we consider the evolution of $\hat{x}_{edge}^{(t)}$. By combining (16) and (22), we have

$$\hat{x}_{edge}^{(t+1)} = (1-s)A^{(t)}D^{(t)}y + sA^{(t)}D^{(t)}\hat{x}_{edge}^{(t)}$$
$$-(1-s)A^{(t)}D^{(t)}CBA^{(t)-1}\hat{x}_{edge}^{(t)} \quad t \geq 0, \tag{23}$$

where the matrix $A^{(t)} \in \mathbb{R}^{|\vec{E}| \times |\vec{E}|}$ is given by

$$A_{ij,uk}^{(t)} = \begin{cases} \dfrac{1}{1-(1-s)^2 J_{ij}^2 \gamma_{ij}^{(t)} \gamma_{ji}^{(t)}} & u = i, k = j \text{ and } [i,j] \in \vec{E} \\ \dfrac{-(1-s)J_{ij}\gamma_{ij}^{(t)}}{1-(1-s)^2 J_{ij}^2 \gamma_{ij}^{(t)} \gamma_{ji}^{(t)}} & u = j, k = i \text{ and } [i,j] \in \vec{E} \\ 0 & \text{otherwise} \end{cases}$$

The matrix $A^{(t)}$ converges to A_s^* as $\gamma^{(t)} \to \gamma_s^*$.

Upon obtaining (23), the remaining work is to study under what conditions $\hat{x}_{edge}^{(t)}$ converges to the optimal solution x_{edge}^*. To achieve this goal, we analyze (23) in two steps. In the first step, we consider the extreme case with D_s^* and A_s^* in (23). In this situation, $\hat{x}_{edge}^{(t)}$ can be alternatively expressed as

$$\hat{x}_{edge}^{(t)} = (1-s)A_s^* \sum_{i=0}^{t} [D_s^*(sA_s^* - (1-s)CB)]^i D_s^* y. \tag{24}$$

It is immediate from (24) that if the spectral radius of the matrix $D_s^*(sA_s^* - (1-s)CB)$ is less than 1 (i.e., $\rho(D_s^*(sA_s^* - (1-s)CB)) < 1$), $\hat{x}_{edge}^{(t)}$ converges to a fixed point as $t \to \infty$. We note that at this moment it is unclear if the fixed point $\hat{x}_{edge}^{(\infty)}$ is identical to x_{edge}^*.

In the second step, we consider the overall convergence specified by (23). We assume J_s is walk-summable and the spectral radius of $D_s^*(sA_s^* - (1-s)CB)$ is less than 1. By using the result of Lemma 1, it is known that there exists an integer K such that when $t \geq K$, the spectral radius of $D^{(t)}(sA^{(t)} - (1-s)CB)$ is less than one. This implies that $\hat{x}_{edge}^{(t)}$ in (23) also converges to a fixed point provided with sufficient time. In fact, both (23) and (24) converge to the same fixed point. Due to limited space, we will not provide the proof here. One can refer to Section VI of [6] for a detailed argument on proving a similar result.

Based on the above analysis, we summarize the result in a lemma below.

Lemma 2. *Under the initialization $\gamma^{(0)} = 0$, if the matrix J_s in (19) is walk-summable and the spectral radius of $D_s^*(sA_s^* - (1-s)CB)$ is less than 1, the estimation vector $\hat{x}_{edge}^{(t)}$ converges to a fixed point. In particular, the fixed point is given by*

$$\lim_{t \to \infty} \hat{x}_{edge}^{(t)} = (1-s)A_s^*\left(I - sD_s^*A_s^* + (1-s)D_s^*CB\right)^{-1}D_s^*y. \quad (25)$$

Lemma 2 provides a general sufficient convergence condition for the min-sum-min algorithm. For the situation that the algorithm converges, one natural question is if the fixed point $\hat{x}_{edge}^{(\infty)}$ is identical to the optimal solution x_{edge}^*. To clarify, x_{edge}^* is constructed from x^*, and is of dimension $|\vec{E}|$. We show in the following that $\hat{x}_{edge}^{(\infty)} = x_{edge}^*$ when the algorithm converges. We let $\gamma^{(0)} = \gamma_s^*$ to simplify the argument.

Lemma 3. *Under the initialization $\gamma^{(0)} = \gamma_s^*$, if the matrix J_s in (19) is walk-summable and the spectral radius of $D_s^*(sA_s^* - (1-s)CB)$ is less than 1, the fixed point $\hat{x}_{edge}^{(\infty)}$ in (25) is the same as x_{edge}^**

$$x_{edge}^* = (1-s)A_s^*\left(I - sD_s^*A_s^* + (1-s)D_s^*CB\right)^{-1}D_s^*y. \quad (26)$$

Proof. From Lemma 2, it is clear that when the algorithm converges, the fixed point $\hat{x}_{edge}^{(\infty)}$ is independent of the initial vector $\hat{x}_{edge}^{(0)}$. In other words, any initialization would result in the same fixed point. In order to prove the lemma, we consider a special initialization for the estimation vector. That is $\hat{x}_{edge}^{(0)} = x_{edge}^*$. It is immediate from (10)-(11) and (16) that $\hat{x}_{edge}^{(t)} = \hat{x}_{edge}^*$ for any $t \geq 0$. The optimal solution x_{edge}^* is the fixed point. The proof is complete. □

Remark 3. In fact, one can generalize Lemma 2 by considering more general initializations. See [6] for how to initialize $\gamma^{(0)}$ and $z^{(0)}$. In this paper, we consider the special initialization for simplicity.

3.2 Convergence Analysis

We have known from (6) that the parameter s determines the amount of feedback in computing new messages and estimates. We show in the following that when s approaches to 1, the min-sum-min algorithm converges. We use the Taylor expansions in the argument.

As indicated in Lemma 3, the key point in proving the algorithm convergence is to study the spectral radius of the matrix $D_s^*(sA_s^* - (1-s)CB)$. Note that the two matrices D_s^* and A_s^* take complicated forms while the matrix CB is much simple. We now study the properties of A_s^* and D_s^* in detail. Due to the special structure of A_s^*, its inverse can be easily computed:

$$A_s^{*-1} = I + (1-s)D_s^*H, \tag{27}$$

where

$$H_{\{ij,uk\}} = \begin{cases} J_{ij} & u = j, k = i \text{ and } [i,j] \in \vec{E} \\ 0 & \text{otherwise} \end{cases}.$$

By using (27), the matrix A_s^* can be represented by an infinite series in terms of D_s^*H, which is given by $A_s^* = \sum_{i=0}^{\infty}(-1)^i(1-s)^i(D_s^*H)^i$. By using algebra on the infinite series, we obtain

$$A_s^* = I - (1-s)A_s^*D_s^*H. \tag{28}$$

Similarly, by applying the Taylor expansion on D_s^*, we have

$$D_s^* = I + (1-s)^2 D_s^* P_s, \tag{29}$$

where the matrix P_s is given by

$$P_{\{s,ij,uk\}} = \begin{cases} \sum_{v \in N(i)\backslash j} J_{vi}^2 \gamma_{\{s,vi\}}^* & [i,j] = [u,k] \text{ and } [i,j] \in \vec{E} \\ 0 & \text{otherwise} \end{cases}.$$

Now we are ready to study the matrix $D_s^*(sA_s^* - (1-s)CB)$. By applying (28)-(29), the matrix can be rewritten as

$$\begin{aligned}
D_s^*(sA_s^* &- (1-s)CB) \\
&= D_s^*(sI - s(1-s)A_s^*D_s^*H - (1-s)CB) \\
&= D_s^*\big[sI - (1-s)CB - (1-s)D_s^*H \\
&\quad +(1-s)^2 D_s^*H + s(1-s)^2 A_s^*(D_s^*H)^2\big] \\
&= sI - (1-s)(CB + H) \\
&\quad +(1-s)^2 g(A_s^*, D_s^*, P_s, H, CB),
\end{aligned} \tag{30}$$

where $g(\cdot)$ is a matrix function in terms of the matrices $\{A_s^*, D_s^*, P_s, H, CB\}$. Note that the last term in (30) is of second order of $(1-s)$. Also, as $s \to 1$, γ_s^* converges to $\mathbf{1}$. This implies that the matrices A_s^*, D_s^* and P_s are bounded when $s \in (\lfloor 1 - 1/\rho(\bar{R})\rfloor_+, 1)$. Thus, as $s \to 1$, the last term in (30) can be ignored, which results in

$$D_s^*(sA_s^* - (1-s)CB) \approx sI - (1-s)(CB + H), \quad \text{as } s \to 1.$$

To facilitate the analysis in the following, we denote $Q_s = sI - (1-s)(CB+H)$.

Next we derive the eigenvalues of the matrix Q_s. Denote the eigenvalues of J as $\{\lambda_i > 0, i = 1, \ldots, |V|\}$. We first note that the matrix $CB + H$ takes the form

$$(CB + H)_{ij,uk} = \begin{cases} J_{ui} & k = i \text{ and } [i,j], [u,k] \in \overrightarrow{E} \\ 0 & \text{otherwise} \end{cases}.$$

By relating the matrix $CB + H$ with $R = I - J$, one can show that all the non-zero eigenvalues of $CB + H$ are $\{\lambda_i - 1, i = 1, \ldots, |V|\}$. Finally, the eigenvalues of Q_s are give by

$$\{s + (1 - s)(1 - \lambda_i), i = 1, \ldots, |V|\} \bigcup \{s\}.$$

Using the fact that $\lambda_i > 0$ for all i, it can be shown that when $1 > s > \left\lfloor \frac{\rho(R)-1}{\rho(R)+1} \right\rfloor_+$ (i.e., $R = I - J$), the spectral radius of Q_s is less than 1. Further, as $s \to 1$, all the eigenvalues of Q_s approach to 1. As $\rho(R) \leq \rho(\bar{R})$ (see Corollary 6.3 in [1]), it is immediate that

$$\left\lfloor \frac{\rho(R) - 1}{\rho(R) + 1} \right\rfloor_+ \leq \lfloor 1 - 1/\rho(\bar{R}) \rfloor_+.$$

Thus, we can safely say that when $1 > s > \lfloor 1 - 1/\rho(\bar{R}) \rfloor_+$, the spectral radius of Q_s is less than 1.

The above analysis shows that if s is sufficiently close to 1, the min-sum-min algorithm converges, which we summarize in a theorem below.

Theorem 1. *If the parameter s is sufficiently close to 1 from below, the spectral radius of the matrix $D_s^*(sA_s^* - (1-s)CB)$ is less than 1. Consequently, the min-sum-min algorithm converges to the optimal solution.*

Remark 4. We point out that the matrix Q_s can be used to construct the message-updating expression of the Jacobi-relaxation algorithm [1]. In particular, the expression takes the form

$$\hat{x}_{edge}^{(t)} = \sum_{k=0}^{t-1} Q_s^k y + Q_s^t \hat{x}_{edge}^{(0)}.$$

Compared with Jacobi-relaxation algorithm, the min-sum-min algorithm updates the estimates nonlinearly in terms of the elements of J (see (23)), resulting in the last term in (30).

4   Dynamic Adaption of the Weighting Factor s

We have known thus far that when the weighting factor s is sufficiently close to 1, the min-sum-min algorithm converges to the optimal solution. Right now we cannot provide a fixed support region for s with guaranteed convergence. On the other hand, in practice, we have to choose some value for s. Intuitively

Table 2. The min-sum-min algorithm with dynamic parameter s

Initialization: $\gamma^{(0)} = 0$, $\hat{x}_{edge}^{(0)} = 0$, Flag $= 0$,
$\quad s = \lfloor 1 - 1/\rho(\|R\|_1)\rfloor_+$, $s_{best} = s$, $r_{best} = 1$
repeat{min-sum-min iteration: t=1,2,...}
$\quad$ **if** $r^{(t)}$ is stable AND Flag=0 **then**
$\quad\quad$ **if** $r^{(t)} < r_{best}$ **then**
$\quad\quad\quad$ $r_{best} = r^{(t)}$, $s_{best} = s$, $s = s + 0.1$
$\quad\quad$ **else**
$\quad\quad\quad$ Flag=1, $s = s_{best}$
$\quad\quad$ **end if**
$\quad$ **end if**
until it terminates

---

speaking, if the parameter s is chosen to be very close to 1, the min-sum-min algorithm may take many iterations to reach the stoping criterion, making the algorithm less valuable. This motivate us to dynamically adjust the weighting factor s when running the min-sum-min algorithm.

We now explain how we adjust the weighting factor s in the algorithm. we first compress the estimation vector $\hat{x}_{edge}^{(t)}$ from dimension $\overrightarrow{E}$ to $|V|$. In particular, we compute an estimate $\hat{x}_i^{(t)}$ for each optimal component x_i^* by using $\{\hat{x}_i^{u,(t)}, u \in N(i)\}$:

$$x_i^{(t)} = \frac{1}{|N(i)|} \sum_{u \in N(i)} \hat{x}_i^{u,(t)}. \tag{31}$$

We denote the resulting estimation vector as $\hat{x}^{(t)} = [\hat{x}_i^{(t)}, \ldots, \hat{x}_{|V|}^{(t)}]$, which is of dimension $|V|$.

With the vector $\hat{x}^{(t)}$, we then define a new sequence $\{r^{(t)}, t \geq 2\}$:

$$r^{(t)} = \frac{\|\hat{x}^{(t)} - \hat{x}^{(t-1)}\|^2}{\|\hat{x}^{(t-1)} - \hat{x}^{(t-2)}\|^2}. \tag{32}$$

For a fixed parameter s, the sequence $\{r^{(t)}, t \geq 2\}$ would become stable after a number of iterations. We search for a value of s in $[\lfloor 1 - 1/\|R\|_1\rfloor_+, 1)$ such that the corresponding stable value of the sequence $\{r^{(t)}, l \geq 2\}$ is as small as possible, which we denote as s_{best}. We note that once the value s_{best} is found after a number iterations, it will remain the same in the following iterations.

The pseudo-code of the min-sum-min algorithm with dynamic parameter s is provided in Table 2. The stepsize Δs for searching for the value s_{best} is set to be 0.1. The parameter "Flag" is used to indicate if the value s_{best} has been found or not.

Table 3. Numbers of iterations of the two algorithms for seven pairs of (J, h)

	1	2	3	4	5	6	7
$(|V|, |E|)$	(10,26)	(10,16)	(15,75)	(15,39)	(20,56)	(20,70)	(25, 182)
Jacobi-relaxation	1011	4635	4239	3498	6078	24130	15005
GLiCD	587	398	3050	2276	4568	18775	11516
min-sum-min	330	188	2474	1513	3691	16047	9801

4.1   Experiments with Synthetic Data

We tested the min-sum-min algorithm with the synthetic data. In the implementation of the min-sum-min algorithm in Table 2, we measured the error $|r^{(t)} - r^{(t-1)}|$ for checking the stability of the sequence $\{r^{(t)}, t \geq 2\}$. The threshold for $|r^{(t)} - r^{(t-1)}|$ was set as 10^{-4}.

We also implemented the Jacobi-relaxation and GLiCD algorithms for comparison (see Table 1). The GLiCD also has a free parameter s required to be adjusted in order to guarantee its convergence (see [10]). We adapted a similar procedure of Table 2 to adjust the parameter s for GLiCD. For the Jacobi-relaxation algorithm, it is known that when $s = \lfloor 1 - 1/\|R\|_1 \rfloor_+$, it converges to the optimal solution [1]. Therefore, we fixed the parameter $s = \lfloor 1 - 1/\|R\|_1 \rfloor_+$ in implementing the Jacobi-relaxation algorithm. To terminate the iterations of the three algorithms, the infinite norm between an estimate and the optimal solution was measured. The convergence threshold was set as 10^{-5}.

Seven pairs of (J, h) were randomly generated and tested by the three algorithms. The experimental results are displayed in Table 3. It is seen that the Jacobi-relaxation algorithm performs the worst in terms of number of iterations for all the seven optimization problems. Conversely, the min-sum-min algorithm performs the best. This might be because the quadratic messages carry more information than the linear messages (see Table 1 for the algorithmic comparison).

4.2   Experiments with Real Data

We also tested the three algorithms for the J matrices downloaded from the Matrix Market website repository [11], where the matrices originated from some real applications. The vector h in (1) were randomly generated. The implementation of the three algorithms were the same as for the synthetic data.

Fig. 1 displays the performance results of the three algorithms for two particular J matrices (one is of size 48×48 and the other one is of size 468×468). The min-sum-min and GLiCD algorithms converges significantly faster than the Jacobi-relaxation algorithm. This may be because the Jacobi-relaxation algorithm only involves linear updates of the estimates while the other two algorithms apply nonlinear updates of the estimate (See Remark 4 and [10]). Also we have observed that for the J matrix of size 48×48, the convergence speeds

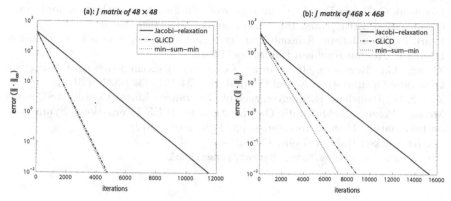

Fig. 1. Performance comparison for J matrices downloaded from [11]

of the min-sum-min and GLiCD algorithms are quite similar. For the above particular case, the GLiCD algorithm is favorable because it only transmits linear messages within the graph, thus saving half of transmission energy required for the min-sum-min algorithm. Other matrices were also tested and similar results were obtained.

5   Conclusion

In this paper, we have proposed the min-sum-min algorithm for the quadratic optimization problem. The min-sum-min algorithm parallels with the Jacobi-relaxation and GLiCD algorithms (See Table 1). Also we have studied the convergence of the min-sum-min algorithm. We have shown that if the feedback signal is set to be large enough (i.e., the parameter s is close to 1), the min-sum-min algorithm converges to the optimal solution. Experimental results show that the min-sum-min algorithm is advantageous over the Jacobi-relaxation and GLiCD algorithms in terms of the convergence speed.

References

1. Bertsekas, D.P., Tsitsikis, J.N.: Parallel and distributed Computation: Numerical Methods. Athena Scientific, Belmont (1997)
2. Bishop, C.M.: Pattern Recognition and Machine Learning. Springer (2007)
3. Johnson, J.K., Malioutov, D.M., Willsky, A.S.: Walk-sum Interpretation and Analysis of Gaussian Belief Propagation. In: Advances in Neural Information Processing Systems, vol. 18. MIT Press, Cambridge (2006)
4. Johnson, J.K., Bickson, D., Dolev, D.: Fixing Convergence of Gaussian Belief Propagation. In: The International Symposium on Information Theory (2009)
5. Malioutov, D.M., Johnson, J.K., Willsky, A.S.: Walk-Sums and Belief Propagation in Gaussian Graphical Models. J. Mach. Learn. Res. 7, 2031–2064 (2006)
6. Moallemi, C.C., Van Roy, B.: Convergence of Min-Sum Message Passing for Quadratic Optimization. IEEE Trans. Inf. Theory 55(5), 2413–2423 (2009)

368 G. Zhang and R. Heusdens

7. Moallemi, C.C., Van Roy, B.: Convergence of Min-Sum Message Passing for Convex Optimization. IEEE Trans. Inf. Theory 56(4), 2041–2050 (2010)
8. Pearl, J.: Probabilistic Reasoning in Intelligent Systems: Networks of Plausible Inference. Morgan Kaufman Publishers (1988)
9. Zhang, G., Heusdens, R.: Linear Coordinate-Descent Message-Passing for Quadratic Optimization. Neural Computation 24(12), 3340–3370 (2012)
10. Zhang, G., Heusdens, R.: Convergence of Generalized Linear Coordinate-Descent Message-Passing for Quadratic Optimization. In: IEEE International Symposium on Information Theory Proceedings, pp. 1997–2001 (2012)
11. Matrix Market: Harwell Boeing Collection, http://math.nist.gov/MatrixMarket/index.html

Clustering Image Search Results by Entity Disambiguation

Kaiqi Zhao, Zhiyuan Cai, Qingyu Sui, Enxun Wei, and Kenny Q. Zhu

Department of Computer Science & Engineering
Shanghai Jiao Tong University, China
{kaiqi_zhao,luckyvega}@163.com,
{sqybilly,weienxun}@gmail.com, kzhu@cs.sjtu.edu.cn**

Abstract. Existing key-word based image search engines return images whose title or immediate surrounding text contains the search term as a keyword. When the search term is ambiguous and means different things, the results often come in a mixed bag of different entities. This paper proposes a novel framework that understands the context and thus infers the most likely entity in the given image by disambiguating the terms in the context into the corresponding concepts from external knowledge in a process called conceptualization. The images can subsequently be clustered by the most likely associated entities. This approach outperforms the best competing image clustering techniques by 29.2% in NMI score. In addition, the framework automatically annotates each cluster of images by its key entities which allows users to quickly identify the images they want.

1 Introduction

Images are one of the most abundant multimedia resources on the Web. Most commercial search engines offer image search today, which enables the user to retrieve images by search terms. By default, all existing image search engines rank the returned images by the relevance of their contexts (i.e. the web pages they are embedded in) to the query keywords. Fig. 1 shows the result for searching "bean" on *Google Image* in October 2013.The result appears to be a random mix of many different entities related to the keyword "bean", e.g., "Mr. Bean (comedian)", "Sean Bean (actor)", "beans (crop)", etc. Ambiguous search terms like this are not rare: Google Image returns at least two different entities for "kiwi", three for "explorer", and over ten different persons named "Jerry Hobbs"!

This paper is concerned with the problem of clustering web images according to the entity or concept they represent. Once the images are clustered, the search engine can return the *original* set of search results classified by distinct entities, offering easier accessibility and more diversity. Note that a separate but different problem [21,22] is mapping images to an entity in a knowledge base like Wikipedia or YAGO [20]. That is a different problem because 1) the entity is unique and known in advance, so its features in the knowledge base can be used for retrieving images whereas our problem does not

---

** Kenny Q. Zhu is the contact author and is supported by NSFC Grants 61100050, 61033002, 61373031 and Google Faculty Research Award.

T. Calders et al. (Eds.): ECML PKDD 2014, Part III, LNCS 8726, pp. 369–384, 2014.

Fig. 1. Search Result of "bean" on Google Image

assume known entities *a priori*; 2) the goal is to rank the relevant images to an entity while our problem is a clustering problem.

In the past, there have been numerous research efforts on image clustering. These efforts can be roughly divided into three categories: visual-based, context-based and hybrid approaches.

Visual-based methods only take into account visual features such as SIFT descriptors, edge histogram, color and contrast[11,27], and these are often insufficient for distinguishing real entities. For example, some images of *Mr. Bean* in Fig. 1 are very different by the look, while other images of *Mr. Bean* and *Sean Bean* are fairly similar as they both wear suits. On the other hand, high level visual object recognition techniques[17,15] focus on detecting objects like bottle, dog, grass, etc. in an image, but are not powerful enough to distinguish entities.

Context-based methods use only textual information in the context of the image. Here context refers to URL, descriptive tags for the image, the surrounding text and even search result snippets [14]. To represent the context, all previous work uses bag-of-words or n-grams model [14]. The bag-of-words (BOW) model can not capture the semantics of the context in an accurate way for three reasons. First, limited length of context provide insufficient signals in words model. Second, terms with one or more words are sometimes better semantic units than single words but they are not handled properly by BOW models. Finally, words can be ambiguous. "Apple" may refer to an IT company or a kind of fruit, but BOW model treats all "apple" terms equally. Similar arguments hold for n-gram models.

Hybrid approaches attempt to combine the visual features with textual features. However, semantic gaps between the visual and textual features make it difficult to directly combine them into one uniform similarity measure. Some hybrid algorithms therefore resort to co-clustering on visual and text simultaneously such as MMCP [11]. But such approach is iterative, time consuming and thus not suitable for online applications such as image search.

Fig. 2. Partial Search Result for "bean" on Prototype System

In this paper, we propose a new context-based approach that emphasizes on understanding textual signals. The reason to focus on text is that, we believe, unlike visual signals, textual signals from the right context explicitly reveal the semantics of the image. Our approach is different from the existing context-based image clustering in three aspects. First, we explicitly disambiguate the context text by converting each phrase to an unambiguous concept from an external knowledge source such as Wikipedia. We call this process "conceptualization". Conceptualization has been previously shown to be a better way to understand textual signals than bag-of-words model[19]. Second, our method provides concept labels to annotate each cluster of images by accumulating the concepts in the contexts from the clusters. With these labels, users can conveniently grasp what each cluster is about. Third, we propose a modified version of hierarchical agglomerative clustering (HAC) in a tri-stage clustering framework, which is more robust to noise. This framework guarantees the purity of each cluster while improving the inverse purity, i.e. forming as large clusters as possible. The experimental result shows that our approach significantly outperforms competing algorithms, and achieves very high purity, F-measure and NMI scores. A partial result of searching for "bean" on our prototype image search system is shown in Fig. 2. Every cluster shows the most relevant images about a distinct entity, and each cluster is labeled with the 5 concepts which are most related to the entity. The four clusters in Fig. 2 have been correctly identified as *Mr Bean, Sean Bean, Frances Bean Cobain* and *Phaseolus vulgaris* (the official name for "common bean").

The rest of the paper is organized as follows. Section 2 presents the structure and each component of our framework; Section 3 demonstrates the experimental results; Section 4 introduces some related work while Section 5 concludes the paper.

2   Framework

In this section, we introduce a novel image clustering framework based on conceptualization of contexts. Our input is an image search query and a set of images returned by this query along with their hosting HTML pages. Our output is a number of clusters of images, each containing images of the same entity and each tagged with a concise list of most relevant concepts. For example, the first cluster of Fig. 2 is tagged with "Mr. Bean", "Rowan Atkinson", etc.

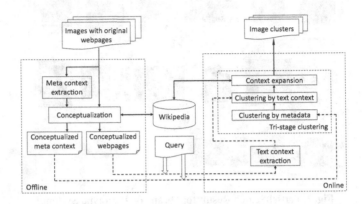

Fig. 3. The Architecture of Image Clustering by Conceptualization

The architecture of our framework is shown in Fig. 3. The framework is divided into two parts: online and offline components. The offline components extract the meta data of the image and conceptualize all of the text in the source page. Online components 1) extract the surrounding text context of the image and query from the conceptualized source page and then use concepts in the context to construct the concept vector representation of the image context; and 2) cluster the images using a tri-stage clustering algorithm. The context extraction process is online because it cannot be done before the query is known. Next, we present each component in more detail.

2.1   Context Extraction

This paper concerns two kinds of image context, meta data context and text context. Meta data context extraction is an offline process while text context is extracted online.

Meta data context (or meta context in short) are all intrinsic attributes of the image, such as the anchor text of the image (i.e., ALT attribute in image tags) in the web page, the URL of the image. The domain and the file extension in the URLs are ignored because they are less relevant to entity in the image. For example, images from Flickr share the same domain but are not the same entity. We split the URL into "words" by directory separators, special characters or letter case conversion (e.g., from lower to upper case) to get context from URL. In some cases, the URL may contain randomly generated strings:

http://domain.com/**53C316-C2oJ5/AppleInc_2012**.jpg

contains these words: "53C316", "C2oJ5", "Apple", "Inc" and "2012". Here, "53C316" and "C2oJ5" has no clear meanings, while "Apple", "Inc" and "2012" are understandable. We extract all 3-grams in each word, such as "C2o", "2oJ" and "oJ5" in "C2oJ5", and "App", "ppl" and "ple" in "Apple". Each 3-gram corresponds to one feature of this word. Then we learn an L2-SVM model using LIBLINEAR [8] to classify these words and filter out meaningless ones with an accuracy of 95.69%. Note that, using a lexicon such as Wikipedia only does not work because simple strings like "5" or "J" are also valid terms.

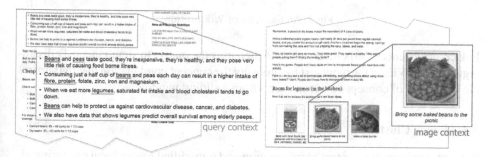

Fig. 4. Image Context and Query Context

Text context is the surrounding plain text of both the image and the query terms in the web page. The reason we employ query context in addition is that the context surrounding the image is likely to be an accurate description of that image but not always enough to distinguish different entities. As Fig. 4 shows, the image context contains limited amount of information. A great deal of signals for identifying "bean" such as "pea (a kind of bean)", "legume (the family that bean belongs to)", "fibre (major ingredient of bean)" and "protein (major ingredient of bean)" can otherwise be found in the query context part. We extract the relevant context by a sibling based method [1]. It retrieves all text nodes which contain the query terms, as well as their sibling nodes in the Document Object Model (DOM) tree of the page.

2.2 Conceptualization of Context

Wikipedia is a rich and comprehensive knowledge source of concepts. Each concept (e.g. *Mr. Bean* or *Phaseolus vulgaris*) has a descriptive article. The goal of conceptualization based on Wikipedia is to convert a piece of plain text into a set of Wikipedia concepts. To achieve this, we need to recognize the multi-word expressions (MWEs)[1] in the text and then disambiguate them by linking each of them to a corresponding Wikipedia article/concept. Fig. 5 shows an example of conceptualization, where "Polar Bear" is recognized as an MWE and correctly linked to the "Snow Patrol" [2] article.

In this paper, we adopt a conceptualization approach known as *wikification* [5] which is based on link co-occurrence in Wikipedia corpus. The technique first constructs a

---

[1] MWE is any term that contains one or more words.
[2] Snow Patrol is a Scottish rock band.

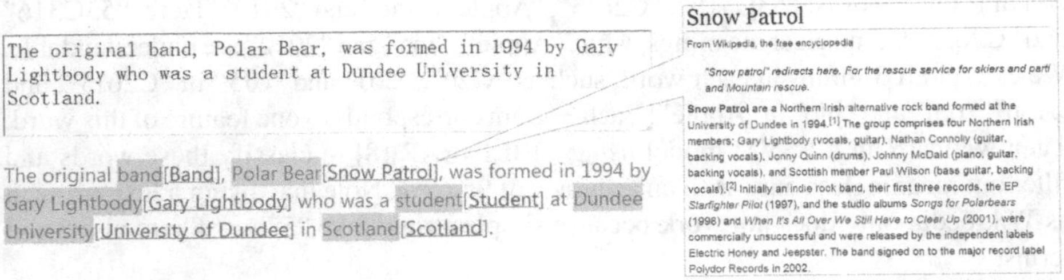

The original band, Polar Bear, was formed in 1994 by Gary
Lightbody who was a student at Dundee University in
Scotland.

The original band[Band], Polar Bear[Snow Patrol], was formed in 1994 by
Gary Lightbody[Gary Lightbody] who was a student[Student] at Dundee
University[University of Dundee] in Scotland[Scotland].

Snow Patrol

From Wikipedia, the free encyclopedia

"Snow patrol" redirects here. For the rescue service for skiers and parti and Mountain rescue.

Snow Patrol are a Northern Irish alternative rock band formed at the University of Dundee in 1994.[1] The group comprises four Northern Irish members; Gary Lightbody (vocals, guitar), Nathan Connolly (guitar, backing vocals), Jonny Quinn (drums), Johnny McDaid (piano, guitar, backing vocals), and Scottish member Paul Wilson (base guitar, backing vocals).[2] Initially an indie rock band, their first three records, the EP Starfighter Pilot (1997), and the studio albums Songs for Polarbears (1996) and When It's All Over We Still Have to Clear Up (2001), were commercially unsuccessful and were released by the independent labels Electric Honey and Jeepster. The band signed on to the major record label Polydor Records in 2002.

Fig. 5. An Example of Wikification

link co-occurrence matrix iteratively, and then uses the matrix to simultaneously disambiguate all MWEs in the input text by choosing the concept combination that maximizes the likelihood of concept co-occurrence within a sliding window.

2.3 Image Clustering

We first introduce the context representation and a modified hierarchical clustering algorithm. We then propose a tri-stage clustering framework.

Context Representation. With concepts extracted from the context, we can draw a concept histogram for each image, which represents the image's semantic information. We use the *vector space model* (VSM) to represent the context. We define a CF-IDF score for each dimension in the concept vector of a textual context. The CF-IDF score of the concept c in context d's concept vector is adapted from the well-known TF-IDF score in information retrieval, and is defined as:

$$\text{CF-IDF}(c,d) = CF(c,d) \times log \frac{|D|}{DF(c)}, \tag{1}$$

where $CF(c,d)$ is the concept frequency of c in d, $|D|$ is the total number of Wikipedia articles from which we compute the document frequency of each concept while $DF(c)$ is document frequency of c. We compute the document frequency of c by counting the number of documents which have links to c.

HAC with Cluster Conceptualization. We apply *cosine similarity* to compute the pairwise similarity of contexts. We use a modified HAC algorithm to cluster the contexts. There are two reasons for using HAC: First, we don't know the exact number of clusters in advance, but we can specify a threshold for minimal similarity within a cluster. Second, HAC is an agglomerative algorithm that merges similar clusters incrementally. Therefore we are able to extend the algorithm by incorporating different features at any step of the clustering process.

There are four common ways to compute similarity between two clusters in HAC: *Single-link, Complete-link, Group Average, Centroid*. These methods compare the individual data points in each cluster without considering each cluster as a whole. This paper adopts a new method to compute cluster similarity. It summarizes the semantic information in each cluster by building a concept histogram for each cluster. Specifically, given a cluster C with n image contexts, $d_1 \ldots d_n$, the weight of concept c in the concept vector for C is

$$V(C)\{c\} = \sum_{d \in C} \text{CF-IDF}(c, d) \qquad (2)$$

To restrict the size of this concept vector and to avoid noise, we keep only top K concepts with the highest weights. The selected concepts and their weights thus represent the semantics of the cluster. This process is called *cluster conceptualization*. The complete HAC with cluster conceptualization (HAC_CC) is shown in Algorithm 1. D is the set of images, Π is the set of resulting clusters, N is the number of images, C_i is an image cluster, $V(C)$ is the concept vector of a cluster C, Sim is the function computing the cosine similarity of the two vectors, S is the similarity matrix of images, and τ_t is the threshold that controls the clustering granularity. Line 9 to 15 merge two most similar clusters each time.

Algorithm 1. HAC with Cluster Conceptualization (HAC_CC)

Input: Set of images D
Output: Image cluster Π
1: **function** HAC_CC(D)
2:     $\Pi \leftarrow \{C_i = \{d_i\} \,|\, d_i \in D\}$
3:     **for** $i \leftarrow 1$ to N **do**
4:       **for** $j \leftarrow i + 1$ to N **do**
5:         $S[i, j] \leftarrow Sim(V(C_i), V(C_j))$
6:       **end for**
7:     **end for**
8:     **for** $iter \leftarrow 1$ to $N - 1$ **do**
9:       $max\_sim = \max_{i<j} S[C_i, C_j]$
10:       **if** $max\_sim < \tau_t$ **then**
11:         **return** Π
12:       **end if**
13:       $C_i, C_j \leftarrow argmax_{C_i \neq C_j} S[C_i, C_j]$
14:       $C_i \leftarrow$ **Combine**(C_i, C_j, S)
15:       $C_j \leftarrow \emptyset$
16:     **end for**
17:     **return** Π
18: **end function**
19: **function** COMBINE(C_i, C_j, S)
20:     $V \leftarrow V(C_i) + V(C_j)$
21:     $V(C_i) \leftarrow top\ K\ concepts\ of\ V$
22:     **for** $m \leftarrow 1$ to N **do**
23:       **if** $m > i$ and $m \neq j$ **then**
24:         $S[i, m] \leftarrow Sim(V(C_i), V(C_m))$
25:       **else if** $m < i$ and $m \neq j$ **then**
26:         $S[m, i] \leftarrow Sim(V(C_i), V(C_m))$
27:       **end if**
28:     **end for**
29:     **return** $C_i \cup C_j$
30: **end function**

The advantage of this method is, we can boost the important signals while ignoring noisy ones. On the other hand, since we just keep K concepts, both cluster similarity and the generation of cluster histogram can be computed in constant time, while HAC using *Group Average* or *Centroid* has a quadratic time complexity to the cluster size.

Similar to the original HAC algorithm, Algorithm 1 has a time complexity of $O(N^3)$ [3]. We can further optimize it to $O(N^2 \log N)$ by using a sorted priority queue to store the rows of the semantic matrix S in line 5, With this optimization, the operation of finding two most similar clusters (line 9) is reduced from N^2 to constant time, and the overall complexity only depends on the sorting process which costs $O(N^2 \log N)$.

---

[3] Strictly speaking, it is $O(K^2 N^3)$, but $K \ll N$ so it is treated as a constant.

Tri-stage Clustering. Generally speaking, meta context is the most reliable image context since it is guaranteed to be related to the image, whereas the text context may contain noise. As such, we use these two kinds of context at different stages of clustering. Further, to remedy insufficient signals, we expand the contexts by using additional information from Wikipedia, and perform the third stage of clustering. The above stages form a tri-stage clustering algorithm which includes *meta context clustering*, *text context clustering* and *expansion clustering*.

In the first stage, we construct the concept vector of each image using the concepts extracted from the URL and anchor texts, and apply the HAC_CC algorithm on the images. Although the signals from meta data are reliable, useful signals are limited. Thus, many small clusters are formed with very high purity.

In the second stage, we merge the concept vector extracted from the text context into the concept vector of meta context for each image and combine all the vectors for each cluster from stage one to obtain the cluster vectors (Eq. (2)). We again apply HAC_CC algorithm on these new cluster vectors. Only top 50 concepts in each resulting cluster are kept to filter out the noise.

The final stage takes as input the clusters formed in the second stage, and expands the context of each cluster in an attempt to merge some of the clusters which should have been together. For each of the top K concepts in a cluster, we extract the top 50 concepts (ranked by CF-IDF) from the Wikipedia article of that concept, and replace the concepts in the previous stage with them. The weight of the concept c in the new vector $V'(C)$ is defined as:

$$V'(C)\{c\} = \sum_{c_i \in V_C} (V(C)\{c_i\} \times \text{CF-IDF}(c, d_{c_i})), \tag{3}$$

where V_C is the previous concept vector of cluster C, c_i is one of the concept in V_C, and d_{c_i} is the Wikipedia article of c_i. After reconstructing the new concept vector, HAC_CC is again applied to form the final clusters.

When the third stage finishes, we rank the concepts (dimensions) in the aggregated concept vector of each cluster by the values and use top concepts to represent the semantics of that image cluster. The complexity of the tri-stage clustering algorithm remains the same as HAC_CC algorithm because the input size of each stage is bounded by the total number of images.

2.4 Use Scenario

Our framework has an online component because the query terms, which are important signals for context extraction, must be processed at runtime. Although the clustering algorithm presented earlier has a non-linear time complexity, the following use case of our framework is typical and practical. User enters a search term and the search engine returns a number of relevant images on page-by-page display. On any given page, the user can choose to "order by entity", and the clustering framework will re-organize the results on that page (typically a few tens to several hundred images) by entities, as shown in Fig. 2. This is practical because, as we will show later, the online part of the algorithm completes within a second for 100 images.

3   Experimental Results

This section evaluates the image clustering system. We first present the experiment set-up and evaluation metrics. Then, we show four experiments. The first experiment evaluates the performance of each key component of our system. The second one gives an end-to-end comparison between our approach and the state-of-the-art systems. The third one illustrates the accuracy of concepts generated by our system for each cluster. The last one evaluates the time efficiency of the system.

3.1   Experiment Setup

We prepare an image data set from Google Image Search, sorted by relevance. We select a list of 50 ambiguous queries as shown in Fig. 6 (10 for parameter training and 40 for testing). For each query, we query in Google Image and download the top 100 images returned by Google with the original web pages of the images. This data set contains a total of 5,000 web pages/images. We then ask two human judges to manually cluster the collected data to create two label sets. All evaluation metrics computed in subsequent experiments are the averaging values over these two sets. All experiments were run on a dual-core Intel i5 machine with 14GB memory.

barcelona, berry, curve, david walker, diff, george foster, john smith, longhorn, manchester, puma

acrobat, adam, amazon, anderson, andrew appel, apple, arthur morgan, bean, british india, carrier, champion, eclipse, emirates, explorer, focus, friends, jaguar, jerry hobbs, jobs, kiwi, lotus, malibu, morgan, nut, palm, patriot, perfume, pluto, polo, santa fe, shell, sigma, studio one, subway, taurus, tick, tucson, venus, visa, wilson

Fig. 6. Queries for training(above) and testing(below)

3.2   Evaluation Metrics

We adopt three well-known metrics to measure the result of image/document clustering: *Purity*, *NMI* and F_1. Purity measures the intra-cluster accuracy. It has an obvious drawback that if we create one cluster for each document, the Purity will be 1, and this is not useful at all. Therefore, Purity should not be viewed independently. NMI (Normalized mutual information) is a better measure that balances the purity of the clusters with the number of clusters. It measures the amount of common information between the computed clusters and the ground truth. Another measure of clustering is F_1 score, which combines Purity and *Inverse Purity*. Inverse purity exchanges the position of the result and the ground truth in the the purity computation, and determines how much of each cluster in the ground truth is correctly clustered together. Similar to the F_1 score used in information retrieval task, F_1 score is computed as:

$$F_1(C, L) = \frac{2 \cdot Purity(C, L) \cdot Purity(L, C)}{Purity(C, L) + Purity(L, C)}, \qquad (4)$$

where C is the clustering result and L is the ground truth clusters. In many studies of clustering algorithms, NMI is more important and sometimes the only measure, because it's extremely difficult to achieve high NMI scores.

3.3 Threshold of Tri-stage Clustering

The tri-stage clustering (TSC) algorithm is based on HAC_CC algorithm. Similar to traditional HAC algorithm, HAC_CC has a threshold to control the granularity of the clustering result. We tune different threshold τ_t of HAC_CC on a training data collected from top 100 images of 10 different queries. Cluster labels are assigned to each image by human judges. Fig. 7 shows the clustering result on different thresholds of HAC_CC. We prefer to choose a threshold which can ensure high purity, F1 and NMI at the same time. NMI reaches a peak value at $\tau_t = 0.15$. At this threshold, the purity is significantly higher than when $\tau_t = 0.1$ and F1 score is relatively high, too. Consequently, in this system, we set τ_t to be 0.15.

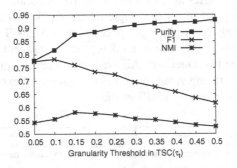

Fig. 7. Clustering Result on Different τ_t

3.4 Evaluation on Key Components

In this sub-section, we experiment on different variants of our system. First, we investigate the effects of different context extraction methods. Then we show the performance of concept representation based on conceptualization. Finally, we show the benefits of tri-stage hierarchical clustering.

Context Extraction: There are three variants of context: the whole page (Page), surrounding text of the image (Image) and surrounding text of both the image and query terms (I & Q). The window size of the surrounding text is empirically set to 200 words (100 words before and after the query/image respectively). Table 1(a) compares the end-to-end results of image clustering on 20 different queries using these three types of context. One can stipulate that the noise in whole page contexts adversely affect the purity of the clusters. Even though the surrounding text of the images already gives rise to very pure clusters, adding the query context gives better F1 and NMI. Overall, the text context of both image and query terms wins because of superior cluster accuracy at limited computation overhead.

Context Representation: We implement two baseline systems to compare with our concept vector (CV) model. One of them uses bag-of-words(BOW) model and the other one uses bag-of-phrase(BOP) model. The latter is a minor enhancement to BOW, and uses (possibly ambiguous) MWEs instead of single words to represent the context. Different from these two baselines, our system disambiguates MWEs in the context to

Table 1. Comparison on Key Components

(a) Diff. Contexts

	Purity	F1	NMI
Page	0.71	0.78	0.35
Image	**0.91**	0.80	0.59
I & Q	0.90	**0.81**	**0.62**

(b) Diff. Representations

	Purity	F1	NMI
BOW	0.92	0.54	0.48
BOP	**0.94**	**0.62**	0.50
CV	**0.94**	**0.62**	**0.55**

(c) Diff. Algorithms

	Purity	F1	NMI	Time
AP	0.92	0.55	0.50	1.9s
HAC	**0.94**	0.62	0.55	0.9s
HAC_CC	**0.94**	0.76	0.59	0.7s
TSC	0.90	**0.81**	**0.62**	1.1s

generate a more accurate representation. To make the end-to-end results comparable, we apply HAC on all three types of representations, since the tri-stage clustering algorithm is only applicable to our CV model. Table 1(b) shows the comprehensive clustering results. This experiment shows that the BOP/CV representations are much more effective than BOW, with particular improvement in F1 score. Phrases are more accurate to identify the semantics of text than single words. CV beats BOP on NMI because it disambiguates the MWEs in the context and thus makes the similarity computation between two images more accurate.

Tri-stage Clustering: We compare HAC_CC and TSC with HAC and Affinity propagation (AP), two very popular clustering algorithms. In this experiment, all algorithms use the concept vector representation. Except for TSC which clusters in three stages, all other algorithms run one time only. The threshold τ_t of HAC and HAC_CC is set to 0.15, while the preference of AP is set to the average similarity between the data points. We also report the time cost of the algorithms by averaging 5 independent runs in the same setting. The result of these three algorithms are shown in Table 1(c). HAC_CC algorithm outperforms AP and HAC due to the enhancement of strong signals and removal of noise in cluster conceptualization process. TSC further improves HAC_CC with concept expansion because 1) we make use of meta context, and 2) the previous clustering stage provides accurate cluster vectors as input to the next stage to further reduce the influence of noise. The experiment demonstrates TSC's capability of boosting important semantic signals which substantially helps improve the accuracy of web image clustering.

3.5 End-to-end Accuracy

We compare our approach (TSC) with two image clustering systems and two text clustering systems from the literature (See Table 2). The first image clustering system is implemented following Cai's [3] approach, which extracts image context using VIPS [4]. The second image clustering system is the multi-modal constraint propagation approach (MMCP) [11]. We also compare with text clustering systems as baselines because our approach only extracts text features from the image context and therefore can be considered as text clustering as well. The two text-based methods that we compare with are HAC clustering on bag-of-words (BOW) and HAC clustering of topics extracted by LDA[2], and both are input with the same text context used in our algorithm, i.e., meta context and text context concatenated in one blob.

Cai's system used visual features, textual features (context), and an image link graph. They used Color Texture Moments[25] as visual features and bag-of-words in the visual context as textual features. We replicate the link graph from a subset of source pages without obtaining the entire set of web pages, according to the property described by Cai. For MMCP, we apply the same modalities mentioned by Fu: local visual, global visual and text. Fu used tags of the images in Flickr as the textual features. However, without available tags, we instead use the bag-of-words in the source page of the image.

The two text clustering systems use different representations for the text context (i.e., BOW and topics) to compute the similarity between two image contexts, and then use HAC algorithm to cluster the contexts. In the LDA system, we directly extract topics in the test data. The parameters of each system are tuned to the one that maximizes the NMI score in the training data. The clustering threshold τ_t is set to 0.2 in BOW baseline and 0.25 in LDA. The number of topics for LDA is set to 150.

The four competing systems generally do not have a good way of handling noise, which is often seen in the contexts of web images. The noise usually dilutes the positive impact of the important signals, especially when the context is of limited size. Our conceptualization and tri-stage clustering method can help remove some of the noise. Some systems like MMCP intends to obtain high NMI score, but their purity is very low. The BOW system achieves the highest purity because of the exact match of the words in the context, but otherwise has a low F1 score. In contrast, the LDA system has some degree of generalization which makes it perform better than BOW in F1 scores. However, LDA failed to capture high quality topics for images that have very short and noisy contexts. Consequently, it has relatively poor purity. Over all, our approach outperforms other systems by producing bigger clusters while preserving the high purity in each clusters. It defeats the best of the peers by significant margins: **17.4%** by F_1 and **29.2%** by NMI score.

Table 2. Results of End-to-End Image Clustering

	Purity	F1	NMI
Cai	0.60	0.71	0.10
MMCP	0.74	0.58	0.34
BOW+HAC	**0.92**	0.54	0.48
LDA+HAC	0.88	0.60	0.44
TSC	0.90	**0.81**	**0.62**

3.6   Cluster Conceptualization Accuracy

In this subsection, we show the conceptualization result on the test queries. To quantify the accuracy of conceptualization on all 40 test queries, we manually label the results in the following manner. For the top 5 clusters of each query, we pick top ten ranked concepts for each cluster and judge whether the concept is relevant to the images in the cluster by human. This results in around 2000 concepts to be labeled. Each query is labeled by three persons and the accuracy for each image clusters is averaged on the judgement from the three persons. Formally, the accuracy of conceptualization of an image cluster is defined in Eq. (5).

$$Accuracy(C) = \frac{1}{M} \sum_{i=1}^{M} \frac{1}{|C|} * \sum_{c \in C} f_i(c), \qquad (5)$$

where C is the set of concepts for an image cluster, M is the number of the human judges ($M = 3$ in our experiment), and f_i is the judgement of the ith judge. If concept c is labeled as relevant to the cluster, $f_i(c) = 1$, otherwise $f_i(c) = 0$. We average the accuracy of all clusters on the test queries, and the final result is **71.82%**.

Table 3 shows some examples of our conceptualization results. For each query, we show only the first two clusters as well as the most related concepts generated from different entities. Terms listed under the images are 5 top-ranking Wikipedia concepts that are conceptualized from each image cluster. Each of the concept has a corresponding Wikipedia article. For example, the concept "Kiwi" in Wikipedia is the bird kiwi, while "Kiwifruit" refers to the fruit kiwi.

Table 3. Conceptualization of Image Clusters (Adam, Eclipse, Kiwi)

Query	Cluster
Adam	Adam Lambert, American Idol, God, Kris Allen, Privacy policy
	Adam Levine, Hijab, Mehndi, Fashion, Hairstyle
Eclipse	Solar eclipse, Sun, Moon, Lunar, Umbra
	Twilight (series), Bella, David Slade, Vampire, Stephenie Meyer
Kiwi	Kiwifruit, Fruit, Recipe, Health benefit, New Zealand
	Kiwi, Bird, New Zealand, Egg, Smithsonian National Zoological Park

3.7   Time Efficiency

First, We evaluate the time cost of the online and offline components in our system. The results are averaged over 5 independent runs, on the 40 test queries. The average execution time per query (with 100 images to cluster) of offline and online components are 471 seconds and 1 second, respectively. The off-line component consists of image context extraction, chunking, and conceptualization, of which conceptualization is the most expensive process. The current offline-online split of the system effectively pushes the most time consuming work to the preprocessing stage and thus makes the online part more efficient and practical.

Second, we compare the average online clustering time of our system (1121 ms) with MMCP (5021 ms) and Cai's system (194 ms). All timing results are averaged over 5 independent runs. MMCP propagates the constraints among modalities. This process clusters on each modality for several times, which explains its long execution time (5 seconds). With all features extracted off-line, Cai's system only need spectral clustering on the images online, which explains why it is the winner here. However, the VIPS extraction module of Cai's system relies on the browser rendering module and crashes frequently. It is almost impossible to automate the context extraction process without human intervention. Our prototype system, which is not optimized, runs for around 1 second per query on average. It is slower than Cai's since we need to extract the query context online, and the expansion of concepts is also time consuming. However, with accuracy, efficiency and reliability all considered, our system is an overall winner in practical web image search tasks.

4   Related Work

We divide existing image clustering methods into three categories: content-based, context-based and the combined approaches.

Content-based image clustering approaches [10,7,13] rely on visual signals. For example, Fu et al.[11] gave a constraint propagation framework for multi-model situations. They constructed multiple graphs, one for each visual modalities such as color histogram, SIFT descriptors [18], etc. The nodes are images while the edges are similarities between the images by a particular visual modality. A random walk process is employed on these graphs. All of the above work uses low-level visual signals of images such colors, gray scales, contrasts, patterns, etc. These signals are insufficient to capture high level semantics of the images. This is evident from our experiments on Fu's algorithm which heavily relies on basic visual signals. There has been some development on high level visual object recognition and semantic annotation [17], but even the state-of-the-art techniques in this area suffer from low accuracy and unreliability.

With the difficulty in content-based clustering, some researchers turn to signals coming from the context of the images, such as file name, alternate text and surrounding text. Cai et al. made some progress in this respect. They represented a web page segmentation algorithm named VIPS [4], which works by rendering the web page visually and detecting the important visual blocks in the page. And they subsequently proposed three kinds of representations for images [3]: visual feature based representation, textual feature based representation and link graph based representation, and proposed a two-level clustering algorithm which combined the latter two. Jing et al. [14] introduced a novel method named IGroup for image clustering. Instead of clustering on returned images directly, they first search the query on normal web search engine, and cluster the titles and snippets from the search results. They then construct a new query string to represent each of the cluster, and send these query strings to the image search engine to get images for each cluster. To construct the query string, they used an algorithm proposed by

Zeng[26]. These bag-of-words approaches are inadequate for understanding the semantics of the context. Relying on bag-of-words or n-grams can easily confuse noise with meaningful signals. Our approach, on the other hand, leverages co-occurrence information on high level concepts mined from Wikipedia, a comprehensive knowledge source, and most importantly, is able to disambiguate entities using this knowledge. Hence, we are able to achieve better results.

Recently there are many attempts on combining visual features and textual features in image clustering. Feng et al.[9] used the surrounding text of images and a visual-based classifier to build a co-training framework. Gao et al.[12] represented the relationship among low-level visual features, images and the surrounding texts in a tripartite graph. Wang et al.[24] reinforced visual and textual features via inter-type links and inversely uses those features to update these links. The visual features, text features and inter-type links are represented as three matrices. Three linear formulas is defined to iteratively update the three matrices. Ding et al.[6] proposed a hierarchical clustering framework. Leuken et al.[16] investigated three methods for visual diversification of image search results in their paper. Tsai et al.[23] proposed a technique based on visual synset for web image annotation. They applied affinity propagation clustering on a set of images associated with a query term based on both visual and textual features. Each cluster represents a visual synset, and is labeled by related query terms. However, this query-based/term-based labeling approach has two limitations: 1) it cannot produce related concepts to the clusters like our system does (e.g. "Teddy" for Cluster 1 in Fig. 2); 2) the related query terms themselves can be ambiguous and are not suitable for representing a visual synset. In our paper, we represent each cluster with high related concepts which are Wikipedia concepts without ambiguity. The main challenge with the above hybrid approaches is the semantic gap between visual signals and textual signals. There is no easy way to combine the two kinds of similarity measures into one unifying measure.

5 Conclusion

In this paper, we proposed a novel framework for clustering web images by their contexts. The novelty lies in that our framework seeks to "understand" a context by converting words and phrases in the context into high level concepts in an external knowledge base such as Wikipedia. Moreover, it performs a tri-stage modified HAC algorithm utilizing information of various reliability. Our experiments show that on 40 "ambiguous" query terms, the purity, F-measure and NMI of our clustering results are consistently better than other recently developed image clustering systems. Our prototype system is practical as it is able to cluster a page of 100 images within 1 second.

References

1. Alcic, S., Conrad, S.: Measuring performance of web image context extraction. In: MDMKDD, vol. 8, p. 8 (2010)
2. Blei, D.M., Ng, A.Y., Jordan, M.I.: Latent dirichlet allocation. J. Mach. Learn. Res. 3, 993–1022 (2003)

3. Cai, D., He, X., Ma, W.Y., Wen, J.R., Zhang, H.: Organizing www images based on the analysis of page layout and web link structure. In: ICME, pp. 113–116 (2004)
4. Cai, D., Yu, S., Wen, J.R., Ma, W.Y.: VIPS: a vision-based page segmentation algorithm. In: Microsoft Technical Report, MSR-TR-2003-79 (2003)
5. Cai, Z., Zhao, K., Zhu, K.Q., Wang, H.: Wikification via link co-occurrence. In: CIKM, CIKM 2013, pp. 1087–1096 (2013)
6. Ding, H., Liu, J., Lu, H.: Hierarchical clustering-based navigation of image search results. In: MM, pp. 741–744 (2008)
7. Fan, J., Gao, Y., Luo, H.: Hierarchical classification for automatic image annotation. In: SIGIR, pp. 111–118 (2007)
8. Fan, R.E., Chang, K.W., Hsieh, C.J., Wang, X.R., Lin, C.J.: LIBLINEAR: A library for large linear classification. Journal of Machine Learning Research 9, 1871–1874 (2008)
9. Feng, H., Shi, R., Chua, T.S.: A bootstrapping framework for annotating and retrieving www images. In: MM, pp. 960–967 (2004)
10. Fergus, R., Li, F.F., Perona, P., Zisserman, A.: Learning object categories from google's image search. In: ICCV, pp. 1816–1823 (2005)
11. Fu, Z., Ip, H.H.S., Lu, H., Lu, Z.: Multi-modal constraint propagation for heterogeneous image clustering. In: MM, pp. 143–152 (2011)
12. Gao, B., Liu, T.Y., Qin, T., Zheng, X., Cheng, Q., Ma, W.Y.: Web image clustering by consistent utilization of visual features and surrounding texts. In: MM, pp. 112–121 (2005)
13. Gao, Y., Fan, J., Luo, H., Satoh, S.: A novel approach for filtering junk images from google search results. In: MMM, pp. 1–12 (2008)
14. Jing, F., Wang, C., Yao, Y., Deng, K., Zhang, L., Ma, W.Y.: IGroup: web image search results clustering. In: MM, pp. 377–384 (2006)
15. Krizhevsky, A., Sutskever, I., Hinton, G.E.: Imagenet classification with deep convolutional neural networks. In: NIPS (2012)
16. van Leuken, R.H., Pueyo, L.G., Olivares, X., van Zwol, R.: Visual diversification of image search results. In: WWW, pp. 341–350 (2009)
17. Li, L.J., Socher, R., Li, F.F.: Towards total scene understanding: Classification, annotation and segmentation in an automatic framework. In: CVPR, pp. 2036–2043 (2009)
18. Lowe, D.G.: Object recognition from local scale-invariant features. In: ICCV, pp. 1150–1157 (1999)
19. Song, Y., Wang, H., Wang, Z., Li, H., Chen, W.: Short text conceptualization using a probabilistic knowledgebase. In: IJCAI (2011)
20. Suchanek, F.M., Kasneci, G., Weikum, G.: YAGO: a core of semantic knowledge. In: WWW, pp. 697–706 (2007)
21. Taneva, B., Kacimi, M., Weikum, G.: Gathering and ranking photos of named entities with high precision, high recall, and diversity. In: WSDM, pp. 431–440 (2010)
22. Taneva, B., Kacimi, M., Weikum, G.: Finding images of difficult entities in the long tail. In: CIKM, CIKM 2011, pp. 189–194 (2011)
23. Tsai, D., Jing, Y., Liu, Y., Rowley, H., Ioffe, S., Rehg, J.: Large-scale image annotation using visual synset. In: ICCV, pp. 611–618 (2011)
24. Wang, X.J., Ma, W.Y., Zhang, L., Li, X.: Iteratively clustering web images based on link and attribute reinforcements. In: MM, pp. 122–131 (2005)
25. Yu, H., Li, M., Zhang, H.J., Feng, J.: Color texture moments for content-based image retrieval. In: International Conference on Image Processing, pp. 24–28 (2003)
26. Zeng, H.J., He, Q.C., Chen, Z., Ma, W.Y., Ma, J.: Learning to cluster web search results. In: SIGIR, pp. 210–217 (2004)
27. Zhong, S., Liu, Y., Liu, Y.: Bilinear deep learning for image classification. In: MM, pp. 343–352 (2011)

Accelerating Model Selection with Safe Screening for L_1-Regularized L_2-SVM

Zheng Zhao, Jun Liu, and James Cox

SAS Institute Inc., 600 Research Drive, Cary, NC 27513, USA
{zheng.zhao,jun.liu,james.cox}@sas.com

Abstract. The L_1-regularized support vector machine (SVM) is a powerful predictive learning model that can generate sparse solutions. Compared to a dense solution, a sparse solution is usually more interoperable and more effective for removing noise and preserving signals. The L_1-regularized SVM has been successfully applied in numerous applications to solve problems from text mining, bioinformatics, and image processing. The regularization parameter has a significant impact on the performance of an L_1-regularized SVM model. Therefore, model selection needs to be performed to choose a good regularization parameter. In model selection, one needs to learn a solution path using a set of predefined parameter values. Therefore, many L_1-regularized SVM models need to be fitted, which is usually very time consuming. This paper proposes a novel safe screening technique to accelerate model selection for the L_1-regularized L_2-SVM, which can lead to much better efficiency in many scenarios. The technique can successfully identify most inactive features in an optimal solution of the L_1-regularized L_2-SVM model and remove them before training. To achieve safe screening, the technique solves a minimization problem for each feature on a convex set that is formed by the intersection of a tight n-dimensional hyperball and the upper half-space. An efficient algorithm is designed to solve the problem based on zero-finding. Every feature that is removed by the proposed technique is guaranteed to have zero weight in the optimal solution. Therefore, an L_1-regularized L_2-SVM solver achieves exactly the same result by using only the selected features as when it uses the full feature set. Empirical study on high-dimensional benchmark data sets produced promising results and demonstrated the effectiveness of the proposed technique.

Keywords: Screening, sparse support vector machine, model selection.

1 Introduction

Feature selection is an effective technique for dimensionality reduction and relevance detection [1]. The L_1-regularized support vector machine (SVM) is a powerful feature selection algorithm [3, 4, 5, 6] that is in the embedded model [2]. It can simultaneously fit a model by margin maximization and remove noisy features by soft-thresholding. It has been successfully applied to solve many problems in text mining, bioinformatics, and image processing. The L_1-regularized

T. Calders et al. (Eds.): ECML PKDD 2014, Part III, LNCS 8726, pp. 385–400, 2014.

SVM enjoys two major advantages compared to other variances of sparse SVM models [7, 8, 9]: first, it solves a convex problem; therefore, an optimal solution can always be obtained without any relaxation of the original problem. Second, it is efficient. A well-implemented L_1-regularized SVM solver can readily handle problems that have tens of millions samples and features [6].

The value of the regularization parameter λ has a significant impact on the performance of an L_1-regularized SVM model. Model selection can be used to select a good parameter value. During model selection, a series of L_1-regularized SVM models need to be fit for a set of predefined regularization parameter values. The best regularization parameter value can be chosen by using a pre-specified criterion, such as the accuracy or the area under the curve (AUC) that is achieved by the resulting models on holdout samples. When data are huge, the computational cost of model selection can be prohibitive. Assume that k regularization parameter values, $\lambda_1 > \lambda_2 > \ldots > \lambda_k$, need to be tried in a model selection process. It is easy to see that this process can be greatly accelerated if the solution obtained for λ_i can be used to speed up the computation of the solution for λ_{i+1}. Based on this idea, highly efficient screening techniques are recently proposed for Lasso [10] to accelerate its model selection. The key idea is that, given a solution w_1^* for $\lambda = \lambda_1$, many features that have zero coefficients in w_2^* when $\lambda = \lambda_2$ can be identified. By removing these "inactive" features, the cost for computing w_2^* can be significantly reduced. Although effective screening algorithms have been designed for Lasso [11, 12, 13, 14, 15], research into screening for the L_1-regularized SVM is largely untouched.

In this paper, a novel screening technique is proposed to speed up model selection for an L_1-regularized L_2-SVM.[1] The technique makes use of the variational inequality [16] and the nonnegative constraint on the dual variables of the L_1-regularized L_2-SVM model for constructing a tight convex set, which can be used to compute bounds for screening features. A prescreening strategy and a fast zero-finding algorithm are designed and implemented to ensure the efficiency of the screening process. Features that are removed by the technique are guaranteed to be inactive in the optimal solution. Therefore, the screening technique is "safe," because an L_1-regularized L_2-SVM solver can achieve exactly the same result when it uses the features selected by the technique as when it uses the full feature set. To the best knowledge of the authors, this is the first screening technique that is proposed for accelerating the speed of model selection for the L_1-regularized L_2-SVM. Empirical study on five high-dimensional benchmark data sets produced promising results and demonstrated that the proposed screening technique can greatly speed up model selection for an L_1-regularized L_2-SVM by efficiently removing a large number of inactive features.

---

[1] Our ongoing work will extend the technique proposed in this paper to screen features for the L_1-regularized L_1-SVM.

2 L_1-Regularized L_2-SVM

Assume that $\mathbf{X} \in \mathbb{R}^{m \times n}$ is a data set that contains n samples, $\mathbf{X} = (\mathbf{x}_1, \ldots, \mathbf{x}_n)$, and m features, $\mathbf{X} = (\mathbf{f}_1^\top, \ldots, \mathbf{f}_m^\top)^\top$. Assume also that $\mathbf{y} = (y_1, \ldots, y_n)$ contains n class labels, $y_i \in \{-1, +1\}$, $i = 1, \ldots, n$. Let $\mathbf{w} \in \mathbb{R}^m$ be the m-dimensional weight vector, let $\xi_i \geq 0, i = 1, \ldots, n$ be the n slack variables, and let $b \in \mathbb{R}$ and $\lambda \in \mathbb{R}^+$ be the bias and the regularization parameter, respectively. The primal form of the L_1-regularized L_2-SVM is defined as:

$$\min_{\boldsymbol{\xi}, \mathbf{w}} \frac{1}{2} \sum_{i=1}^n \xi_i^2 + \lambda \|\mathbf{w}\|_1, \tag{1}$$

$$\text{s.t.} \quad y_i \left(\mathbf{w}^\top \mathbf{x}_i + b \right) \geq 1 - \xi_i, \quad \xi_i \geq 0.$$

Eq. (1) specifies a convex problem that has a non-smooth L_1 regularizer, which enforces that the solution is sparse. Let $\boldsymbol{w}^\star(\lambda)$ be the optimal solution of Eq. (1) for a given λ. All the features that have nonzero values in $\boldsymbol{w}^\star(\lambda)$ are called active features, and the other features are called inactive. Let $\boldsymbol{\alpha} \in \mathbb{R}^n$ be the n-dimensional dual variable. By applying the Lagrangian multiplier [17], the dual of the problem defined in Eq. (1) can be obtained as:

$$\min_{\boldsymbol{\alpha}} \|\boldsymbol{\alpha} - \mathbf{1}\|_2^2 \tag{2}$$

$$\text{s.t.} \quad \|\hat{\mathbf{f}}_j^\top \boldsymbol{\alpha}\| \leq \lambda, \quad j = 1, \ldots, m, \quad \sum_{i=1}^n \alpha_i y_i = 0, \; \boldsymbol{\alpha} \succcurlyeq \mathbf{0}.$$

Here, $\hat{\mathbf{f}} = \mathbf{Y}\mathbf{f}$, and $\mathbf{Y} = diag(\mathbf{y})$ is a diagonal matrix. By defining $\boldsymbol{\alpha} = \lambda\boldsymbol{\theta}$, Eq. (2) can be reformulated as:

$$\min_{\boldsymbol{\theta}} \|\boldsymbol{\theta} - \frac{1}{\lambda}\|_2^2 \tag{3}$$

$$\text{s.t.} \quad \|\hat{\mathbf{f}}_j^\top \boldsymbol{\theta}\| \leq 1, \quad j = 1, \ldots, m, \quad \sum_{i=1}^n \theta_i y_i = 0, \; \boldsymbol{\theta} \succcurlyeq \mathbf{0}.$$

In the primal formulation for the L_1-regularized L_2-SVM, the primal variables are b, $\mathbf{w}$, and $\boldsymbol{\xi}$. And in the dual formulation, the dual variables are $\boldsymbol{\alpha}$ or $\boldsymbol{\theta}$. When b and $\mathbf{w}$ are known, $\boldsymbol{\xi}$, $\boldsymbol{\alpha}$, and $\boldsymbol{\theta}$ can be obtained as:

$$\xi_i = \alpha_i = \lambda\theta_i = \max\left(0, 1 - y_i \left(\mathbf{w}^\top \mathbf{x}_i + b\right)\right). \tag{4}$$

The relation between $\boldsymbol{\theta}$ and $\mathbf{w}$ can be expressed as:

$$\boldsymbol{\theta}^\top \hat{\mathbf{f}}_j = \begin{cases} \text{sign}\,(w_j), & \text{if } w_j \neq 0 \\ [-1, +1], & \text{if } w_j = 0 \end{cases}, \quad j = 1, \ldots, m. \tag{5}$$

$\lambda_{\max}$ is defined as the smallest λ value that leads to $\mathbf{w} = \mathbf{0}$ when it is used in Eq. (1). Given a data set $(\mathbf{X}, \mathbf{y})$, $\lambda_{\max}$ can be obtained in a closed form as:

$$\lambda_{\max} = \left\| \sum_{i=1}^n \left(y_i - \frac{n_+ - n_-}{n} \right) \mathbf{x}_i \right\|_\infty, \tag{6}$$

where n_+ and n_- denote the number of positive and negative samples, respectively. And when $\lambda \geq \lambda_{max}$, the optimal solution of the problem defined in Eq. (1) can be written as:

$$\mathbf{w}^\star = \mathbf{0}, \quad b^\star = \frac{(n_+ - n_-)}{n}. \tag{7}$$

Denote $\mathbf{m} = \sum_{i=1}^{n} \left(y_i - \frac{n_+ - n_-}{n} \right) \mathbf{x}_i$. The first feature to enter the model is the one that corresponds to the element that has the largest magnitude in $\mathbf{m}$.

3   Safe Screening for L_1-Regularized L_2-SVM

Eq. (5) shows that the necessary condition for a feature $\mathbf{f}$ to be active in an optimal solution is $|\boldsymbol{\theta}^\top \hat{\mathbf{f}}| = 1$. On the other hand, for any feature $\mathbf{f}$, if $|\boldsymbol{\theta}^\top \hat{\mathbf{f}}| < 1$, it must be inactive in the optimal solution. Given a λ value, this condition can be used to develop a rule for screening inactive features to speed up training for the L_1-regularized L_2-SVM. The key is to compute the upper bound of $|\boldsymbol{\theta}^\top \hat{\mathbf{f}}|$ for features. A feature can be safely removed if its upper bound value is less than 1. The cost of computing the upper bounds can be much lower than training L_1-regularized L_2-SVM. Therefore, screening can greatly lower the computational cost by removing many inactive features before training.

To bound the value of $|\boldsymbol{\theta}^\top \hat{\mathbf{f}}|$, it is necessary to construct a closed convex set $\mathbf{K}$ that contains $\boldsymbol{\theta}$. The upper bound value can be then computed by maximizing $|\boldsymbol{\theta}^\top \hat{\mathbf{f}}|$ over $\mathbf{K}$, which defines a convex problem with a unique solution.

3.1   Constructing the Convex Set K

Given $\lambda_1, \ldots, \lambda_k$, k models need to be trained for model selection. Let $\boldsymbol{\theta}_i$ be the solution that corresponds to λ_i, this section shows that $\boldsymbol{\theta}_i$ can be used to construct a convex set that contains $\boldsymbol{\theta}_{i+1}$ for bounding the value of $|\boldsymbol{\theta}_{i+1}^\top \hat{\mathbf{f}}|$. When λ_i is close to λ_{i+1}, this convex set can be very tight.

Assume that $\boldsymbol{\theta}^\star$ is the optimal solution of Eq. (3) and $t \geq 0$. It is easy to verify that $\boldsymbol{\theta}^\star$ is also the optimal solution of the following problem:

$$\min_{\boldsymbol{\theta}} \left\| \boldsymbol{\theta} - \left(t\frac{1}{\lambda} + (1-t)\boldsymbol{\theta}^\star \right) \right\|_2^2 \quad , \tag{8}$$

$$s.t. \quad \|\hat{\mathbf{f}}_j^\top \boldsymbol{\theta}\| \leq 1, \quad j = 1, \ldots, m, \quad \sum_{i=1}^{n} \theta_i y_i = 0, \quad \boldsymbol{\theta} \succcurlyeq \mathbf{0}.$$

In the following, Eq. (8) and the variational inequality [16] are used to construct a closed convex set $\mathbf{K}$ to bound $|\boldsymbol{\theta}^\top \hat{\mathbf{f}}|$. Proposition 1 introduces the variational inequality for a convex optimization problem.

Proposition 1. *Let $\theta^\star$ be an optimal solution of a convex problem:*

$$\min g(\theta), \quad s.t. \quad \theta \in \mathbf{K},$$

where g is continuously differentiable and $\mathbf{K}$ is closed and convex. Then the following variational inequality holds:

$$\nabla g\left(\theta^\star\right)^\top \left(\theta - \theta^\star\right) \geq 0, \quad \forall \theta \in \mathbf{K}.$$

Let θ_1 and θ_2 be the optimal solutions of the problem defined in Eq. (3) and Eq. (8) for λ_1 and λ_2, respectively. Assume that $\lambda_1 > \lambda_2$ and that θ_1 is known[2]. The following results can be obtained by applying Proposition 1 to the problem defined in Eq. (8) for θ_1 and θ_2, respectively:

$$\left(\theta_1 - \left(t_1 \frac{1}{\lambda_1} + (1 - t_1)\,\theta_1\right)\right)^\top \left(\theta - \theta_1\right) \geq 0, \tag{9}$$

$$\left(\theta_2 - \left(t_2 \frac{1}{\lambda_2} + (1 - t_2)\,\theta_2\right)\right)^\top \left(\theta - \theta_2\right) \geq 0. \tag{10}$$

Let $t = \frac{t_1}{t_2} \geq 0$. By substituting $\theta = \theta_2$ and $\theta = \theta_1$ into Eq. (9) and Eq. (10), respectively, and then combining the two inequalities, it leads to:

$$\mathbf{B}_t = \left\{\theta_2 : (\theta_2 - \mathbf{c})^\top (\theta_2 - \mathbf{c}) \leq l^2\right\}, \tag{11}$$

$$\mathbf{c} = \frac{1}{2}\left(t\theta_1 - t\frac{1}{\lambda_1} + \frac{1}{\lambda_2} + \theta_1\right), l = \frac{1}{2}\left\|t\theta_1 - t\frac{1}{\lambda_1} + \frac{1}{\lambda_2} - \theta_1\right\|_2.$$

As the value of t changes from 0 to ∞, Eq. (11) generates a series of hyperballs that contains θ_2. The following theorem studies when the radius of the hyperball generated by Eq. (11) reaches its minimum:

Theorem 1. *Let* $\mathbf{a} = \frac{\frac{1}{\lambda_1} - \theta_1}{\left\|\frac{1}{\lambda_1} - \theta_1\right\|_2}$. *The radius of the hyperball generated by Eq. (11) reaches it minimum when*

$$t = 1 + \left(\frac{1}{\lambda_2} - \frac{1}{\lambda_1}\right)\frac{\mathbf{a}^\top \mathbf{1}}{\left\|\frac{1}{\lambda_1} - \theta_1\right\|_2}. \tag{12}$$

Let $\mathbf{c}$ be the center of the ball and l be the radius. When the minimum is reached, they can be computed as:

$$\mathbf{c} = \frac{1}{2}\left(\frac{1}{\lambda_2} - \frac{1}{\lambda_1}\right) P_\mathbf{a}\left(\mathbf{1}\right) + \theta_1, \ l = \frac{1}{2}\left(\frac{1}{\lambda_2} - \frac{1}{\lambda_1}\right)\left\|P_\mathbf{a}\left(\mathbf{1}\right)\right\|. \tag{13}$$

Here, $P_\mathbf{u}\left(\mathbf{v}\right) = \mathbf{v} - \frac{\mathbf{v}^\top \mathbf{u}}{\|\mathbf{u}\|_2^2}\mathbf{u}$ is an operator that projects $\mathbf{v}$ to the null-space of $\mathbf{u}$. Since $\|\mathbf{a}\|_2 = 1$, $P_\mathbf{a}\left(\mathbf{1}\right) = \mathbf{1} - \left(\mathbf{a}^\top \mathbf{1}\right)\mathbf{a}$.

---

[2] When $\lambda_1 = \lambda_{max}$, $\mathbf{w} = 0$ and θ_1 is given in Eq. (4).

Proof. The theorem can be proved by minimizing the r defined in Eq. (11).

$\square$

Theorem 1 suggests that when $t = 1 + \left(\frac{1}{\lambda_2} - \frac{1}{\lambda_1}\right) \mathbf{a}^\top \mathbf{1} \left\|\frac{1}{\lambda_1} - \boldsymbol{\theta}_1\right\|_2^{-1}$, the volume of $\mathbf{B}_t$ is minimized, which forms a good basis for constructing $\mathbf{K}$. The nonnegative constraint on the dual variable confines $\boldsymbol{\theta}$ in the upper half-space: $\boldsymbol{\theta} \succcurlyeq \mathbf{0}$, and can be used to further reduce the volume of $\mathbf{K}$:

$$\mathbf{K} = \left\{ \boldsymbol{\theta} : (\boldsymbol{\theta} - \mathbf{c})^\top (\boldsymbol{\theta} - \mathbf{c}) \leq l^2, \boldsymbol{\theta} \succcurlyeq \mathbf{0} \right\}, \qquad (14)$$

$$\mathbf{c} = \frac{1}{2}\left(\frac{1}{\lambda_2} - \frac{1}{\lambda_1}\right) P_{\mathbf{a}}(1) + \boldsymbol{\theta}_1, \quad l = \frac{1}{2}\left(\frac{1}{\lambda_2} - \frac{1}{\lambda_1}\right) \|P_{\mathbf{a}}(1)\|.$$

3.2 Computing the Upper Bound

Given the convex set $\mathbf{K}$ defined in Eq. (14), the maximum value of $\left|\boldsymbol{\theta}_2^\top \hat{\mathbf{f}}\right|$ can be computed by solving the problem:

$$\max \left|\boldsymbol{\theta}^\top \hat{\mathbf{f}}\right|, \ \ s.t. \ (\boldsymbol{\theta} - \mathbf{c})^\top (\boldsymbol{\theta} - \mathbf{c}) \leq l^2, \ \boldsymbol{\theta} \succcurlyeq \mathbf{0}. \qquad (15)$$

Since the following equation holds:

$$\max |x| = \max \left\{-\min(x), \max(x)\right\} = \max \left\{-\min(x), -\min(-x)\right\}.$$

The computation of $\max \left|\boldsymbol{\theta}^\top \hat{\mathbf{f}}\right|$ can be decomposed to the following two subproblems: $m_1 = -\min \boldsymbol{\theta}^\top \hat{\mathbf{f}}$, $m_2 = -\min \boldsymbol{\theta}^\top (-\hat{\mathbf{f}})$. And $\max \left|\boldsymbol{\theta}^\top \hat{\mathbf{f}}\right| = \max(m_1, m_2)$. This suggests that the key to bound $\left|\boldsymbol{\theta}^\top \hat{\mathbf{f}}\right|$ is to compute:

$$\min \boldsymbol{\theta}^\top \hat{\mathbf{f}}, \ \ s.t. \ (\boldsymbol{\theta} - \mathbf{c})^\top (\boldsymbol{\theta} - \mathbf{c}) \leq l^2, \ \boldsymbol{\theta} \succcurlyeq \mathbf{0}. \qquad (16)$$

Its Lagrangian $L(\boldsymbol{\theta}, \alpha, \boldsymbol{\nu})$ can be written as:

$$L(\boldsymbol{\theta}, \alpha, \boldsymbol{\nu}) = \boldsymbol{\theta}^\top \hat{\mathbf{f}} + \frac{1}{2}\alpha \left(\|\boldsymbol{\theta} - \mathbf{c}\|_2^2 - l^2\right) + \boldsymbol{\nu}^\top \boldsymbol{\theta}, \ \alpha \geq 0, \ \boldsymbol{\nu} \succcurlyeq \mathbf{0}. \qquad (17)$$

Since $\|\boldsymbol{\theta} - \mathbf{c}\|_2^2 \leq l^2$, the problem specified in Eq. (16) is bounded from below by $-(\|\mathbf{c}\|_2 + l) \|\mathbf{f}\|_2$. Thus, $\min_{\boldsymbol{\theta}} L(\boldsymbol{\theta}, \alpha, \boldsymbol{\nu})$ is also bounded from below. Since the minimum achieves on the boundary, it must hold that $\alpha > 0$. It is also easy to verify that $\alpha = 0 \Rightarrow \left|\boldsymbol{\theta}^\top \hat{\mathbf{f}}\right| = 0$.

Setting the derivative of $L(\boldsymbol{\theta}, \alpha, \boldsymbol{\nu})$ to be zero leads to the equation:

$$\mathbf{f} + \alpha(\boldsymbol{\theta} - \mathbf{c}) - \boldsymbol{\nu} = 0 \Rightarrow \boldsymbol{\theta} = \frac{1}{\alpha}\boldsymbol{\nu} - \frac{1}{\alpha}\mathbf{f} + \mathbf{c}.$$

Therefore, $\theta_i = \frac{1}{\alpha}\nu_i - \frac{1}{\alpha}f_i + c_i$, $i = 1, \ldots, n$. According to the complementary slackness condition, $\boldsymbol{\nu}^\top\boldsymbol{\theta} = 0$. Also since $\boldsymbol{\nu} \succeq 0$ and $\boldsymbol{\theta} \succeq 0$. It must hold that $\nu_i\theta_i = 0$, $i = 1, \ldots, n$. These conditions lead to the following equations:

$$\theta_i = \max\left(c_i - \frac{1}{\alpha}f_i, \ 0\right). \tag{18}$$

This suggests that when α is know, $\boldsymbol{\theta}$ can be computed by using Eq. (18). In the following, it shows that the value of α can be efficiently computed by solving a zero finding problem through binary search.

Computing α via zero finding According to the complementary slackness condition, $\alpha\left(\|\boldsymbol{\theta} - \mathbf{c}\|_2^2 - l^2\right) = 0$. Because $\alpha > 0$, it must hold that:

$$\|\boldsymbol{\theta} - \mathbf{c}\|_2^2 - l^2 = 0 \Rightarrow \boldsymbol{\theta}^\top\boldsymbol{\theta} - 2\mathbf{c}^\top\boldsymbol{\theta} - l^2 + \mathbf{c}^\top\mathbf{c} = 0. \tag{19}$$

Let $\mathcal{A} = \{i : \theta_i > 0\}$. The following equation can be obtained.

$$\boldsymbol{\theta}^\top\boldsymbol{\theta} - 2\mathbf{c}^\top\boldsymbol{\theta} - l^2 + \mathbf{c}^\top\mathbf{c} = \sum_{i\in\mathcal{A}}\theta_i^2 - 2\sum_{i\in\mathcal{A}}c_i\theta_i - l^2 + \mathbf{c}^\top\mathbf{c}. \tag{20}$$

By plugging Eq. (18) into Eq. (20). A function of α can be obtained as:

$$g\left(\frac{1}{\alpha}\right) = \frac{1}{\alpha^2}\sum_{i\in\mathcal{A}}f_i^2 - \sum_{i\in\mathcal{A}}c_i^2 - l^2 + \mathbf{c}^\top\mathbf{c}. \tag{21}$$

And the α value can be obtained by solving the zero finding problem:

$$g\left(\frac{1}{\alpha}\right) = 0 \tag{22}$$

The following theorem suggests that $g\left(\frac{1}{\alpha}\right)$ monotonically increases as $\frac{1}{\alpha}$ increases. Therefore this problem can be solved efficiently via binary search.

Theorem 2. *The function $g\left(\frac{1}{\alpha}\right)$ monotonically increases as $\frac{1}{\alpha}$ increases.*

Proof. Assume that $g_i\left(\frac{1}{\alpha}\right)$ is defined as:

$$g_i\left(\frac{1}{\alpha}\right) = \begin{cases} i \in \mathcal{A}, & \frac{1}{\alpha^2}f_i^2 - c_i^2 \\ i \notin \mathcal{A}, & 0 \end{cases}. \tag{23}$$

$g\left(\frac{1}{\alpha}\right)$ can be rewritten as:

$$g\left(\frac{1}{\alpha}\right) = \sum_{i=1}^n g_i\left(\frac{1}{\alpha}\right) - l^2 + \mathbf{c}^\top\mathbf{c}.$$

The theorem can be proved by showing that for $\forall i \in \{1, \ldots, n\}$, $g_i\left(\frac{1}{\alpha}\right)$ either increases monotonically as $\frac{1}{\alpha}$ increases, or is a constant. Let $\epsilon > 0$, this can be proved by comparing $g_i\left(\frac{1}{\alpha}\right)$ to $g_i\left(\frac{1}{\alpha} + \epsilon\right)$ in the following four cases.

1. $c_i > 0$, $f_i \leq 0$: $c_i > 0$, $f_i \leq 0 \Rightarrow \theta_i = c_i - \frac{1}{\alpha} f_i$, $i \in \mathcal{A}$, for $\forall \frac{1}{\alpha} \in \mathbb{R}^+$. In this case $g_i\left(\frac{1}{\alpha}\right)$ can be written as:

$$g_i\left(\frac{1}{\alpha}\right) = \frac{1}{\alpha^2} f_i^2 - c_i^2. \tag{24}$$

And it can be verify that $g_i\left(\frac{1}{\alpha} + \epsilon\right) > g_i\left(\frac{1}{\alpha}\right)$ when $c_i > 0$, $f_i \leq 0$.

2. $c_i \leq 0$, $f_i > 0$: $c_i \leq 0$, $f_i > 0 \Rightarrow \theta_i = 0$, $i \notin \mathcal{A}$, for $\forall \frac{1}{\alpha} \in \mathbb{R}^+$. In this case $g_i\left(\frac{1}{\alpha}\right)$ can be written as:

$$g_i\left(\frac{1}{\alpha}\right) = 0. \tag{25}$$

Therefore, $g_i\left(\frac{1}{\alpha}\right)$ is a constant when $c_i \leq 0$, $f_i > 0$.

3. $c_i > 0$, $f_i > 0$: In this case $g_i\left(\frac{1}{\alpha}\right)$ can be written as:

$$g_i\left(\frac{1}{\alpha}\right) = \begin{cases} \frac{1}{\alpha} \in \left(0, \frac{c_i}{f_i}\right) \Rightarrow \theta_i = c_i - \frac{1}{\alpha} f_i, i \in \mathcal{A}, & \frac{1}{\alpha^2} f_i^2 - c_i^2 \\ \frac{1}{\alpha} \in \left[\frac{c_i}{f_i}, +\infty\right) \Rightarrow \theta_i = 0, i \notin \mathcal{A} & , \qquad 0 \end{cases} \tag{26}$$

And it can be verify that $g_i\left(\frac{1}{\alpha} + \epsilon\right) > g_i\left(\frac{1}{\alpha}\right)$ when $c_i > 0, f_i > 0$.

4. $c_i < 0$, $f_i \leq 0$: In this case $g_i\left(\frac{1}{\alpha}\right)$ can be written as:

$$g_i\left(\frac{1}{\alpha}\right) = \begin{cases} \frac{1}{\alpha} \in \left(0, \frac{c_i}{f_i}\right] \Rightarrow \theta_i = 0, i \notin \mathcal{A} & , \qquad 0 \\ \frac{1}{\alpha} \in \left(\frac{c_i}{f_i}, +\infty\right) \Rightarrow \theta_i = c_i - \frac{1}{\alpha} f_i, i \in \mathcal{A}, & \frac{1}{\alpha^2} f_i^2 - c_i^2 \end{cases} \tag{27}$$

It can also be verify that $g_i\left(\frac{1}{\alpha} + \epsilon\right) > g_i\left(\frac{1}{\alpha}\right)$ when $c_i < 0$, $f_i \leq 0$.

This finishes the proof of the theorem.

$\square$

When a value is given to α, $\mathcal{A}$ can be determined via computing θ_i by using one of the four equations $\left(\text{Eq. }(24) - \text{Eq. }(27)\right)$ provided in Theorem 2 according to the value of c_i and f_i. And the obtained $\mathcal{A}$ can be used to compute the value of $g\left(\frac{1}{\alpha}\right)$. When $\mathcal{A}$ is determined, solving Eq. (22) leads to the following equation:

$$\frac{1}{\alpha'} = \sqrt{\frac{l^2 - \mathbf{c}^\top \mathbf{c} + \sum\limits_{i \in \mathcal{A}} c_i^2}{\sum\limits_{i \in \mathcal{A}} f_i^2}} \tag{28}$$

Let an index set $\mathcal{B}$ is defined as $\mathcal{B} = \{i : (c_i > 0, f_i > 0) \text{ or } (c_i < 0, f_i \leq 0)\}$. Assume that $\mathcal{B}$ contains k members. A sorted index set $\mathcal{B}_{sorted} = \{i_1, \ldots, i_k\}$ can be obtained by sorting the value of $\frac{c_i}{f_i}$, $i \in \mathcal{B}$. The following theorem provides the stopping condition for using binary search to solve the zero finding problem.

Theorem 3. *Let $\mathcal{T} = \left\{0, \frac{c_{i_1}}{f_{i_1}}, \ldots, \frac{c_{i_k}}{f_{i_k}}, +\infty\right\} = \{t_1, \ldots, t_{k+2}\}$, where $i_1, \ldots, i_k$ are the k sorted indices in $\mathcal{B}_{sorted}$. Given $\frac{1}{\alpha}$, assume that $t_j < \frac{1}{\alpha} \leq t_{j+1}$. The binary search stops when the $\frac{1}{\alpha'}$ computed by using Eq. (28) also satisfies that $t_j < \frac{1}{\alpha'} \leq t_{j+1}$. In this case, set $\frac{1}{\alpha} = \frac{1}{\alpha'}$ and it can verified that $g\left(\frac{1}{\alpha}\right) = 0$.*

Proof. The theorem can be proved by using the fact that when the value of $\frac{1}{\alpha}$ varies in $(t_j, t_{j+1}]$, $\mathcal{A}$ keeps unchanged.

$\square$

Theorem 3 suggests that t_{k+1} can be used as the starting point for binary search. If $g\left(\frac{1}{\alpha}\right) > 0$, decrease $\frac{1}{\alpha}$. If $g\left(\frac{1}{\alpha}\right) < 0$, increase $\frac{1}{\alpha}$. The search stops when the condition specified in Theorem 3 is satisfied. And the obtained $\frac{1}{\alpha}$ and $\mathcal{A}$ can be used to compute $\boldsymbol{\theta}^\top \hat{\mathbf{f}}$ by using the following equation:

$$\boldsymbol{\theta}^\top \hat{\mathbf{f}} = \sum_{i \in \mathcal{A}} c_i f_i - \frac{1}{\alpha} \sum_{i \in \mathcal{A}} f_i^2. \tag{29}$$

3.3 Computing the Upper Bound without Using $\boldsymbol{\theta} \succcurlyeq 0$

When $\boldsymbol{\theta} \succcurlyeq 0$ is not used to construct $\mathbf{K}$, $\max \left| \boldsymbol{\theta}^\top \hat{\mathbf{f}} \right|$ has a closed form solution on the hyper-ball defined in Theorem 1.

Theorem 4. *The optimization problem:*

$$\max \left| \boldsymbol{\theta}^\top \hat{\mathbf{f}} \right|, \ s.t. \ (\boldsymbol{\theta} - \mathbf{c})^\top (\boldsymbol{\theta} - \mathbf{c}) \leq l^2, \tag{30}$$

has a closed form solution:

$$\max \left| \boldsymbol{\theta}^\top \hat{\mathbf{f}} \right| = \left| \mathbf{c}^\top \hat{\mathbf{f}} \right| + l \left\| \hat{\mathbf{f}} \right\|_2. \tag{31}$$

Proof. The theorem can be proved by using the method of Lagrange multipliers.

$\square$

Let m be the bound computed by solving Eq. (15), and m' be the bound computed by solving Eq. (30). It is easy to see that $m < m'$, since the $\mathbf{K}$ used in Eq. (15) is tighter. However, since m' can be computed in closed form, its computational cost is low . Therefore, it can be used for pre-screening features. More specifically, If $m' < 1$, there is no need to compute m by solving Eq. (15), since $m < m' < 1$. Computing m requires to solve a zero finding problem using binary search which is usually more expensive than computing Eq. (31).

3.4 The Screening Algorithm

Algorithm 1 shows the procedure of screening features for L_1-regularized L_2-SVM. Given λ_1, λ_2, and $\boldsymbol{\theta}_1$, the algorithm returns a list $\mathcal{L}$, which contains the indices of the features that are potentially active in the optimal solution that corresponds to λ_2. The algorithm first weights a feature using $\mathbf{Y}$ in Line 3. It then computes a bound for $|\hat{\mathbf{f}}^\top \boldsymbol{\theta}|$ using Eq. (31) in Line 4. If this bound is less than 1, the algorithm goes to test the next feature. This is the pre-screening step for improving algorithm's efficiency by using a bound that is cheaper to compute. If a feature passes the pre-screening, the algorithm computes a tighter bound for the feature in Line 8 and Line 9. If the bound is larger than 1, it adds the index of the feature to $\mathcal{L}$ in Line 11. The function neg_min($\hat{\mathbf{f}}$) computes $-\min \boldsymbol{\theta}_2^\top \hat{\mathbf{f}}$. It first solves a zero finding problem for $\hat{\mathbf{f}}$ in Line 17, then uses the obtained $\frac{1}{\alpha}$ and $\mathcal{A}$ to compute $\min \boldsymbol{\theta}_2^\top \hat{\mathbf{f}}$ in Line 18. It returns $-\min \boldsymbol{\theta}_2^\top \hat{\mathbf{f}}$ in Line 19. The function zero_finding($\hat{\mathbf{f}}$) solves the zero finding problem. This function first uses $max\left(\frac{c_j}{f_j}, j \in \mathcal{B}\right)$ as the starting value for $\frac{1}{\alpha}$. If $g\left(\frac{1}{\alpha}\right) < 0$, it must hold that $\frac{1}{\alpha} \leq \frac{1}{\alpha'} < \infty$. Therefore the stopping condition specified in Theorem 3 is satisfied. The algorithm returns $\frac{1}{\alpha}$ and $\mathcal{A}$ in Line 28. Otherwise it setups the *low* and *high* variables for binary search. The binary search is performed in Line 32 to Line 45. The stopping condition is tested in Line 36. If this condition is satisfied, the function stops searching and returns $\frac{1}{\alpha}$ and $\mathcal{A}$.

The algorithm needs to be implemented carefully to ensure efficiency. First, each step of the computation needs to be decomposed to many small substeps, so that the intermediate results obtained in the preceding substeps can be used by the following substeps to accelerate computation. Second, the substeps need to be organized and ordered properly so that no redundant computation is performed. It turns out the procedure listed in Algorithm 1 can be very efficient.

The pre-screening step requires to compute $\mathbf{Yf}$, $\mathbf{f}^\top \mathbf{y}$, and $\mathbf{f}^\top \mathbf{f}$. Since these computations are independent of $\boldsymbol{\theta}_1$, λ_1, and λ_2. Therefore, they can been precomputed before training[3], and the cost is $O\left(mp\right)$ for m features. Here p is the average feature length[4]. The pre-screening step also requires to compute $\boldsymbol{\theta}_1^\top \mathbf{1}$ and $\boldsymbol{\theta}_1^\top \boldsymbol{\theta}_1$. They are shared by all the features. So they can be computed at the beginning of screening, and the cost is $O\left(n\right)$. For each feature, the pre-screening step requires to compute $\boldsymbol{\theta}_1^\top \mathbf{f}$, and its cost is $O\left(mp\right)$ for m features. However, when a solver fits a L_1-regularized L_2-SVM model, it might have already computed $\hat{\mathbf{f}}^\top \boldsymbol{\theta}_1$ as an intermediate result for all the features. In this case, $\hat{\mathbf{f}}^\top \boldsymbol{\theta}_1$ can be obtained from the solver for screening features at no cost. Given these intermediate results, the bound in the pre-screening step can be obtained in $O\left(m\right)$ for m feature. Therefore, the total computational cost for pre-screening m features is $O\left(mp\right)$. And if $\hat{\mathbf{f}}^\top \boldsymbol{\theta}_1$ can be obtained from the intermediate results generated by the L_1-regularized L_2-SVM solver and $\mathbf{Yf}$, $\mathbf{f}^\top \mathbf{y}$, and $\mathbf{f}^\top \mathbf{f}$ are precomputed before training, the total cost can decrease to just $O\left(m + n\right)$.

Assume that q features passed the pre-screening[5]. To compute the tighter bounds for these features, the algorithm requires to compute $\mathbf{c}$ and l. The cost is $O\left(n\right)$. For each feature, it can be verified that the algorithm takes at most $O\left(\log\left(p\right)\right)$ steps to solve the zero finding problem. In each step, it takes $O\left(p\right)$ to determine $\mathcal{A}$ and compute $g\left(\frac{1}{\alpha}\right)$. Thus, cost for solving the zero-finding problem is $O\left(p\log\left(p\right)\right)$. In the process of solving the zero-finding problem, $\sum_{i \in \mathcal{A}} c_i f_i$ and $\sum_{i \in \mathcal{A}} f_i^2$ are computed as the intermediate results. Given them as well as the $\frac{1}{\alpha}$

---

[3] They can also be used by the L_1-regularized L_2-SVM solver.
[4] For dense data $p = n$, for sparse data usually $p \ll n$.
[5] Usually, $q \ll m$.

Input: $\mathbf{X} \in \mathbb{R}^{n \times m}$, $\mathbf{y} \in \mathbb{R}^n$, $\lambda_1, \lambda_2, \boldsymbol{\theta}_1 \in \mathbb{R}^n$.
Output: $\mathcal{L}$, the retained feature list.

```
 1  𝕃 = ∅, i = 1, Y = diag (y);
 2  for i ≤ m do
 3  │  f̂ = Yfᵢ;
 4  │  m = |cᵀf̂| + l ‖f̂‖₂;
 5  │  if m < 1 then
 6  │  │  continue;
 7  │  end
 8  │  m₁ = neg_min(f̂), m₂ = neg_min(−f̂);
 9  │  m = max {m₁, m₂};
10  │  if m ≥ 1 then
11  │  │  𝓛 = 𝓛 ∪ {i};
12  │  end
13  │  i = i + 1;
14  end
15  return 𝓛;

16  Function neg_min(f̂)
17  │  {1/α, 𝒜} = zero_finding(f̂);
18  │  m = Σ_{i∈𝒜} cᵢfᵢ − 1/α Σ_{i∈𝒜} fᵢ²;
19  │  return −m;
20  end

21  Function zero_finding(f̂)
22  │  ℬ = {i : (cᵢ > 0, fᵢ > 0) or (cᵢ < 0, fᵢ ≤ 0)};
23  │  search = true, 1/α = max ( cⱼ/fⱼ , j ∈ ℬ );
24  │  compute 𝒜 and g (1/α);
25  │  if g (1/α) < 0 then
26  │  │  compute 1/α′ using Eq. (28);
27  │  │  1/α = 1/α′;
28  │  │  return {1/α, 𝒜};
29  │  else
30  │  │  low = 0, high = 1/α;
31  │  end
32  │  while search do
33  │  │  1/α = ½ (low + high);
34  │  │  compute 𝒜 and g (1/α), compute 1/α′ using Eq. (28);
35  │  │  if the condition specified in Theroem 3 is satisfied then
36  │  │  │  1/α = 1/α′, search = false;
37  │  │  else
38  │  │  │  if g (1/α) < 0 then
39  │  │  │  │  low = tⱼ₊₁, tⱼ₊₁ is as defied in Theroem 3;
40  │  │  │  else
41  │  │  │  │  high = tⱼ, tⱼ is as defied in Theroem 3;
42  │  │  │  end
43  │  │  end
44  │  end
45  │  return {1/α, 𝒜};
46  end
```

Algorithm 1. Screening for L_1-regularized L_2-SVM

determined by zero finding, $\boldsymbol{\theta}^\top \hat{\mathbf{f}}$ can be computed in $O(1)$. Therefore, the total cost for computing the tighter bounds for q features is $O(n + qp \log(p))$.

In summary, in the worst case of the proposed procedure, the total computational cost for screening a data set that has m features is $O(mp + qp \log(p))$. And if $\hat{\mathbf{f}}^\top \boldsymbol{\theta}_1$ can be obtained from the intermediate results generated by the L_1-regularized L_2-SVM solver and $\mathbf{Yf}$, $\mathbf{f}^\top \mathbf{y}$, and $\mathbf{f}^\top \mathbf{f}$ are precomputed before training, the total cost can decrease to just $O(m + n + qp \log(p))$.

4   Empirical Study

The screening approach presented in Algorithm 1 was implemented in the C language. This section evaluates its power for accelerating model selection for L_1-regularized L_2-SVM. Experiments are performed on a Windows Server 2008 R2 with two Intel Xeon® L5530 (2.40GHz) CPUs and 72GB memory.

4.1   Experiment Setup

Five benchmark data sets are used in the experiment. One is a microarray data set: gli_85. Three are text data sets: rcv1.binary(rcv1b), real-sim, and news20.binary (news20b). And one is a educational data mining data set: kdd2010 bridge-to-algebra (kddb). The gli_85 data set is downloaded from Gene Expression Omnibus,[6] and the other four data sets are downloaded from the LIBSVM data repository.[7] According to the feature-to-sample ratio (m/n), the five data sets fall into three groups: (1) the $m \gg n$ group, including the gli_85 and news20b data sets; (2) the $m \approx n$ group, including the rcv1b and kddb data sets; and (3) the $m \ll n$ group, including the real-sim data set. Table 1 shows detailed information about the five benchmark data sets.

Table 1. Summary of the benchmark data sets

Data Set	sample (n)	feature (m)	m/n
gli_85	85	22,283	262.15
rcv1b	20,242	47,236	2.33
real-sim	72,309	20,958	0.29
news20b	19,996	1,355,191	67.77
kddb	19,264,097	29,890,095	1.55

A L_1-regularized L_2-SVM solver based on the coordinate gradient descent (cgd) algorithm [18] is implemented in the C language for training the L_1-regularized L_2-SVM model. A similar solver is also implemented in the liblinear

---

[6] www.ncbi.nlm.nih.gov/geo/query/acc.cgi?acc=GSE4412
[7] www.csie.ntu.edu.tw/ cjlin/libsvmtools/datasets/

package [6]. The difference is that in liblinear, the bias term b is also penalized by the L_1 regularizer and is inactive in most cases. In contrast, the solver that is implemented for this paper solves the problem specified in Eq. (1) exactly. Therefore, the bias term is not penalized and is alway active.

For each benchmark data set, the L_1-regularized L_2-SVM solver is used to fit model along a sequence of 20 λ values: $\left\{ \lambda_k = \frac{1}{k} \lambda_{max} - \epsilon, k = 1, \ldots, 20, \epsilon = 10^{-8} \right\}$. When $\lambda = \lambda_{max} - \epsilon$, only one feature is active. Denote n_+ and n_- as the number of positive and negative samples, respectively. And let $\mathbf{m} = \sum_{i=1}^{n} \left(y_i - \frac{n_+ - n_-}{n} \right) \mathbf{x}_i$. This feature corresponds to the largest element in $\mathbf{m}$.

For each given benchmark data set, the L_1-regularized L_2-SVM solver runs in four different configurations: (1) In **org**, the solver runs without any accelerating technique. (2) In **warm**, the solver runs with warm-start. In the kth iteration, the $\mathbf{w}_{k-1}$ obtained in the $(k-1)$th iteration is used as the initial $\mathbf{w}_k$ for fitting model. When λ_k and λ_{k-1} are close, warm-start can effectively speed up training by reducing the number of iterations for the solver to converge. (3) In **scr**, the solver runs with the screening technique. (4) In **warm_scr**, the solver runs with both warm-start and the screening technique. Both warm-start and screening can be used to speed up model selection. The main purpose of running the L_1-regularized L_2-SVM solver with different configurations is not only to compare screening with warm-start, but also to provide a sensitivity study to explore that whether better performance can be achieved by combining two techniques.

Table 2. Total run time (in sec.) of the L_1-regularized L_2-SVM solver when different combinations of accelerating techniques are used to speed up model selection.

Alg.	gli_85	rcv1b	real-sim	news20b	kddb
org	284.08	19.04	20.73	1040.22	9071.73
warm	259.20	11.54	14.06	786.44	5770.12
scr	1.89	4.09	8.53	25.97	947.01
warm_scr	**1.83**	**2.70**	**5.90**	**18.22**	**643.34**

Table 3. Total number of iterations for the L_1-regularized L_2-SVM solver to converge when different combinations of accelerating techniques are used

Alg.	gli_85	rcv1_trainb	real-sim	news20b	kddb
org	16,176	1004	548	2,501	737
warm	14,772	578	361	1,908	483
scr	16,028	995	591	2,857	809
warm_scr	15,227	606	369	2,035	499

4.2   Results

Table 2 and Table 3 show the results of the total run time and the total number of iterations for the L_1-regularized L_2-SVM solver to converge when different combinations of accelerating techniques are used. The total run time and total number of iterations are obtained by aggregating the time and number of iterations used by the L_1-regularized L_2-SVM solver when it fits models using different regularization parameters. In terms of total running time, screening with warm-start (**warm_scr**) provides the best performance. Compared to **org**, for the $m \gg n$ group, the speed-up ratio is about 155.5 for the gli_85 data and 57.1 for the news20b data. For the $m \approx n$ group, the speed-up ratio is about 7.1 for the rcv1b data and 14.1 for the kddb data. And for the $m \ll n$ group, the speed-up ratio is about 3.5 for the real-sim data. The result shows that **warm_scr** is more effective when the number of features is larger than the number of samples. A similar trend is observed on **scr**. In terms of the total iteration number, the best performance is achieved by **warm** and **warm_scr**. This suggests that warm-start can effectively speed up convergence by providing a good start point for optimization. When **org** is compared to **scr**, the result suggests that the proposed screening technique can significantly improve the efficiency of the L_1-regularized L_2-SVM solver. This justifies that screening can effectively reduce the computational cost of training by removing most inactive features.

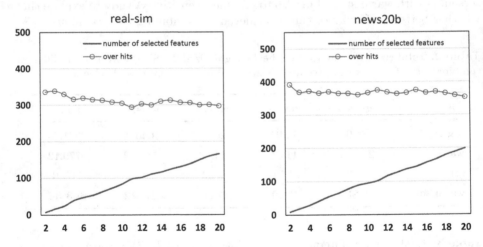

Fig. 1. Detailed information about the "over hits" on two benchmark data sets when λ decreases from λ_{max} to $\frac{1}{20}\lambda_{max}$. "Over hits" is the number of inactive features that are not removed by screening. The results show that the number of leftover inactive features is stable, and is small when compared to the size of the original feature set.

Figure 1 shows detailed information about the number of leftover inactive features on the real-sim and news20b data sets when λ decreases from λ_{max} to $\frac{1}{20}\lambda_{max}$. The result shows that this number is very stable during model selection. Let k be the number of active features. The proposed screening technique keeps to retain about $k + 400$ features for training the L_1-regularized L_2-SVM model. This number is much smaller than the dimensionality of the original data sets. Similar trends are also observed on other data sets and are not presented in this paper because of the space limit. Table 4 compares the time used by screening to the time used by training. Compared to training time, the screening time is marginal, especially when $m \gg n$. Notice that for training, the solver uses only the features that are selected by screening. The training time can be much longer if screening is not used to eliminate inactive features.

Table 4. Comparison of screening to training time. For training, the solver uses only the features that are selected by the proposed screening technique. The training time can be much longer if screening is not used to eliminate inactive features.

Tech.	gli_85	rcv1b	real-sim	news20b	kddb
			scr		
scr	0.01	0.73	1.79	1.29	35.29
tr	1.89	3.35	6.74	24.68	911.72
ratio	0.01	0.22	0.27	0.05	0.04
			warm_scr		
scr	0.03	0.75	1.75	1.31	34.93
tr	1.79	1.95	4.15	16.91	608.41
ratio	0.02	0.38	0.42	0.08	0.06

The results indicate that the proposed screening technqiue is effective for removing inactive features. And with warm-start they form a powerful combination for accelerating model selection for the L_1-regularized L_2-SVM.

5   Conclusion

Screening is an effective technique for accelerating model selection for L_1-regularized sparse learning model by eliminating features that are guaranteed to be inactive. This paper proposes a novel technique to screen features for L_1-regularized L_2-SVM by bounding $|\hat{\mathbf{f}}^\top \boldsymbol{\theta}|$ on a tight convex set formed by the interaction of an n-dimensional hyper-ball and the upper half-space. An efficient binary search algorithm is designed and implemented to compute this bound for features. Empirical study shows that the proposed technique can greatly improve model selection efficiency by stably eliminating a large portion of the inactive features. Our ongoing work will extend the technique to screen features for the L_1-regularized L_1-SVM model and provide support for distributed computing in a massively parallel processing (MPP) environment.

Acknowledgments. The authors would like to thank Anne Baxter, Russell Albright, and the anonymous reviewers for their valuable suggestions to improve this paper.

References

[1] Liu, H., Motoda, H.: Feature Selection for Knowledge Discovery and Data Mining. Kluwer Academic Publishers, Boston (1998)

[2] Guyon, I., Elisseeff, A.: An introduction to variable and feature selection. JMLR 3, 1157–1182 (2003)

[3] Bradley, P.S., Mangasarian, L.O.: Feature selection via concave minimization and support vector machines. In: ICML (1998)

[4] Zhu, J., Rosset, S., Hastie, T., Tibshirani, R.: 1-norm support vector machines. In: NIPS (2003)

[5] Bi, J., Embrechts, M., Breneman, C.M., Song, M.: Dimensionality reduction via sparse support vector machines. JMLR 3, 1229–1243 (2003)

[6] Fan, R.E., Chang, K.W., Hsieh, C.J., Wang, X.R., Lin, C.J.: Liblinear: A library for large linear classification. JMLR 9, 1871–1874 (2008)

[7] Weston, J., Elisseff, A., Schoelkopf, B., Tipping, M.: Use of the zero norm with linear models and kernel methods. JMLR 3, 1439–1461 (2003)

[8] Guyon, I., Weston, J., Barnhill, S., Vapnik, V.: Gene selection for cancer classification using support vector machines. Machine Learning 46, 389–422 (2002)

[9] Li Wang, M.T., Tsang, I.W.: Learning sparse svm for feature selection on very high dimensional datasets. In: ICML (2010)

[10] Tibshirani, R.: Regression shrinkage and selection via the lasso. Journal of the Royal Statistical Society, Series B 58, 267–288 (1996)

[11] Ghaoui, L., Viallon, V., Rabbani, T.: Safe feature elimination in sparse supervised learning. Pacific Journal of Optimization 8, 667–698 (2012)

[12] Wang, J., Lin, B., Gong, P., Wonka, P., Ye, J.: Lasso screening rules via dual polytope projection. In: NIPS (2013)

[13] Zhen, J.X., Hao, X., Peter, J.R.: Learning sparse representations of high dimensional data on large scale dictionaries. In: NIPS (2011)

[14] Liu, J., Zhao, Z., Wang, J., Ye, J.: Safe screening with variational inequalities and its applicaiton to lasso. arXiv:1307.7577 (2013)

[15] Tibshirani, R., Bien, J., Friedman, J.H., Hastie, T., Simon, N., Taylor, J., Tibshirani, R.J.: Strong rules for discarding predictors in lasso-type problems. Journal of the Royal Statistical Society: Series B 74, 245–266 (2012)

[16] Lions, J.L., Stampacchia, G.: Variational inequalities. Communications on Pure and Applied Mathematics 20(3), 493–519 (1967)

[17] Boyd, S., Vandenberghe, L.: Convex Optimization. Cambridge University Press (2004)

[18] Tseng, P., Yun, S.: A coordinate gradient descent method for nonsmooth separable minimization. Mathematical Programming 117, 387–423 (2009)

Kernel Alignment Inspired Linear Discriminant Analysis

Shuai Zheng and Chris Ding

Department of Computer Science and Engineering,
University of Texas at Arlington, TX, USA
zhengs123@gmail.com, chqding@uta.edu

Abstract. Kernel alignment measures the degree of similarity between two kernels. In this paper, inspired from kernel alignment, we propose a new Linear Discriminant Analysis (LDA) formulation, kernel alignment LDA (kaLDA). We first define two kernels, data kernel and class indicator kernel. The problem is to find a subspace to maximize the alignment between subspace-transformed data kernel and class indicator kernel. Surprisingly, the kernel alignment induced kaLDA objective function is very similar to classical LDA and can be expressed using between-class and total scatter matrices. This can be extended to multi-label data. We use a Stiefel-manifold gradient descent algorithm to solve this problem. We perform experiments on 8 single-label and 6 multi-label data sets. Results show that kaLDA has very good performance on many single-label and multi-label problems.

Keywords: Kernel Alignment, LDA.

1 Introduction

Kernel alignment [2] is a way to incorporate class label information into kernels which are traditionally directly constructed from data without using class labels. Kernel alignment can be viewed as a measurement of consistency between the similarity function (the kernel) and class structure in the data. Improving this consistency helps to enforce data become more separated when using the class label aligned kernel. Kernel alignment has been applied to pattern recognition and feature selection recently [3,28,10,11,4].

In this paper, we find that if we use the widely used linear kernel and a kernel built from class indicators, the resulting kernel alignment function is very similar to the widely used linear discriminant analysis (LDA), using the well-known between-class scatter matrix S_b and total scatter matrix S_t. We call this objective function as kernel alignment induced LDA (kaLDA). If we transform data into a linear subspace, the optimal solution is to maximize this kaLDA.

We further analyze this kaLDA and propose a Stiefel-manifold gradient descent algorithm to solve it. We also extend kaLDA to multi-label problems. Surprisingly, the scatter matrices arising in multi-label kernel alignment are identical those matrices developed in Multi-label LDA [21].

T. Calders et al. (Eds.): ECML PKDD 2014, Part III, LNCS 8726, pp. 401–416, 2014.

We perform extensive experiments by comparing kaLDA with other approaches on 8 single-label datasets and 6 multi-label data sets. Results show that kernel alignment LDA approach has good performance in terms of classification accuracy and F1 score.

2   From Kernel Alignment to LDA

Kernel Alignment is a similarity measurement between a kernel function and a target function. In other words, kernel alignment evaluates the degree of fitness between the data in kernel space and the target function. For this reason, we usually set the target function to be the class indicator function. The other kernel function is the data matrix. By measuring the similarity between data kernel and class indicator kernel, we can get a sense of how easily this data can be separated in kernel subspace. The alignment of two kernels $\mathcal{K}_1$ and $\mathcal{K}_2$ is given as [2]:

$$A(\mathcal{K}_1, \mathcal{K}_2) = \frac{\mathrm{Tr}(\mathcal{K}_1 \mathcal{K}_2)}{\sqrt{\mathrm{Tr}(\mathcal{K}_1 \mathcal{K}_1)} \sqrt{\mathrm{Tr}(\mathcal{K}_2 \mathcal{K}_2)}}. \tag{1}$$

We first introduce some notations, and then present Theorem 1 and kernel alignment projective function.

Let data matrix be $X \in \Re^{p \times n}$ and $X = (\mathbf{x}_1, \cdots, \mathbf{x}_n)$, where p is data dimension, n is number of data points, $\mathbf{x}_i$ is a data point. Let normalized class indicator matrix be $Y \in \Re^{n \times K}$, which was used to prove the equivalence between PCA and K-means clustering [26,5], and

$$Y_{ik} = \begin{cases} \frac{1}{\sqrt{n_k}}, & \text{if point } i \text{ is in class } k. \\ 0, & \text{otherwise.} \end{cases} \tag{2}$$

where K is total class number, n_k is the number of data points in class k. Class mean is $\mathbf{m}_k = \sum_{\mathbf{x}_i \in k} \mathbf{x}_i / n_k$ and total mean of data is $\mathbf{m} = \sum_i \mathbf{x}_i / n$.

Theorem 1. *Define data kernel $\mathcal{K}_1$ and class label kernel $\mathcal{K}_2$ as follows:*

$$\mathcal{K}_1 = X^T X, \quad \mathcal{K}_2 = Y Y^T, \tag{3}$$

we have

$$A(\mathcal{K}_1, \mathcal{K}_2) = c \frac{TrS_b}{\sqrt{TrS_t^2}} \tag{4}$$

where $c = 1/\sqrt{Tr(YY^T)^2}$ is a constant independent of X.

Furthermore, let $G \in \Re^{p \times k}$ be a linear transformation to a k-dimensional subspace

$$\tilde{X} = G^T X, \quad \tilde{\mathcal{K}}_1 = \tilde{X}^T \tilde{X}, \tag{5}$$

we have

$$A(\widetilde{\mathcal{K}}_1, \mathcal{K}_2) = c \frac{Tr(G^T S_b G)}{\sqrt{Tr(G^T S_t G)^2}} \tag{6}$$

where

$$S_b = \sum_{k=1}^{K} n_k (\mathbf{m}_k - \mathbf{m})(\mathbf{m}_k - \mathbf{m})^T, \tag{7}$$

$$S_t = \sum_{i=1}^{n} (\mathbf{x}_i - \mathbf{m})(\mathbf{x}_i - \mathbf{m})^T, \tag{8}$$

Theorem 1 shows that kernel alignment can be expressed using scatter matrices S_b and S_t. In applications, we adjust G such that kernel alignment is maximized, i.e., we solve the following problem:

$$\max_{G} \frac{Tr(G^T S_b G)}{\sqrt{Tr(G^T S_t G)^2}}. \tag{9}$$

In general, columns of G are assumed to be linearly independent.

A striking feature of this kernel alignment problem is that it is very similar to classic LDA.

2.1   Proof of Theorem 1 and Analysis

Here we note a useful lemma and then prove Theorem 1.

In most data analysis, data are centered, i.e., $\sum_i \mathbf{x}_i = \mathbf{0}$. Here we assume data is already centered. The following results remain correct if data is not centered. We have the following relations:

Lemma 1. *Scatter matrices S_b, S_t can be expressed as:*

$$S_b = XYY^T X^T, \tag{10}$$

$$S_t = XX^T. \tag{11}$$

These results are previously known, for example, Theorem 3 of [5].

Proof of Theorem 1. To prove Eq.(4), we substitute $\mathcal{K}_1, \mathcal{K}_2$ into Eq.(1) and obtain, noting $Tr(AB) = Tr(BA)$,

$$A(\mathcal{K}_1, \mathcal{K}_2) = \frac{Tr(XYY^T X^T)}{\sqrt{Tr(XX^T)^2}\sqrt{Tr(YY^T)^2}} = c \frac{TrS_b}{\sqrt{TrS_t^2}}.$$

where we used Lemma 1. $c = 1/\sqrt{Tr(YY^T)^2}$ is a constant independent of data X.

To prove Eq.(6),

$$A(\widetilde{\mathcal{K}}_1, \mathcal{K}_2) = c \frac{Tr(G^T XYY^T X^T G)}{\sqrt{Tr(G^T XX^T G)^2}} = c \frac{Tr(G^T S_b G)}{\sqrt{Tr(G^T S_t G)^2}},$$

thus we obtain Eq.(6) using Lemma 1.

2.2   Relation to Classical LDA

In classical LDA, the between-class scatter matrix S_b is defined as Eq.(7), and the within-class scatter matrix S_w and total scatter matrix S_t are defined as:

$$S_w = \sum_{k=1}^{K} \sum_{\mathbf{x}_i \in k} (\mathbf{x}_i - \mathbf{m}_k)(\mathbf{x}_i - \mathbf{m}_k)^T, \ S_t = S_b + S_w, \tag{12}$$

where $\mathbf{m}_k$ and $\mathbf{m}$ are class means. Classical LDA finds a projection matrix $G \in \Re^{p \times (K-1)}$ that minimizes S_w and maximizes S_b using the following objective:

$$\max_G \operatorname{Tr} \frac{G^T S_b G}{G^T S_w G}, \tag{13}$$

or

$$\max_G \frac{\operatorname{Tr}(G^T S_b G)}{\operatorname{Tr}(G^T S_w G)}. \tag{14}$$

Eq.(14) is also called trace ratio (TR) problem [22]. It is easy to see [1] that Eq.(14) can be expressed as

$$\max_G \frac{\operatorname{Tr}(G^T S_b G)}{\operatorname{Tr}(G^T S_t G)}. \tag{15}$$

As we can see, kernel alignment LDA objective function Eq.(9) is very similar to Eq.(15). Thus kernel alignment provides an interesting alternative explanation of LDA. In fact, we can similarly show that in Eq.(9), S_w is also maximized as in the standard LDA. First, Eq.(9) is equivalent to

$$\max_G \operatorname{Tr}(G^T S_b G) \ s.t. \ \operatorname{Tr}(G^T S_t G)^2 = \eta,$$

where η is a fixed-value. The precise value of η is unimportant, since the scale of G is undefined in LDA: if G^* is an optimal solution, and r is any real number, $G^{**} = rG^*$ is also an optimal solution with the same optimal objective function value. The above optimization is approximately equivalent to

$$\max_G \operatorname{Tr}(G^T S_b G) \ s.t. \ \operatorname{Tr}(G^T S_t G) = \eta,$$

This is same as

$$\max_G \operatorname{Tr}(G^T S_b G) \ s.t. \ \operatorname{Tr}(G^T S_w G) = \eta - \operatorname{Tr}(G^T S_b G),$$

In other words, S_b is maximized while S_w is minimized — recovering the LDA main theme.

---

[1] Eq.(14) is equivalent to $\min \frac{\operatorname{Tr}(G^T S_w G)}{\operatorname{Tr}(G^T S_b G)}$, which is $\min \left(\frac{\operatorname{Tr}(G^T S_w G)}{\operatorname{Tr}(G^T S_b G)} + 1 \right)$. Reversing to maximization and using $S_t = S_b + S_w$, we obtain Eq.(15).

3   Computational Algorithm

In this section, we develop efficient algorithm to solve kaLDA objective function Eq.(9):

$$\max_{G} J_1 = \frac{\text{Tr}(G^T S_b G)}{\sqrt{\text{Tr}(G^T S_t G)^2}}, \quad s.t. \ G^T G = I. \tag{16}$$

The condition $G^T G = I$ ensures different columns of G mutually independent. The gradient of $J_1(G)$ is

$$\nabla J_1 \triangleq \frac{\partial J_1}{\partial G} = 2\frac{A}{\sqrt{\text{Tr}D^2}} - 2\frac{\text{Tr}B}{(\text{Tr}D^2)^{\frac{3}{2}}} CD, \tag{17}$$

where $A = S_b G$, $B = G^T A$, $C = S_t G$, $D = G^T C$.

Constraint $G^T G = I$ enforces G on the Stiefel manifold. Variations of G on this manifold is parallel transport, which gives some restriction to the gradient. This has been been worked out in [6]. The gradient that reserves the manifold structure is

$$\nabla J_1 - G[\nabla J_1]^T G. \tag{18}$$

Thus the algorithm computes the new G is given as follows:

$$G \leftarrow G - \eta(\nabla J_1 - G[\nabla J_1]^T G). \tag{19}$$

The step size η is usually chosen as:

$$\eta = \tau \|G\|_1 / \|\nabla J_1 - G(\nabla J_1)^T G\|_1, \quad \tau = 0.001 \sim 0.01. \tag{20}$$

where $\|G\|_1 = \sum_{ij} |G_{ij}|$.

Occasionally, due to the loss of numerical accuracy, we use projection $G \leftarrow G(G^T G)^{-\frac{1}{2}}$ to restore $G^T G = I$. Starting with the standard LDA solution of G, this algorithm is iterated until the algorithm converges to a local optimal solution. In fact, objective function will converge quickly when choosing η properly. Figure 1 shows that J_1 converges in about 200 iterations when $\tau = 0.001$, for datasets ATT, Binalpha, Mnist, and Umist (more details about the datasets will be introduced in experiment section). In summary, kernel alignment LDA (kaLDA) procedure is shown in Algorithm 1.

To show the effectiveness of proposed kaLDA, we visualize a real dataset in 2-D subspace in Figure 2. In this example, we take 3 classes of 644-dimension Umist data, 18 data points in each class. Figure 2a shows the original data projected in 2-D PCA subspace. Blue points are in class 1; red circle points are in class 2; black square points are in class 3. Data points from the three classes are mixed together in 2-D PCA subspace. It is difficult to find a linear boundary to separate points of different classes. Figure 2b shows the data in 2-D standard LDA subspace. We can see that data points in different classes have been projected into different clusters. Figure 2c shows the data projected in 2-D kaLDA subspace. Compared to Figure 2b, the within-class distance in Figure 2c is much smaller. The distance between different classes is larger.

Algorithm 1. $[G] = kaLDA(X, Y)$

Input: Data matrix $X \in \Re^{p \times n}$, class indicator matrix $Y \in \Re^{n \times K}$
Output: Projection matrix $G \in \Re^{p \times k}$
1: Compute S_b and S_t using Eq.(10) and Eq.(11)
2: Initialize G using classical LDA solution
3: **repeat**
4:     Compute gradient using Eq.(17)
5:     Update G using Eq.(19)
6: **until** J_1 Converges

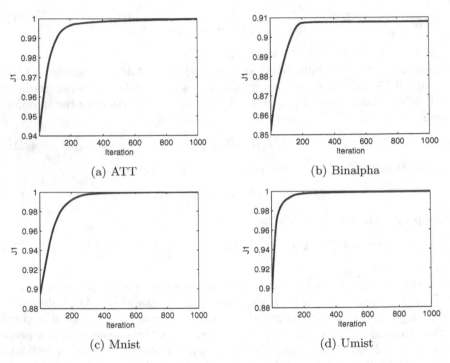

(a) ATT

(b) Binalpha

(c) Mnist

(d) Umist

Fig. 1. Objective J1 converges using Stiefel-manifold gradient descent algorithm ($\tau = 0.001$)

4   Extension to Multi-label Data

Multi-label problem arises frequently in image and video annotations, multi-topic text categorization, music classification. etc.[21]. In multi-label data, a data point could have several class labels (belonging to several classes). For example, an image could have "cloud", "building", "tree" labels. This is different from the case of single-label problem, where one point can have only one class label. Multi-label is very natural and common in our everyday life. For example, a film can be simultaneously classified as "drama", "romance", "historic" (if it

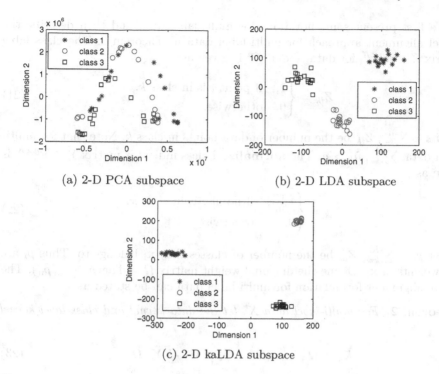

(a) 2-D PCA subspace

(b) 2-D LDA subspace

(c) 2-D kaLDA subspace

Fig. 2. Visualization of Umist data in 2-D PCA, 2-D LDA and 2-D kaLDA subspace

is about a true story). A news article can have topic labels such as "economics", "sports", etc.

Kernel alignment approach can be easily and naturally extended to multi-label data, because the class label kernel can be clearly and unambiguously defined using class label matrix Z on both single label and multi-label data sets. The data kernel is defined as usual. In the following we further develop this approach.

One important result of our kernel alignment approach for single label data is that it has close relationship with LDA. For multi-label data, each data point could belong to several classes. The standard scatter matrices S_b, S_w are ambiguous, because S_b, S_w are only defined for single label data where each data point belongs to one class only. However, our kernel alignment approach on multi-label data leads to new definitions of scatter matrices and similar objective function; this can be viewed as the generalization of LDA from single-label data to multi-label data via kernel alignment approach.

Indeed, the new scatter matrices we obtained from kernel alignment approach are identical to the so-called "multi-label LDA" [21] developed from a class-separate, probabilistic point of view, very different from our point of view. The fact that these two approaches lead to the same set of scatter matrices show that the resulting multi-label LDA framework has a broad theoretical basis.

We first present some notations for multi-label data and then describe the kernel alignment approach for multi-label data in Theorem 2. The class label matrix $Z \in \Re^{n \times K}$ for data $X \in \Re^{p \times n}$ is given as:

$$Z_{ik} = \begin{cases} 1, & \text{if point } i \text{ is in class } k. \\ 0, & \text{otherwise.} \end{cases} \tag{21}$$

Let $\tilde{n}_k = \sum_{i=1}^n Z_{ik}$ be the number of data points in class k. Note that for multi-label data, $\sum_{k=1}^K \tilde{n}_k > n$. The **normalized** class indicator matrix $\tilde{Y} \in \Re^{n \times K}$ is given as:

$$\tilde{Y}_{ik} = \begin{cases} \frac{1}{\sqrt{\tilde{n}_k}}, & \text{if point } i \text{ is in class } k. \\ 0, & \text{otherwise.} \end{cases} \tag{22}$$

Let $\rho_i = \sum_{k=1}^K Z_{ik}$ be the number of classes that $\mathbf{x}_i$ belongs to. Thus ρ_i are the weights of $\mathbf{x}_i$. Define the diagonal weight matrix $\Omega = \mathrm{diag}(\rho_1, \cdots, \rho_n)$. The kernel alignment formulation for multi-label data can be stated as

Theorem 2. *For multi-label data X, let the data kernel and class label kernel be*

$$\mathcal{K}_1 = \Omega^{\frac{1}{2}} X^T X \Omega^{\frac{1}{2}}, \quad \mathcal{K}_2 = \Omega^{-\frac{1}{2}} \tilde{Y} \tilde{Y}^T \Omega^{-\frac{1}{2}}. \tag{23}$$

We have the alignment

$$A(\mathcal{K}_1, \mathcal{K}_2) = c \frac{Tr S_b}{\sqrt{Tr S_t^2}} \tag{24}$$

where $c = 1/\sqrt{Tr(\Omega^{-1} \tilde{Y} \tilde{Y}^T)^2}$ is a constant independent of data X, and S_b, S_t are given in Eqs.(27, 28).

Furthermore, let $G \in \Re^{p \times k}$ be the linear transformation to a k-dimensional subspace,

$$\tilde{X} = G^T X, \quad \tilde{\mathcal{K}}_1 = \Omega^{1/2} \tilde{X}^T \tilde{X} \Omega^{1/2}, \tag{25}$$

we have

$$A(\tilde{\mathcal{K}}_1, \mathcal{K}_2) = c \frac{Tr(G^T S_b G)}{\sqrt{Tr(G^T S_t G)^2}} \tag{26}$$

The matrices S_b, S_t in Theorem 2 are defined as:

$$S_b = \sum_{k=1}^K \tilde{n}_k (\mathbf{m}_k - \mathbf{m})(\mathbf{m}_k - \mathbf{m})^T, \tag{27}$$

$$S_t = \sum_{k=1}^K \sum_{i=1}^n Z_{ik} (\mathbf{x}_i - \mathbf{m})(\mathbf{x}_i - \mathbf{m})^T, \tag{28}$$

where $\mathbf{m}_k$ is the mean of class k and $\mathbf{m}$ is global mean, defined as:

$$\mathbf{m}_k = \frac{\sum_{i=1}^{n} Z_{ik}\mathbf{x}_i}{\tilde{n}_k}, \quad \mathbf{m} = \frac{\sum_{i=1}^{n} \rho_i \mathbf{x}_i}{\sum_{k=1}^{K} \tilde{n}_k}. \tag{29}$$

Therefore, we can seek an optimal subspace for multi-label data by solving Eq.(16) with S_b, S_t given in Eqs.(27,28).

4.1 Proof of Theorem 2 and Equivalence to Multi-label LDA

Here we note a useful lemma for multi-label data and then prove Theorem 2. We consider the case the data is centered, i.e., $\sum_{i=1}^{n} \rho_i \mathbf{x}_i = \mathbf{0}$. The results also hold when data is not centered, but the proofs are slightly complicated.

Lemma 2. *For multi-label data, S_b, S_t of Eqs.(27,28) can be expressed as*

$$S_b = X\widetilde{Y}\widetilde{Y}^T X^T \tag{30}$$
$$S_t = X\Omega X^T \tag{31}$$

Proof. From the definition of $\mathbf{m}_k$ and $\widetilde{Y}$ in multi-label data, we have

$$X\widetilde{Y} = (\mathbf{m}_1, \cdots, \mathbf{m}_K) \begin{pmatrix} \sqrt{\tilde{n}_1} & & \\ & \ddots & \\ & & \sqrt{\tilde{n}_K} \end{pmatrix}.$$

Thus $X\widetilde{Y}\widetilde{Y}^T X^T = \sum_{k=1}^{K} \tilde{n}_k \mathbf{m}_k \mathbf{m}_k^T$ recovers S_b of Eq.(27).

To prove Eq.(31), note that $X\Omega = (\rho_1 \mathbf{x}_1, \cdots, \rho_n \mathbf{x}_n)$, thus $X\Omega X^T = \sum_{i=1}^{n} \rho_i \mathbf{x}_i \mathbf{x}_i^T$.

Proof of Theorem 2. Using Lemma 2, to prove Eq.(24),

$$A(\mathcal{K}_1, \mathcal{K}_2) = c\frac{\mathrm{Tr}(X\widetilde{Y}\widetilde{Y}^T X^T)}{\sqrt{\mathrm{Tr}(X\Omega X^T)^2}} = c\frac{\mathrm{Tr}S_b}{\sqrt{\mathrm{Tr}S_t^2}},$$

where $c = 1/\sqrt{\mathrm{Tr}(\Omega^{-1}\widetilde{Y}\widetilde{Y}^T)^2}$ is independent of X.

To prove Eq.(26),

$$A(\widetilde{\mathcal{K}}_1, \mathcal{K}_2) = c\frac{\mathrm{Tr}(G^T X\widetilde{Y}\widetilde{Y}^T X^T G)}{\sqrt{\mathrm{Tr}(G^T X\Omega X^T G)^2}} = c\frac{\mathrm{Tr}(G^T S_b G)}{\sqrt{\mathrm{Tr}(G^T S_t G)^2}}.$$

For single-label data, $\rho_i = 1$, $\Omega = I$, $\tilde{n}_k = n_k$, Eqs.(30, 31) reduce to Eqs.(10, 11), and Theorem 2 reduces to Theorem 1.

As we can see, surprisingly, the scatter matrices S_b, S_t of Eqs.(27, 28) arising in Theorem 2 are identical to that in Multi-label LDA proposed in [21].

Table 1. Single-label datasets attributes

Data	n	p	k
Caltec07	210	432	7
Caltec20	1230	432	20
MSRC	210	432	7
ATT	400	644	40
Binalpha	1014	320	26
Mnist	150	784	10
Umist	360	644	20
Pie	680	1024	68

Table 2. Classification accuracy on Single-label datasets ($K - 1$ dimension)

Data	kaLDA	LDA	TR	sdpLDA	MMC	RLDA	OCM
Caltec07	0.7524	0.6619	0.6762	0.5619	0.6000	**0.7952**	0.7619
Caltec20	**0.7068**	0.6320	0.4465	0.3386	0.5838	0.6812	0.6696
MSRC	**0.7762**	0.6857	0.5714	0.5952	0.5667	0.7333	0.7286
ATT	**0.9775**	0.9750	0.9675	0.9750	0.9750	0.9675	0.9675
Binalpha	0.7817	0.6078	0.4620	0.2507	0.7638	0.7983	**0.8204**
Mnist	**0.8800**	0.8733	0.8667	0.8467	0.8467	0.8667	0.8467
Umist	0.9900	0.9900	**0.9917**	0.9133	0.9633	0.9800	0.9783
Pie	0.8765	**0.8838**	0.8441	0.8632	0.8676	0.6515	0.6515

5   Related Work

Linear Discriminant Analysis (LDA) is a widely-used dimension reduction and subspace learning algorithm. There are many LDA reformulation publications in recent years. Trace Ratio problem is to find a subspace transformation matrix G such that the within-class distance is minimized and the between-class distance is maximized. Formally, Trace Ratio maximizes the ratio of two trace terms, $\max_G \mathrm{Tr}(G^T S_b G)/\mathrm{Tr}(G^T S_t G)$ [22,13], where S_t is total scatter matrix and S_b is between-class scatter matrix. Other popular LDA approach includes, regularized LDA(RLDA) [9], Orthogonal Centroid Method (OCM) [18], Uncorrelated LDA(ULDA) [23], Orthogonal LDA (OLDA) [23], etc.. These approaches mainly compute the eigendecomposition of matrix $S_t^{-1} S_b$, but use different formulation of total scatter matrix S_t [24].

Maximum Margin Criteria (MMC) [17] is a simpler and more efficient method. MMC finds a subspace projection matrix G to maximize $\mathrm{Tr}(G^T(S_b - S_w)G)$. Though in a different way, MMC also maximizes between-class distance while minimizing within-class distance. Semi-Definite Positive LDA (sdpLDA) [14]

Table 3. Multi-label datasets attributes

Data	n	p	k
MSRC-MOM	591	384	23
Barcelona	139	48	4
Emotion	593	72	6
Yeast	2,417	103	14
MSRC-SIFT	591	240	23
Scene	2,407	294	6

Table 4. Classification accuracy on Multi-label datasets ($K - 1$ dimension)

Data	kaLDA	MLSI	MDDM	MLLS	MLDA
MSRC-MOM	**0.9150**	0.8962	0.9044	0.8994	0.9036
Barcelona	**0.6579**	0.6436	0.6470	0.6524	0.6290
Emotion	**0.7634**	0.7397	0.7540	0.7529	0.7619
Yeast	**0.7405**	0.7317	0.7371	0.7364	0.7368
MSRC-SIFT	0.8839	0.8762	0.8800	0.8807	**0.8858**
Scene	**0.8870**	0.8534	0.8713	0.8229	0.8771

solves the maximization of $\mathrm{Tr}(G^T(S_b - \lambda_1 S_w)G)$, where λ_1 is the largest eigenvalue of $S_w^{-1}S_b$. sdpLDA is derived from the maximum margin principle.

Multi-label problem arise frequently in image and video annotations and many other related applications, such as multi-topic text categorization [21]. There are many Multi-label dimension reduction approaches, such as Multi-label Linear Regression (MLR), Multi-label informed Latent Semantic Indexing (MLSI) [25], Multi-label Dimensionality reduction via Dependence Maximization (MDDM) [27], Multi-Label Least Square (MLLS) [12], Multi-label Linear Discriminant Analysis (MLDA) [21].

6 Experiments

In this section, we first compare kernel alignment LDA (kaLDA) with other six different methods on 8 single label data sets and compare kaLDA multi-label version with four other methods on 6 multi-label data sets.

6.1 Comparison with Trace Ratio w.r.t. Subspace Dimension

Eight single-label datasets are used in this experiment. These datasets come from different domains, such as image scene Caltec [8] and MSRC [16], face datasets ATT, Umist, Pie [19], and digit datasets Mnist [15] and Binalpha. Table 1 summarizes the attributes of those datasets.

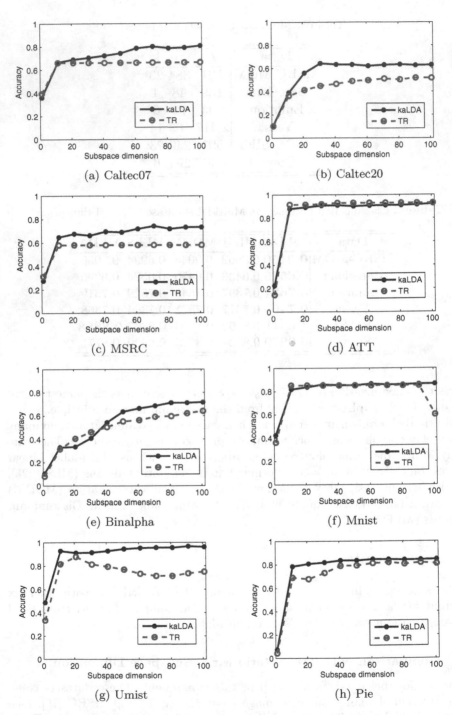

Fig. 3. Classification accuracy w.r.t. dimension of the subspace

Table 5. Macro F1 score on Multi-label datasets ($K - 1$ dimension)

Dataset	kaLDA	MLSI	MDDM	MLLS	MLDA
MSRC-MOM	**0.6104**	0.5244	0.5593	0.5426	0.5571
Barcelona	**0.7377**	0.7286	0.7301	0.7341	0.7169
Emotion	**0.6274**	0.5873	0.6101	0.6041	0.6200
Yeast	**0.5757**	0.5568	0.5696	0.5691	0.5693
MSRC-SIFT	0.4712	0.4334	0.4522	0.4544	**0.4773**
Scene	**0.6851**	0.5911	0.6411	0.5048	0.6568

Caltec07 and **Caltec20** are subsets of Caltech 101 data. Only the HOG feature is used in this paper.

MSRC is a image scene data, includes tree, building, plane, cow, face, car and so on. It has 210 images from 7 classes and each image has 432 dimension.

ATT data contains 400 images of 40 persons, with 10 images for each person. The images has been resized to 28×23.

Binalpha data contains 26 binary hand-written alphabets. It has 1014 images in total and each image has 320 dimension.

Mnist is a handwritten digits dataset. The digits have been size-normalized and centred. It has 10 classes and 150 images in total, with 784 dimension each image.

Umist is a face image dataset (Sheffield Face database) with 360 images from 20 individuals with mixed race, gender and appearance.

Pie is a face database collected by Carnegie Mellon Robotics Institute between October and December 2000. In total, it has 68 different persons.

In this part, we compare the classification accuracy of kaLDA and Trace Ratio [22] with respect to subspace dimension. The dimension of the subspace that kaLDA can find is not restricted to $K - 1$. After subspace projection, KNN classifier ($knn = 3$) is applied to perform classification. Results are shown in Figure 3. Solid line denotes kaLDA accuracy and dashed line denotes Trace Ratio accuracy. As we can see, in Figures 3a, 3b, 3c, 3g, and 3h, kaLDA has higher accuracy than Trace Ratio when using the same number of reduced features. In Figures 3d, 3e, 3f, kaLDA has competitive classification accuracy with Trace Ratio. However, kaLDA is more stable than Trace Ratio. For example, in Figure 3f and 3g, we observe a decrease in accuracy when feature number increases using Trace Ratio.

6.2 Comparison with other LDA Methods

We compare kaLDA with six other different methods, including LDA, Trace Ratio (TR), spdLDA, Maximum Margin Criteria (MMC), regularized LDA (RLDA), and Orthogonal Centroid Method (OCM). All LDA will reduce data to $K - 1$ dimension. KNN ($knn = 3$) will be applied to do the classification after data is

Table 6. Micro F1 score on Multi-label datasets ($K - 1$ dimension)

Dataset	kaLDA	MLSI	MDDM	MLLS	MLDA
MSRC-MOM	**0.5138**	0.4064	0.4432	0.4370	0.4448
Barcelona	**0.6969**	0.6891	0.6861	0.6904	0.6772
Emotion	**0.6203**	0.5779	0.6030	0.5961	0.6151
Yeast	**0.4249**	0.4026	0.4205	0.4216	0.4213
MSRC-SIFT	0.3943	0.3510	0.3637	0.3667	**0.3959**
Scene	**0.6966**	0.6006	0.6493	0.5062	0.6643

projected into the selected subspace. The other algorithms have already been introduced in related work section. The final classification accuracy is the average of 5-fold cross validation, and is reported in Table 2. The first column "kaLDA" reports kaLDA classification accuracy. kaLDA has the highest accuracy on 4 out of 8 datasets, including Caltec20, MSRC-MOM, ATT and Mnist. For Umist and Pie, kaLDA results are very close to the highest accuracy. Overall, kaLDA performs better than all other methods.

6.3 Multi-label Classification

Six multi-label datasets are used in this part. These datasets include images features, music emotion and so on. Table 3 summarizes the attributes of those datasets.

MSRC-MOM and **MSRC-SIFT** data set is provided by Microsoft Research in Cambridge. It includes 591 images of 23 classes. **MSRC-MOM** is the Moment invariants (MOM) feature of images and each image has 384 dimensions. **MSRC-SIFT** is the SIFT feature and each image has 240 dimensions. About 80% of the images are annotated with at least one classes and about three classes per image on average.

Barcelona data set contains 139 images with 4 classes, i.e., "building", "flora", "people" and "sky". Each image has at least two labels.

Emotion [20] is a music emotion data, which comprises 593 songs with 6 emotions. The dimension of Emotion is 72.

Yeast [7] is a multi-label data set which contains functional classes of genes in the Yeast Saccharomyces cerevisiae.

Scene [1] contains images of still scenes with semantic indexing. It has 2407 images from 6 classes.

We use 5-fold cross validation to evaluate classification performance of different algorithms. K-Nearest Neighbour (KNN) classifier is used after the subspace projection. The algorithms we compared in this section includes Multi-label informed Latent Semantic Indexing (MLSI), Multi-label Dimensionality reduction via Dependence Maximization (MDDM), Multi-Label Least Square (MLLS), Multi-label Linear Discriminant Analysis (MLDA). These algorithms have been introduced in related work section.

We compare the performance of kaLDA and other algorithms using macro accuracy (Table 4), macro-averaged F1-score (Table 5) and micro-averaged (Table 6) F1-score. Accuracy and F1 score are computed using standard binary classification definitions. In multi-label classification, macro average is a standard class-wise average, and it is related to number of samples in each class. However, micro average gives equal weight to all classes [21]. kaLDA achieves highest classification accuracy on 5 out of 6 datasets. On the remaining MSRC-SIFT dataset, kaLDA result is very close to the best method MLDA and beat all rest methods. kaLDA achieves highest macro and micro F1 score on 5 out of 6 datasets. Furthermore, kaLDA has the second highest macro and micro F1 score on dataset MSRC-SIFT. Overall, kaLDA outperforms other multi-label algorithms in terms of classification accuracy and macro and micro F1 score.

7 Conclusions

In this paper, we propose a new kernel alignment induced LDA (kaLDA). The objective function of kaLDA is very similar to classical LDA objective. The Stifel-manifold gradient descent algorithm can solve kaLDA objective efficiently. We have also extended kaLDA to multi-label problems. Extensive experiments show the effectiveness of kaLDA in both single-label and multi-label problems.

Acknowledgment. This work is partially supported by US NSF CCF-0917274 and NSF DMS-0915228.

References

1. Boutell, M.R., Luo, J., Shen, X., Brown, C.M.: Learning multi-label scene classification. Pattern Recognition 37(9), 1757–1771 (2004)
2. Cristianini, N., Shawe-taylor, J., Elisseeff, A., Kandola, J.S.: On kernel target alignment. Advances in Neural Information Processing Systems 14, 367 (2002)
3. Cristianini, N., et al.: Method of using kernel alignment to extract significant features from a large dataset. US Patent 7,299,213 (2007)
4. Cuturi, M.: Fast global alignment kernels. In: Proceedings of the 28th International Conference on Machine Learning (ICML 2011), pp. 929–936 (2011)
5. Ding, C., He, X.: K-means clustering via principal component analysis. In: Proc. of International Conference on Machine Learning, ICML 2004 (2004)
6. Edelman, A., Arias, T.A., Smith, S.T.: The geometry of algorithms with orthogonality constraints. SIAM Journal on Matrix Analysis and Applications 20(2), 303–353 (1998)
7. Elisseeff, A., Weston, J.: A kernel method for multi-labelled classification. In: NIPS, vol. 14, pp. 681–687 (2001)
8. Fei-Fei, L., Fergus, R., Perona, P.: Learning generative visual models from few training examples: An incremental bayesian approach tested on 101 object categories. Computer Vision and Image Understanding 106(1), 59–70 (2007)
9. Guo, Y., Hastie, T., Tibshirani, R.: Regularized linear discriminant analysis and its application in microarrays. Biostatistics 8(1), 86–100 (2007)

10. Hoi, S.C., Lyu, M.R., Chang, E.Y.: Learning the unified kernel machines for classification. In: Proceedings of the 12th ACM SIGKDD International Conference on Knowledge Discovery and Data Mining, pp. 187–196. ACM (2006)
11. Howard, A., Jebara, T.: Transformation learning via kernel alignment. In: International Conference on Machine Learning and Applications, ICMLA 2009, pp. 301–308. IEEE (2009)
12. Ji, S., Tang, L., Yu, S., Ye, J.: Extracting shared subspace for multi-label classification. In: Proceedings of the 14th ACM SIGKDD International Conference on Knowledge Discovery and Data Mining, pp. 381–389. ACM (2008)
13. Jia, Y., Nie, F., Zhang, C.: Trace ratio problem revisited. IEEE Transactions on Neural Networks 20(4), 729–735 (2009)
14. Kong, D., Ding, C.: A semi-definite positive linear discriminant analysis and its applications. In: 2012 IEEE 12th International Conference on Data Mining (ICDM), pp. 942–947. IEEE (2012)
15. LeCun, Y., Bottou, L., Bengio, Y., Haffner, P.: Gradient-based learning applied to document recognition. Proceedings of the IEEE 86(11), 2278–2324 (1998)
16. Lee, Y.J., Grauman, K.: Foreground focus: Unsupervised learning from partially matching images. International Journal of Computer Vision 85(2), 143–166 (2009)
17. Li, H., Jiang, T., Zhang, K.: Efficient and robust feature extraction by maximum margin criterion. IEEE Transactions on Neural Networks 17(1), 157–165 (2006)
18. Park, H., Jeon, M., Rosen, J.B.: Lower dimensional representation of text data based on centroids and least squares. BIT Numerical Mathematics 43(2), 427–448 (2003)
19. Sim, T., Baker, S., Bsat, M.: The cmu pose, illumination, and expression (pie) database of human faces. Tech. Rep. CMU-RI-TR-01-02, Robotics Institute, Pittsburgh, PA (January 2001)
20. Trohidis, K., Tsoumakas, G., Kalliris, G., Vlahavas, I.P.: Multi-label classification of music into emotions. In: ISMIR, vol. 8, pp. 325–330 (2008)
21. Wang, H., Ding, C., Huang, H.: Multi-label linear discriminant analysis. In: Daniilidis, K., Maragos, P., Paragios, N. (eds.) ECCV 2010, Part VI. LNCS, vol. 6316, pp. 126–139. Springer, Heidelberg (2010)
22. Wang, H., Yan, S., Xu, D., Tang, X., Huang, T.: Trace ratio vs. ratio trace for dimensionality reduction. In: IEEE Conference on Computer Vision and Pattern Recognition, CVPR 2007, pp. 1–8. IEEE (2007)
23. Ye, J.: Characterization of a family of algorithms for generalized discriminant analysis on undersampled problems. Journal of Machine Learning Research, 483–502 (2005)
24. Ye, J., Ji, S.: Discriminant analysis for dimensionality reduction: An overview of recent developments. Biometrics: Theory, Methods, and Applications. Wiley-IEEE Press, New York (2010)
25. Yu, K., Yu, S., Tresp, V.: Multi-label informed latent semantic indexing. In: Proceedings of the 28th Annual International ACM SIGIR Conference on Research and Development in Information Retrieval, pp. 258–265. ACM (2005)
26. Zha, H., Ding, C., Gu, M., He, X., Simon, H.: Spectral relaxation for K-means clustering. In: Advances in Neural Information Processing Systems 14 (NIPS 2001), pp. 1057–1064 (2001)
27. Zhang, Y., Zhou, Z.H.: Multilabel dimensionality reduction via dependence maximization. ACM Transactions on Knowledge Discovery from Data (TKDD) 4(3), 14 (2010)
28. Zhu, X., Kandola, J., Ghahramani, Z., Lafferty, J.D.: Nonparametric transforms of graph kernels for semi-supervised learning. In: Advances in Neural Information Processing Systems, pp. 1641–1648 (2004)

Transfer Learning with Multiple Sources via Consensus Regularized Autoencoders

Fuzhen Zhuang[1], Xiaohu Cheng[1,2], Sinno Jialin Pan[3],
Wenchao Yu[1,2], Qing He[1], and Zhongzhi Shi[1]

[1] Key Laboratory of Intelligent Information Processing, Institute of Computing Technology,
Chinese Academy of Sciences, Beijing 100190, China
[2] University of Chinese Academy of Sciences, Beijing 100049, China
[3] Institute for Infocomm Research, Singapore
{zhuangfz,chengxh,yuwc,heq,shizz}@ics.ict.ac.cn
sinnocat@gmail.com

Abstract. Knowledge transfer from multiple source domains to a target domain is crucial in transfer learning. Most existing methods are focused on learning weights for different domains based on the similarities between each source domain and the target domain or learning more precise classifiers from the source domain data jointly by maximizing their consensus of predictions on the target domain data. However, these methods only consider measuring similarities or building classifiers on the original data space, and fail to discover a more powerful feature representation of the data when transferring knowledge from multiple source domains to the target domain. In this paper, we propose a new framework for transfer learning with multiple source domains. Specifically, in the proposed framework, we adopt autoencoders to construct a feature mapping from an original instance to a hidden representation, and train multiple classifiers from the source domain data jointly by performing an entropy-based consensus regularizer on the predictions on the target domain. Based on the framework, a particular solution is proposed to learn the hidden representation and classifiers simultaneously. Experimental results on image and text real-world datasets demonstrate the effectiveness of our proposed method compared with state-of-the-art methods.

Keywords: Transfer Learning, Multiple Sources, Consensus Regularization, Feature Representation.

1 Introduction

Transfer learning or domain adaptation aims to extract common knowledge across domains such that a model trained on one domain can be adapted effectively to other domains [16]. In the past decade, a number of transfer learning methods have been proposed, most of which are focused on the 1vs1 transfer learning setting, where only one source domain and one target domain are assumed to be available when knowledge is transferred. However, in many real-world scenarios, given a target domain, there may be more than one source domain available for building classifiers. In this case, how to fully utilize multiple sources to ensure effective knowledge transfer is crucial.

T. Calders et al. (Eds.): ECML PKDD 2014, Part III, LNCS 8726, pp. 417–431, 2014.

So far, there are several works proposed for transfer learning with multiple source domains [8,27,7,4,9]. Most of them are focused on learning weights for different domains based on the similarities between each source domain and the target domain or learning more precise classifiers from the source domain data jointly by maximizing their consensus of predictions on the target domain data. For instance, Gao et al. [8] proposed a lazy ensemble method for multi-source transfer learning. Specifically, a number of supervised classifiers are trained from the source domains, then given an instance in the target domain, its local structure constructed in the source domains is used to estimate the weights for different source-domain classifiers to make predictions. Zhuang et al. [27] proposed a consensus regularization framework for multi-source transfer learning, where classifiers trained on multiple source domains are optimized jointly not only to achieve high prediction results on the corresponding domains, but also to make consistent predictions on target domain data. Similarly, Chattopadhyay et al. [4] introduced a transfer learning framework based on the multi-source domain adaptation methodology for detecting different stages of fatigue using surface electromyography signals. The works [7,4,9] need a few labeled data in the target domain, while in our work there are only labeled data in the source domains.

A common characteristic of most transfer learning methods with multiple domains is that knowledge transfer is performed on the original data space. However, in many applications, the supports of features of different domains may not be the same. In other words, there may exist domain-specific features in different domains, e.g., different product domains have their specific opinion words [1,14]. In this case, adapting models on the original data space may not be able to transfer knowledge effectively. Moreover, in many other applications, the data observed may be very complex, e.g., sensor signals. In this case, measuring similarity or dissimilarity between domains on the original data space may not be precise, which may limit the transferability across domains [13,15]. To address these issues, another branch of methods, which is referred to as the feature-based transfer learning approach, has been proposed in the 1vs1 transfer learning setting. The motivation of this approach is to learn a feature mapping or transformation to map the original data to a new feature space where the difference or distance between different domains can be reduced implicitly or explicitly.

Motivated by the idea of the feature-based methods in the 1vs1 transfer learning setting, in this paper, we propose an embedding-based framework for multi-source transfer learning. Specifically, in the proposed framework, we first adopt autoencoders [10] to construct a feature mapping to map an original instance to a hidden representation. Note that this mapping is shared by all the source and target domain data. We then train multiple classifiers on different source domain labeled data with the hidden representation jointly by introducing an entropy-based consensus regularizer on the predictions on the target domain data with the hidden representation. Based on the framework, a particular solution is proposed to learn the hidden representation and consensus regularized classifiers simultaneously. Different from the existing work proposed by Zhuang et al. [27], where a consensus regularizer is performing on the original data space, our model instead of a hidden feature space. We believe the great success of representation learning of autoencoders can lead to better transferability of our framework. As will be shown in the Experimental section, extensive experiments on image and text datasets verify our hypotheses and demonstrate the superiority of our proposed framework over a variety of state-of-the-art methods.

2 Notations and Preliminaries

In this section, we first introduce some frequently used notations as presented in Table 1, and some preliminaries which will be used in our proposed framwork.

Table 1. The Notation and Denotation

$\mathcal{D}^{(i)}$	A data domain i
r	The number of source domains
m	The number of original features of a data domain
n_i	The number of instances of a data domain i
k	The number of hidden features
x	An original instance
y	A class label
$\hat{x}$	The reconstruction of x
z	An embedded instance
W, b	A weight matrix and bias vector of encoding
W', b'	A weight matrix and bias vector of decoding
θ_i	A vector of parameters of a classifier i
$\top$	The transposition of a matrix
$\circ$	The dot product of vectors or matrixes

2.1 Logistic Regression

In our proposed framework, we adopt logistic regression [6] as the base classifier. Note that the proposed framework is general, thus other types of classifiers can also be plugged into our framework. The goal of logistic regression is to estimate a conditional probability $P(y|x)$ in terms of a vector of parameters $\theta \in \mathbb{R}^{m \times 1}$ by solving the following maximization problem,

$$\min_{\theta} \sum_{i=1}^{n} \log \sigma(y_i \theta^\top x_i) - \lambda \theta^\top \theta, \tag{1}$$

over a set of labeled data $\{x_i, y_i\}_{i=1}^{n}$, where $x_i \in \mathbb{R}^{m \times 1}$ is an input instance, y_i is its correspondingly discrete output, e.g., for binary classification $y_i \in \{-1, 1\}$, and $\sigma(u)$ is a sigmoid function defined as follows,

$$\sigma(u) = \frac{1}{1 + e^{-u}}. \tag{2}$$

The second term in (1) is a regularization term to avoid overfitting, where the trade-off parameter λ is a small positive constant. After θ is estimated, the conditional probability of y given x can be computed by

$$p(y|x; \theta) = \sigma(y\theta^\top x), \tag{3}$$

which is used to classify target domain data, i.e., the predicted label of x is $\max_y p(y|x; \theta)$.

2.2   Autoencoders

An autoencoder first maps an input instance x to a hidden representation z through an encoding mapping:

$$z = h(\mathbf{W}x + b),$$

where h is a nonlinear activation function, $\mathbf{W} \in \mathbb{R}^{k \times m}$ is a weight matrix, and $b \in \mathbb{R}^{k \times 1}$ is a bias vector. The resulting latent representation z is then mapped back to a reconstruction $\hat{x}$ through a decoding mapping:

$$\hat{x} = g(\mathbf{W}'z + b'),$$

where g is a nonlinear activation function, $\mathbf{W}' \in \mathbb{R}^{m \times k}$ is a weight matrix, and $b' \in \mathbb{R}^{m \times 1}$ is a bias vector. Given a set of inputs $\{x_i\}_{i=1}^n$, the parameters of an autoencoder are optimized by minimizing the reconstruction error as follows,

$$\min_{\mathbf{W}, b, \mathbf{W}', b'} = \sum_{i=1}^n \|x_i - \hat{x}_i\|^2. \tag{4}$$

Note that, in this paper we adopt the sigmoid function σ defined in (2), which is widely used in constructing autoencoders, as the nonlinear activation functions g and h for encoding and decoding respectively.

2.3   Consensus Measure

Given r classifiers in terms of their parameter vectors $(\theta_1, \theta_2, \cdots, \theta_r)$ and an instance x, for a specific class c, we denote $(p_1(c), p_2(c), \cdots, p_r(c))$ a vector of the predicted probabilities $P(y = c|x)$ of the r classifiers accordingly. Then the consensus measure of the predictions of the r classifiers on x is given by

$$\psi(x; \{\theta_i\}_{i=1}^r) = -\sum_{c \in \mathcal{C}} \bar{p}(c) \log \frac{1}{\bar{p}(c)}, \tag{5}$$

where $\bar{p}(c) = \frac{1}{r} \sum_{i=1}^r p_i(c)$, and $\mathcal{C}$ is the total set of classes. As shown in [27], maximizing (5) is equivalent to enforcing the r classifiers to make consistent predictions on x as well as minimizing the entropy of the predictions of each classifier on x. For binary classification, (5) can be rewritten as

$$\psi(x; \{\theta_i\}_{i=1}^r) = (\bar{p} - (1 - \bar{p}))^2 = (2\bar{p} - 1)^2. \tag{6}$$

Note that we say (5) and (6) are equivalent for binary classification in the sense that they have the same effect that: when maximizing them, the predictions on any instance from all the classifiers (from the different domains) are similar. Thus, their effects on making the prediction consensus are similar, though their value scales are not the same. In this paper, we focus on binary classification, thus adopt (6) as the consensus measure in the following section.

3 Consensus Regularized Autoencoders

3.1 Problem Formalization

Given r source domains $\mathcal{D}_S^{(1)}, \cdots, \mathcal{D}_S^{(r)}$, where for each source domain $j \in \{1, \cdots, r\}$, there are n_j labeled data, i.e., $\mathcal{D}_S^{(j)} = \left\{ \boldsymbol{x}_{S_i}^{(j)}, y_{S_i}^{(j)} \right\}_{i=1}^{n_j}$, where $y_{S_i}^{(j)} \in \{-1, 1\}$, and a target domain $\mathcal{D}_T$ without any labeled data, i.e, $\mathcal{D}_T = \{\boldsymbol{x}_{T_i}, y_{T_i}\}_{i=1}^{n}$, the goal is to train a classifier f to make precise predictions on $\mathcal{D}_T$ or previously unseen instances in the target domain. Note that, in our transfer scenario there is not any labeled data in the target domain.

Our proposed optimization problem for multi-source transfer learning is formulated as follows,

$$\min_{\Theta, \Theta', \{\boldsymbol{\theta}_j\}} \mathcal{J} = \epsilon(\boldsymbol{x}_S, \hat{\boldsymbol{x}}_S, \boldsymbol{x}_T, \hat{\boldsymbol{x}}_T) + \gamma \Omega(\Theta, \Theta')$$
$$+ \alpha \ell(\boldsymbol{z}_S, y_S; \{\boldsymbol{\theta}_j\}) - \beta \psi(\boldsymbol{z}_T; \{\boldsymbol{\theta}_j\}), \tag{7}$$

where the first term in the objective is the reconstruction error of the source and target domain data, which can be written as follows,

$$\epsilon(\boldsymbol{x}_S, \hat{\boldsymbol{x}}_S, \boldsymbol{x}_T, \hat{\boldsymbol{x}}_T) = \sum_{j=1}^{r} \sum_{i=1}^{n_j} \|\boldsymbol{x}_{S_i} - \hat{\boldsymbol{x}}_{S_i}\|^2 + \sum_{i=1}^{n} \|\boldsymbol{x}_{T_i} - \hat{\boldsymbol{x}}_{T_i}\|^2,$$

and

$$\boldsymbol{z}_{S_i}^{(j)} = \sigma(\mathbf{W}\boldsymbol{x}_{S_i}^{(j)} + \boldsymbol{b}), \ \boldsymbol{z}_{T_i} = \sigma(\mathbf{W}\boldsymbol{x}_{T_i} + \boldsymbol{b}),$$
$$\hat{\boldsymbol{x}}_{S_i}^{(j)} = \sigma(\mathbf{W}'\boldsymbol{z}_{S_i}^{(j)} + \boldsymbol{b}'), \ \hat{\boldsymbol{x}}_{T_i} = \sigma(\mathbf{W}'\boldsymbol{z}_{T_i} + \boldsymbol{b}').$$

The second term in the objective is a regularization term on the parameters $\Theta = \{\mathbf{W}, \boldsymbol{b}\}$ and $\Theta' = \{\mathbf{W}', \boldsymbol{b}'\}$, which can be written as

$$\Omega(\Theta, \Theta') = (\|\mathbf{W}\|^2 + \|\boldsymbol{b}\|^2 + \|\mathbf{W}'\|^2 + \|\boldsymbol{b}'\|^2).$$

The third term in (7) is the total loss of each source classifiers over the corresponding source label data with the hidden representation, which can be written as

$$\ell(\boldsymbol{z}_S, y_S; \{\boldsymbol{\theta}_j\}) = \sum_{j=1}^{r} \left(-\sum_{i=1}^{n_j} \log \sigma(y_{S_i}^{(j)} \boldsymbol{\theta}_j^\top \boldsymbol{z}_{S_i}^{(j)}) + \lambda \boldsymbol{\theta}_j^\top \boldsymbol{\theta}_j \right),$$

where $\boldsymbol{\theta}_j \in \mathbb{R}^{k \times 1}$. The last term in (7) is the consensus regularization terms of the predictions of the source classifiers on the target domain data, which can be written as

$$\psi(\boldsymbol{z}_T; \{\boldsymbol{\theta}_j\}) = \sum_{i=1}^{n} \left\| 2\frac{\sum_{j=1}^{r} \sigma(\boldsymbol{\theta}_j^\top \boldsymbol{z}_{T_i})}{r} - 1 \right\|^2.$$

The trade-off parameters α, β, γ and λ are small positive contents to balance the effect of different terms to the overall objective (7).

3.2   A Particular Solution

The optimization problem (7) is an unconstrained optimization with five types of variables $\mathbf{W}$, $\mathbf{b}$, $\mathbf{W}'$, $\mathbf{b}'$ and $\{\boldsymbol{\theta}_j\}$'s to be optimized, and does not have closed form solutions. To derive the solutions of the five types of variables, we propose to use gradient descent methods. To simplify the math expressions, we first introduce the following intermediate variables.

$$
A_{S_i}^{(j)} = \left(\hat{\boldsymbol{x}}_{S_i}^{(j)} - \boldsymbol{x}_{S_i}^{(j)} \right) \circ \hat{\boldsymbol{x}}_{S_i}^{(j)} \circ \left(1 - \hat{\boldsymbol{x}}_{S_i}^{(j)} \right),
$$

$$
A_{T_i} = (\hat{\boldsymbol{x}}_{T_i} - \boldsymbol{x}_{T_i}) \circ \hat{\boldsymbol{x}}_{T_i} \circ (1 - \hat{\boldsymbol{x}}_{T_i}),
$$

$$
B_{S_i}^{(j)} = \boldsymbol{z}_{S_i}^{(j)} \circ \left(1 - \boldsymbol{z}_{S_i}^{(j)} \right),
$$

$$
B_{T_i} = \boldsymbol{z}_{T_i} \circ (1 - \boldsymbol{z}_{T_i}),
$$

$$
C_{T_i}^{(j)} = \sigma(\boldsymbol{\theta}_j^\top \boldsymbol{z}_{T_i}) \left(1 - \sigma(\boldsymbol{\theta}_j^\top \boldsymbol{z}_{T_i}) \right).
$$

Then, it can be shown that the partial derivatives of the objective $\mathcal{J}$ in (7) with respect to $\mathbf{W}$, $\mathbf{b}$, $\mathbf{W}'$, $\mathbf{b}'$ and $\{\boldsymbol{\theta}_j\}$'s can be computed as follows respectively,

$$
\begin{aligned}
\frac{\partial \mathcal{J}}{\partial \mathbf{W}} =\ & 2\mathbf{W}'^\top \left(\sum_{j=1}^{r}\sum_{i=1}^{n_j} A_{S_i}^{(j)} \circ B_{S_i}^{(j)} \boldsymbol{x}_{S_i}^{(j)\top} + \sum_{i=1}^{n} A_{T_i} \circ B_{T_i} \boldsymbol{x}_{T_i}^\top \right) \\
& - \alpha \sum_{j=1}^{r}\sum_{i=1}^{n_j} \left(1 - \sigma(y_{S_i}^{(j)} \boldsymbol{\theta}_j^\top \boldsymbol{z}_{S_i}^{(j)}) \right) y_{S_i}^{(j)} \boldsymbol{\theta}_j \circ B_{S_i}^{(j)} \boldsymbol{x}_{S_i}^{(j)\top} \\
& - \frac{4\beta}{r^2} \sum_{i=1}^{n} \left(\left(2\sum_{j=1}^{r} \sigma(\boldsymbol{\theta}_j^\top \boldsymbol{z}_{T_i}) - r \right) \sum_{j=1}^{r} (C_{T_i}^{(j)} \boldsymbol{\theta}_j \circ B_{S_i}^{(j)} \boldsymbol{x}_{T_i}^\top) \right) \\
& + 2\gamma \mathbf{W},
\end{aligned}
\tag{8}
$$

$$
\begin{aligned}
\frac{\partial \mathcal{J}}{\partial \mathbf{b}} =\ & 2\mathbf{W}'^\top \left(\sum_{j=1}^{r}\sum_{i=1}^{n_j} A_{S_i}^{(j)} \circ B_{S_i}^{(j)} + \sum_{i=1}^{n} A_{T_i} \circ B_{T_i} \right) \\
& - \alpha \sum_{j=1}^{r}\sum_{i=1}^{n_j} \left(1 - \sigma(y_{S_i}^{(j)} \boldsymbol{\theta}_j^\top \boldsymbol{z}_{S_i}^{(j)}) \right) y_{S_i}^{(j)} \boldsymbol{\theta}_j \circ B_{S_i}^{(j)} \\
& - \frac{4\beta}{r^2} \sum_{i=1}^{n} \left(\left(2\sum_{j=1}^{r} \sigma(\boldsymbol{\theta}_j^\top \boldsymbol{z}_{T_i}) - r \right) \sum_{j=1}^{r} (C_{T_i}^{(j)} \boldsymbol{\theta}_j \circ B_{S_i}^{(j)}) \right) \\
& + 2\gamma \mathbf{b},
\end{aligned}
\tag{9}
$$

$$
\frac{\partial \mathcal{J}}{\partial \mathbf{W}'} = \sum_{j=1}^{r}\sum_{i=1}^{n_j} 2A_{S_i}^{(j)} \boldsymbol{z}_{S_i}^{(j)\top} + \sum_{i=1}^{n} 2A_{T_i} \boldsymbol{z}_{T_i}^\top + 2\gamma \mathbf{W}',
\tag{10}
$$

$$\frac{\partial \mathcal{J}}{\partial b'} = \sum_{j=1}^{r} \sum_{i=1}^{n_j} 2A_{S_i}^{(j)} + \sum_{i=1}^{n} 2A_{T_i} + 2\gamma b', \tag{11}$$

$$\frac{\partial \mathcal{J}}{\partial \theta_j} = \alpha \left(-\sum_{i=1}^{n_j} \left(1 - \sigma(y_{S_i}^{(j)} \theta_j^{\mathsf{T}} z_{S_i}^{(j)})\right) y_{S_i}^{(j)} z_{S_i}^{(j)\mathsf{T}} + 2\lambda \theta_j^{\mathsf{T}} \right)$$
$$-\frac{4\beta}{r^2} \sum_{i=1}^{n} \sum_{j=1}^{r} \left(2\sigma(\theta_j^{\mathsf{T}} z_{T_i}) - r\right) C_{T_i}^{(j)} z_{T_i}^{\mathsf{T}}. \tag{12}$$

Based on the above partial derivatives, with an initialization of $\mathbf{W}$, b, $\mathbf{W}'$, b' and $\{\theta_j\}$'s, we can update them alternatively and iteratively by applying the following rules till the solutions are converged,

$$\mathbf{W} \leftarrow \mathbf{W} - \eta \frac{\partial \mathcal{J}}{\partial \mathbf{W}}, \qquad b \leftarrow b - \eta \frac{\partial \mathcal{J}}{\partial b},$$
$$\mathbf{W}' \leftarrow \mathbf{W}' - \eta \frac{\partial \mathcal{J}}{\partial \mathbf{W}'}, \qquad b' \leftarrow b' - \eta \frac{\partial \mathcal{J}}{\partial b'}, \tag{13}$$
$$\theta_j \leftarrow \theta_j - \eta \frac{\partial \mathcal{J}}{\partial \theta_j},$$

where η is a learning rate. That is, in each iteration, we alteratively fix four of the five types of the variables and optimize the rest one.

3.3 Target Classifier Construction

After the solutions of $\mathbf{W}$, b, $\mathbf{W}'$, b' and $\{\theta_j\}$'s are obtained, one can construct a classifier f_T in terms of θ_T for the target domain in two ways. One way is to construct the classifier combining all source classifiers $\{\theta_j\}$'s based on a voting scheme. That is, for any instance x_T from the target domain, which can be either from the observed unlabeled sample $\mathcal{D}_T$ or unseen data sample, the classifier f_T make a prediction on it based on

$$f_T(x_T) = \frac{1}{r} \sum_{j=1}^{r} \sigma \left(\theta_j^{\mathsf{T}} (\sigma(\mathbf{W} x_T + b)) \right).$$

Alternatively, another way to construct a target classifier is to first map instances from all the source domains to their corresponding hidden representations by $z_{S_i}^{(j)} = \sigma \left(\mathbf{W} x_{S_i}^{(j)} + b \right)$, and then apply standard classification algorithms, e.g., logistic regression or Support Vector Machine (SVM) [2], on the labeled data, $\{z_{S_i}^{(j)}, y_{S_i}^{(j)}\}_{i=1,\cdots,n_j}^{j=1,\cdots,r}$, to train a unified classifier f_T in terms of a vector of parameter θ_T. For any instance x_T from the target domain, one can first map it to an hidden representation by $z_T = \sigma(\mathbf{W} x_T + b)$, and then use θ_T to make an prediction. In the sequel, we denote Consensus Regularized Autoencoders (CRA) for our proposed framework and the particular solution. The overall algorithm of CRA is summarized in Algorithm 1.

Algorithm 1. Consensus Regularized Autoencoders (CRA)

Input: Given r source domains $\mathcal{D}_S^{(1)}, \cdots, \mathcal{D}_S^{(r)}$, where $\mathcal{D}_S^{(j)} = \{x_{S_i}^{(j)}, y_{S_i}^{(j)}\}_{i=1}^{n_j}$, a target domain $D_T = \{x_{T_i}\}_{i=1}^n$, trade-off parameters $\alpha, \beta, \gamma, \lambda$, and the number of hidden features k.
Output: A classifier on the target domain.

1. Initialize $\mathbf{W}, b, \mathbf{W}'$, and b' by performing an autoencoder algorithm on instances of all the domains, and train $\{\theta_j\}$'s on the corresponding domain data independently.
2. Fix $\{\theta_j\}$'s, update $\mathbf{W}, b, \mathbf{W}'$, and b' alteratively based on the update rules in (13) and the corresponding derivatives in (8), (9), (10) and (11).
3. Fix $\mathbf{W}, b, \mathbf{W}'$, and b', update $\{\theta_j\}$'s based on the update rules in (13) and the corresponding derivative in (12).
4. If the solutions are converged, construct a target classifier as described in Section 3.3, otherwise, go to Step 2.

Table 2. Description of the image dataset

	flower				*traffic*			
	sunflower	*rose*	*lotus*	*tulip*	*aviation*	*bus*	*boat*	*dogsled*
No. of instance	85	100	66	100	100	100	100	100

4 Experiments

In this section, we conduct extensive experiments on two real-world datasets to systemically evaluate the effectiveness of our proposed method for multi-source transfer learning.

4.1 Datasets

Image Dataset. We conduct experiments on the image dataset of multi-source transfer learning problems used in [27]. The dataset contains two main categories, *flower* and *traffic*, selected from the COREL collection[1]. Each main category further contains four subcategories. The *flower* category can be further classified into *sunflower*, *rose*, *lotus* and *tulip*, while the *traffic* category can be further classified into *aviation*, *bus*, *boat* and *dogsled*. Figure 1 shows one example of each subcategory respectively. Following the same preprocessing proposed in [27], we randomly select one subcategory from *flower* and one subcategory from *traffic* to construct a domain, thus can construct 24 (4!) different groups of domains, where each group contains 4 different domains and each subcategory appears once and only once in each group. In each group, we then randomly select one domain as the target domain, and the rest 3 domains as the source domains. Finally, we can construct 96 (4 × 4!) multi-source (3 source domains) image classification problems. Each image is represented by 87 features, which include 36 features are based on color histogram [25] and 51 features are based on SILBP texture histogram [19]. The description of the image dataset is summarized in Table 2.

---

[1] http://archive.ics.uci.edu/ml/datasets/Corel+Image+Features

Fig. 1. Examples of the eight subcategories of the dataset

Sentiment Dataset. We use the Multi-domain sentiment benchmark dataset generated by [1] for experiments. The dataset contains reviews of 4 types of products, books, dvd, electronics, and kitchen, crawled from Amazon.com. Each product review is annotated as positive or negative based on its overall sentiment polarity. Each type of products is considered as a domain, and each domain contains 2,000 reviews, of which 1,000 are positive and the other 1,000 are negative. Each review is represented as a vector of 3126 word features. Following similar preprocessing used in [1], we randomly select one of the 4 domains as the target domain, and the rest 3 domains as the source domains. Therefore, we can conduct four multi-source sentiment classification problems.

4.2   Baseline Methods and Implementation Details

Baseline Methods. We compare our proposed method CRA with various baseline methods, including the standard logistic regression (LR) and SVM without transfer learning, an embedding method based on autoencoders (EAER) [23], a dimensionality reduction method for 1vs1 transfer learning problems, Transfer Component Analysis (TCA) [15], the Centralized Consensus Regularization (CCR_3) [27] for multi-source transfer learning problems on the original data space, and a recently proposed 1vs1 transfer learning method based on autoencoders, marginalized Stacked Denoising Autoencoders (mSDA) [5].

Note that the methods EAER and TCA only map original data to a latent space, where a classifier needs to be further specified for final classification problems. Here, we consider LR or SVM as the base classifier for EAER and TCA. Moreover, except for CCR_3, all the other baselines are not proposed for multi-source transfer learning problems, to conduct experiments with multiple source domains, we can either apply them on each pair of a source domain and the target domain or apply them on the pair of a *unified* source domain which simply combines all source domains and the target domain to learn a target classifier. For each of these baselines, i.e., LR, SVM, EAER, TCA, and mSDA, we report the mean, the maximum as well as the minimum accuracies of their corresponding target classifiers based on pairwise domains. For our proposed CRA, as we discussed, there are two ways to construct a target classifier. One is a voting-based combination of the multiple learned source classifiers, the other is to learn

Table 3. Average results (in %) on the 96 multi-source image classification problems

	LR	SVM	LR		SVM		mSDA	CCR$_3$	CRA$_v$	LR CRA$_u$	SVM CRA$_u$
			EAER	TCA	EAER	TCA					
Max	83.9	81.7	83.2	84.2	85.6	85.2	83.1	87.5	**89.2**	**89.4**	88.9
Min	65.0	56.0	62.3	66.8	71.3	69.8	64.6	83.5			
Mean	76.1	69.6	74.9	77.0	79.4	79.1	73.5	85.9			

a unified classifier from hidden representations of all source domains. We denote CRA$_v$ and CRA$_u$ the target classifiers built in these two ways respectively.

Implementation Details. For the trade-off parameters in CRA, the settings are listed as follows, $\alpha = 1$, $\beta = 0.5$, $k = 10$, $\gamma = 0.0001$, $\lambda = 1$ for the image dataset, and $\alpha = 100$, $\beta = 20$, $k = 80$, $\gamma = 0.0001$, $\lambda = 1$ for sentiment dataset. In experiments, we also study the parameter sensitivity of the parameters. For the parameters in TCA, EAER and mSDA, we carefully tune the number of dimensions k, and report the best results (e.g., in TCA, k varies from 10 to 80 with interval 10 for image data). We set the parameters of CCR$_3$ as the those published in [27], in which the parameter θ controlling the importance of consensus is sampled from $[0.05, 0.25]$ with interval 0.05. Thus the three values of minimum, mean and maximum for CCR$_3$ are also reported.

4.3 Experimental Results

Results on Image Data. We show the detailed mean accuracies of 96 image classification problems in Figure 2, and their average results in Table 3. From these results, we have some attractive observations: 1) CRA is significantly better than the traditional machine learning algorithms LR and SVM, which validate the effectiveness of the proposed transfer learning framework. 2) CRA outperforms TCA, EAER and mSDA, which shows that CRA can benefit from discovering a more powerful feature representation and incorporating consensus regularization from multiple source domains. 3) CRA performs better than CCR$_3$, which indicates the superiority of representation learning of autoencoders. Furthermore, the t-test with 95% confidence shows that CRA is significantly better than all the compared baselines.

Results on Sentiment Data. To further verify the effectiveness of the proposed framework CRA, we also make comparisons of all algorithms on sentiment classification problems. The detailed results are recorded in Table 4. Except mSDA is slightly better than CRA according to the maximum accuracies, CRA outperforms all the baselines. These results again validate the effectiveness of CRA, which can take full advantage of autoencoders and consensus regularization from multiple source domains simultaneously in a unified optimization framework.

4.4 Parameter Sensitivity

Here, we also investigate the parameter influence of three important trade-off parameters on image data, i.e., the relative importance of incorporating labeled information

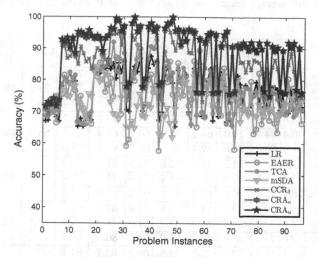

(a) The mean accuracies of 96 multi-source image classification problems using LR as the base classifier

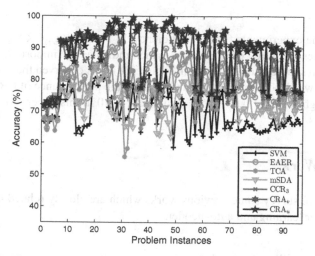

(b) The mean accuracies of 96 multi-source image classification problems using SVM as the base classifier

Fig. 2. The mean accuracies of 96 multi-source image classification problems

from source domains, the effect of considering consensus regularization and the number of hidden nodes for autoencoder. When we consider one parameter, the rest parameters are fixed. α and β are sampled from the value set $\{0.01, 0.1, 0.5, 1, 5, 10, 50, 100\}$, and k is sampled from the value set $\{5, 10, 20, 30, 40, 50, 60, 70, 80\}$. Six problems are randomly selected from 96 ones, and all the results of CRA_v are shown in Figure 3. We find that CRA is not sensitive to the number of hidden nodes k from Figure 3(c), so we

Table 4. Detailed and average results (in %) on the 4 multi-source sentiment classification problems

Tasks		LR	SVM	LR		SVM		mSDA	CCR$_3$	CRA$_v$	LR CRA$_u$	SVM CRA$_u$
				EAER	TCA	EAER	TCA					
tar.book	Max	79.3	78.4	67.8	68.5	73.0	66.2	**82.3**	78.6			
	Min	71.0	71.5	57.0	58.9	69.3	59.3	77.6	78.2	79.2	79.2	79.1
	Mean	75.7	74.9	63.0	64.2	70.9	62.8	**79.9**	78.4			
tar.kitchen	Max	85.6	85.4	78.9	75.2	77.5	73.1	84.7	**86.1**			
	Min	76.4	74.9	71.0	64.2	75.9	64.7	81.4	85.6	85.9	**86.3**	85.8
	Mean	81.0	80.5	76.6	69.4	76.7	68.7	83.5	85.9			
tar.elec.	Max	83.9	83.1	74.2	72.9	72.8	70.5	**85.2**	79.3			
	Min	73.5	73.0	68.5	60.7	69.4	59.4	74.4	75.4	84.1	**84.7**	82.4
	Mean	78.7	78.9	70.8	67.1	71.2	65.2	81.0	75.6			
tar.dvd	Max	79.7	79.5	69.5	68.5	70.8	67.4	**82.3**	80.2			
	Min	73.6	72.2	56.5	61.4	67.7	61.3	78.2	79.7	80.6	**81.1**	80.8
	Mean	77.0	75.9	65.1	65.2	69.0	64.3	80.3	80.1			
Average	Max	82.1	81.6	72.6	71.3	73.5	69.3	**83.7**	81.1			
	Min	73.6	72.9	63.2	61.3	70.6	61.2	77.9	79.7	82.5	**82.8**	82.0
	Mean	78.1	77.5	68.9	66.5	72.0	65.3	81.2	80.5			

set $k = 10$ in the experiments for high efficiency. In Figure 3(a), CRA gets very low performance when the value of α is small, which indicates the importance of labeled information from source domains. Also in Figure 3(b), it is observed that the setting of large value of β will lead to over-fitting and degrade the performance of CRA. According to these insights, we set $\alpha = 1$, $\beta = 0.5$ and $k = 10$ in this paper to achieve good and stable results.

5   Related Work

In this section, we survey some previous works which are closely related to our work, including transfer learning and autoencoder.

5.1   Embedding with Autoencoder

Autoencoders are primarily seen as a dimensionality reduction technique and thus use a bottleneck, namely the lower dimensional hidden layer of autoencoder, to learn a compressed representation which is represented by the hidden layer [3,10]. Currently variants of autoencoders have been investigated. Sparse autoencoders [17] use the idea of introducing a form of sparsity regularization to restrict the capacity of hidden units. Denoising autoencoders [21,22] learn to reconstruct the clean input from a artificially corrupted input and capture the structure of the input distribution. Sparse coding [12] can be viewed as a kind of autoencoder that uses a linear decoder tends to favor learning over-complete representations. These are often called regularized autoencoders, where some regularization terms are proposed to improve the data reconstruction performance.

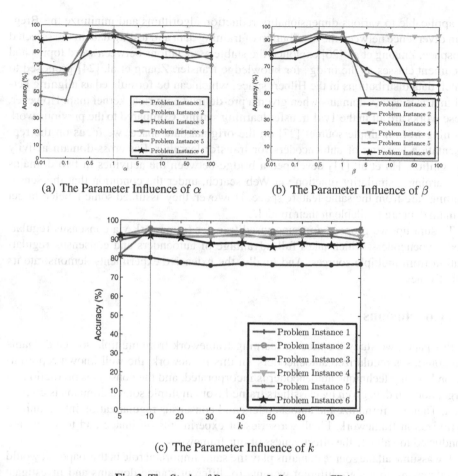

(a) The Parameter Influence of α (b) The Parameter Influence of β

(c) The Parameter Influence of k

Fig. 3. The Study of Parameter Influence on CRA

Contractive autoencoders [18], which shares a similar motivation with Denoising autoencoders, learn robust representations by adding an analytic contractive penalty term to the basic autoencoder. Marginalized Stacked Denoising Autoencoders (mSDA) [5] can be seen as the first try to use autoencoding technique for domain adaptation. However they have not considered consensus regularization from multiple sources.

5.2 Transfer Learning

Recent years have witnessed numerous research in transfer learning [16]. Here we only list some closely related works, i.e., transfer embedding and subspace learning (or learning on topic level). Pan et al. [13] proposed a dimensionality reduction approach to find out such latent feature space that supervised learning algorithms can be applied to train classification models and obtain satisfying results. After that, they also proposed a transfer component analysis (TCA) algorithm to learn some transfer components across domains [15]. Si et al. [20] developed a transfer subspace learning framework, which can be applicable to various dimensionality reduction algorithms and minimize the Bregman divergence between the distribution of training data and testing data in the selected subspace. Zhuang et al. [26] exploited the stable associations between word topics and document classes as the bridge for knowledge transfer. Zhang et al. [24] proposed to match data distributions in the Hilbert space, which can be formulated as aligning kernel matrices across domains when given a pre-defined empirical kernel map. However, these works are all in the 1vs1 transfer learning setting. Compared to the previous work learning from multiple sources [27] on the original data space, we focus on the representation learning of autoencoders for transfer learning. For cross-domain activity recognition, Hu et al. [11] developed a bridge between the activities in two domains by learning a similarity function via Web search, under the condition that the sensor readings are from the same feature space. However, they assumed some labeled target domain data are available in their model.

To sum up, we propose a unsupervised transfer framework via consensus regularized autoencoders, which takes full advantage of autoenders and consensus regularization from multiple sources. And finally, the extensive experiments demonstrate its effectiveness.

6   Conclusions

In this paper, we study the transfer learning framework from multiple source domains via consensus regularized autoencoders. In this framework, the well known representation learning technique autoencoder is incorporated, and the consensus prediction on target domain data given by classifiers trained from multiple source domains is considered. Then we formalize the autoencoders and consensus regularization into a unified optimization framework. Finally, a series of experiments on image and text data are conducted to validate the effectiveness of our framework.

We assume all the source domains play the same important role in this paper. It would be interesting to assign different weights to different source domains and investigate their importance in the future work.

Acknowledgments. This work is supported by the National Natural Science Foundation of China (No. 61175052, 61203297, 61035003), National High-tech R&D Program of China (863 Program) (No.2014AA012205, 2013AA01A606, 2012AA011003).

References

1. Blitzer, J., Dredze, M., Pereira, F.: Biographies, bollywood, boom-boxes and blenders: Domain adaptation for sentiment classification. In: Proceedings of the 45th ACL (2007)
2. Boser, B.E., Guyon, I.M., Vapnik, V.N.: A training algorithm for optimal margin classifiers. In: Proceedings of the 5th AWCLT (1992)
3. Bourlard, H., Kamp, Y.: Auto-association by multilayer perceptrons and singular value decomposition. Biological Cybernetics (1988)
4. Chattopadhyay, R., Ye, J.P., Panchanathan, S., Fan, W., Davidson, I.: Multi-source domain adaptation and its application to early detection of fatigue. In: Proceedings of the 17th ACM SIGKDD, pp. 717–725. ACM (2011)

5. Chen, M.M., Xu, Z.X., Weinberger, K., Sha, F.: Marginalized denoising autoencoders for domain adaptation. In: Proceedings of the 29th ICML (2012)
6. David, H., Stanley, L.: Applied Logistic Regression. Wiley, New York (2000)
7. Duan, L., Tsang, I.W., Xu, D., Chua, T.S.: Domain adaptation from multiple sources via auxiliary classifiers. In: Proceedings of the 26th ICML (2009)
8. Gao, J., Fan, W., Jiang, J., Han, J.W.: Knowledge transfer via multiple model local structure mapping. In: Proceedings of the 14th ACM SIGKDD (2008)
9. Ge, L., Gao, J., Zhang, A.D.: Oms-tl: a framework of online multiple source transfer learning. In: Proceedings of the 22nd ACM CIKM, pp. 2423–2428. ACM (2013)
10. Hinton, G.E., Zemel, R.S.: Autoencoders, minimum description length, and helmholtz free energy. In: Advances in NIPS (1994)
11. Hu, D.H., Zheng, V.W., Yang, Q.: Cross-domain activity recognition via transfer learning. Pervasive and Mobile Computing 7(3), 344–358 (2011)
12. Olshausen, B.A., Field, D.J.: Sparse coding with an overcomplete basis set: A strategy employed by v1? Vision Research (1997)
13. Pan, S.J., Kwok, J.T., Yang, Q.: Transfer learning via dimensionality reduction. In: Proceedings of the 23rd AAAI (2008)
14. Pan, S.J., Ni, X., Sun, J.T., Yang, Q., Chen, Z.: Cross-domain sentiment classification via spectral feature alignment. In: Proceedings of the 19th WWW (2010)
15. Pan, S.J., Tsang, I.W., Kwok, J.T., Yang, Q.: Domain adaptation via transfer component analysis. IEEE TNN (2011)
16. Pan, S.J., Yang, Q.: A survey on transfer learning. IEEE TKDE (2010)
17. Poultney, C., Chopra, S., Cun, Y.L.: Efficient learning of sparse representations with an energy-based model. In: Advances in NIPS (2006)
18. Rifai, S., Vincent, P., Muller, X., Glorot, X., Bengio, Y.: Contractive auto-encoders: Explicit invariance during feature extraction. In: Proceedings of the 28th ICML (2011)
19. Shi, Z.P., Ye, F., He, Q., Shi, Z.Z.: Symmetrical invariant lbr texture descriptor and application for image retrieval. In: Congress on Image and Signal Processing (2008)
20. Si, S., Tao, D.C., Geng, B.: Bregman divergence-based regularization for transfer subspace learning. IEEE TKDE (2010)
21. Vincent, P., Larochelle, H., Bengio, Y., Manzagol, P.: Extracting and composing robust features with denoising autoencoders. In: Proceedings of the 25th ICML (2008)
22. Vincent, P., Larochelle, H., Lajoie, I., Bengio, Y., Manzagol, P.: Stacked denoising autoencoders: Learning useful representations in a deep network with a local denoising criterion. JMLR (2010)
23. Yu, W., Zeng, G., Luo, P., Zhuang, F., He, Q., Shi, Z.: Embedding with autoencoder regularization. In: Blockeel, H., Kersting, K., Nijssen, S., Železný, F. (eds.) ECML PKDD 2013, Part III. LNCS, vol. 8190, pp. 208–223. Springer, Heidelberg (2013)
24. Zhang, K., Zheng, V., Wang, Q.J., Kwok, J., Yang, Q., Marsic, I.: Covariate shift in hilbert space: A solution via sorrogate kernels. In: Proceedings of The 30th ICML, pp. 388–395 (2013)
25. Zhang, L.: The Research on Human-computer Cooperation in Content-based Image Retrieval. Ph.D. thesis, Tsinghua University, Beijing (2001) (in Chinese)
26. Zhuang, F.Z., Luo, P., Xiong, H., He, Q., Xiong, Y.H., Shi, Z.Z.: Exploiting associations between word clusters and document classes for cross-domain text categorization. Statistical Analysis and Data Mining (2011)
27. Zhuang, F.Z., Luo, P., Xiong, H., Xiong, Y.H., He, Q., Shi, Z.Z.: Cross-domain learning from multiple sources: A consensus regularization perspective. IEEE TKDE (2010)

Branty: A Social Media Ranking Tool for Brands

Alexandros Arvanitidis*, Anna Serafi*,
Athena Vakali, and Grigorios Tsoumakas

Dept of Informatics, Aristotle University of Thessaloniki, Thessaloniki, Greece
{arvanian,annasera,avakali,greg}@csd.auth.gr

Abstract. In the competitive world of popular brands, strong presence in social media is of major importance for customer engagement and products advertising. Up to now, many such tools and applications enable end-users to observe and monitor their company's web profile, their statistics, as well as their market outreach and competition status. This work goes beyond the individual brands statistics since it automates a brand ranking process based on opinions emerging in social media users' posts. Twitter streaming API is exploited to track micro-blogging activity for a number of famous brands with emphasis on users' opinions and interactions. The social impact is captured from 3 different perspectives (objective counts, opinion reckoning, influence analysis), which estimate a score assigned to each brand via a multi-criteria algorithm. The results are then exposed in a Web application as a list of the most social brands on Twitter. But, are conventional metrics, such as followers, enough in order to measure the social impact of a brand? Different usage scenarios of our application reveal that the social presence of a brand is more complex than current social impact frameworks care to admit.

Keywords: social media analytics, brand ranking, multiple criteria decision analysis, sentiment classification, visualization.

1 Introduction

Twitter has become a valuable tool for extracting public opinion. Indeed, a trend towards replacing conventional surveys by opinion mining over popular social media has already been highlighted in the literature [1]. Large scale vendors have always spent a lot of money to gain information about their products and services. Exploiting social media statistics, sentiment analysis and further metrics is today gaining a momentum and significantly impacting brands' marketing strategy. Such tools exist, a popular one being *Sysomos* [2], which started as a research project and is now a large commercial company that comes with a price, whereas free-of-charge platforms provide nearly enough data for someone to recognize what he has to change for his brand to improve its social web profile.

BRANTY[1] is a partly open-source social media monitoring platform that analyzes, ranks and visualizes the social presence of brands on Twitter. BRANTY

---

* Authors contributed equally to this work.
[1] Branty web application http://branty.org/

T. Calders et al. (Eds.): ECML PKDD 2014, Part III, LNCS 8726, pp. 432–435, 2014.

ranks brands based on social media analytics. Users can specify their preferences via an easy Web interface which is facilitating users adjustments and results summarizations. BRANTY is characterized by its openness with an aim to provide the BRANTY rating as an external service, which would be pluggable by other social media tools, product review sites and e-shops.

2   Ranking Brands with Branty

BRANTY formulates the problem of brand ranking as a multiple criteria decision analysis (MCDA) problem. Each brand is characterized with respect to the 16 criteria summarized in Table 1. BRANTY employs the Technique for Order of Preference by Similarity to Ideal Solution (TOPSIS) [3] for ranking the monitored brands. TOPSIS assumes two extra brands which have the best and worst possible score in each criterion and assigns the best rate to the brand closest to the optimal and furthermost to the worst one. It proceeds to ranking by receiving input weights by the user, showing the relative importance of each criterion.

Table 1. Criteria each brand is characterized with

Objective Counts Criteria	
Overall time	**Current week**
Number of brand followers	Number of tweets by users
Friends-Followers ratio	Number of tweets by the brand
Verified Twitter page or not	Appearances of the brand's website
Twitter lists the brand is member of	Times the brand has been mentioned
Average number of brand's posts per month, since its registration	Sum of re-tweets of the brand's tweets
	Sum of favorites of the brand's tweets
Opinion Reckoning Criteria	
Overall time	**Current week**
-	Number of users' positive tweets
	Number of users' negative tweets
	Number of users' neutral tweets
Influence Analysis Criteria	
Overall time	**Current week**
Klout Score	Positive tweets on trending topics

These criteria assess three different views of brands' social impact on Twitter and span either a small recent time period (e.g. week) or a more archival time period (the whole dataset). *Objective Counts Criteria* are statistics which assess the social presence of a brand on Twitter while *Opinion Reckoning Criteria* assess the current opinion Twitter users have for this brand. Positive tweets on trending topics show the strength a brand has among popular topics, meaning that a @brand is met along with at least one of the most popular hashtags.

We track the other *Influence Analysis Criterion* in a comparative to our score manner, using Klout service[2].

MongoDB[3] was used to store the data, while the implementation was based on the Java Twitter4J[4] library in order to utilize Twitter API. The current version of BRANTY is outlined with data for approximately 400 brands in 4 distinct categories: Auto, Fashion, Food/Beverages, Technology, collected from December 2013 until today.

2.1   Classifying Tweets by Sentiment

Opinion Reckoning Criteria require the classification of tweets by sentiment as positive, negative or neutral. This is achieved via a linear support vector machine classifier, trained on a number of tweets (approximately 1000 tweets manually annotated in this case study). Particular pre-processing has cleaned text by removing URLs, references, punctuation, hash-tag symbols and all other non-alphanumeric characters, while question and exclamation marks were converted to words. We then used a standard tf-idf representation for the pre-processed tweets. Also, SentiWordNet (SWN) [4] was utilized to derive 4 additional features corresponding to the fraction of positive, negative and neutral words within a sentence (tweet text in our case), as well as to the overall sentiment score of the sentence (the result of the SWN lexicon). The inclusion of these SWN attributes increased the accuracy of our approach by 5%. We managed to achieve a prediction accuracy of around 80%, similarly to [5,6].

3   Visualizing Data on a Web Application

Figure 3 shows the interface of BRANTY. On the left, a folding tab hosts horizontal sliders that allow users to weigh each of the criteria according to their preferences. The calculate button at the bottom of this tab refreshes the brand ranking according to the current weight settings. Brands are ranked in descending order according to their score, which is displayed as both a number and a horizontal bar. At the top right, clickable icons can be used to filter the ranking by brand category. At the same time, the top 10 hash-tags within the tweets of the selected category's brands are displayed above the ranking.

Experimenting with BRANTY, we noticed its functionality since differentiating weights in partial or all criteria has resulted in deviating brand ranking results. Its innovative contribution is that it enables tunable weighting for the end-users, using criteria which have different emphasis on the brands' ranking. Utilizing Twitter is justified by the fact that its presence is dominant, which is why it is used by BRANTY, integrating all of its features into a user-friendly interface. Via BRANTY's fine grained tunable and automated brand ranking,

---

[2] Klout Service, an external framework which ranks its users after measuring their social influence http://klout.com/

[3] http://www.mongodb.org/

[4] http://twitter4j.org/

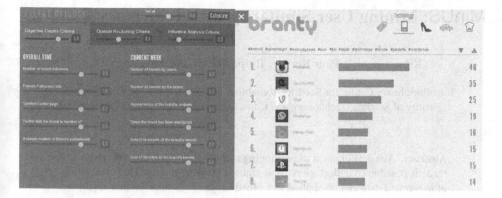

Fig. 1. BRANTY's interface. The top 8 Technology brands ranking snaphot

companies can monitor their social presence relative to that of their competitors with respect to different criteria, discover their weak points and adopt strategies for improvements and effective decision making.

4   Conclusion and Future Work

We have developed a ranking framework based on social media for evaluating brands in a comparative manner. Our first hypothesis was that a ranking analysis is defined by many factors and demands a deeper analysis. On our framework, a parameterized view of the current picture in social media can be created, giving brands a powerful tool to play with and analyze the results near real-time.

In the future we plan to extend BRANTY with criteria coming from other social media, such as Facebook, Google+ and LinkedIn. We also plan to increase the number of monitored brands by replacing our single server system with a distributed infrastructure solution.

References

1. O'Connor, B., Balasubramanyan, R., Routledge, B.R., Smith, N.A.: From tweets to polls: Linking text sentiment to public opinion time series. ICWSM 11, 122–129 (2010)
2. Cheng, A., Evans, M., Singh, H.: Inside twitter (2009)
3. Hwang, C., Yoon, K.: Multiple Attribute Decision Making: Methods and Applications. Springer, New York (1981)
4. Esuli, A., Sebastiani, F.: Sentiwordnet: A publicly available lexical resource for opinion mining. In: Proceedings of LREC, vol. 6, pp. 417–422 (2006)
5. Go, A., Bhayani, R., Huang, L.: Twitter sentiment classification using distant supervision. Processing, 1–6 (2009)
6. Chamlertwat, W., Bhattarakosol, P., Rungkasiri, T., Haruechaiyasak, C.: Discovering consumer insight from twitter via sentiment analysis. Journal of Universal Computer Science 18(8), 973–992 (2012)

MinUS: Mining User Similarity with Trajectory Patterns

Xihui Chen[1,*], Piotr Kordy[1], Ruipeng Lu[2], and Jun Pang[1,2,**]

[1] Interdisciplinary Centre for Security, Reliability and Trust, University of Luxembourg
[2] Faculty of Science, Technology and Communication, University of Luxembourg

Abstract. The development of positioning systems and wireless connectivity has made it possible to collect users' fine-grained movement data. This availability of movement data can be applied in a broad range of services. In this paper, we present a novel tool for calculating users' similarity based on their movements. This tool, MinUS, integrates the technologies of trajectory pattern mining with the state-of-the-art research on discovering user similarity with trajectory patterns. Specifically, with MinUS, we provide a platform to manage movement datasets, and construct and compare users' trajectory patterns. Tool users can compare results given by a series of user similarity metrics, which allows them to learn the importance and limitations of different similarity metrics and promotes studies in related areas, e.g., location privacy. Additionally, MinUS can also be used by researchers as a tool for preliminary process of movement data and parameter tuning in trajectory pattern mining.

1 Introduction

Due to the free access to GPS (global positioning system), people have access to their precise whereabouts. This access in turn leads to the collection of enormous amount of movement data, which offers us a new source of information to study human being's behaviour. For instance, we can check whether humans possess *swarm patterns* as birds and other animals during their movement. We can also learn the mutual influences between users' movements and their social relationships [1]. Among all interesting patterns, periodicity is naturally inherited in people's movement [2]. In other words, people have the intention to repeat some of their routes, possibly with similar temporal patterns. For instance, as a student of University of Luxembourg, Pierre has a daily routine from his residence to the train station where he takes bus to Campus Kirchberg. We call such repeated routines *trajectory patterns*. Intuitively, a trajectory pattern is a sequence of places of interests (PoIs) which a user frequently visits. With this interpretation, the daily routine of the student can be expressed as *residence → train station → Campus Kirchberg*. Typical transition time between two consecutive places of interest can also be added as part of a trajectory pattern. The extraction of trajectory patterns from users' travel history has been well studied in the literature and many mining algorithms have been proposed (e.g., [3]).

Trajectory patterns can be explored in many ways [4,5,6], one of which is friend recommendation in social networks. This is inspired by the fact that users' movement

---

* Supported by the FNR Luxembourg under project SECLOC 794361.
** To whom correspondence should be addressed.

T. Calders et al. (Eds.): ECML PKDD 2014, Part III, LNCS 8726, pp. 436–439, 2014.

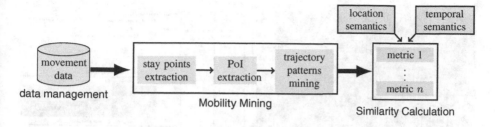

Fig. 1. The architecture of MinUS

reflects their interest. For instance, frequent visits to a reading club can indicate a user as a fan of literature. Thus, by the similarity between two users in terms of their trajectory patterns, we can infer whether they share similar hobbies. To the best of our knowledge, no tools are publicly available to calculate user similarity with trajectory patterns.

In this paper, we present a tool which integrates the technologies of trajectory pattern mining with recent research on user similarity calculation with trajectory patterns. Compared to existing trajectory pattern mining tools, our tool provides a graphical interface and allows users to control the mining process. More specifically, users have access to all intermediate results which can be visualised with our tool. Thus, researchers can use MinUS as a tool to preliminarily process movement data and tune parameters in trajectory pattern mining. With a series of metrics implemented, tool users can measure user similarity from different perspectives and thus learn the limitations and importance of different metrics. MinUS has one distinguishing feature that allows users to use the semantics of spatio-temporal information in trajectory patterns to mine user similarity.

Our tool is implemented with Java and C#. It is available online and can be downloaded from the following link http://satoss.uni.lu/software/MinUS.

2   The MinUS Tool

The MinUS tool has three function modules which are shown in Figure 1. The first module, *data management*, is in charge of managing movement datasets which are collected by different organisations. Such a dataset consists of a number of users whose movement is stored in the form of daily trajectories. This module keeps track of the statistic information about the users in each dataset. The statistics will be updated automatically once the values of the fields are available. The second function module, *mobility mining*, takes users' daily trajectories as input and outputs their trajectory patterns. The third module, *similarity calculation*, calculates the similarity values between users selected by tool users using the chosen similarity metric. In the following discussion, we give more details about the last two modules.

Mobility mining. In this function module, we implement the three sequential steps to calculate users' trajectory patterns according to the methodology given in [7,8]. At each step, a type of information about users' movement is mined. At the first step, we traverse all of the selected users' trajectories and detect their *stay points*, the centroids of small areas where a user stayed for a certain amount of time. A *place of interest* (PoI) is an area where users *frequently* visit, stay for a while and preform certain activities, such as

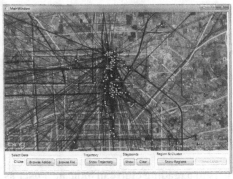

(a) MapView

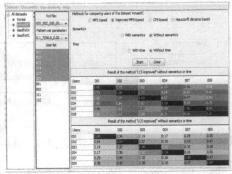

(b) User similarity calculation

supermarkets and theatres. Based on this interpretation, at the second step, we calculate users' PoIs by identifying the regions where stay points are densely located. We implement a hierarchical clustering algorithm to calculate the clusters of stay points which are close to each other. Outlying stay points that are isolated from other points may increase the size of some PoIs. We make use of LOF (local outlier factor) to measure the extent of isolation of a stay point and remove a certain percentage of stay points that are most isolated. With extracted PoIs, users' trajectories are transformed into sequences of PoIs. At the last step, we explore the trajectory pattern mining tool of Giannotti et al. [3] to extract trajectory patterns.

Our tool gives tool users the control (by specifying the parameters required by the underlying algorithms) and the access to all intermediate results, such as users' stay points and PoIs. With a visualisation interface called MapView, tool users can put all the intermediate results on the map. In Figure 2(a), we show the screen shot of the visualisation interface. Users' trajectories are denoted by blue lines, while the yellow dots represent stay points and PoIs are depicted by red rectangles.

Similarity Calculation. This module provides a platform to apply different metrics to measure user similarity with trajectory patterns. Figure 2(b) shows the interface of users similarity calculation. The tool users start with selecting a subset of users and the tool will return the similarity values between any two selected users. The results are visualised by a grid where the grey level of each cell indicates the similarity values between a pair of users. So far, we have implemented three categories of user similarity metrics: *maximal trajectory pattern based* [9,7,8], *common pattern set based* [10] and *Hausdorff distance based* [11]. Our tool also allows for taking into account *location semantics* and *temporal semantics* in the calculation. Location semantics denotes the functionalities of a PoI, e.g., restaurant and school, while temporal semantics represents the information revealed by time, e.g., weekends and weekdays. Since a PoI may correspond to a number of location semantics, we propose to use a probability distribution over all possible location semantics to demonstrate the uncertainty about a user's purpose in a PoI. Sometimes location semantics are used just for the purpose of comparison or validating the effectiveness of different metrics. For such cases, we implement a separate panel for tool users to simulate the distribution over location semantics for a PoI by giving the freedom to choose the number of location semantic tags and other factors.

The MinUS tool has been applied for experimental validation on different trajectory datasets [7,8,10]. We refer readers to those papers for the details.

3   Related Work

There have been many tools developed for mining users' movement patterns of different forms. The trajectory pattern miner by Giannotti et al. [3], used in MinUS, is developed to extract spatio-temporal frequent trajectory patterns from GPS-like trajectories of a set of move objects. Li et al. [4] design the tool MoveMine for scalable analysis on massive movement data, e.g., movement pattern mining and trajectory clustering. Two types of movement patterns are extracted: periodic patterns and swarm patterns.

By applying users' trajectory patterns, a few applications have been implemented. AllAboard by Berlingerio et al. [6] is a software system that makes use of cellphone data to improve existing public transport systems. Using users' call locations stored on cellphones, their trajectory patterns are extracted, with which new routes will be calculated and added to reduce users' waiting and travel time. Pelekis et al. [5] develop a tool called Hermoupolis to generate trajectories by simulating given trajectory patterns. This tool provides a method to synthesise trajectory datasets for researchers when real-life datasets are unavailable or not sufficiently large in experimental validation. So far, MinUS is the first tool which is publicly available to compare users' trajectory patterns.

References

1. Srivatsa, M., Hicks, M.: Deanonymizing mobility traces: using social network as a side-channel. In: Proc. CCS, pp. 628–637. ACM Press (2012)
2. Li, Z., Ding, B., Han, J., Kays, R., Nye, P.: Mining periodic behaviors for moving objects. In: Proc. KDD, pp. 1099–1108. ACM Press (2010)
3. Giannotti, F., Nanni, M., Pinelli, F., Pedreschi, D.: Trajectory pattern mining. In: Proc. KDD, pp. 330–339. ACM Press (2007)
4. Li, Z., Ji, M., Lee, J.G., Tang, L.A., Yu, Y., Han, J., Kays, R.: Movemine: mining moving object databases. In: Proc. SIGMOD, pp. 1203–1206. ACM Press (2010)
5. Pelekis, N., Ntrigkogias, C., Tampakis, P., Sideridis, S., Theodoridis, Y.: Hermoupolis: A trajectory generator for simulating generalized mobility patterns. In: Blockeel, H., Kersting, K., Nijssen, S., Železný, F. (eds.) ECML PKDD 2013, Part III. LNCS, vol. 8190, pp. 659–662. Springer, Heidelberg (2013)
6. Lorenzo, G.D., Sbodio, M.L., Calabrese, F., Berlingerio, M., Nair, R., Pinelli, F.: Allaboard: visual exploration of cellphone mobility data to optimise public transport. In: Proc. IUI, pp. 335–340. ACM Press (2014)
7. Chen, X., Pang, J., Xue, R.: Constructing and comparing user mobility profiles for location-based services. In: Proc. SAC, pp. 261–266. ACM Press (2013)
8. Chen, X., Pang, J., Xue, R.: Constructing and comparing user mobility profiles. ACM TWEB (accepted, 2014)
9. Ying, J.C., Lu, H.C., Lee, W.C., Weng, T.C., Tseng, S.: Mining user similarity from semantic trajectories. In: Proc. LBSN, pp. 19–26. ACM Press (2010)
10. Chen, X., Lu, R., Ma, X., Pang, J.: Measuring user similarity with trajectory patterns: Principles and new metrics. In: Liu, G. (ed.) APWeb 2014. LNCS, vol. 8709, pp. 437–448. Springer, Heidelberg (2014)
11. Lu, R.: New user similarity measures based on mobility profiles. Master's thesis, University of Luxembourg (2013)

KnowNow: A Serendipity-Based Educational Tool for Learning Time-Linked Knowledge

Luigi Di Caro[1], Livio Robaldo[1], and Nicoletta Bersia[2]

[1] University of Turin, Italy
{dicaro,robaldo}@di.unito.it
[2] Telecom Italia, Turin, Italy
nicoletta.bersia@telecomitalia.it

Abstract. In this paper we present the system *KnowNow*, a tool whose aim is to let the users navigate into text corpora through dynamic semantic information networks, created in real-time according to delimited time ranges. In educational scenarios, students are often asked to write short essays on different topics linked by temporal information. This usually involves a combination of several aspects to be evaluated, such as knowledge, imagination, structure and presentation. In the light of this, the introduction of Natural Language Understanding techniques together with cross-topic navigation and visualization tools and considerably help students to retrieve, link, and create well-structured and original contributions, as we demonstrate by using *KnowNow*.

Keywords: Natural Language Understanding, Semantic Search, Education.

1 Introduction

In educational scenarios, it is common to find teachers' requests for short essays that students must elaborate by picking different topics directly connected by temporal constraints. For instance, a student may present a work to combine history, geography, and physics by mentioning the military and political leader *Napoleon Bonaparte* (died in 1821), the *Eyjafjallajokull* volcano in Iceland (that it began to erupt in 1821), and the physician *Elizabeth Blackwell* (born in 1821, who was the first woman to receive a medical degree in the United States).

However, building such *knowledge graph* often results to be "boring" for the following reason: students are usually interested in topics that are likely to be temporally disconnected, so they have to select only one as starting point, and then attach quite unintentional facts that cover other domains.

Nowadays, there is a plenty of freely available resources that can be used for educational purposes, like Wikipedia[1]. Wikipedia is the largest free on-line encyclopedia that includes information of different areas and in different languages that has been already used in this context [3]. Since it contains several historical

---

[1] www.wikipedia.org

T. Calders et al. (Eds.): ECML PKDD 2014, Part III, LNCS 8726, pp. 440–443, 2014.

facts (and so it is full of temporal information) but also hundreds of other topics, it perfectly fits the above-mentioned context.

In the next section we will illustrate the underlying technology of *KnowNow*, which is a combination of advanced Natural Language Techniques, Data Mining, Human-Interaction models, and Data Visualization schemes. *KnowNow* is the result of the project named *KnowYouAll*, that won a national competition for innovative ideas promoted by Telecom Italia[2].

2   KnowNow

KnowNow is made of different modules: a Time Extractor, a Named Entity Analyzer and Semantic Network builder, a Content Summarizer, and an interactive fish-eye visualization tool.

2.1   Data

As already mentioned in the introduction, we directly used Wikipedia as input corpus. For the demonstration, we randomly selected $10,000$ Wikipedia pages, removing metadata information, html tags, links, and Wikipedia-specific texts that are not related to the content. This limit, however, does not reflect technical problems since our syntactic, semantic, and statistical analyses are only applied on small time-delimited document sets.

2.2   Time Extractor

After the cleansing of the input corpus, the system syntactically parses the text using TULE [2], a dependency parser for English and Italian. Since a single document may contain multiple temporal information (related to facts happened in different periods), the system has to extract them in order to build an inverse *temporal map* $< t_k, \{doc_{ids}\} >$ that links time frames[3] t_k with sets of documents $\{doc_{ids}\}$ that contain at least one fact happened in t_k. For recognizing temporal expressions, we used the rule-based techniques proposed in [5].

2.3   Semantic Network

While the syntactic analysis supports the extraction of temporal expressions, the system also includes a semantic analyzer that deals with the identification of *semantic units* for semantic search and access. We define a *semantic unit* as a named entity in the classical NLP task Named Entity Recognition (NER) [4]. A named entity is a type of class of objects, like people, organizations, places, and others. In *KnowNow*, we used the large ontology of semantic information of DBPedia[4], which is a structured version of Wikipedia. It contains several

---

[2] Working Capital (ed. 2012), www.workingcapital.telecomitalia.it
[3] Only timestamps having at least the year and the month are preserved.
[4] www.dbpedia.org

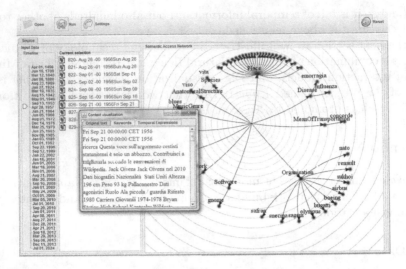

Fig. 1. A screenshot of the *KnowNow*'s main interface. The left panel contains a slider that allows the user to focus on different the time periods automatically extracted from the input corpus. On its right, the interface shows the list of documents that represent the *temporal window* around the selected date. The larger panel on the right shows a fish-eye semantic network calculated in real-time with respect the the left selection. Notice that in this example, an Italian Wikipedia corpus is used, and it is navigable through an English-based semantic network, relying on the Wikipedia interlingual links. Clicking on the button *Run* is then possible to see a hierarchical tag-cloud of the most dominant common terms used in the selected texts, as in [1]. The little window on the top shows the original content of one selected document in the list.

semantic units, organized in a multi-level taxonomy. For example, the instance *Pink Floyd* is associated to the node *Band*, which is a subclass of *Organization*, and so forth. By using these resources, *KnowNow* is able to let the users explore non-English texts with navigable semantic networks written in English. This is done by making use of the interlingual links of the Wikipedia pages, that provides the translation of specific entity names in different languages. This is a powerful feature, since it allows to explore the semantics of texts expressed in several languages by means of an English-based semantic network.

2.4   Content Summarizer

Texts are not only made of semantic units, but they also contain several *common words* that describe the content, and specifically, which named entities are involved in the events, and how. In this case, standard Data Mining techniques applied on texts are useful to allow the users navigate through the content by leveraging on words frequencies and co-occurrences. In particular, *KnowNow* relies on a technique that applies Latent Semantic Analysis on the input texts to construct a navigable tree of *dominant terms* [1].

2.5 Visualization Tool

In *KnowNow*, the information is displayed using different parts of the interface, shown in Figure 1. On the left, a slider permits to observe all the time frames extracted by the time extractor. The user is then able to focus on a particular *temporal window* around the selected date. This parameter, like all the ones mentioned (and not mentioned because of lack of space) in this paper are adjustable through the interface of the system. Once the user selected a temporal window W, *KnowNow* shows a fish-eye tree with all the semantic units found in the texts that have been associated to W, and so that contain facts happened in W. These documents are listed side by side with the slider, and the content can be visualized by clicking on them. The user can do *drag-and-drop* operations on the semantic network to put more visual emphasis on a specific subtree. Then, clicking on a node (or more than one node), *KnowNow* highlights those documents which are related to that relative semantics. Finally, the user may also want to explore the content expressed by common words. In this sense, the *Content Summarizer* extracts a hierarchical tag-cloud of the most dominant terms in the input texts by leveraging on a Latent Semantic Analysis of the term-document matrix. The user can click on the button *Run* to perform such process over the content associated to the current temporal window.

3 Demo Scenario

During the demonstration, we will allow the users to select different time ranges, showing how the semantic network is able to capture and visualize the main semantic information contained in the input texts, in real-time. Then, the tool allows for a number of further interactions, like the selection of specific semantic nodes, the classification and the ranking of the most relevant texts, fish-eye visualization of dominant terms, and the impact of parameters like size of time ranges, amount of data to be displayed, and several others.

References

1. Di Caro, L., Candan, K.S., Sapino, M.L.: Navigating within news collections using tag-flakes. Journal of Visual Languages & Computing 22(2), 120–139 (2011)
2. Lesmo, L.: The Turin University Parser at Evalita 2009. Proceedings of EVALITA 9 (2009)
3. Moy, C.L., Locke, J.R., Coppola, B.P., McNeil, A.J.: Improving science education and understanding through editing wikipedia. vol. 87, pp. 1159–1162. ACS Publications (2010)
4. Nadeau, D., Sekine, S.: A survey of named entity recognition and classification. Lingvisticae Investigationes 30(1), 3–26 (2007)
5. Robaldo, L., Caselli, T., Russo, I., Grella, M.: From italian text to timeML document via dependency parsing. In: Gelbukh, A. (ed.) CICLing 2011, Part II. LNCS, vol. 6609, pp. 177–187. Springer, Heidelberg (2011)

Khiops CoViz: A Tool for Visual Exploratory Analysis of k-Coclustering Results

Bruno Guerraz, Marc Boullé, Dominique Gay, and Fabrice Clérot

Orange Labs
2, avenue Pierre Marzin, F-22307 Lannion Cedex, France
firstname.name@orange.com

Abstract. Identifying and visually analyzing interesting interactions between variables in large-scale data sets through k-coclustering is of high importance. We present Khiops CoViz [1], a tool for visual analysis of interesting relationships between two or more variables (categorical and/or numerical). The visualization of k variables coclustering takes the form of a grid/matrix whose dimensions are partitioned: categorical variables are grouped into clusters and numerical variables are discretized. The tool allows several kinds of visualization at various scales for grid representation of coclustering results by means of several criteria each of which providing different insights into the data. Hereafter, several screen shots describe the main visual components of the tool.

---

[1] http://www.khiops.com, an exhaustive user manual is available when downloading the tool.

T. Calders et al. (Eds.): ECML PKDD 2014, Part III, LNCS 8726, pp. 444–447, 2014.

A Tool for Visual Exploratory Analysis of k-Coclustering Results   445

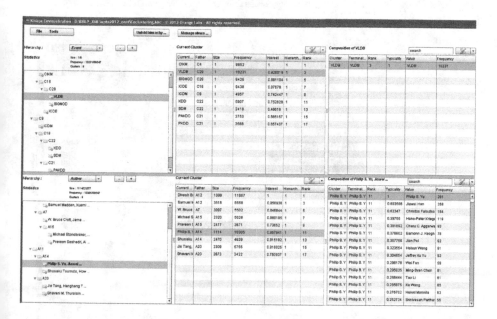

Fig. 1. Case study on a examplary sample of DBLP data set. Three-dimensional data ($Author \times Year \times Event$) is considered. A screen shot of Khiops CoViz: (from left to right) hierarchies of parts for two dimensions ($Author$ and $Event$), list of terminal parts of hierarchies and composition of selected parts.

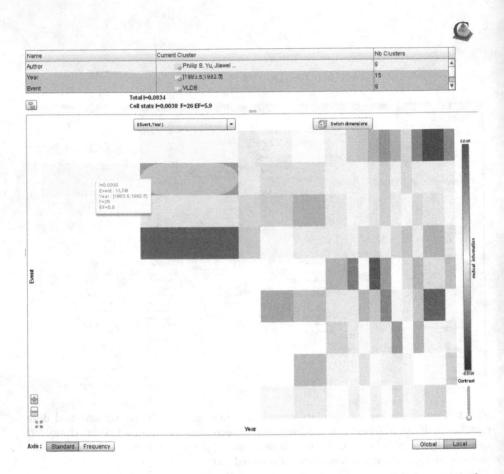

Fig. 2. (Left): Grid/matrix visualization of contribution to mutual information for *Year* × *Event* dimensions. Other criteria for visualization are available: e.g., contrast, frequency, conditional probability and joint probability... The tool allows navigating along the partition of a selected dimension (e.g., *Author*) while the others (*Year, Event*) are fixed and dedicated to the visualization.

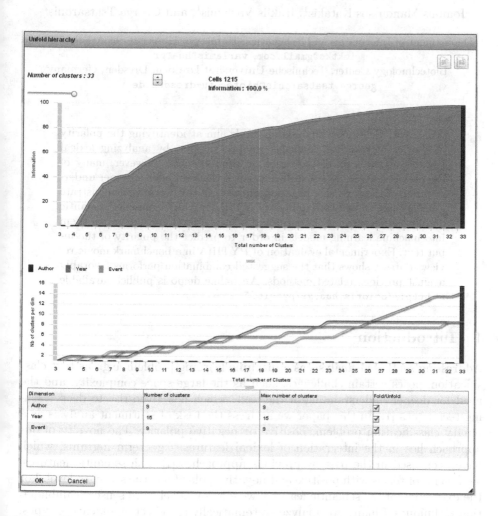

Fig. 3. Choosing the wanted granularity for visualizing the grid is available through the "Unfold Hierarchy" functionality. The user can control either the number of parts of the dimensions or the grid quality (w.r.t. to the optimal grid) by optimal merging or per-dimension customized non-optimal merging.

PYTHIA: Employing Lexical and Semantic Features for Sentiment Analysis

Ioannis Manoussos Katakis[1], Iraklis Varlamis[1], and George Tsatsaronis[2]

[1] Department of Informatics and Telematics, Harokopio University of Athens, Greece
imktks@gmail.com, varlamis@hua.gr
[2] Biotechnology Center, Technische Universität Dresden, Dresden, Germany
george.tsatsaronis@biotec.tu-dresden.de

Abstract. Sentiment analysis methods aim at identifying the polarity of a piece of text, e.g., passage, review, snippet, by analyzing lexical features at the level of the terms or the sentences. However, many of the previous works do not utilize features that can offer a deeper understanding of the text, e.g., negation phrases. In this work we demonstrate a novel piece of software, namely PYTHIA[1], which combines semantic and lexical features at the term and sentence level and integrates them into machine learning models in order to predict the polarity of the input text. Experimental evaluation of PYTHIA in a benchmark movie reviews dataset shows that the suggested combination performs favorably against previous related methods. An online demo is publicly available at http://omiotis.hua.gr/pythia.

1 Introduction

Addressing sentiment analysis as a machine learning problem, e.g., as text classification, poses certain challenges, such as the large space complexity, and the need for deeper understanding of the text, which has given rise to deep learning techniques [1]. In this paper we address the task of sentiment analysis as a binary classification problem; positive or negative polarity. The novelty of our approach lies in the integration of lexical features, e.g., term n-grams, which have been used in the past by previous approaches [1] with semantic features, e.g., count of terms with positive and negative polarity at the sentence level. For the extraction of the semantic features we employ novel word sense disambiguation techniques. Finally, we analyze systematically the effect of all feature types, and we evaluate experimentally our approach by using a variety of machine learners for the task. Comparative evaluation with previous works in a movie review benchmark dataset shows that the suggested approach compares favorably against the previously reported results. The resulting approach is offered as an online system which is publicly available (http://omiotis.hua.gr/pythia) and customizable. The users may select among different disambiguation methods, machine learners and feature types, and can test the approach with any

---

[1] PYTHIA (pronounced pɪθɪə), was the priestess at the Oracle of Delphi. The story says that PYTHIA spoke gibberish, which was then interpreted by the priests.

T. Calders et al. (Eds.): ECML PKDD 2014, Part III, LNCS 8726, pp. 448–451, 2014.
© Springer-Verlag Berlin Heidelberg 2014

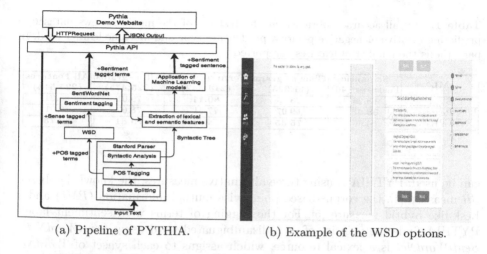

(a) Pipeline of PYTHIA.   (b) Example of the WSD options.

Fig. 1. Overview of PYTHIA pipeline and an example of the user options

piece of text as input. The output of the system is the prediction of the overall polarity of the text, and the annotation and highlighting of the text fragments with information that played important role in the final polarity decision. As a result, the presented online system is of great value to researchers of the field, as well as practitioners who aim at utilizing sentiment analysis approaches in wider text processing software components.

2   Methods, Evaluation and Demonstration of PYTHIA

The overview of the processing pipeline of the PYTHIA system is shown in Figure 1(a). PYTHIA implements five main components: (1) syntactic analysis of input text, (2) word sense disambiguation (WSD), (3) tagging of terms with sentiment labels, (4) extraction of lexical and semantic features, and, (5) application of machine learning models to predict the polarity of the input text. For several of these components the API of the online system offers multiple alternatives that can be customized by the user, e.g., as Figure 1(b) shows for the WSD options.

For the syntactic analysis PYTHIA employs the *Stanford Parser*[2]. The output of this step is a set of trees (one for each sentence) annotated with part of speech (POS) information for each term. The POS information is useful for the WSD step that follows. For the WSD component, PYTHIA implements three options: (1) the first sense heuristic, that always selects the most frequent sense for each word based on *WordNet*, (2) a graph based disambiguation technique, called weighted degree (WDEG) [2], which is a version of Degree Centrality for weighted graphs, and, (3) an Integer Linear Programming approach ILP [3] which solves the ILP problem of maximizing the total pairwise relatedness of the selected senses. For the latter WSD approach, any sense relatedness measure

---

[2] http://nlp.stanford.edu/software/lex-parser.shtml

Table 1. Overall accuracy obtained at the test set of the movie reviews dataset in predicting positive or negative polarity, per feature type and machine learner used. In parenthesis the number of features is reported.

ML	Semantic Features (40)	Char n−grams (11,923)	Term n−grams (214,342)	All n−grams (225,475)	All Features (225,515)
SVM	68.26	73.35	**80.11**	79.01	80.04
Log. Regression	**68.43**	69.07	77.31	78.65	79.01
Naive Bayes	64.66	**75.35**	74.32	**79.81**	**80.73**

can be used; PYTHIA is using three alternative measures: the knowledge-based SR measure [4], the corpus-based point-wise mutual information (PMI) and a Lesk-like hybrid measure [5]. For the tagging of terms with sentiment labels PYTHIA finds the polarity of each disambiguated word using $SentiWordNet$[3]. $SentiWordNet$ is a lexical resource, which assigns to each synset of $WordNet$ sentiment scores for positivity, negativity and objectivity. Next, PYTHIA uses the output of the previous components to extract semantic and lexical features for the input text. The semantic features employed by PYTHIA are 40 and are of two types: (i) at the term level, they capture the number, type and polarity score of the terms that contribute some sentiment to the sentence, and, (ii) at the phrases level they capture the same information as before, but by analyzing whole phrases of the sentence instead of terms. Examples of the semantic features are: the number of nouns with positive polarity in the sentence, the total positive polarity score of verbs, and, the number of noun phrases with negative polarity. In addition to the semantic features, PYTHIA also employs two types of lexical features, namely character and term n−grams, with $n = [1, 3]$. The final step is the application of a machine learning model using some or all of the aforementioned features, in order to predict the polarity of the input text.

The selection of the offered classifiers in PYTHIA is based on the results of the comparative experimental evaluation we conducted. For this purpose, we used a benchmark dataset in sentiment analysis that contains $9,613$ sentences from movie reviews. We used the split into training $(7,792)$ and test $(1,821)$ introduced in [1]. Table 1 shows the results of the evaluation, reporting only on the top-3 tested classifiers (Support Vector Machines, Naive Bayes, and Logistic Regression). The top accuracy obtained for each of the feature types is highlighted, reaching up to 80.73% when all of the features are used, and Naive Bayes is used as a learner. These results are comparable with the SoA results presented in [1], where the authors report 79.4% for the SVM, and 81.8% for the Naive Bayes using BoW representations of the text. The key finding of the experimental evaluation, which constitutes the novelty of the PYTHIA approach, is that the combination of semantic and lexical features leads to the best results.

Finally, in Figure 2 we present screenshots of the PYTHIA demo. Figure 2(a) shows the results of the sentiment analysis of an input sentence, which are presented with a user-friendly GUI. The used model is automatically selected

---

[3] http://sentiwordnet.isti.cnr.it/

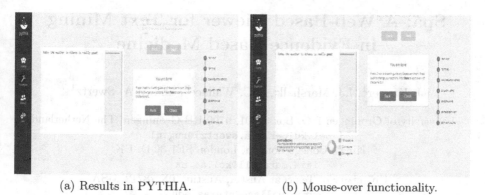

(a) Results in PYTHIA.   (b) Mouse-over functionality.

Fig. 2. Screenshots of the PYTHIA demo

based on the selection of the feature types, e.g., if all features are selected the Naive Bayes classifier is used. Figure 2(b) shows the response of PYTHIA when the user places her mouse over the sentence terms; the polarity score of the specific term is shown. In addition to this, all PYTHIA features are publicly available via an API with GET and POST methods that return *JSON* objects[4].

3 Summary

In this article we presented PYTHIA, a demo for sentiment analysis which employs semantic and lexical features in order to predict the sentiment of an input text. Evaluation of PYTHIA in a benchmark dataset with movie reviews showed that the implemented methods may achieve an accuracy of up to 81%, and that the combination of the semantic and lexical features provided the best performing set up.

References

1. Socher, R., Perelygin, A., Wu, J., Chuang, J., Manning, C.D., Ng, A.Y., Potts, C.: Recursive deep models for semantic compositionality over a sentiment treebank. In: EMNLP, pp. 1631–1642. ACL (2013)
2. Sinha, R., Mihalcea, R.: Unsupervised graph-based word sense disambiguation using measures of word semantic similarity. In: Proc. of the IEEE ICSC, pp. 363–369 (2007)
3. Panagiotopoulou, V., Varlamis, I., Androutsopoulos, I., Tsatsaronis, G.: Word sense disambiguation as an integer linear programming problem. In: Maglogiannis, I., Plagianakos, V., Vlahavas, I. (eds.) SETN 2012. LNCS, vol. 7297, pp. 33–40. Springer, Heidelberg (2012)
4. Tsatsaronis, G., Varlamis, I., Vazirgiannis, M.: Text relatedness based on a word thesaurus. JAIR 37, 1–39 (2010)
5. Nguyen, K., Ock, C.: Word sense disambiguation as a travelling salesman problem. AI Review, 1–23 (2011)

---

[4] http://omiotis.hua.gr/pythia/api.html

Spá: A Web-Based Viewer for Text Mining in Evidence Based Medicine

J. Kuiper[1], I.J. Marshall[2], B.C. Wallace[3], and M.A. Swertz[1]

[1] University of Groningen P.O. Box 30001, 9700 RB Groningen, The Netherlands
{joel.kuiper,m.a.swertz}@rug.nl
[2] King's College London, London SE1 3QD, UK
iain.marshall@kcl.ac.uk
[3] University of Texas at Austin, Austin, TX 78712, USA
byron.wallace@utexas.edu

Abstract. Summarizing the evidence about medical interventions is an immense undertaking, in part because unstructured Portable Document Format (PDF) documents remain the main vehicle for disseminating scientific findings. Clinicians and researchers must therefore manually extract and synthesise information from these PDFs. We introduce Spá,[1][2] a web-based viewer that enables automated annotation and summarisation of PDFs via machine learning. To illustrate its functionality, we use Spá to semi-automate the assessment of bias in clinical trials. Spá has a modular architecture, therefore the tool may be widely useful in other domains with a PDF-based literature, including law, physics, and biology.

1 Introduction

Imposing structure on full-text documents is an important and practical task in natural language processing and machine learning. *Systematic reviews* are an instructive example. Such reviews aim to answer clinical questions by providing an exhaustive synthesis of all the current evidence in published literature. They are fundamental tools in Evidence-Based Medicine (EBM) [2,3]. Data must be manually extracted from the literature to produce the systematic reviews. These extraction tasks are extremely laborious, but could potentially be assisted by machine learning approaches.

As an example we consider risk of bias assessment. Here reviewers assess, e.g., whether study participants and personnel were properly blinded [4]. Assessing risk of bias is a time-consuming task. A single trial typically takes a domain expert ten minutes [5], and a single review typically includes several dozen trials. Making matters worse, due to low rates of reviewer agreement it is regarded as best practice to have each study assessed twice by independent reviewers who later come to a consensus [6].

---

[1] From the Old Norse word spá or spæ referring to prophesying (prophecy)
[2] Source code available under GPLv3 at https://github.com/joelkuiper/spa [1]; demo available at http://spa.clinici.co/

T. Calders et al. (Eds.): ECML PKDD 2014, Part III, LNCS 8726, pp. 452–455, 2014.
© Springer-Verlag Berlin Heidelberg 2014

Machine learning methods could provide the machinery to automate such extractions; as they can effectively impose the desired structure onto PDFs. But if such technologies are to be practically useful, we need tools that visualize these model predictions and annotations. Here we describe Spá, which aspires to realize this aim.

Spá is an open-source, web-based tool that can incorporate machine learning to automatically annotate PDF articles. As a practical demonstration of this technology, we have built a machine learning system that automatically annotates PDFs to aid EBM. This tool is unique in that it leverages state-of-the-art machine learning (ML) models applied to full-text articles to assist practitioners of EBM.

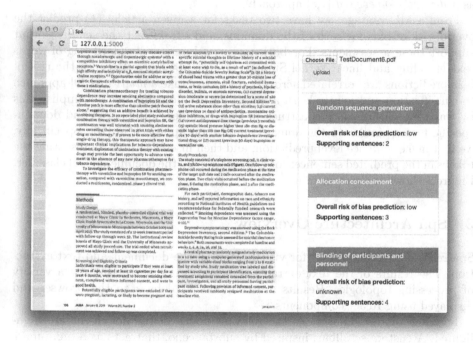

Fig. 1. Screenshot of a PDF with highlighted risk of bias. Here the risk of bias is assessed to be low, for example, and one of the supporting sentences for this assessment describes the randomization procedure (highlighted in green).

While our application of interest is EBM, we emphasize that the visualization tool can be used for any domain in which one wants to annotate PDFs, e.g. genome-wide association studies or jurisprudence. Thus the contribution of this work is two-fold, as we present: (1) a practical tool that incorporates machine learning to help researchers rapidly assess the risk of biases in published biomedical articles, and, (2) a general open-source system for visualizing the predictions of trained models from full-text articles on the web.

2   Case Study: Risk of Bias in Evidence-Based Medicine

2.1   Machine Learning Approaches

To automatically assess the study risk of bias, we have leveraged the Cochrane Database of Systematic Reviews (CDSR) in lieu of manually annotated data, which would be expensive to collect. The CDSR contains descriptions and data about clinical trials reported in existing systematic reviews. We match the full-texts of studies to entries in the CDSR, which contains risk of bias assessment; providing document level labels. The CDSR also contains quotations that reviewers indicated as supporting their assessments. We match these strings to substrings in the PDFs to provide sentence-level supervision. This can be viewed as a *distantly supervised* [7,8] approach.

From a ML vantage, we have two tasks for a given article: (1) predict the overall risk of bias for each of the domains, and (2) extract the sentences that support these assessments. For both tasks we leverage standard bag-of-words text encoding and linear-kernel Support Vector Machines. Because the risk of bias predictions are correlated across domains, we take a *multi-task* [9] approach to classification and jointly learn a model for the domains. We accomplish this by way of a feature space construction that includes both shared and domain-specific terms, similar to the domain adaptation approach in [10]. Specifically, we first make sentence level prediction, and then insert features representing the tokens in the predicted sentences for exploitation by the document level classifier. Figure 1 shows the system in use.

3   Spá Architecture Overview

Spá relies on Mozilla pdf.js[3] for visualization of the document and text extraction. The results of the text extraction are processed server-side by a variety of processing topologies. Results are communicated back to the browser and displayed using React components.[4]

For each of the annotations the relevant nodes in the document are highlighted. A custom scrollbar, inspired by substance.io, that acts as a 'mini-map' is projected to show where annotations reside within the document. The user can interactively activate and inspect specific results.

4   Future Work

We have presented a web-based tool for visualization of annotations and marginalia for PDF documents. Furthermore, we have demonstrated the use of this system within the context of Evidence-Based Medicine by automatically extracting potential risks of bias.

---

[3] http://mozilla.github.io/pdf.js
[4] http://facebook.github.io/react

We believe the tool to be potentially useful for a much wider range of machine learning applications. Currently we are developing a pluggable system for processing topologies, allowing developers to quickly plug in new systems for automated PDF annotation. Furthermore, we are working to allow users to persist annotations and marginalia, possibly embedded within the document itself, for sharing and off-line use. The vision is to have an extensible system for machine assisted data extraction that will greatly increase both the quality and the reproducibility (i.e. data provenance) of current Evidence-Based Medicine.

Acknowledgments. Part of this research was funded by the European Union Seventh Framework Programme (FP7/2007-2013) under grant agreement n° 261433.

References

1. Kuiper, J., Wallace, B.C., Marshall, I.J.: Spa (2014),
   http://figshare.com/articles/Spa/997707
2. Sackett, D.L., Rosenberg, W.M., Gray, J., Haynes, R.B., Richardson, W.S.: Evidence based medicine: what it is and what it isn't. BMJ: British Medical Journal 312(7023), 71–72 (1996)
3. Valkenhoef, G., Tervonen, T., Brock, B., Hillege, H.: Deficiencies in the transfer and availability of clinical trials evidence: a review of existing systems and standards. BMC Medical Informatics and Decision Making 12(1), 95 (2012)
4. Higgins, J., Altman, D., Gotzsche, P., Juni, P., Moher, D., Oxman, A., Savovic, J., Schulz, K., Weeks, L., Sterne, J.: The Cochrane Collaboration's tool for assessing risk of bias in randomised trials. BMJ 343, d5928 (2011)
5. Hartling, L., Bond, K., Vandermeer, B., Seida, J., Dryden, D.M., Rowe, B.H.: Applying the risk of bias tool in a systematic review of combination long-acting beta-agonists and inhaled corticosteroids for persistent asthma. PloS one 6(2), e17242 (2011)
6. Hartling, L., Ospina, M., Liang, Y.: Risk of bias versus quality assessment of randomised controlled trials: cross sectional study. BMJ 339, b4012 (2009)
7. Mintz, M., Bills, S., Snow, R., Jurafsky, D.: Distant supervision for relation extraction without labeled data. In: Proceedings of the Joint Conference of the 47th Annual Meeting of the ACL, Association for Computational Linguistics, pp. 1003–1011 (2009)
8. Nguyen, T., Moschitti, A.: End-to-end relation extraction using distant supervision from external semantic repositories. In: Proceedings of the 49th Annual Meeting of the Association for Computational Linguistics, Association for Computational Linguistics, pp. 277–282 (2011)
9. Evgeniou, T., Pontil, M.: Regularized multi–task learning. In: Proceedings of the Tenth ACM SIGKDD International Conference on Knowledge Discovery and Data Mining, KDD 2004, pp. 109–117. ACM, New York (2004)
10. Daumé III, H.: Frustratingly easy domain adaptation. In: Association for Computatoinal Linguistics (ACL), vol. 1785 (2007)

Propositionalization Online

Nada Lavrač[1,2,3], Matic Perovšek[1,2], and Anže Vavpetič[1,2]

[1] Jožef Stefan Institute, Ljubljana, Slovenia
[2] Jožef Stefan International Postgraduate School, Ljubljana, Slovenia
[3] University of Nova Gorica, Nova Gorica, Slovenia
{nada.lavrac,matic.perovsek,anze.vavpetic}@ijs.si

Abstract. Inductive Logic Programming and Relational Data Mining address the task of inducing models or patterns from multi-relational data. An established relational data mining approach is propositionalization, characterized by transforming a relational database into a single-table representation. The paper presents a propositionalization toolkit implemented in the web-based data mining platform ClowdFlows. As a contemporary integration platform it enables workflow construction and execution, provides open access to Aleph, RSD, RelF and Wordification feature construction engines, and enables RDM performance comparison through cross-validation and ViperCharts results visualization.

Keywords: relational data mining, propositionalization, web access.

1 Introduction

Propositional data mining algorithms induce hypotheses in the form of models or patterns learned from a given data table. In contrast, Inductive Logic Programming (ILP) [6] and Relational Data Mining (RDM) [1] algorithms induce models or patterns from multi-relational data (e.g., relational databases). For relational databases with clearly identifiable instances (i.e., *individual-centered representations* [2], characterized by one-to-many relationships among data tables), propositionalization techniques [3] can be used to transform a relational database into a propositional single-table format, followed by propositional learning, e.g., by using a decision tree or a classification rule learner.

This paper presents an online propositionalization toolkit, which can be used to construct RDM workflows. As completed workflows, data, and results can be made public by the author of the workflow, the platform can serve as an easy-to-access integration platform for various RDM workflows.

2 Clowdflows ILP module

The ClowdFlows platform [4] is an open-source, web-based data mining platform that supports the construction and execution of scientific workflows. This web application can be accessed and controlled from anywhere while the processing is performed in a cloud of computing nodes. A public installation of ClowdFlows

T. Calders et al. (Eds.): ECML PKDD 2014, Part III, LNCS 8726, pp. 456–459, 2014.

is accessible at `http://clowdflows.org`. For a developer, the graphical user interface supports simple operations that enable workflow construction: adding workflow components (widgets) on a canvas and creating connections between the components to form an executable workflow, which can be shared by other users or developers. Upon registration, the user can access, execute, modify, and store the modified workflows, enabling their sharing and reuse. On the other hand, by using anonymous login, the user can execute a predefined workflow, while any workflow modifications would be lost upon logout.

We have extended ClowdFlows with the implementation of an ILP toolkit, including the popular ILP system Aleph [9] together with its feature construction component, as well as RSD [10], RelF [5] and Wordification [7] propositionalization engines. Construction of RDM workflows is supported by other specialized RDM components (e.g., the MySQL package providing access to a relational database by connecting to a MySQL database server), other data mining components (e.g., the Weka classifiers) and other supporting components (including cross-validation), accessible from other ClowdFlows modules. Each public workflow is assigned a unique URL that can be accessed by any user to either repeat the experiment, or use the workflow as a template to design another workflow. Consequently, the incorporated RDM algorithms become handy to use in real-life data analytics, which may therefore contribute to improved accessibility and popularity of ILP and RDM.

Figure 1 shows two simple workflows using the ILP and Weka module components. The first workflow assumes that the user uploads the files required by RSD

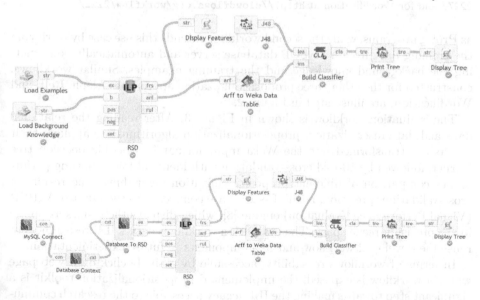

Fig. 1. Above: Simple RSD propositionalization workflow using ILP and Weka components, available online at `http://clowdflows.org/workflow/471/`. Below: The same RSD workflow, extended by accessing the training data using a MySQL database, available at `http://clowdflows.org/workflow/611/`.

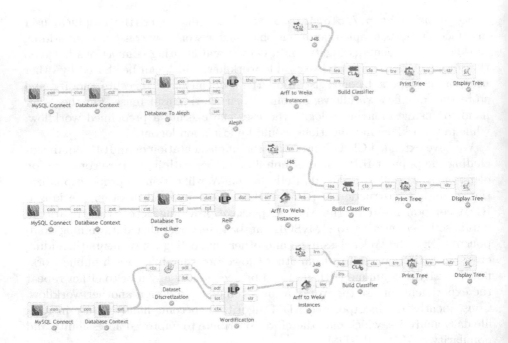

Fig. 2. Propositionalization workflows available online: for Aleph at http://clowdflows.org/workflow/2224/, for RelF at http://clowdflows.org/workflow/2227/ and for Wordification at http://clowdflows.org/workflow/2222/.

as Prolog programs, while the second workflow extends this use case by retrieving the training data from a MySQL database server and automatically constructing the background knowledge and the training examples. Similar workflows, constructed for the other three propositionalization approaches Aleph, RelF and Wordification, are illustrated in Figure 2.

The evaluation workflow is shown in Figure 3. After reading the relational data and data discretization, propositionalization algorithms are applied, their results are transformed into the Weka input format for the J48 decision tree learner, followed by 10-fold cross-validation with identical folds allowing performance comparison of different propositionalization algorithms. The results of cross-validation (precision, recall, F-score) are connected to the input of VIPER (Visual Performance Evaluation) engine [8], which displays the results as points in the precision-recall space. The evaluation workflow enables ILP researchers to reuse the developed workflow and its components in future experimentation.

In terms of workflows reusability, accessible by a single click on a web page where a workflow is exposed, the implemented propositionalization toolkit is a significant step towards making the ILP legacy accessible to the research community in a systematic and user-friendly way. To the best of our knowledge, this is the only workflow-based implementation of ILP and RDM algorithms in a platform accessible through a web browser, enabling simple workflow adaptation to the user's needs.

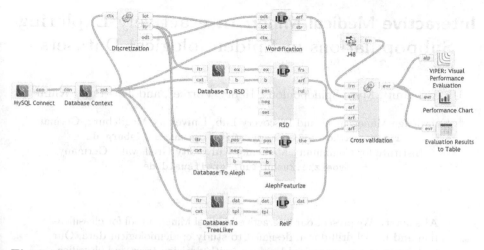

Fig. 3. Performance evaluation workflow, available at http://clowdflows.org/workflow/2210/, comparing the results of J48 after propositionalization by Aleph, RSD, RelF and Wordification.

References

[1] Džeroski, S., Lavrač, N. (eds.): Relational Data Mining. Springer (2001)
[2] Flach, P.A., Lachiche, N.: 1BC: A First-Order Bayesian Classifier. In: Džeroski, S., Flach, P.A. (eds.) ILP 1999. LNCS (LNAI), vol. 1634, pp. 92–103. Springer, Heidelberg (1999)
[3] Kramer, S., Lavrač, N., Flach, P.A.: Propositionalization approaches to relational data mining. In: Džeroski and Lavrač pp. 262–292
[4] Kranjc, J., Podpečan, V., Lavrač, N.: ClowdFlows: A cloud based scientific workflow platform. In: Flach, P.A., De Bie, T., Cristianini, N. (eds.) ECML PKDD 2012, Part II. LNCS, vol. 7524, pp. 816–819. Springer, Heidelberg (2012)
[5] Kuželka, O., Železný, F.: Block-wise construction of tree-like relational features with monotone reducibility and redundancy. Machine Learning 83(2), 163–192 (2011)
[6] Muggleton, S. (ed.): Inductive Logic Programming. Academic Press, London (1992)
[7] Perovšek, M., Vavpetič, A., Cestnik, B., Lavrač, N.: A wordification approach to relational data mining. In: Fürnkranz, J., Hüllermeier, E., Higuchi, T. (eds.) DS 2013. LNCS, vol. 8140, pp. 141–154. Springer, Heidelberg (2013)
[8] Sluban, B., Lavrač, N.: ViperCharts: Visual performance evaluation platform. In: Blockeel, H., Kersting, K., Nijssen, S., Železný, F. (eds.) ECML PKDD 2013, Part III. LNCS, vol. 8190, pp. 650–653. Springer, Heidelberg (2013)
[9] Srinivasan, A.: Aleph manual (March 2007), http://www.cs.ox.ac.uk/activities/machinelearning/Aleph/
[10] Železný, F., Lavrač, N.: Propositionalization-based relational subgroup discovery with RSD. Machine Learning 62(1-2), 33–63 (2006)

Interactive Medical Miner: Interactively Exploring Subpopulations in Epidemiological Datasets

Uli Niemann[1], Myra Spiliopoulou[1], Henry Völzke[2], and Jens-Peter Kühn[2]

[1] Knowledge Management and Discovery Lab, University Magdeburg, Germany
uliniemann@hotmail.com, myra@iti.cs.uni-magdeburg.de
[2] Institute for Community Medicine, University Greifswald, Germany
{voelzke,kuehn}@uni-greifswald.de

Abstract. We present our Interactive Medical Miner, a tool for classification and model drill-down, designed to study epidemiological data. Our tool encompasses supervised learning (with decision trees and classification rules), utilities for data selection, and a rich panel with options for inspecting individual classification rules, and for studying the distribution of variables in each of the target classes. Since some of the epidemiological data available to the medical researcher may be still unlabeled (e.g. because the medical recordings for some part of the cohort are still in progress), our Interactive Medical Miner also supports the juxtaposition of labeled and unlabeled data. The set of methods and scientific workflow supported with our tool have been published in [1].

1 Introduction

High quality decisions of personalized medicine involve identifying subgroups that share some risk factors or symptoms associated with a certain disease. In [1], we have presented mining methods for the discovery of risk factors in subgroups of an epidemiological study's cohort. In [1], we have shown a mining approach for splitting a cohort in subgroups and for discovering factors associated with the outcome for each subgroup. We have thereby demonstrated that different subgroups exhibit different factors. The top-classification rules found by our approach agree with research results in epidemiology publications.

We present here our Interactive Medical Miner[1], as well as utilities for learning and model inspection. The Interactive Medical Miner derives models from epidemiological data containing a nominal target variable, for instance a diagnosis report outcome, and allows the user to drill down to the data of distinct individuals, and to further explore detailed information about summary statistics or class distribution histograms. Contrary to medical research practice which is hypotheses-based, we use a data-driven approach, as practiced e.g. in [2,3]. Our Interactive Medical Miner offers, under a simple interface, several functionalities for medical researches who aim to interactively explore their datasets and inspect classification patterns derived on them. To this purpose, the Interactive Medical

---

[1] The tool can be downloaded at http://kmd.cs.ovgu.de/res/imm/.

T. Calders et al. (Eds.): ECML PKDD 2014, Part III, LNCS 8726, pp. 460–463, 2014.
© Springer-Verlag Berlin Heidelberg 2014

Miner provides a tailored workflow for preparation and classification of epidemiological data, including model drill-down and summary statistics for each class and rule. The tool can also be used in a more general medical (clinical) context.

We use algorithms from the Weka[2] library: We leverage the HotSpot[3] algorithm for classification rule discovery and employ the J48 (equivalent to the C4.5 algorithm [4]) for decision tree induction.

2   Workflow of the Interactive Medical Miner

Our tool takes as input the data of an epidemiological cohort, where the target variable concerns a medical outcome, e.g. the presence of increased fat in the liver [1]. The tool allows the medical researcher to specify that either the complete dataset or a selection of cohort participants (a subpopulation) should be used for learning. On this dataset, the Interactive Medical Miner discovers classification rules and builds decision trees. The decision trees are presented to the expert, while classification rules can be further *explored*. In particular, the expert is shown histograms on the distribution of the participants supporting a rule with respect to each class, and histograms on the distribution of the rule's variables inside each class. Since some of the cohort participants in the dataset we study (SHIP [5]) are not yet labeled, our tool supports the inspection of the values in each classification rule's antecedent on the unlabeled data as well.

3   User Interface

The user interface of the Interactive Medical Miner consists of two areas; each one is comprised of six panels. Subsequently, we describe the layout of the Interactive Medical Miner while referring to Figure 1.

In the upper left "Settings" panel, the user controls the most important algorithmic parameters. For classification rules generation, the user has to specify:

- Minimum value count: the minimum percentage (or number) of instances supporting the rule AND belonging to specified target class,
- Maximum rule length: the number of variables in the antecedent,
- Maximum branching factor: the maximum number of variables that may be added to an existing classification rule,
- Minimum improvement: the minimum relative confidence improvement to be achieved by the addition of a further variable to the classification rule.

For decision tree induction, the following three parameters are required:

- Minimum number of data records in a leaf node,
- Pruning factor: the threshold that must be satisfied if the tree is pruned; if the value falls below this threshold, the tree is not pruned further,

---

[2] http://www.cs.waikato.ac.nz/ml/weka/

[3] http://weka.sourceforge.net/packageMetaData/hotSpot/Latest.html

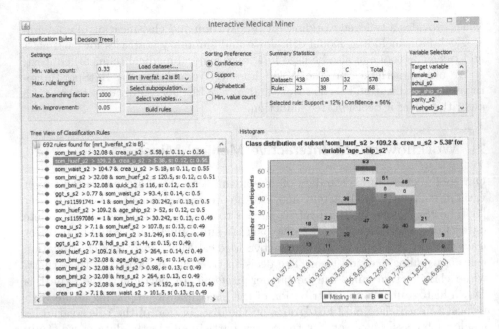

Fig. 1. Screenshot of the Interactive Medical Miner with its six panels

- Only binary splits: a flag indicating, whether nominal variables with a value range of n values should be subjected to binary splits or to n-ary splits.

Further, the tool allows the user to specify a subpopulation of the dataset. For instance, one might filter out the male participants and study only female participants. To this purpose, a new frame pops up where filtering queries in the form of **Variable Operator Value** can be specified. The defined restrictions are shown in a table where the user can optionally select and remove single entries. Next, the Interactive Medical Miner also offers a button "Filter Variables..." which opens a pop-up frame where the user can (de-)select one or more variables for model generation by shifting them from one list to the other. For example, the user might exclude a variable that is known to be highly correlated with another variable that is already considered for model learning.

By clicking on "Build Rules" / "Generate Tree", the resulting model is depicted in the "Tree View" panel. For classification rules, the retrieved rules can be sorted via the radio buttons in the panel "Sorting Preference" (area right to "Settings") according to several criteria, including confidence (default), support, alphabetical and minimum value count. For decision trees, the output tree structure can be visualized with Weka's TreeVisualizer.

When a rule/ node is selected, the top middle area "Summary Statistics" is refreshed. The first row shows the class distribution of the dataset, while the second row shows how the instances supporting the antecedent are distributed among the existing classes. Hence, the user gets insights about the class distribution of a single rule/ node and thus can control the mining process by adapting parameters or selecting specific variables or value ranges. For instance, the tool allows the user to trade of high confidence against high support rules.

The summary statistics table contains information about the labeled instances, while the histogram in the bottom right panel covers the unlabeled instances. The user can choose a further variable from the panel "Variable Selection" (cf. Figure 1, upper right corner) to see how the values of these variables are distributed. Unlabeled data are marked as "Missing". To plot the histograms, we employ the open source chart library JFreeChart[4].

4   Conclusion

The Interactive Medical Miner generates classification models on epidemiological datasets and supports interactive exploration of individual classification rules and decision tree nodes. The tool provides options for filtering cohort participants and selecting a subset of variables. It allows the user to tune algorithm parameters and thus guide the mining process. The visual representation of class distributions and the juxtaposition of labeled and unlabeled cohort participants improves data understanding and might reveal idiosyncrasies of the labeled data. In future work, we intend to extend the tool by adding more classifiers, a more elaborate visualization as well as model and graphic export possibilities.

Acknowledgements. Part of this work was supported by the German Research Foundation project SP 572/11-1 "IMPRINT: Incremental Mining for Perennial Objects".
The data used in this work were made available through the cooperation SHIP/ 2012/06/D "Predictors of Steatosis Hepatis".

References

1. Niemann, U., Völzke, H., Kühn, J.-P., Spiliopoulou, M.: Learning and inspecting classification rules from longitudinal epidemiological data to identify predictive features on hepatic steatosis. Expert Systems with Applications 41(11), 5405–5415 (2014)
2. Zhanga, C., Kodell, R.L.: Subpopulation-specific confidence designation for more informative biomedical classification. Artificial Intelligence in Medicine 58(3), 155–163 (2013)
3. Pinheiro, F., Kuo, M.-H., Thomo, A., Barnett, J.: Extracting association rules from liver cancer data using the FP-growth algorithm. In: 3rd International Conference on Computational Advances in Bio and Medical Sciences, ICCABS (2013)
4. Quinlan, J.R.: Learning with continuous classes. In: 5th Australian Joint Conference on Artificial Intelligence, vol. 92, pp. 343–348 (1992)
5. Völzke, H., Alte, D., Schmidt, C.O., Radke, D., Lorbeer, R., Friedrich, N., et al.: Cohort Profile: The Study of Health in Pomerania. International Journal of Epidemiology 40(2), 294–307 (2011)

---

[4] http://www.jfree.org/jfreechart/

WebDR: A Web Workbench for Data Reduction

Stefanos Ougiaroglou* and Georgios Evangelidis

Department of Applied Informatics, School of Information Sciences,
University of Macedonia, 156 Egnatia str, GR-54006 Thessaloniki, Greece
{stoug,gevan}@uom.gr

Abstract. Data reduction is a common preprocessing task in the context
of the k nearest neighbour classification. This paper presents WebDR, a
web-based application where several data reduction techniques have been
integrated and can be executed on-line. WebDR allows the performance
evaluation of the classification process through a web interface. Therefore,
it can be used by the academia for educational and experimental purposes.

Keywords: k-NN classification, data reduction, web-based application.

1 Introduction

The k Nearest Neighbour (k-NN) classifier [3] is an effective classifier but has
some weaknesses that may render its use inappropriate. The first one is the
high computational cost involved (all distances between each unseen item and
all training data must be computed). In cases of large datasets, this drawback
renders the classification a time-consuming procedure. Another weakness is that
the k-NN classifier must maintain all the training data always available. Thus,
it involves high storage requirements. Moreover, the accuracy achieved by the
classifier depends on the quality of the training set (TS). Noise and mislabelled
data, as well as outliers and overlaps between data regions of different classes
may mislead the algorithm and affect the accuracy.

Data Reduction Techniques (DRTs) can cope with all the weaknesses. They
can be grouped into two main categories: (i) prototype selection algorithms
(PS) [6], and, (ii) prototype abstraction algorithms (PA) [17]. PS algorithms
select representative items (or prototypes) from the initial training set, whereas
PA algorithms generate items by summarizing on similar training items.

PS algorithms are divided into two subcategories. They can be either condensing or editing algorithms. PA and PS-condensing algorithms have the same
motivation. They aim to build a small representative set of the TS. This set is
called the condensing set (CS). Usage of the CS has the benefits of low computational cost and storage requirements, while accuracy is not affected. Editing
algorithms aim to improve accuracy rather than achieve high reduction rates.
For that purpose, they try to improve the quality of the TS by removing outliers,
noise and by smoothing the class decision boundaries.

---

* S. Ougiaroglou is supported by the State Scholarships Foundation of Greece (I.K.Y.)

T. Calders et al. (Eds.): ECML PKDD 2014, Part III, LNCS 8726, pp. 464–467, 2014.

Several papers have been published that present DRTs with the corresponding experimental results. Some of them have been implemented under KEEL [2], an open-source java-based framework. However, to the best of our knowledge there is no software that allows experimentations over the web. This observation is behind the motivation of this work. We introduce WebDR[1] (Web-based Data Reduction), a web-based application that allows the execution and the performance evaluation of several DRTs over the web.

Section 2 outlines k-NN classification through data reduction. Section 3 presents WebDR. Finally, Section 3 concludes the paper and presents our future plans.

2 k-NN Classification through Data Reduction

The reduction rates achieved by many PA and PS-condensing algorithms depend on the level of noise in the TS. The higher the level of noise is, the lower reduction rates are achieved. Hence, their effective application implies the removal of noise, i.e., execution of editing beforehand [4]. Therefore, an editing algorithm should be run in order to either improve accuracy or make more effective the application of a PA or PS-condensing algorithm.

k-NN classification through data reduction is summarized in Figure 1. The process has two stages, preprocessing (optionally) and classification. There are four possible preprocessing types: (i) **No-preprocessing:** If the TS is small and noise-free, no preprocessing is required. (ii) **Only editing:** If the TS is small but contains noise, only editing should be executed during preprocessing. (iii) **Only condensing:** In cases of large and noise-free TSs, data reduction without editing should be executed (i.e., a PA or PS-condensing algorithm). (iv) **Both editing and PA or PS-condensing:** In cases of large TSs that contain noise, both types of preprocessing algorithms must be run.

The goal of a complete data reduction preprocessing procedure is to build a noise-free CS by keeping or generating for each class a sufficient number of prototypes that are essential for the k-NN classification.

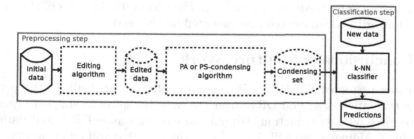

Fig. 1. k-NN classification through data reduction

---

[1] http://dbtech.uom.gr/webdr

3   WebDR

WebDR offers several DRTs available on-line. The user can plan and run experiments and measure the classification performance through an interactive web
interface over several known datasets distributed by the KEEL[2] or/and the UCI[3]
dataset repositories and time-series datasets distributed by the UCR time-series
classification/clustering website[4]. All the available datasets can be explored in
detail using the "dataset explorer" tool that is available in WebDR.

WebDR allows the performance evaluation of the DRTs by measuring three
criteria, namely, (i) **Reduction rate:** the ratio of the discarded items over
the initial items of the TS. The higher the reduction rate, the faster the k-
NN classification (fewer distances are computed); (ii) **Accuracy** achieved by
the k-NN classifier when it runs over the CS; (iii) **Preprocessing cost:** the
computational cost required for the construction of the CS.

The preprocessing costs are estimated by counting the distances computed
by the corresponding DRTs. WebDR adopts the Euclidean distance as the distance metric. The reported performance measurements are averages obtained via
five-fold cross-validation. It is worth mentioning that all datasets built during
preprocessing are available to the users in a five-fold form (five pairs of training and testing sets). They can be downloaded and used by the user locally. Of
course, the number of the nearest neighbours and the DRT specific parameters
(if any) can be adjusted through the interface.

All the possible preprocessing types can be executed by WebDR. Its main
page offers four links. Each one leads to the corresponding type of preprocessing.
Currently, the following DRTs have been integrated in WebDR:

- **Editing algorithms:** ENN-rule [18], All-k-NN [16], Multiedit [5], EHC [14]
- **Condensing algorithms:** CNN-rule [7], IB2 [1], PSC [8]
- **PA algorithms:** RSP3 [15], RHC [11,10], dRHC [10], ERHC [9], AIB2 [13],
  RkM [12]

WebDR is hosted on a Debian GNU/Linux server with two 64-bit Quad-Core
CPUs and 2GB of main memory. All algorithms were coded in C. The web
interface was developed using PHP (server-side programming) and html/CSS
and javascript (client-side programming). The executable binaries of the implemented algorithms are located and executed on the server.

4   Conclusions and Future Work

The paper presented WebDR, a web-based application that allows the performance evaluation of several DRTs over the web. It aspires to support teaching
and research on data reduction. We plan to integrate more DRTs and datasets
in WebDR. Moreover, we will develop a mechanism that will allow users to run
experiments on their own datasets.

---

[2] http://sci2s.ugr.es/keel/datasets.php
[3] http://archive.ics.uci.edu/ml/
[4] http://www.cs.ucr.edu/~eamonn/time_series_data/

References

1. Aha, D.W., Kibler, D., Albert, M.K.: Instance-based learning algorithms. Mach. Learn. 6(1), 37–66 (1991), http://dx.doi.org/10.1023/A:1022689900470
2. Alcala-Fdez, J., Sanchez, L., Garcia, S., del Jesus, M.J., Ventura, S., Garrell, J.M., Otero, J., Romero, C., Bacardit, J., Rivas, V.M., Fernandez, J.C., Herrera, F.: Keel: a software tool to assess evolutionary algorithms for data mining problems. Soft Comput. 13(3), 307–318 (2008)
3. Dasarathy, B.V.: Nearest neighbor (NN) norms: NN pattern classification techniques. IEEE Computer Society Press (1991)
4. Dasarathy, B.V., Sánchez, J.S., Townsend, S.: Nearest neighbour editing and condensing tools synergy exploitation. Pattern Analysis & Applications 3(1), 19–30 (2000)
5. Devijver, P.A., Kittler, J.: On the edited nearest neighbor rule. In: Proceedings of the Fifth International Conference on Pattern Recognition. The Institute of Electrical and Electronics Engineers (1980)
6. Garcia, S., Derrac, J., Cano, J., Herrera, F.: Prototype selection for nearest neighbor classification: Taxonomy and empirical study. IEEE Trans. Pattern Anal. Mach. Intell. 34(3), 417–435 (2012)
7. Hart, P.E.: The condensed nearest neighbor rule. IEEE Transactions on Information Theory 14(3), 515–516 (1968)
8. Olvera-Lopez, J.A., Carrasco-Ochoa, J.A., Trinidad, J.F.M.: A new fast prototype selection method based on clustering. Pattern Anal. Appl. 13(2), 131–141 (2010)
9. Ougiaroglou, S., Evangelidis, G.: Efficient editing and data abstraction by finding homogeneous clusters. In: Submitted, under review
10. Ougiaroglou, S., Evangelidis, G.: RHC: Non-parametric cluster-based data reduction for efficient k-nn classification. Pattern Analysis and Applications pp. (accepted, to appear)
11. Ougiaroglou, S., Evangelidis, G.: Efficient dataset size reduction by finding homogeneous clusters. In: Proceedings of the Fifth Balkan Conference in Informatics, BCI 2012, pp. 168–173. ACM Press, New York (2012)
12. Ougiaroglou, S., Evangelidis, G.: A simple noise-tolerant abstraction algorithm for fast k-NN classification. In: Corchado, E., Snášel, V., Abraham, A., Woźniak, M., Graña, M., Cho, S.-B. (eds.) HAIS 2012, Part II. LNCS, vol. 7209, pp. 210–221. Springer, Heidelberg (2012)
13. Ougiaroglou, S., Evangelidis, G.: AIB2: An abstraction data reduction technique based on ib2. In: Proceedings of the 6th Balkan Conference in Informatics, BCI 2013, pp. 13–16. ACM, New York (2013)
14. Ougiaroglou, S., Evangelidis, G.: EHC: Non-parametric editing by finding homogeneous clusters. In: Beierle, C., Meghini, C. (eds.) FoIKS 2014. LNCS, vol. 8367, pp. 290–304. Springer, Heidelberg (2014)
15. Sánchez, J.S.: High training set size reduction by space partitioning and prototype abstraction. Pattern Recognition 37(7), 1561–1564 (2004)
16. Tomek, I.: An experiment with the edited nearest-neighbor rule. IEEE Transactions on Systems, Man, and Cybernetics 6, 448–452 (1976)
17. Triguero, I., Derrac, J., Garcia, S., Herrera, F.: A taxonomy and experimental study on prototype generation for nearest neighbor classification. Trans. Sys. Man Cyber Part C 42(1), 86–100 (2012)
18. Wilson, D.L.: Asymptotic properties of nearest neighbor rules using edited data. IEEE Trans. on Systems, Man, and Cybernetics 2(3), 408–421 (1972)

GrammarViz 2.0: A Tool for Grammar-Based Pattern Discovery in Time Series

Pavel Senin[1], Jessica Lin[2], Xing Wang[2], Tim Oates[3], Sunil Gandhi[3],
Arnold P. Boedihardjo[4], Crystal Chen[4], Susan Frankenstein[4], and Manfred Lerner[5]

[1] University of Hawaii, Manoa, ICS Dept., CSDL, USA
`senin@hawaii.edu`
[2] George Mason University, Dept. of Computer Science, USA
`{jessica,xwang24}@gmu.edu`
[3] University of Maryland, Baltimore County, Dept. of Computer Science, USA
`oates@cs.umbc.edu, sunilga1@umbc.edu`
[4] U.S. Army Corps of Engineers, Engineer Research and Development Center, USA
`{arnold.p.boedihardjo,crystal.chen,`
`susan.frankenstein}@usace.army.mil`
[5] SAP Germany
`manfred.lerner@sap.com`

Abstract. The problem of frequent and anomalous patterns discovery in time series has received a lot of attention in the past decade. Addressing the common limitation of existing techniques, which require a pattern length to be known in advance, we recently proposed grammar-based algorithms for efficient discovery of variable length frequent and rare patterns. In this paper we present GrammarViz 2.0, an interactive tool that, based on our previous work, implements algorithms for grammar-driven mining and visualization of variable length time series patterns.[1]

1 Introduction

The ability to efficiently detect frequent and anomalous patterns in time series allows for the exploration, summation, and compression of data. In addition, such information is crucial to a variety of application domains where these patterns convey critical and actionable information, such as health care, equipment safety, and security. Furthermore, these patterns are often used as input features for data mining tasks, such as association rule mining and classification.

Previously, we defined time series motifs (frequent patterns) [1] and time series discords (anomalous patterns) [2], and proposed efficient *exact* solutions for their discovery based on Symbolic Aggregate Approximation (SAX) [3]. While there has been a great amount of follow-up work on the discovery of both pattern types [4], one common limitation of currently available techniques is that they require the length of a potential

---

[1] This research is partially supported by the National Science Foundation under Grant No. 1218325 and 1218318.

T. Calders et al. (Eds.): ECML PKDD 2014, Part III, LNCS 8726, pp. 468–472, 2014.

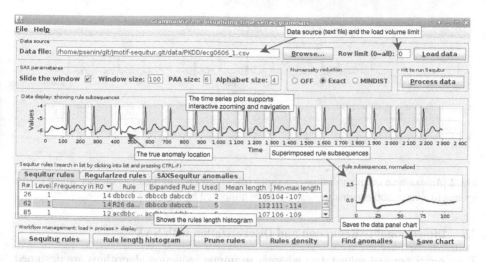

Fig. 1. An example of a recurrent grammar rule (i.e. *motif*) discovery in the ECG dataset using GrammarViz 2.0. Note, that the highlighted motif does not cover an anomalous heartbeat and that rule-corresponding subsequences vary in length.

motif or discord to be specified as input. This is unreasonable for most real-world problems as such information may not be known in advance, and patterns of different lengths may co-exist in the data.

Addressing this limitation, we recently proposed an alternative solution for the discovery of variable-length motifs [5] and anomalies [6] based on SAX discretization and the Sequitur grammar inference algorithm [7]. We showed that our algorithm is able to efficiently discover *co-existing variable-length approximate motifs and anomalies* without any prior knowledge about their length, shape, or minimal occurrence frequency. In this work, we present a time series pattern discovery application called GrammarViz 2.0 that can simultaneously discover variable-length motifs and anomalies.

2   Our Approach and the Tool for Time Series Patterns Mining

Our approach is built on a three phase process: time series discretization, context free grammar induction, and motif/anomaly detection. The first step is to model the time series as discretized elements and convert it into a symbolic representation. The second step is to parse the symbolic series and decompose it into a context free grammar [5, 6]. Since rules of a context free grammar are hierarchically organized, it is possible to establish the probability of occurrence of a time series subsequence using its corresponding rule hierarchy and rule counts in the entire time series. Intuitively, since each grammar rule represents a discretized subsequence *pattern* of the input time series, frequently used rules are likely to correspond to recurrent subsequences, while infrequently used rules are likely to correspond to rare subsequences.

Next, we discuss the detailed steps of the above approach and its implementation in our grammar-driven workflow for time series patterns discovery.

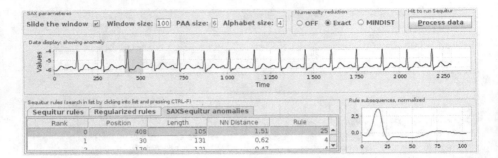

Fig. 2. An example of an anomalous grammar rule discovered in the ECG dataset which corresponds to a very subtle anomaly in the ST wave annotated by an expert [2].

2.1 Dimensionality Reduction and Discretization with SAX

Time series are real-valued data whereas grammar induction algorithms are designed for discrete values. We rely on SAX [3] to discretize the input time series. For time series T of length m, SAX obtains a lower-dimensional representation by first performing a z-normalization then dividing the time series into w equal-sized segments. Next, for each segment, SAX computes a mean value and maps it to a symbol according to a predefined set of breakpoints dividing the data space into α equiprobable regions, where α is the user specified alphabet size. While dimensionality reduction is a desirable feature for exploring global patterns, the high compression ratio (m/w) significantly affects performance in cases where localized phenomena are of interest. Thus, for the local pattern discovery, and specifically for motif and anomaly detection, SAX is typically applied to a set of subsequences that represent local features – a technique called subsequence discretization [1] which is implemented via a sliding window.

Our tool implements both global and local discretization and allows an interactive tuning of discretization parameters using "SAX parameters" panel (Fig. 1). In addition, next to the SAX parameters selection, users can toggle the numerosity reduction strategy, which not only mitigates for trivial and degenerate pattern discovery [2,3], but enables an essential feature of our technique – the discovery of variable-length co-existing patterns [5, 6].

2.2 Context Free Grammar Induction with Sequitur

For grammar inference, we rely on Sequitur - a linear time and space algorithm that derives a context-free grammar from a string incrementally [7]. By identifying frequent subsequences in the input string, the algorithm builds a compact context-free grammar reflecting the input string specificity. In addition, we are currently extending our application with mSequitur algorithm implementation that introduces a merging operator and is capable of further grammar reduction by generalization [11].

Since Sequitur requires no input parameters, in a single "Process data" step (Fig. 1) our tool performs both discretization and grammar induction procedures. Once grammar is built, its rules are presented to the user in a table format enabling efficient examination and exploration of rules and their corresponding subsequences. GrammarViz 2.0

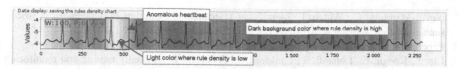

Fig. 3. An example of the "Data display" panel showing the "Rule density" plot used for highly efficient approximate anomaly discovery through visual examination.

shows rule locations on the original time series and superimposes all rule subsequences on a separate panel. This allows visual evaluation of the results from selected parameters as well as their interactive tuning (Fig.1).

2.3 Exploiting Context-Free Grammar for Pattern Discovery

Motif Discovery. With the capability to sort the rule table by the rule usage frequency, as well as the effective visual presentation of grammar rules, GrammarViz 2.0 allows user to navigate the rules and visually inspect their corresponding subsequences ("motifs").

Discord discovery. GrammarViz 2.0 enables anomaly detection in two ways: by integrating grammar induction in the HOTSAX discord discovery framework [2] (Fig.2), and by visualization of the grammar rule density (Fig.3). Both approaches allow the user to visually evaluate potential anomalous rules and their corresponding subsequences.

3 Target Audience and Similar Applications

As time series are often used as a proxy to represent a large variety of wide ranging real-life phenomena, the GrammarViz 2.0 application targets diverse audiences including researchers, practitioners, engineers, medical specialists, and safety and security personnel. While other time series pattern visualization tools exist [9, 10], we are not aware of any tool that has the same capabilities as GrammarViz 2.0; namely, the discovery of hierarchical patterns and variable-length motifs and discords.

References

1. Lin, J., Keogh, E., Patel, P., Lonardi, S.: Finding Motifs in Time Series. In: The 2nd Workshop on Temporal Data Mining, the 8th ACM Int'l Conference on KDD, pp. 53–68 (2002)
2. Keogh, E., Lin, J., Fu, A.: HOT SAX: Efficiently Finding the Most Unusual Time Series Subsequence. In: Proc. ICDM, pp. 226–233 (2005)
3. Patel, P., Keogh, E., Lin, J., Lonardi, S.: Mining Motifs in Massive Time Series Databases. In: Proc. ICDM (2002)
4. Chandola, V., Cheboli, D., Kumar, V.: Detecting Anomalies in a Time Series Database. CS Technical Report 09–004 (2009)
5. Li, Y., Lin, J., Oates, T.: Visualizing variable-length time series motifs. In: Proc. of the 2012 SIAM International Conference on Data Mining, pp. 895–906 (2012)
6. Senin, P., Lin, J., Wang, X., Oates, T., Boedihardjo, A.P., Chen, C., Frankenstein, S., Gandhi, S.: Grammar-driven anomaly discovery in time series. CSDL Techreport 14-05 (2014)

7. Nevill-Manning, C., Witten, I.: Identifying Hierarchical Structure in Sequences: A linear-time algorithm. Journal of Artificial Intelligence Research 7, 67–82 (1997)
8. Paper authors. Supporting webpage: https://code.google.com/p/jmotif/
9. Lin, J., Keogh, E., Lonardi, S., Lankford, J., Nystrom, D.: Visually mining and monitoring massive time series. In: Proc. 10th ACM SIGKDD Intl. Conf. on KDD, pp. 460–469 (2004)
10. Hao, M., Marwah, M., Janetzko, H., Dayal, U., Keim, D., Patnaik, D., Ramakrishnan, N., Sharma, R.K.: Visual Exploration of Frequent Patterns in Multivariate Time Series. Information Visualization 11(1), 71–83 (2012)
11. Oates, T., Boedihardjo, A., Lin, J., Chen, C., Frankenstein, S., Gandhi, S.: Motif discovery in spatial trajectories using grammar inference. In: Proc. of ACM Intl. Conf. on Information and Knowledge Management, CIKM (2013)

Insight4News: Connecting News to Relevant Social Conversations

Bichen Shi, Georgiana Ifrim, and Neil Hurley

Insight Centre for Data Analytics,
University College Dublin
Dublin, Ireland
{bichen.shi,georgiana.ifrim,neil.hurley}@insight-centre.org

Abstract. We present the **Insight4News** system that connects news articles to social conversations, as echoed in microblogs such as Twitter. **Insight4News** tracks feeds from mainstream media, e.g., BBC, Irish Times, and extracts relevant topics that summarize the tweet activity around each article, recommends relevant hashtags, and presents complementary views and statistics on the tweet activity, related news articles, and timeline of the story with regard to Twitter reaction. The user can track their own news article or a topic-focused Twitter stream. While many systems tap on the social knowledge of Twitter to help users stay on top of the information wave, none is available for connecting news to relevant Twitter content on a large scale, in real time, with high precision and recall. **Insight4News** builds on our award winning Twitter topic detection approach and several machine learning components, to deliver news in a social context.

Keywords: news tracking, social media, Twitter, summarization.

1 Introduction

Famously in August 2011, news of an earthquake in Virginia, USA, reached New York by Twitter before the tremors were felt. Media stories such as the death of Michael Jackson have spread rapidly on social media in advance of breaking on the mainstream media. Nowadays, more often than not, news stories break online long before appearing in newspapers. The landscape of news delivery and dissemination has changed dramatically in less than a decade since the widespread take-up of social media. Writer and entrepreneur Chris Anderson captures this through his statement *"The ants have megaphones now."* [5].

Insight4News links news articles from mainstream media (e.g., BBC), to relevant Twitter conversations, as delivered by tweets, hashtags and automatically detected events. It builds on our prior work [1] for automatically mining social media streams to provide users with a set of headlines summarizing the most important topics discussed over a given time period. Our topic detection approach was assessed by practicing journalists and ranked first as the most effective "news miner" with regards to several evaluation criteria, amongst which

T. Calders et al. (Eds.): ECML PKDD 2014, Part III, LNCS 8726, pp. 473–476, 2014.
© Springer-Verlag Berlin Heidelberg 2014

were precision and recall. Furthermore, **Insight4News** provides social context for news articles via a machine learning algorithm that classifies and ranks hashtags [2] and provides a timeline of each article with respect to relevant tweets, hashtags, topics and photos.

Since Twitter has become very popular as a channel for citizen-driven media, many systems aim to tap this resource. Storyful [7] is a social media news agency that tracks and curates Twitter content for breaking news and potential stories for newsrooms. The focus is mostly on content curation and licensing. The *headlines* feature of Twitter provides links to articles that relate to a specific tweet. The Tweeted Times promises a personal newspaper, by aggregating news from the Twitter stream and ranking them by popularity among the users' friends. Storify [8] provides a service where users can manually search for topics of interests on several platforms (Twitter, Facebook, YouTube) in order to add social context to a story. Blews [4] focuses on political news by tracking blogs and the articles they cite, tagging each article with the number of blogs citing it, political orientation and emotional charge of those blogs. Hash2News [9] takes a hashtag as input and presents relevant news articles for that hashtag. Most of these systems go from the social media to the news, via the urls shared in tweets. This drastically reduces recall, since many related tweets do not post the url explicitly. We propose a system that combines both directions, by connecting mainstream media news articles to the relevant social feeds on Twitter, while allowing the user to track in real-time the newsworthy topics directly from the Twitter stream. Most readers are interested in other people's opinion on the topics, the connections between articles and the development of stories, *"users crave more relevant news with deeper contextualization"* [6]. **Insight4News** aims to fulfil this need. This is also an important step in the development of new digital journalism support tools. It is expected that such tools will become commonplace in the newsroom of the near future [10].

2   The Insight4News System

In this section, we present the key components of **Insight4News** [3], as illustrated in Figure 1. The system is written in Python3 with the Django web framework and is deployed on an Apache web server. Celery, a distributed task queue, is used for back end data collection and processing.

Data Collection and Processing. We poll 14 RSS news feeds every 15 minutes (currently from BBC and Irish Times), covering international and local news. We extract the urls and retrieve the articles (around 400 daily). We automatically extract representative keywords for each article, pool keywords for all articles, and feed them to the Twitter Streaming API, constraining each retrieved tweet to contain at least two article keywords (for more details see [2]). On average, we get about 500k tweets per day. Each article's keywords are streamed for 24 hours, by updating the all-article-keywords-list every 5 minutes. This step aims to retrieve a large set of relevant tweets without being restricted to a set of manually curated user lists, locations or article urls. Via shallow matching of

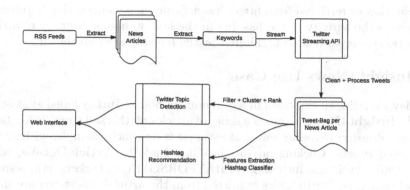

Fig. 1. High-level overview of the **Insight4News** system

tweet and article keywords, we get a local tweet-bag per article which we use for topic detection and hashtag recommendation (recomputed every 5 mins).

Topic Extraction. Based on our award winning aproach [1], this stage relies on tweet-clustering combined with a few layers of filtering, aggregation and ranking. The detailed steps include hierarchical clustering of tweets, time-dependent n-gram and cluster ranking and headlines re-clustering. For each article tweet-bag, we execute these steps to obtain a set of headlines or topics that summarize the tweet activity relevant to the article. This approach also works on an arbitrary Twitter stream (not article-focused).

Hashtag Recommendation. We pose this as a learning problem. Using the tweets-bag per article, we form article-hashtag pairs, and compute four features for each pair that capture the global (whole stream) and local (article tweet-bag) profile of the hashtag (wrt popularity and relevance). To train a hashtag classifier, we use 2,500 manually labeled article-hashtag pairs. A good source of relevant hashtags that does not require manual effort, are user tweets that post the article url and hashtags for that article. We use a Logistic Regression classifier with 87% Precision and 79% Recall from [2]. The classifier provides a score describing the likelihood that a hashtag is relevant to the article, which we use to rank hashtags for each article, and recommend the top10 hashtags with classification score above 0.5.

The **web interface** currently has 6 views. **TrackedNews** shows the latest news articles, with headline, number of tweets retrieved and published time. **ArticleDetails** shows content and social context, recommended hashtags and extracted topics from related tweets. The default hashtag view is the classifier result, while 3 other views show top10 hashtags based on frequency, recency, and cosine similarity between hashtag and article profiles. **TrackYourArticleHere** allows the user to track their own news article by providing a valid url or the full text. **TrackYourTopicsHere** allows the user to track events by providing keywords (e.g., Russia, Ukraine, EU). **PopularHashtags** lists the top50 most popular hashtags in the last 24 hours. **HashtagDetails** shows a specific hashtag, its definition as retrieved from http://tagdef.com, and co-occurring hashtags. A plot shows the activity of the hashtag in the last 30 hours. A group of articles and a tweet stream related-to/filtered-by the hashtag are also shown.

3 Insight4News Use Case

On May 13, 2014, an explosion in a coal mine in Soma, Turkey killed at least 280 people. **Insight4News** captures a series of articles on this sad story. *BBC: Turkey coal mine disaster: Desperate search at Soma pit* is one such example, with around 5k related tweets. Clicking on this headline leads to **ArticleDetails**, which shows top10 hashtags, including #PRAYFORSOMA, #Turkey, #turkeymine, #WorkersRights. Top10 topics extracted from the article's tweet-bag are shown (e.g., **1.** *Mourners by miners' graves in the western town of Soma. The toll from Turkey's worst mining accident is now 282.*, **2.** *Bayram Ilki poured water on the grave of his son Saban, as many other coal miners were buried nearby.*). A photo is shown beside each topic, which can be expanded to show the tweets summarized by this topic-headline. Top10 learned hashtags are shown by default. Clicking on #Soma leads to the hashtag definition and related hashtags (#PrayForTurkey, #Protests). A plot shows the activity for #Soma in the last 30 hours. Below the plot, there are articles related to #Soma. One of the articles *Turkish mine disaster prompts violent protests* is a follow up news. It is therefore useful to look at news articles related to the same story by following related hashtags. On the right side of the plot we show the most recent 100 tweets for the current hashtag.

In the future we intend to scale **Insight4News** to more RSS news feeds and tap into additional social media (e.g., Facebook, Reddit) as sources of social context for news.

Acknowledgements. This work was supported by Science Foundation Ireland under grant 07/CE/I1147 (Insight Centre for Data Analytics).

References

1. Ifrim, G., Shi, B., Brigadir, I.: Event Detection in Twitter using Aggressive Filtering and Hierarchical Tweet Clustering. In: Proceedings of SNOW 2014 Data Challenge, WWW (2014)
2. Shi, B., Ifrim, G., Hurley, N.: Be in The Know: Connecting News Articles and Relevant Twitter Conversations. arXiv:1405.3117 (2014)
3. http://insight4news.ucd.ie/insight4news/
4. Gamon, M., Basu, S., Belenko, D., Fisher, D., Hurst, M., König, A.C.: BLEWS: Using Blogs to Provide Context for News Articles. In: ICWSM. AAAI Press (2008)
5. http://www.thetimes.co.uk/tto/business/article2118052.ece
6. http://www.forbes.com/sites/oreillymedia/2012/05/09/the-future-of-the-newspaper/
7. http://storyful.com/products/newsrooms
8. https://storify.com/tour
9. http://hujo.deri.ie/hujo-newshack-ii/
10. http://www.goatmustbefed.com/resources/pdf/goat-must-be-fed.pdf

BestTime:
Finding Representatives in Time Series Datasets

Stephan Spiegel, David Schultz, and Sahin Albayrak

DAI-Lab, Berlin Institute of Technology,
Ernst-Reuter-Platz 7, 10587 Berlin, Germany
{spiegel,schultz,albayrak}@dai-lab.de
http://www.dai-lab.de/~spiegel/besttime.html

Abstract. Given a set of time series, we aim at finding representatives which best comprehend the recurring temporal patterns contained in the data. We demonstrate BestTime, a Matlab application that uses recurrence quantification analysis to find time series representatives.

1 Introduction

This work presents BestTime, a platform-independent Matlab application with graphical user interface, which enables us to find representatives that best comprehend the recurring temporal patterns contained in a certain time series dataset. Although BestTime was originally designed to analyze vehicular sensor data and identify characteristic operational profiles that comprise frequent behavior patterns [6], our extended version [7] can be used to find representatives in arbitrary sets of single- or multi-dimensional time series of variable length.

Our approach to find representatives in time series datasets is based on agglomerative hierarchical clustering [3]. We define a representative as the time series that is closest to the corresponding cluster center of gravity [5]. Since we want a representative to comprehend the recurring temporal patterns contained in the time series of the respective cluster, we need a distance measure that accounts for similar subsequences regardless of their position in time [6].

However, traditional time series distance measures, such as the Euclidean distance (ED) and Dynamic Time Warping (DTW), are not suitable to match similar subsequences that occur in arbitrary order [1,2]. Hence, we propose to employ Recurrence Plots (RPs) and corresponding Recurrence Quantification Analysis (RQA) [4,9] to measure the pairwise (dis)similarity of time series with similar patterns at arbitrary positions. In earlier work [8] we introduced a novel Recurrence Plot-based distance measure, which is used by our BestTime tool to cluster time series and find representatives.

The following section describes the operation of our BestTime application and illustrates the identification of representatives on a small set of sample time series. We furthermore provide supplementary online material [7], including the executable Matlab code of BestTime, real-life data for testing, a video demonstration of BestTime, and a technical report with an introduction to the formal problem statement and employed recurrence plot-based distance measure.

T. Calders et al. (Eds.): ECML PKDD 2014, Part III, LNCS 8726, pp. 477–480, 2014.
© Springer-Verlag Berlin Heidelberg 2014

2   BestTime

BestTime is a platform-independent Matlab application which provides an user-friendly interface. It enables a user to find representatives in arbitrary time series datasets by clustering the data sequences according to co-occurring patterns. In the following we briefly describe the operation of our BestTime application and illustrate the data processing for a small set of sample time series in Figure 1. Please feel free to download our BestTime tool [7] to follow the stepwise operating instructions given below.

Input Data. BestTime is able to analyze multivariate time series with same dimensionality and of variable length. Each individual time series needs to be stored in an independent csv (comma separated values) file, where rows correspond to observations and columns correspond to variables. Optionally, the first row may specify the names of the variables. The user selects an input folder that should contain all time series in specified csv format. A small set of sample time series that we use as input is illustrated in Figure 1(a).

Minimum Number of Observations. Depending on the application, the user can optionally reduce the size of the dataset by specifying the minimum length of the time series which should be consider for further processing.

Data Reduction Rate. Since the cost of our pairwise distance calculations is quadratic in the length of the time series, we offer the possibility to reduce the length via piecewise aggregate approximation [2]. Given a time series of length n and a reduction rate r, the approximate time series is of length n/r.

Minimum Pattern Length. The predetermined minimum pattern length directly influences the time series similarity. This parameter strongly depends on the application and needs to be chosen by a domain expert.

Variable Selection. In case of time series datasets with multiple dimensions, the user interface of our tool offers the possibility to select the variables that should be considered for further analysis.

Similarity Threshold. This parameter is usually very sensitive and directly influences the clustering result. Since it may be challenging to determine an appropriate similarity threshold for each variable, our tool can alternatively recommend (estimated) thresholds.

Parallel Computing. Calculating the distance matrix is costly for large datasets. However, this step is fully parallelized and runs almost n_{CPU}-times faster than serial processing. Up to twelve parallel workers are supported.

Quality Control. Our tool presents a colored plot of the computed distance matrix and a histogram of the distance distribution in order to ensure appropriate parameter settings as well as clusters that preserve the time series characteristics. Since both plots are updated iteratively during distance calculations, we can abort computation anytime the preview suggests undesired results. For the distance matrix, a high variance in the distances/colors indicates an appropriate parameter setting, and a low variance in the distances/colors may result in poor clustering. In general, good clustering results can be achieved when the distances do not accumulate at either end of the interval (all close to zero or one). Figure 1(b) shows the quality control for our sample dataset.

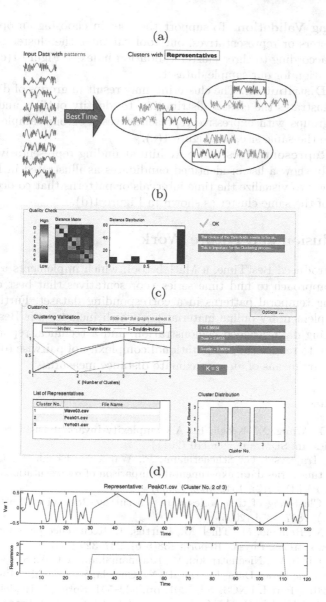

Fig. 1. BestTime operation and data processing for finding representatives in a certain time series dataset. (a) Given a set of time series with previously unknown patterns, we aim to cluster the data and find a representative (highlighted) for each group. (b) Visualization of computed distance matrix and distance distribution, which are used to ensure both appropriate parameter settings and clusters that preserve the time series characteristics. (c) Clustering results which show various validation indexes for a changing number of clusters, the list of identified representatives for a selected number of clusters, and the cardinality of the individual clusters. (d) Detailed view of a representative with corresponding pattern frequency regarding the selected cluster.

Clustering Validation. To support the user in choosing an optimal number of k clusters or representatives, our tool validates the cluster goodness for changing k according to three cluster validation indexes. Figure 1(c) shows the cluster validation for our sample dataset.

Cluster Distribution. The clustering may result in groups of different size. Our tools illustrates the cluster distribution to identify outliers and emphasize prominent groups with expressive representatives. For our sample dataset all clusters have the same size, see Figure 1(c).

List of Representatives. Since we aim at finding representatives, our tool does not only show a list of identified candidates as illustrated in Figure 1(c), but also allows to visualize the time intervals or patterns that co-occur in other time series of the same cluster as shown in Figure 1(d).

3   Conclusion and Future Work

We have introduced BestTime, a Matlab tool, which implements a recurrence-plot based approach to find time series representatives that best comprehend the recurring temporal patterns in a corresponding dataset. Furthermore, we provide supplementary online material [7], which includes our BestTime tool, real-life testing data, a video demonstration, and a technical report. In future work we plan to reduce the computational complexity of pairwise (dis)similarity comparisons by means of an approximate distance measure.

References

1. Batista, G., Wang, X., Keogh, E.: A Complexity-Invariant Distance Measure for Time Series. In: SDM, pp. 699–710 (2011)
2. Ding, H., Trajcevski, G., Scheuermann, P., Wang, X., Keogh, E.: Querying and mining of time series data: experimental comparison of representations and distance measures. PVLDB 1(2), 1542–1552 (2008)
3. Liao, T.: Clustering of time series data - a survey. Journal on Pattern Recognition 38(11), 1857–1874 (2005)
4. Marwan, N., Romano, M., Thiel, M., Kurths, J.: Recurrence plots for the analysis of complex systems. Physics Reports 438(5-6), 237–329 (2007)
5. Meesrikamolkul, W., Niennattrakul, V., Ratanamahatana, C.A.: Shape-Based Clustering for Time Series Data. In: Tan, P.-N., Chawla, S., Ho, C.K., Bailey, J. (eds.) PAKDD 2012, Part I. LNCS, vol. 7301, pp. 530–541. Springer, Heidelberg (2012)
6. Spiegel, S., Albayrak, S.: An Order-invariant Time Series Distance Measure. In: KDIR, pp. 264–268 (2012)
7. Spiegel, S., Schultz, D., Schacht, M., Albayrak, S.: Supplementary Online Material - BestTime App, Test Data, Video Demonstration, Technical Report, http://www.dai-lab.de/~spiegel/besttime.html
8. Spiegel, S., Jain, B.J., Albayrak, S.: A Recurrence Plot-based Distance Measure. Springer Proceedings in Mathematics (to appear, 2014)
9. Webber, C., Marwan, N., Facchini, A., Giuliani, A.: Simpler methods do it better: Success of Recurrence Quantification Analysis as a general purpose data analysis tool. Physics Letters A 373(41), 3753–3756 (2009)

BMaD – A Boolean Matrix Decomposition Framework

Andrey Tyukin, Stefan Kramer, and Jörg Wicker

Johannes Gutenberg-Universität Mainz, Staudingerweg 9, D-55128 Mainz, Germany
tyukiand@students.uni-mainz.de, {kramer,wicker}@informatik.uni-mainz.de

Abstract. Boolean matrix decomposition is a method to obtain a compressed representation of a matrix with Boolean entries. We present a modular framework that unifies several Boolean matrix decomposition algorithms, and provide methods to evaluate their performance. The main advantages of the framework are its modular approach and hence the flexible combination of the steps of a Boolean matrix decomposition and the capability of handling missing values. The framework is licensed under the GPLv3 and can be downloaded freely at http://projects.informatik.uni-mainz.de/bmad.

1 Introduction

The goal of a Boolean matrix decomposition (BMD) is to represent a given Boolean matrix as a product of two or more Boolean factor matrices. It is a well-known and researched problem with a wide range of applications [2], e.g. in multi-label classification [9], clustering [7], bioinformatics [10], or pattern mining [6]. In this demo, we introduce BMaD system, that understands BMD as a three step process. This division into three steps is inspired by the work of Pauli Miettinen [3,5,4] and uses and extends this work.

The presented implementation differs from previous ones in three points. First, the BMD is implemented as a modular algorithm, where each step can be carried out using multiple algorithms. Hence, different modules from different previous publications can be freely combined to a new BMD method. Second, all implemented matrix decomposition methods support missing values in the data. Published algorithms for BMD can be easily extended to support this. In most cases, no modification of the algorithms were necessary, yet no previous implementation supported missing values in the data. This became necessary in previous work, when BMD was applied to multi-label classification, where some labels can be set to unknown [9]. Third, BMaD is implemented in Java, which makes it easy to use with WEKA [1] and run on many systems out of the box. Methods to load WEKA instances directly into BMaD are provided.

While BMaD does not introduce a completely new algorithm, it provides an easy way to combine state-of-the-art steps of BMD to use established methods for new ways of BMD. Due to its capability to handle missing values, it was already used in previous publications [9] and has the potential to be used in

T. Calders et al. (Eds.): ECML PKDD 2014, Part III, LNCS 8726, pp. 481–484, 2014.

future research, e.g. for developing machine learning algorithms using BMD, or any applications using BMD that benefit from a Java implementation.

Previous implementations did not provide such a wide range of BMD methods in one framework. Additionally, so far there is no implementation capable of handling missing values and no implementation available in Java.

2   Boolean Matrix Decomposition

Let $\mathbb{B}$ be the two-element Boolean algebra, i.e. the set $\{0, 1\}$ equipped with binary operations $\wedge$ (AND), $\vee$ (OR), and the unary operation $\neg$. First, we define the *Boolean matrix product*. Let $A \in \mathbb{B}^{h \times m}$ and $B \in \mathbb{B}^{m \times w}$ be two Boolean matrices for $h, w, m \in \mathbb{N}$. Their Boolean product $A \otimes B \in \mathbb{B}^{h \times w}$ is defined as:

$$A \otimes B := \left[\bigvee_{k=1}^{m} A_{i,k} \wedge B_{k,j} \right]_{i,j}.$$

Real-world data often contains missing values (which we denote by '?'). BMAD accepts Boolean matrices with missing values as input for the decomposition algorithms. Given a Boolean matrix $A \in (\mathbb{B} \cup \{?\})^{h \times w}$ and a parameter $d \in \{1, \ldots, h\}$, the goal is to find factor matrices $C \in \mathbb{B}^{h \times d}$ and $B \in \mathbb{B}^{d \times w}$, such that the reconstruction $C \otimes B$ is as close to the original matrix A as possible. More precisely, the *reconstruction error* E is defined as:

$$E(A, \tilde{A}) := \#\{(i, j) \in \{1, \ldots, h\} \times \{1, \ldots, w\} : A_{i,j} \neq ? \text{ and } A_{i,j} \neq \tilde{A}_{i,j}\}$$

for $A \in (\mathbb{B} \cup \{?\})^{h \times w}$ and $\tilde{A} \in \mathbb{B}^{h \times w}$, that is, the number of entries in $\tilde{A}$ that differ from *known* values of the matrix A. If all entries of A are known, this error is just the L^1 norm of the real-valued difference between $\tilde{A}$ and A. Using these definitions, a more precise formulation of the BMD problem is as follows: given $A \in (\mathbb{B} \cup \{?\})^{h \times w}$, and compression dimension $d \in \{1, \ldots, h\}$, find matrices $C \in \mathbb{B}^{h \times d}$ and $B \in \mathbb{B}^{d \times w}$, such that the reconstruction error $E(A, C \otimes B)$ is minimized.

The problem is known to be NP-complete for Boolean matrices without unknown values [3,5]. By obvious reduction, this also holds for the problem presented here. Hence, we do not attempt to solve the problem exactly, but instead consider a family of heuristics suitable for finding approximate solutions.

3   Algorithms and Implementation

We provide an implementation of a family of modular algorithms, which include several algorithms previously proposed by Miettinen *et al.* [3,5,4]. As previously, let $A \in \mathbb{B}^{w \times h}$, $d \in \{1, \ldots, h\}$ and $c > 0$ be the input parameters. Algorithms representable in BMAD consist of three subalgorithms, which are more or less independent of each other.

Table 1. Example program using BMAD. First, the single modules are created, then they are used to decompose the matrix. Finally, the reconstruction error is calculated on the decomposition.

```
1    // initialize the modules
2    CandidateGenerator generator = new AssociationGenerator(0.2);
3    BasisSelector selector = new GreedySelector();
4    Combinator combinator = new DensityGreedyCombinator();
5    // import matrix from WEKA instances objet
6    BooleanMatrix original = new BooleanMatrix(instances);
7    // decompose matrix using variable dim as dimension
8    BooleanMatrixDecomposition bmd = new
         BooleanMatrixDecomposition(generator, selector, combinator);
9    Tuple<BooleanMatrix, BooleanMatrix> t = bmd.decompose(original,
         dim);
10   BooleanMatrix c = t._1, b = t._2;
11   // generate reconstruction by Boolean multiplication
12   BooleanMatrix reconstruction = c.booleanProduct(b);
13   // calculate the relative reconstruction error
14   double reconstructionError =
         original.relativeReconstructionError(reconstruction,
         onesWeight);
```

Candidate generation First, a set of *potential* basis patterns is generated from the matrix A. It consists of rows of same width as matrix A.

Identity All rows of A are declared to be candidates [5].

Association Candidates are generated using pairwise associations [5].

Intersection For each pair of rows of A, the entrywise minimum is a candidate [8].

Basis selection The size of the set generated in the first step is usually much larger than the parameter d. Hence we have to sort out the less meaningful patterns and retain exactly d candidate patterns that are included into the basis matrix B (basis, second factor). Most of the subalgorithms discussed here also generate a coarse approximation of the matrix C at this step. In the next step, one can obtain the final version of the matrix C by either refining the approximation, or building C from scratch.

Greedy Algorithm The error is minimized in a greedy manner [5].

Local Search (with minor variations) Similar to the Greedy Algorithm but it iterates over $k \in \{1, \ldots, d\}$, replacing the k-th basis row of B [5].

Boolean combination In this step, the matrix C (combination, first factor) is constructed. The goal is to represent rows of the original matrix A as *Boolean combinations* of the basis patterns from the matrix B. Clearly, each row of the matrix A can be represented independently, hence it is enough to specify how to calculate entries of one single row of C.

Iter Iterates multiple times over entries of C, and uses the change in reconstruction error to check if flipping an entry decreases the error [5].

Cover-greedy algorithm Start with an empty matrix C and repeatedly
search a basis row ρ that maximizes the change in reconstruction error.
Density-greedy algorithm Basis rows with fewer 1s are preferred to rows
with more 1s.

The API of BMAD is straightforward (an example call is given in Table 1).
For each module, a class exists, for each step, a class must be initialized and
used. The classes provide the appropriate methods to perform the given step
of the BMD. Additionally, methods to compute errors or visualize matrices and
errors are implemented (examples of visualizations are shown on the web site
`http://projects.informatik.uni-mainz.de/bmad` and in the demo).

4   Conclusion

This demo presents a modular framework for BMD, implementing it as a modular
algorithm. Each step of the algorithm can be carried out by several modules,
providing a flexible implementation of the BMD. It is implemented in Java and
provides support for missing values in the data set and a WEKA interface. In
the demo, we will present a step-by-step tutorial how to use BMAD, showing
the possibilities of it and visualizing the results.

References

1. Hall, M., Frank, E., Holmes, G., Pfahringer, B., Reutemann, P., Witten, I.H.: The
WEKA data mining software: an update. ACM SIGKDD Explorations Newsletter 11(1), 10–18 (2009)
2. Lu, H.: Boolean matrix decomposition and extension with applications. PhD thesis,
Rutgers University (2011)
3. Miettinen, P.: The Boolean column and column-row matrix decompositions. Data
Mining and Knowledge Discovery 17(1), 39–56 (2008)
4. Miettinen, P., et al.: Matrix decomposition methods for data mining: Computational complexity and algorithms. PhD thesis, University of Helsinki (2009)
5. Miettinen, P., Mielikainen, T., Gionis, A., Das, G., Mannila, H.: The discrete
basis problem. IEEE Transactions on Knowledge and Data Engineering 20(10),
1348–1362 (2008)
6. Shen, B.-H., Ji, S., Ye, J.: Mining discrete patterns via binary matrix factorization.
In: Proceedings of the 15th ACM SIGKDD International Conference on Knowledge
Discovery and Data Mining, pp. 757–766. ACM (2009)
7. Streich, A.P., Frank, M., Basin, D., Buhmann, J.M.: Multi-assignment clustering
for Boolean data. In: Proceedings of the 26th Annual International Conference on
Machine Learning, pp. 969–976. ACM (2009)
8. Vaidya, J.: Boolean matrix decomposition problem: Theory, variations and applications to data engineering. In: 2012 IEEE 28th International Conference on Data
Engineering (ICDE), pp. 1222–1224. IEEE (2012)
9. Wicker, J., Pfahringer, B., Kramer, S.: Multi-label classification using Boolean
matrix decomposition. In: Proceedings of the 27th Annual ACM Symposium on
Applied Computing, pp. 179–186. ACM (2012)
10. Zhang, Z.-Y., Li, T., Ding, C., Ren, X.-W., Zhang, X.-S.: Binary matrix factorization for analyzing gene expression data. Data Mining and Knowledge Discovery 20(1), 28–52 (2010)

Analyzing and Grounding Social Interaction in Online and Offline Networks

Martin Atzmueller

University of Kassel,
Knowledge and Data Engineering Group,
Wilhelmshöher Allee 73, 34121 Kassel, Germany
atzmueller@cs.uni-kassel.de

Abstract. In social network analysis, there are a variety of options for investigating social interactions. This paper reviews our recent work on analyzing and grounding social interactions in online and offline networks considering distributional semantics, structural network correlation and network inter-dependencies. Specifically, we focus on the analysis of user relatedness, community structure, and relations on online and offline networks. We discuss findings and results that justify the use of even implicitly accruing social interaction networks for the analysis of user-relatedness, community structure, etc. Furthermore, we provide insights into recent work on analyzing and grounding offline social networks.

Keywords: social network analysis, social interaction networks, mining social media, distributional semantics, community structure, social distributional hypothesis.

1 Introduction

The analysis of user relatedness [10, 13, 14], community structure [11, 12, 15], and the relation between online and offline networks [7, 16] are prominent research topics in data mining and social network analysis. In this context, this paper summarizes our recent work on analyzing and grounding social interaction. We analyze user interaction formalized in so-called social interaction networks [2, 14]: These refer to user-related social networks in social media that are capturing social relations inherent in social interactions, social activities and other social phenomena which act as proxies for social user-relatedness. Essentially, social interaction networks focus on *interaction* relations between *people*, see [19, p. 37 ff.], that are the corresponding actors.

First, we present the *social distributional hypothesis* [13] – a pragmatic proxy for homophily [10] – stating that users with similar interaction characteristics are related, and provide supporting evidence. Second, we extend this to the analysis of *communities* [11, 12] showing structural correlations between implicit networks of user interactions. Third, we investigate the *structural grounding* considering both online and offline network properties [5, 16]. In this way, we provide novel insights into the *grounding of offline behavior*. Our analysis results justify the analysis of even implicitly accruing social interaction networks with respect to user-relatedness, semantics and community structure, and provide for valuable insights, e. g., for the development of link analysis methods, community detection, and the connection of online and offline information.

T. Calders et al. (Eds.): ECML PKDD 2014, Part III, LNCS 8726, pp. 485–488, 2014.

2   Analysis of Social Interaction Networks

With the rise of social software and social media, a wealth of user-generated data and user interactions is being created in online social networks. We adopt an intuitive definition of social media, regarding it as online systems and services in the ubiquitous web, which create and provide social data generated by human interaction and communication [1, 3]. We consider social interactions in an online and offline context, that is, connections and relations in online systems as well as real-world face-to-face contacts.

In the following, we focus on the analysis of such social interaction networks. Figure 1 provides an overview on the analysis and grounding approaches, while Table 1 further summarizes the methods, applied techniques and results which we discuss below in more detail, reviewing our recent work. In particular, we propose the *social distributional hypothesis* [13] stating, that users with similar interaction characteristics tend to be related. Considering users as (social) entities, their distributional characteristics can be observed utilizing social interaction networks. The social distributional hypothesis is postulated similar to the *distributional hypothesis* [8] in linguistics; it states that words with similar distributional characteristics tend to be semantically related, i. e., that words occurring in similar contexts have a similar meaning.

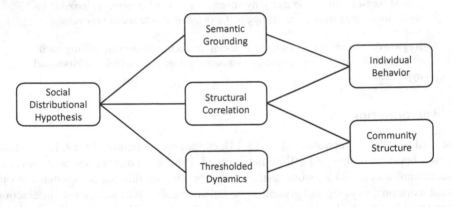

Fig. 1. Overview on the analysis and grounding setup: Starting with the social distributional hypothesis, we apply several methods for analyzing individual behavior and community structure

In [13, 14] we conduct a series of experiments on social interaction networks from Twitter, Flickr and BibSonomy and investigate the user-relatedness concerning the interactions, their frequency, and the specific interaction characteristics. The results indicate interrelations between structural similarity of interaction characteristics and semantic relatedness of users, supporting the social distributional hypothesis. This also grounds methods for analyzing social interaction networks in general.

On a structural level, we investigate two further issues in [11, 12] on the social interaction networks: Are there interrelations and correlations between the interaction networks? Furthermore, can these be applied for the analysis and data-driven assessment of communities? We analyze general structural properties of the obtained networks and comparatively discuss major structural characteristics in order to show that

Table 1. Overview on the applied methods and specific analysis techniques

Analysis	Method	Results
Semantic Grounding [13, 14]	Similarity Covariate Analysis (tag-based, location-based)	Interrelations between structural similarity of interaction characteristics and semantic user-relatedness
Structural Correlation [11, 12]	Degree Correlation, Neighborhood, Graph Covariance	Structural inter-network correlations; consistent community structure and ranking accross networks
Time-based Link Patterns [5, 9, 16]	Community Analysis, Role Analysis, Link Prediction	Semantically consistent community and role structures; indicators for complementing network structures

there are structural and semantic inter-network correlations between the different evidence networks. In particular, we examine several general structural properties, the degree distribution and the degree correlation, indicating significant similarities of the networks. Furthermore, we analyze dependencies of the networks' neighborhood, and inter-network correlations. The results indicate strong correlations and interrelations between the considered social interaction networks, that are strong enough for inferring reciprocal conclusions between the networks. Based on these results, we propose an approach for (relative) community assessment based on the idea of *reconstructing existing social structures* [18] for the assessment and evaluation of a given clustering.

Furthermore, for analyzing and grounding offline networks we focus on real-world offline networks of *human contacts*, that is, *face-to-face* proximity contacts between persons in [5, 9, 16]. In contrast to virtual networks, the involved contacts were collected using the social conference guidance system CONFERATOR [4] – a ubiquitous RFID-based system that allows us to collect face-to-face contact data [6]. Thus, we can observe and analyze (offline) social interaction at a very detailed level, including the specific event sequences and durations. Also, we complement the analysis of the offline social interaction networks with additional node-level properties and further networks, e.g., utilizing the DBLP co-authorship relations. In this context, we analyze different time-based link patterns using offline and online information. We ground user-interaction and community structure accordingly using different online and offline properties in [5]. In a threshold-based analysis, e.g., using different minimal contact durations of the contact data, we analyze general structural properties of the contact network, investigate the stability and dynamics of community structures, and examine different explicit and implicit roles [17] of conference participants. Furthermore, we analyze the predictability of links grounded using different online and offline information [16]. Our results show semantically grounded consistent community and role structure. In addition, we observe that different online and offline networks can complement each other well for improving link analysis methods, e.g., concerning link prediction.

3 Conclusions and Outlook

We proposed the social distributional hypothesis as one foundational issue for the analysis of social interaction networks and presented supporting experimental results. Furthermore, we successfully investigated structural correlation and time-based link patterns on online and offline social interaction networks. Overall, our analysis results are not only relevant for gaining justifications and important insights into structural and semantic relations for social interaction networks. They can also help, e. g., for implementing new link mining, community detection or user recommendation algorithms.

References

1. Atzmueller, M.: Mining Social Media: Key Players, Sentiments, and Communities. WIREs: Data Mining and Knowledge Discovery 2, 411–419 (2012)
2. Atzmueller, M.: Data Mining on Social Interaction Networks. JDMDH 1 (June 2014)
3. Atzmueller, M.: Social Behavior in Mobile Social Networks: Characterizing Links, Roles and Communities. In: Mobile Social Networking, pp. 65–78. Springer, Berlin (2014)
4. Atzmueller, M., Benz, D., Doerfel, S., Hotho, A.: Enhancing Social Interactions at Conferences. IT 53(3) (2011)
5. Atzmueller, M., Doerfel, S., Hotho, A., Mitzlaff, F., Stumme, G.: Face-to-Face Contacts at a Conference: Dynamics of Communities and Roles. In: Atzmueller, M., Chin, A., Helic, D., Hotho, A. (eds.) MUSE 2011 and MSM 2011. LNCS, vol. 7472, pp. 21–39. Springer, Heidelberg (2012)
6. Barrat, A., Cattuto, C., Colizza, V., Pinton, J., den Broeck, W.V., Vespignani, A.: High Resolution Dynamical Mapping of Social Interactions with Active RFID. PLoS ONE 5(7) (2010)
7. Barrat, A., Cattuto, C., Szomszor, M., Van den Broeck, W., Alani, H.: Social Dynamics in Conferences: Analyses of Data from the Live Social Semantics Application. In: Patel-Schneider, P.F., Pan, Y., Hitzler, P., Mika, P., Zhang, L., Pan, J.Z., Horrocks, I., Glimm, B. (eds.) ISWC 2010, Part II. LNCS, vol. 6497, pp. 17–33. Springer, Heidelberg (2010)
8. Harris, Z.S.: Distributional Structure. Word (1954)
9. Macek, B.-E., Scholz, C., Atzmueller, M., Stumme, G.: Anatomy of a Conference. In: Proc. Hypertext, pp. 245–254. ACM, New York (2012)
10. McPherson, M., Smith-Lovin, L., Cook, J.M.: Birds of a Feather: Homophily in Social Networks. Annual Review of Sociology 27(1), 415–444 (2001)
11. Mitzlaff, F., Atzmueller, M., Benz, D., Hotho, A., Stumme, G.: Community Assessment Using Evidence Networks. In: Atzmueller, M., Hotho, A., Strohmaier, M., Chin, A. (eds.) MUSE/MSM 2010. LNCS, vol. 6904, pp. 79–98. Springer, Heidelberg (2011)
12. Mitzlaff, F., Atzmueller, M., Benz, D., Hotho, A., Stumme, G.: User-Relatedness and Community Structure in Social Interaction Networks. CoRR/abs 1309.3888 (2013)
13. Mitzlaff, F., Atzmueller, M., Hotho, A., Stumme, G.: The Social Distributional Hypothesis. Journal of Social Network Analysis and Mining, accepted) (2014)
14. Mitzlaff, F., Atzmueller, M., Stumme, G., Hotho, A.: Semantics of User Interaction in Social Media. In: Ghoshal, G., Poncela-Casasnovas, J., Tolksdorf, R. (eds.) Complex Networks IV. Studies in Computational Intelligence, vol. 476, pp. 13–25. Springer, Heidelberg (2013)
15. Newman, M.E.J.: Detecting Community Structure in Networks. Europ Physical J. 38 (2004)
16. Scholz, C., Atzmueller, M., Barrat, A., Cattuto, C., Stumme, G.: New Insights and Methods For Predicting Face-To-Face Contacts. In: Proc. ICWSM. AAAI, Palo Alto (2013)
17. Scripps, J., Tan, P.-N., Esfahanian, A.-H.: Exploration of Link Structure and Community-Based Node Roles in Network Analysis. In: Proc. ICDM, pp. 649–654 (October 2007)
18. Siersdorfer, S., Sizov, S.: Social Recommender Systems for Web 2.0 Folksonomies. In: Proc. Hypertext, pp. 261–270. ACM, New York (2009)
19. Wasserman, S., Faust, K.: Social Network Analysis: Methods and Applications, 1st edn. Structural Analysis in the Social Sciences, vol. (8). Cambridge University Press (1994)

Be *Certain* of How-to
before Mining *Uncertain* Data

Francesco Gullo[1], Giovanni Ponti[2], and Andrea Tagarelli[3]

[1] Yahoo Labs, Barcelona, Spain
gullo@yahoo-inc.com
[2] ENEA, Portici Research Center (NA), Italy
giovanni.ponti@enea.it
[3] DIMES, University of Calabria, Italy
tagarelli@dimes.unical.it

Abstract. The purpose of this technical note is to introduce the problems of *similarity detection* and *summarization* in uncertain data. We provide the essential arguments that make the problems relevant to the data-mining and machine-learning community, stating major issues and summarizing our contributions in the field. Further challenges and directions of research are also issued.

1 Uncertainty: What We Have to Face

The term *uncertainty* describes an ubiquitous status of the information as being produced, transmitted, and acquired in real-world data sources. Exemplary scenarios are related to the use of location-based services for tracking moving objects and sensor networks, which normally produce data whose representation (attributes) is imprecise at a certain degree. Imprecision arises from the presence of noisy factors in the device or transmission medium, but also from a high variability in the measurements (e.g., locations of a moving object) that obviously prevents an exact representation at a given time. This is the case virtually for any field in scientific computing, and consequently for a plethora of application fields, including: pattern recognition (e.g., image processing), bioinformatics (e.g., gene expression microarray), computational fluid dynamics and geophysics (e.g., weather forecasting), financial planning (e.g., stock market analysis), GIS applications to distributed network analysis [1].

For data management purposes, uncertainty has been traditionally treated at the attribute level, as this is particularly appealing for inductive learning tasks [6]. In general, attribute-level uncertainty is handled based on a probabilistic representation approach that exploits probability distributions describing the likelihood that any given data tuple appears at each position in a multidimensional domain region; the term *uncertain objects* is commonly used to refer to such data tuples described in terms of probability distributions defined over multidimensional domain regions.

Uncertainty in data representation needs to be carefully handled in order to produce meaningful knowledge patterns. Consider for instance the scenario depicted in Fig. 1— uncertain objects are represented in terms only of their domain regions for the sake of simplicity (probability distribution assumed to be uniform for all the objects). The

T. Calders et al. (Eds.): ECML PKDD 2014, Part III, LNCS 8726, pp. 489–493, 2014.

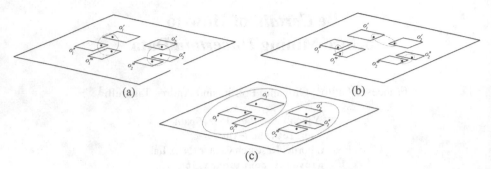

Fig. 1. Grouping uncertain data: (a) true representations of objects and their desired grouping, (b) observed representations which may lead to unexpected groupings, (c) desired grouping identified by considering the object uncertainty (domain regions).

"true" representation of each uncertain object (black circles in Fig. 1(a)) corresponds to a point within its domain region and can be in general far away from its "observed" representation (black circles in Fig. 1(b)). Thus, considering only the observed representations may lead to discover groups of similar objects (i.e., $\{o'_1, o'_2\}$, $\{o''_1, o'''_1\}$, $\{o''_2, o'''_2\}$ in Fig. 1(b)) that are substantially different from the ideal ones which would be identified by considering the true representations (i.e., $\{o'_1, o''_1, o'''_1\}$, $\{o'_2, o''_2, o'''_2\}$ in Fig. 1(a)). Instead, considering the whole domain regions (and pdfs) of the uncertain objects, may help to recognize the correct grouping (Fig. 1(c)).

The computation of proximity between uncertain objects is a fundamental primitive needed in many data-management tasks. Existing approaches fall into two main categories: (*i*) computing the distance between aggregated values extracted from the probability distributions of the uncertain objects (e.g., expected values), or (*ii*) computing the *expected distance* (ED) between distributions, which involves the whole information available from the distributions. The first approach is efficient as it has a time complexity linear in the number of statistical samples used for representing distributions, but it also has an evident accuracy issue since all the information available from the distributions is collapsed into a single numerical value; conversely, the ED-based approach is more accurate but also inefficient (it takes quadratic time). Within this view, our major contribution presented in [4] was to define a novel distance function that achieves a good tradeoff between accuracy and efficiency, by being able to capitalize on the whole information provided by the object distributions while keeping linear-time complexity.

Summarization of uncertain objects is another critical task, which is generally required in scenarios where a more compact representation is essential to analyze and/or further process a large set of (uncertain) objects that would be hard to manage otherwise. Surprisingly, a common trend in the early state-of-the-art was to employ a simple average of the expected values of the set members, which is clearly ineffective in most cases. Our contribution on this topic was to account also for the *variance* of the individual set members. In particular, we proposed a model based on a random variable derived from the realizations of the uncertain objects to be summarized [5], as well as a mixture-model-based summarization method [3,2].

Similarity detection in uncertain data is obviously central in a variety of mining tasks. Analogous consideration holds for uncertain data summarization, as it impacts on how proximity can conveniently be computed between any uncertain object and a "prototype" object summarizing a set (e.g., cluster) of uncertain objects. Next we informally articulate the approaches we have proposed in the aforementioned contexts.

2 Uncertainty: How We Can Deal With

Similarity Detection in Uncertain Data. *Information-theory* (IT) has represented a fruitful research area to devise measures for comparing probability distributions accurately and, in most cases, in linear time with respect to the number of distribution samples. However, none of the prominent existing IT measures can be directly used to define distances for uncertain objects, mainly because of the assumption that the distributions need to share a common event space, which does not necessarily hold for distributions associated to uncertain objects. For this purpose, in [4] we developed a distance measure between uncertain objects that is able to exploit the full information stored in the object distributions, while being fast to compute. A major feature of our proposed distance is a combination of an IT measure with a measure based on aggregated information (i.e., expected value) extracted from the object distributions.

Besides representing a good tradeoff between effectiveness and efficiency, a further nice feature of the proposed distance is that the frequency of occurrence of the no-intersection event, and thus the overall accuracy of the measure, can be statistically controlled. Specifically, the width of the domain region shared between the uncertain objects to be compared represents a useful indicator of the feasibility of the distance calculation by means of the IT term only, and hence of the limited need for comparing the object distributions by also considering the expected-value-based term. This reasoning can profitably be exploited in tasks where the distances are to be computed for objects whose domain regions become larger as their processing goes on. An exemplary task where this happens is *prototype-based agglomerative hierarchical clustering*, where each cluster of uncertain objects is represented according to some notion of prototype whose domain region is ensured to increase with later steps of the clustering process.

Uncertain Data Summarization. As previously mentioned, the naïve notion of uncertain prototype as average of the expected values of the objects in a set has been widely used in the literature. Notwithstanding, it might easily result in limited accuracy, as (*i*) it has a deterministic representation, and (*ii*) it expresses only the central tendency of the objects to be summarized. This prompted us to investigate better ways of summarizing uncertain data. As a first attempt towards this matter, we proposed in [3,2] a notion of uncertain prototype as a mixture model of the set of random variables representing the uncertain objects to be summarized; that is, a notion that enables an uncertain representation while, at the same time, accounting for the variance of the individual objects rather than their central tendency only. A significant part of our work also focused on how to exploit our summarization approach in a classic data-mining task like clustering. Particularly, we demonstrated that a clustering objective criterion can be defined based on the minimization of the variance of the cluster mixture models, and that both efficiency and accuracy requirements can be satisfied. A major remark in this regard is that the proposed criterion enables the definition of fast heuristics that do not require any distance measure between uncertain objects.

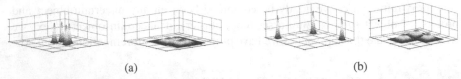

(a)   (b)

Fig. 2. Impact of the use of central tendency and variance of the individual objects on uncertain data summarization. (a) The two groups of objects have the same central tendency, but different variances: considering only central tendency leads to mistakenly summarize the two sets by the same prototype. (b) Considering only the variance is however not enough: even though the objects on the left have lower variance, they evidently form a less compact group than the one on the right.

Although the mixture-model-based approach has the merit of introducing a definition of uncertain prototype that is an uncertain object itself, a criterion based only on the minimization of the variance of the uncertain objects may still lead to unsatisfactory results. (Figure 2). Thus, in [5] we introduced a novel notion of uncertain prototype, named *U-centroid*: an uncertain object that is defined in terms of a random variable whose realizations correspond to all possible deterministic representations deriving from the uncertain objects to be summarized. Besides deriving the analytical expressions of domain region and probability distribution of the proposed U-centroid, a major contribution in this regard was the definition of a closed-form-computable compactness criterion that, coupled with the proposed U-centroid, naturally defines an effective yet efficient objective criterion for grouping uncertain objects.

3   Challenges and Future Directions

We provided a short review of problems related to similarity detection and summarization in uncertain data, and how we addressed them in our previous studies [3,5,2,4], which the interested reader is referred to for any technical details, including further developments we envisaged. Here we rather conclude raising a couple of concerns regarding the current trends of representing uncertainty and evaluating the induced mining results. First, we argue that a more complete treatment of uncertainty in data mining and machine learning could be obtained by integrating attribute-level with tuple-level notions of uncertainty, which have been long studied in database theory and management fields. This could imply the specification of a new yet more expressive class of models and algorithms for mining uncertain data. Second, we believe there is a strong need for the construction of benchmarks for assessing the mining results, which would avoid to bias the performance evaluation often due to the artificial, non-standardized methods for the generation of uncertainty in the selected test data. Another open problem that is worth to be addressed is the design of new assessment criteria to evaluate the many aspects inherent the quality of knowledge patterns induced from uncertain datasets.

References

1. Aggarwal, C.C.: Managing and Mining Uncertain Data. Springer (2009)
2. Gullo, F., Ponti, G., Tagarelli, A.: Minimizing the variance of cluster mixture models for clustering uncertain objects. In: IEEE ICDM, pp. 839–844 (2010)

3. Gullo, F., Ponti, G., Tagarelli, A.: Minimizing the variance of cluster mixture models for clustering uncertain objects. Statistical Analysis and Data Mining 6(2), 116–135 (2013)
4. Gullo, F., Ponti, G., Tagarelli, A., Greco, S.: A Hierarchical Algorithm for Clustering Uncertain Data via an Information-Theoretic Approach. In: IEEE ICDM, pp. 821–826 (2008)
5. Gullo, F., Tagarelli, A.: Uncertain Centroid based Partitional Clustering of Uncertain Data. PVLDB 5(7), 610–621 (2012)
6. Sarma, A.D., Benjelloun, O., Halevy, A.Y., Nabar, S.U., Widom, J.: Representing uncertain data: models, properties, and algorithms. The VLDB Journal 18(5), 989–1019 (2009)

Active Learning Is Planning: Nonmyopic
ε-Bayes-Optimal Active Learning of Gaussian Processes

Trong Nghia Hoang[1], Kian Hsiang Low[1], Patrick Jaillet[2], and Mohan Kankanhalli[1]

[1] National University of Singapore, Singapore
{nghiaht,lowkh,mohan}@comp.nus.edu.sg
[2] Massachusetts Institute of Technology, USA
jaillet@mit.edu

Abstract. A fundamental issue in active learning of Gaussian processes is that of the exploration-exploitation trade-off. This paper presents a novel nonmyopic *ε-Bayes-optimal active learning* (ε-BAL) approach [4] that jointly optimizes the trade-off. In contrast, existing works have primarily developed greedy algorithms or performed exploration and exploitation separately. To perform active learning in real time, we then propose an anytime algorithm [4] based on ε-BAL with performance guarantee and empirically demonstrate using a real-world dataset that, with limited budget, it outperforms the state-of-the-art algorithms.

1 Introduction

Active learning/sensing has become an increasingly important focal theme in environmental sensing and monitoring applications (e.g., precision agriculture [7], monitoring of ocean and freshwater phenomena). Its objective is to derive an optimal sequential policy that plans the most informative locations to be observed for minimizing the predictive uncertainty of the unobserved areas of a spatially varying environmental phenomenon given a sampling budget (e.g., number of deployed sensors, energy consumption). To achieve this, many existing active sensing algorithms [1,2,3,6,7,8] have modeled the phenomenon as a *Gaussian process* (GP), which allows its spatial correlation structure to be formally characterized and its predictive uncertainty to be formally quantified (e.g., based on entropy, or mutual information criterion). However, they have assumed the spatial correlation structure (specifically, the parameters defining it) to be known, which is often violated in real-world applications. The predictive performance of the GP model in fact depends on how informative the gathered observations are for both parameter estimation and spatial prediction given the true parameters.

Interestingly, as revealed in [9], policies that are efficient for parameter estimation are not necessarily efficient for spatial prediction with respect to the true model parameters. Thus, active learning/sensing involves a potential trade-off between sampling the most informative locations for spatial prediction given the current, possibly incomplete knowledge of the parameters (i.e., exploitation) vs. observing locations that gain more information about the parameters (i.e., exploration). To address this trade-off, one principled approach is to frame active sensing as a sequential decision problem that jointly optimizes the above exploration-exploitation trade-off while maintaining a Bayesian belief over the model parameters. Solving this problem then results in an induced policy that is guaranteed to be optimal in the expected active sensing performance [4].

T. Calders et al. (Eds.): ECML PKDD 2014, Part III, LNCS 8726, pp. 494–498, 2014.

Unfortunately, such a nonmyopic *Bayes-optimal active learning* (BAL) policy cannot be derived exactly due to an uncountable set of candidate observations and unknown model parameters. As a result, existing works advocate using greedy policies [10] or performing exploration and exploitation separately [5] to sidestep the difficulty of solving for the exact BAL policy. But, these algorithms are sub-optimal in the presence of budget constraints due to their imbalance between exploration and exploitation [4].

This paper presents a novel nonmyopic active learning algorithm [4] that can still preserve and exploit the principled Bayesian sequential decision problem framework for jointly optimizing the exploration-exploitation trade-off (Section 2.2) and consequently does not incur the limitations of existing works. In particular, although the exact BAL policy cannot be derived, we show that it is in fact possible to solve for a nonmyopic ϵ-*Bayes-optimal active learning* (ϵ-BAL) policy (Section 2.3) given an arbitrary loss bound ϵ. To meet real-time requirement in time-critical applications, we then propose an asymptotically ϵ-optimal anytime algorithm based on ϵ-BAL with performance guarantee (Section 2.4). We empirically demonstrate using a real-world dataset that, with limited budget, our approach outperforms state-of-the-art algorithms (Section 3).

2 Nonmyopic ϵ-Bayes-Optimal Active Learning

2.1 Modeling Spatial Phenomena with Gaussian Processes

Let $\mathcal{X}$ denote a set of sampling locations representing the domain of the phenomenon such that each location $x \in \mathcal{X}$ is associated with a realized (random) measurement z_x (Z_x) if x is observed (unobserved). Let $Z_{\mathcal{X}} \triangleq \{Z_x\}_{x \in \mathcal{X}}$ denote a GP [4]. The GP is fully specified by its *prior* mean $\mu_x \triangleq \mathbb{E}[Z_x]$ and covariance $\sigma_{xx'|\lambda} \triangleq \mathrm{cov}[Z_x, Z_{x'}|\lambda]$ for all locations $x, x' \in \mathcal{X}$; its model parameters are denoted by λ. When λ is known and a set $z_{\mathcal{D}}$ of realized measurements is observed for $\mathcal{D} \subset \mathcal{X}$, the GP prediction for any unobserved location $x \in \mathcal{X} \setminus \mathcal{D}$ is given by $p(z_x|z_{\mathcal{D}}, \lambda) = \mathcal{N}(\mu_{x|\mathcal{D},\lambda}, \sigma_{xx|\mathcal{D},\lambda})$ [4]. However, since λ is not known, a probabilistic belief $b_{\mathcal{D}}(\lambda) \triangleq p(\lambda|z_{\mathcal{D}})$ is maintained over all possible λ and updated using Bayes' rule to the posterior belief $b_{\mathcal{D} \cup \{x\}}(\lambda) \propto p(z_x|z_{\mathcal{D}}, \lambda) \, b_{\mathcal{D}}(\lambda)$ given a new measurement z_x. Then, using belief $b_{\mathcal{D}}$, the predictive distribution is obtained by marginalizing out λ: $p(z_x|z_{\mathcal{D}}) = \sum_{\lambda \in \Lambda} p(z_x|z_{\mathcal{D}}, \lambda) \, b_{\mathcal{D}}(\lambda)$.

2.2 Problem Formulation

To cast active sensing as a Bayesian sequential decision problem, we define a sequential active sensing policy $\pi \triangleq \{\pi_n\}_{n=1}^{N}$ that is structured to sequentially decide the next location $\pi_n(z_{\mathcal{D}}) \in \mathcal{X} \setminus \mathcal{D}$ to be observed at each stage n based on the current observations $z_{\mathcal{D}}$ over a finite planning horizon of N stages (i.e., sampling budget). To measure the predictive uncertainty over unobserved areas of the phenomenon, we use the entropy criterion and define the value under a policy π to be the joint entropy of its selected observations when starting with some prior observations $z_{\mathcal{D}_0}$ and following π thereafter [4].

The work of [7] has established that minimizing the posterior joint entropy (i.e., predictive uncertainty) remaining in unobserved locations of the phenomenon is equivalent to maximizing the joint entropy of π. Thus, solving the active sensing problem entails choosing a sequential BAL policy $\pi_n^*(z_{\mathcal{D}}) = \arg\max_{x \in \mathcal{X} \setminus \mathcal{D}} Q_n^*(z_{\mathcal{D}}, x)$ induced from the following N-stage Bellman equations, as formally derived in [4]:

$$
\begin{aligned}
V_n^*(z_{\mathcal{D}}) &\triangleq \max_{x \in \mathcal{X} \setminus \mathcal{D}} Q_n^*(z_{\mathcal{D}}, x) \\
Q_n^*(z_{\mathcal{D}}, x) &\triangleq \mathbb{E}\left[-\log p(Z_x | z_{\mathcal{D}})\right] + \mathbb{E}\left[V_{n+1}^*(z_{\mathcal{D}} \cup \{Z_x\}) | z_{\mathcal{D}}\right]
\end{aligned}
\tag{1}
$$

for stage $n = 1, \ldots, N$ where $p(z_x | z_{\mathcal{D}})$ is defined in Section 2.1 and the second expectation term is omitted from right-hand side expression of Q_N^* at stage N. Unfortunately, since the BAL policy π^* cannot be derived exactly, we instead consider solving for an ϵ-BAL policy π^ϵ whose joint entropy approximates that of π^* within $\epsilon > 0$.

2.3 ϵ-BAL Policy

The key idea of our proposed nonmyopic ϵ-BAL policy π^ϵ is to approximate the expectation terms in (1) at every stage using truncated sampling. Specifically, given realized measurements $z_{\mathcal{D}}$, a finite set of τ-truncated, i.i.d. observations $\{z_x^i\}_{i=1}^S$ [4] is generated and exploited for approximating V_n^* (1) through the following Bellman equations:

$$
\begin{aligned}
V_n^\epsilon(z_{\mathcal{D}}) &\triangleq \max_{x \in \mathcal{X} \setminus \mathcal{D}} Q_n^\epsilon(z_{\mathcal{D}}, x) \\
Q_n^\epsilon(z_{\mathcal{D}}, x) &\triangleq \frac{1}{S} \sum_{i=1}^S -\log p\left(z_x^i | z_{\mathcal{D}}\right) + V_{n+1}^\epsilon\left(z_{\mathcal{D}} \cup \{z_x^i\}\right)
\end{aligned}
\tag{2}
$$

for stage $n = 1, \ldots, N$. The use of truncation is motivated by a technical necessity for theoretically guaranteeing the *expected* active sensing performance (specifically, ϵ-Bayes-optimality) of π^ϵ relative to that of π^* [4].

2.4 Anytime ϵ-BAL ($\langle \alpha, \epsilon \rangle$-BAL) Algorithm

Although π^ϵ can be derived exactly, the cost of deriving it is exponential in the length N of planning horizon since it has to compute the values $V_n^\epsilon(z_{\mathcal{D}})$ (2) for all $(S|\mathcal{X}|)^N$ possible states $(n, z_{\mathcal{D}})$. To ease this computational burden, we propose an anytime algorithm based on ϵ-BAL that can produce a good policy fast and improve its approximation quality over time. The key intuition behind our *anytime ϵ-BAL algorithm* ($\langle \alpha, \epsilon \rangle$-BAL) is to focus the simulation of greedy exploration paths through the most uncertain regions of the state space (i.e., in terms of the values $V_n^\epsilon(z_{\mathcal{D}})$) instead of evaluating the entire state space like π^ϵ. Interested readers are referred to [4] for more details.

3 Experiments and Discussion

This section evaluates the active sensing performance and time efficiency of our $\langle \alpha, \epsilon \rangle$-BAL policy $\pi^{\langle \alpha, \epsilon \rangle}$ empirically under using a real-world dataset of a large-scale traffic

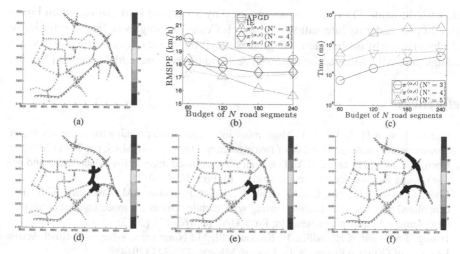

Fig. 1. (a) Traffic phenomenon (i.e., speeds (km/h) of road segments) over an urban road network, graphs of (b) root mean squared prediction error of APGD, IE, and $\langle \alpha, \epsilon \rangle$-BAL policies with horizon length $N' = 3, 4, 5$ and (c) total online processing cost of $\langle \alpha, \epsilon \rangle$-BAL policies with $N' = 3, 4, 5$ vs. budget of N segments, and (d-f) road segments observed (shaded in black) by respective APGD, IE, and $\langle \alpha, \epsilon \rangle$-BAL policies ($N' = 5$) with $N = 60$.

phenomenon (i.e., speeds of road segments) over an urban road network; refer to [4] for additional experimental results on a simulated spatial phenomenon. Fig. 1a shows the urban road network $\mathcal{X}$ comprising 775 road segments in Tampines area, Singapore during lunch hours on June 20, 2011. Each road segment $x \in \mathcal{X}$ is specified by a 4-dimensional vector of features: length, number of lanes, speed limit, and direction. More details of our experimental setup can be found in [4]. The performance of our $\langle \alpha, \epsilon \rangle$-BAL policies with planning horizon length $N' = 3, 4, 5$ are compared to that of APGD and IE policies [5] by running each of them on a mobile robotic probe to direct its active sensing along a path of adjacent road segments according to the road network topology. Fig. 1 shows results of the tested policies averaged over 5 independent runs: It can be observed from Fig. 1b that our $\langle \alpha, \epsilon \rangle$-BAL policies outperform APGD and IE policies due to their nonmyopic exploration behavior. Fig. 1c shows that $\langle \alpha, \epsilon \rangle$-BAL incurs < 4.5 hours given a budget of $N = 240$ road segments, which can be afforded by modern computing power. To illustrate the behavior of each policy, Figs. 1d-f show, respectively, the road segments observed (shaded in black) by the mobile probe running APGD, IE, and $\langle \alpha, \epsilon \rangle$-BAL policies with $N' = 5$ given a budget of $N = 60$. Interestingly, Figs. 1d-e show that both APGD and IE cause the probe to move away from the slip roads and highways to low-speed segments whose measurements vary much more smoothly; this is expected due to their myopic exploration behavior. In contrast, $\langle \alpha, \epsilon \rangle$-BAL nonmyopically plans the probe's path and direct it to observe the more informative slip roads and highways with highly varying traffic measurements (Fig. 1f) to achieve better performance.

Acknowledgments. This work was supported by Singapore National Research Foundation in part under its International Research Center @ Singapore Funding Initiative and administered by the Interactive Digital Media Programme Office and in part through the Singapore-MIT Alliance for Research & Technology Subaward Agreement No. 52.

References

1. Cao, N., Low, K.H., Dolan, J.M.: Multi-robot informative path planning for active sensing of environmental phenomena: A tale of two algorithms. In: Proc. AAMAS, pp. 7–14 (2013)
2. Chen, J., Low, K.H., Tan, C.K.Y.: Gaussian process-based decentralized data fusion and active sensing for mobility-on-demand system. In: Proc. RSS (2013)
3. Chen, J., Low, K.H., Tan, C.K.Y., Oran, A., Jaillet, P., Dolan, J.M., Sukhatme, G.S.: Decentralized data fusion and active sensing with mobile sensors for modeling and predicting spatiotemporal traffic phenomena. In: Proc. UAI. pp. 163–173 (2012)
4. Hoang, T.N., Low, K.H., Jaillet, P., Kankanhalli, M.: Nonmyopic ϵ-Bayes-Optimal Active Learning of Gaussian Processes. In: Proc. ICML, pp. 739–747 (2014)
5. Krause, A., Guestrin, C.: Nonmyopic active learning of Gaussian processes: An exploration-exploitation approach. In: Proc. ICML, pp. 449–456 (2007)
6. Low, K.H., Dolan, J.M., Khosla, P.: Adaptive multi-robot wide-area exploration and mapping. In: Proc. AAMAS, pp. 23–30 (2008)
7. Low, K.H., Dolan, J.M., Khosla, P.: Information-theoretic approach to efficient adaptive path planning for mobile robotic environmental sensing. In: Proc. ICAPS (2009)
8. Low, K.H., Dolan, J.M., Khosla, P.: Active Markov information-theoretic path planning for robotic environmental sensing. In: Proc. AAMAS, pp. 753–760 (2011)
9. Martin, R.J.: Comparing and contrasting some environmental and experimental design problems. Environmetrics 12(3), 303–317 (2001)
10. Ouyang, R., Low, K.H., Chen, J., Jaillet, P.: Multi-robot active sensing of non-stationary Gaussian process-based environmental phenomena. In: Proc. AAMAS, pp. 573–580 (2014)

Generalized Online Sparse Gaussian Processes with Application to Persistent Mobile Robot Localization

Kian Hsiang Low[1], Nuo Xu[1], Jie Chen[2], Keng Kiat Lim[1], and Etkin Barış Özgül[1]

[1] Nat'l Univ. of Singapore, Singapore
{lowkh,xunuo,kengkiat,ebozgul}@comp.nus.edu.sg
[2] Singapore-MIT Alliance for Research and Technology, Singapore
chenjie@smart.mit.edu

Abstract. This paper presents a novel online sparse *Gaussian process* (GP) approximation method [3] that is capable of achieving *constant* time and memory (i.e., independent of the size of the data) per time step. We theoretically guarantee its predictive performance to be equivalent to that of a sophisticated offline sparse GP approximation method. We empirically demonstrate the practical feasibility of using our online sparse GP approximation method through a real-world persistent mobile robot localization experiment.

1 Introduction

Gaussian process (GP) models are a rich class of Bayesian non-parametric models that can perform probabilistic regression by providing Gaussian predictive distributions with formal measures of the predictive uncertainty. Unfortunately, the expressive power of a full GP model comes at a cost of poor scalability (i.e., cubic time) in the size of the data, which hinders its practical use for performing real-time predictions necessary in many time-critical applications and decision support systems (e.g., ocean sensing, traffic monitoring, geographical information systems) that need to process and analyze huge quantities of data streaming in over time (e.g., in astronomy, internet traffic, meteorology, surveillance). When the data stream is expected to be (possibly indefinitely) long, it is also computationally impractical to repeatedly use existing offline sparse GP approximation methods [2] or online GP model [1] for training at each time step because they incur, respectively, linear and quadratic time in the data size per time step.

This paper presents a novel online sparse GP approximation method [3] (Section 3) that, in contrast to existing works mentioned above, is capable of achieving *constant* time and memory (i.e., independent of the size of the data/observations) per time step. We provide a theoretical guarantee on its predictive performance to be equivalent to that of the offline sparse *partially independent training conditional* (PITC) approximation method. Our proposed method [3] generalizes the sparse online GP model of [1] by relaxing its conditional independence assumption significantly, hence potentially improving the predictive performance. We empirically demonstrate the practical feasibility of using our generalized online sparse GP approximation method [3] through a real-world persistent mobile robot localization experiment described in Section 4.

T. Calders et al. (Eds.): ECML PKDD 2014, Part III, LNCS 8726, pp. 499–503, 2014.
© Springer-Verlag Berlin Heidelberg 2014

2 Background

A Gaussian process (GP) model can be used to perform probabilistic regression as follows: Let $\mathcal{X}$ be a set representing the input domain such that each input $x \in \mathcal{X}$ denotes a d-dimensional feature vector and is associated with a realized output value z_x (random output variable Z_x) if it is observed (unobserved). Let $\{Z_x\}_{x \in \mathcal{X}}$ denote a GP, that is, every finite subset of $\{Z_x\}_{x \in \mathcal{X}}$ has a multivariate Gaussian distribution. The GP is fully specified by its *prior* mean $\mu_x \triangleq \mathbb{E}[Z_x]$ and covariance $\sigma_{xx'} \triangleq \text{cov}[Z_x, Z_{x'}]$ for all $x, x' \in \mathcal{X}$. Supposing a column vector $z_{\mathcal{D}}$ of realized outputs is observed for some set $\mathcal{D} \in \mathcal{X}$ of inputs, the full GP model can exploit these observations to predict the unobserved measurement for any input $x \in \mathcal{X} \setminus \mathcal{D}$ as well as provide its predictive uncertainty using a Gaussian predictive distribution $p(z_x|x, \mathcal{D}, z_{\mathcal{D}}) = \mathcal{N}(\mu_{x|\mathcal{D}}, \sigma_{xx|\mathcal{D}})$ with the following *posterior* mean and variance, respectively:

$$\mu_{x|\mathcal{D}} \triangleq \mu_x + \Sigma_{x\mathcal{D}}\Sigma_{\mathcal{D}\mathcal{D}}^{-1}(z_{\mathcal{D}} - \mu_{\mathcal{D}}) \quad \text{and} \quad \sigma_{xx|\mathcal{D}} \triangleq \sigma_{xx} - \Sigma_{x\mathcal{D}}\Sigma_{\mathcal{D}\mathcal{D}}^{-1}\Sigma_{\mathcal{D}x} \quad (1)$$

where $\mu_{\mathcal{D}}$ is a column vector with mean components $\mu_{x'}$ for all $x' \in \mathcal{D}$, $\Sigma_{x\mathcal{D}}$ is a row vector with covariance components $\sigma_{xx'}$ for all $x' \in \mathcal{D}$, $\Sigma_{\mathcal{D}x}$ is the transpose of $\Sigma_{x\mathcal{D}}$, and $\Sigma_{\mathcal{D}\mathcal{D}}$ is a matrix with components $\sigma_{x'x''}$ for all $x', x'' \in \mathcal{D}$.

The key limitation hindering the practical use of the full GP model is that computing (1) requires inverting the covariance matrix $\Sigma_{\mathcal{D}\mathcal{D}}$, which incurs $\mathcal{O}(|\mathcal{D}|^3)$ time and $\mathcal{O}(|\mathcal{D}|^2)$ memory. To improve its scalability, the sparse *partially independent training conditional* (PITC) [2] approximation method is the most general form of a class of reduced-rank covariance matrix approximation methods in [2] exploiting the notion of a support set $\mathcal{S} \subset \mathcal{X}$. PITC computes a Gaussian predictive distribution of the unobserved measurement for any $x \in \mathcal{X} \setminus \mathcal{D}$ with the following posterior mean and variance:

$$\mu_{x|\mathcal{D}}^{\text{PITC}} \triangleq \mu_x + \Gamma_{x\mathcal{D}}(\Gamma_{\mathcal{D}\mathcal{D}} + \Lambda)^{-1}(z_{\mathcal{D}} - \mu_{\mathcal{D}}) \quad \text{and} \quad \sigma_{xx|\mathcal{D}}^{\text{PITC}} \triangleq \sigma_{xx} - \Gamma_{x\mathcal{D}}(\Gamma_{\mathcal{D}\mathcal{D}} + \Lambda)^{-1}\Gamma_{\mathcal{D}x} \quad (2)$$

where $\Gamma_{\mathcal{A}\mathcal{A}'} = \Sigma_{\mathcal{A}\mathcal{S}}\Sigma_{\mathcal{S}\mathcal{S}}^{-1}\Sigma_{\mathcal{S}\mathcal{A}'}$ for all $\mathcal{A}, \mathcal{A}' \subset \mathcal{X}$ and Λ is a block-diagonal matrix constructed from the N diagonal blocks of $\Sigma_{\mathcal{D}\mathcal{D}|\mathcal{S}}$, each of which is a matrix $\Sigma_{\mathcal{D}_n\mathcal{D}_n|\mathcal{S}}$ for $n = 1, \cdots, N$ where $\mathcal{D} = \bigcup_{n=1}^{N} \mathcal{D}_n$. The covariance matrix $\Sigma_{\mathcal{D}\mathcal{D}}$ in (1) is approximated by a reduced-rank matrix $\Gamma_{\mathcal{D}\mathcal{D}}$ summed with the resulting sparsified residual matrix Λ in (2). So, computing either $\mu_{x|\mathcal{D}}^{\text{PITC}}$ or $\sigma_{xx|\mathcal{D}}^{\text{PITC}}$ (2), which requires inverting the approximated covariance matrix $\Gamma_{\mathcal{D}\mathcal{D}} + \Lambda$, incurs $\mathcal{O}(|\mathcal{D}|(|\mathcal{S}|^2 + (|\mathcal{D}|/N)^2))$ time and $\mathcal{O}(|\mathcal{S}|^2 + (|\mathcal{D}|/N)^2)$ memory. The sparse *fully independent training conditional* (FITC) approximation method is a special case of PITC where Λ is a diagonal matrix constructed from $\sigma_{x'x'|\mathcal{S}}$ for all $x' \in \mathcal{D}$ (i.e., $N = |\mathcal{D}|$).

3 Generalized Online Sparse GP (GOSGP) Approximation

The key idea of our GOSGP approximation method [3] is to summarize the newly gathered data/observations at regular time intervals/slices, assimilate the summary information of the new data with that of all the previously gathered data/observations, and then exploit the resulting assimilated summary information to compute a Gaussian predictive distribution of the unobserved measurement for any input. Let $x_{1:t-1} \triangleq \{x_1, \ldots, x_{t-1}\}$

denote a set of inputs from time steps 1 to $t - 1$, each time slice n span time steps $(n - 1)\tau + 1$ to $n\tau$ for some user-defined slice size $\tau \in \mathbb{Z}^+$, and the number of time slices available thus far up until time step t be denoted by N (i.e., $N\tau < t$).

Definition 1 (Slice Summary). *Given a support set* $\mathcal{S} \subset \mathcal{X}$, *a subset* $\mathcal{D}_n \triangleq x_{(n-1)\tau+1:n\tau} \in x_{1:t-1}$ *of inputs associated with time slice n, and the column vector* $z_{\mathcal{D}_n} = z_{(n-1)\tau+1:n\tau}$ *of corresponding realized measurements, the slice summary of time slice n is defined as a tuple* $(\mu_\circledcirc^n, \Sigma_\circledcirc^n)$ *for* $n = 1, \ldots, N$ *where* $\mu_\circledcirc^n \triangleq \Sigma_{\mathcal{S}\mathcal{D}_n} \Sigma_{\mathcal{D}_n \mathcal{D}_n | \mathcal{S}}^{-1} (z_{\mathcal{D}_n} - \mu_{\mathcal{D}_n})$ *and* $\Sigma_\circledcirc^n \triangleq \Sigma_{\mathcal{S}\mathcal{D}_n} \Sigma_{\mathcal{D}_n \mathcal{D}_n | \mathcal{S}}^{-1} \Sigma_{\mathcal{D}_n \mathcal{S}}$ *such that* $\mu_{\mathcal{D}_n}$ *is defined in a similar manner as* $\mu_{\mathcal{D}}$ *in* (1) *and* $\Sigma_{\mathcal{D}_n \mathcal{D}_n | \mathcal{S}}$ *is a posterior covariance matrix with components* $\sigma_{xx'|\mathcal{S}}$ *for all* $x, x' \in \mathcal{D}_n$, *each of which is defined in a similar way as* (1).

Definition 2 (Assimilated Summary). *Given* $(\mu_\circledcirc^n, \Sigma_\circledcirc^n)$, *the assimilated summary* $(\mu_\circledcirc^n, \Sigma_\circledcirc^n)$ *of time slices 1 to n is updated from the assimilated summary* $(\mu_\circledcirc^{n-1}, \Sigma_\circledcirc^{n-1})$ *of time slices 1 to $n-1$ using* $\mu_\circledcirc^n \triangleq \mu_\circledcirc^{n-1} + \mu_\circledcirc^n$ *and* $\Sigma_\circledcirc^n \triangleq \Sigma_\circledcirc^{n-1} + \Sigma_\circledcirc^n$ *for* $n = 1, \ldots, N$ *where* $\mu_\circledcirc^0 \triangleq 0$ *and* $\Sigma_\circledcirc^0 \triangleq \Sigma_{\mathcal{S}\mathcal{S}}$.

Remark 1. After constructing and assimilating $(\mu_\circledcirc^n, \Sigma_\circledcirc^n)$ with $(\mu_\circledcirc^{n-1}, \Sigma_\circledcirc^{n-1})$ to form $(\mu_\circledcirc^n, \Sigma_\circledcirc^n)$, $\mathcal{D}_n = x_{(n-1)\tau+1:n\tau}$, $z_{\mathcal{D}_n} = z_{(n-1)\tau+1:n\tau}$, and $(\mu_\circledcirc^n, \Sigma_\circledcirc^n)$ (Definition 1) are no longer needed and can be removed from memory. As a result, at time step t where $N\tau + 1 \leq t \leq (N + 1)\tau$, only $(\mu_\circledcirc^N, \Sigma_\circledcirc^N)$, $x_{N\tau+1:t-1}$, and $z_{N\tau+1:t-1}$ have to be kept in memory, thus requiring only constant memory (i.e., independent of t).

Remark 2. The slice summaries are constructed and assimilated at a regular time interval of τ, specifically, at time steps $N\tau + 1$ for $N \in \mathbb{Z}^+$.

Theorem 1. *Given* $\mathcal{S} \subset \mathcal{X}$ *and* $(\mu_\circledcirc^N, \Sigma_\circledcirc^N)$, *our GOSGP approximation method computes a Gaussian predictive distribution* $p(z_t | x_t, \mu_\circledcirc^N, \Sigma_\circledcirc^N) = \mathcal{N}(\tilde{\mu}_{x_t}, \tilde{\sigma}_{x_t x_t})$ *of the measurement for any* $x_t \in \mathcal{X}$ *at time step t (i.e., $N\tau + 1 \leq t \leq (N + 1)\tau$) where*

$$\tilde{\mu}_{x_t} \triangleq \mu_{x_t} + \Sigma_{x_t \mathcal{S}} \left(\Sigma_\circledcirc^N \right)^{-1} \mu_\circledcirc^N \text{ and } \tilde{\sigma}_{x_t x_t} \triangleq \sigma_{x_t x_t} - \Sigma_{x_t \mathcal{S}} \left(\Sigma_{\mathcal{S}\mathcal{S}}^{-1} - \left(\Sigma_\circledcirc^N \right)^{-1} \right) \Sigma_{\mathcal{S} x_t}. \tag{3}$$

If $t = N\tau + 1$, $\tilde{\mu}_{x_t} = \mu_{x_t | x_{1:t-1}}^{\text{PITC}}$ *and* $\tilde{\sigma}_{x_t x_t} = \sigma_{x_t x_t | x_{1:t-1}}^{\text{PITC}}$.

Remark 1. Theorem 1 implies that our GOSGP approximation method [3] is in fact equivalent to an online learning formulation/variant of the offline PITC (Section 2). Supposing $\tau < |\mathcal{S}|$, the $\mathcal{O}(t|\mathcal{S}|^2)$ time incurred by offline PITC can then be reduced to $\mathcal{O}(\tau|\mathcal{S}|^2)$ time (i.e., time independent of t) incurred by GOSGP [3] at time steps $t = N\tau + 1$ for $N \in \mathbb{Z}^+$ when slice summaries are constructed and assimilated. Otherwise, GOSGP [3] only incurs $\mathcal{O}(|\mathcal{S}|^2)$ time per time step.

Remark 2. The above equivalence result allows the structural property of GOSGP [3] to be elucidated using that of offline PITC: The measurements $Z_{\mathcal{D}_1}, \ldots, Z_{\mathcal{D}_N}, Z_{x_t}$ between different time slices are assumed to be conditionally independent given $Z_\mathcal{S}$. Such an assumption enables the data gathered during each time slice to be summarized independently of that in other time slices. Increasing slice size τ (i.e., less frequent assimilations of larger slice summaries) relaxes this conditional independence assumption (hence, potentially improving the predictive performance), but incurs more time at time steps when slice summaries are constructed and assimilated (see Remark 1).

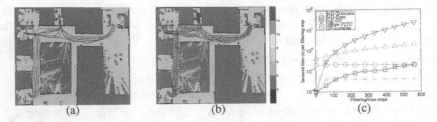

Fig. 1. (a) Pioneer 3-DX mobile robot trajectory of about 280 m in SMART FM IRG office/lab generated by AMCL package in ROS, along which (b) 561 relative light (%) observations/data are gathered at locations denoted by small colored circles. (c) Graphs of incurred time (s) per time step vs. number of time steps comparing different GP localization algorithms.

Remark 3. Since offline PITC generalizes offline FITC, our GOSGP approximation method [3] generalizes the online learning variant of FITC (i.e., $\tau = 1$) [1].

When $N\tau+1 < t \le (N+1)\tau$ (i.e., before the next slice summary of time slice $N+1$ is constructed and assimilated), the most recent observations (i.e., $\mathcal{D}' \triangleq x_{N\tau+1:t-1}$ and $z_{\mathcal{D}'} = z_{N\tau+1:t-1}$), which are often highly informative, are not used to update $\widetilde{\mu}_{x_t}$ and $\widetilde{\sigma}_{x_t x_t}$ (3). This may hurt the predictive performance when τ is large. To resolve this, we exploit incremental update formulas of Gaussian posterior mean and variance [3] to update $\widetilde{\mu}_{x_t}$ and $\widetilde{\sigma}_{x_t x_t}$ with the most recent observations, thereby yielding a Gaussian predictive distribution $p(z_t | x_t, \mu_@^N, \Sigma_@^N, \mathcal{D}', z_{\mathcal{D}'}) = \mathcal{N}(\widetilde{\mu}_{x_t | \mathcal{D}'}, \widetilde{\sigma}_{x_t x_t | \mathcal{D}'})$ where

$$\widetilde{\mu}_{x_t | \mathcal{D}'} \triangleq \widetilde{\mu}_{x_t} + \widetilde{\Sigma}_{x_t \mathcal{D}'} \widetilde{\Sigma}_{\mathcal{D}' \mathcal{D}'}^{-1} (z_{\mathcal{D}'} - \widetilde{\mu}_{\mathcal{D}'}) \text{ and } \widetilde{\sigma}_{x_t x_t | \mathcal{D}'} \triangleq \widetilde{\sigma}_{x_t x_t} - \widetilde{\Sigma}_{x_t \mathcal{D}'} \widetilde{\Sigma}_{\mathcal{D}' \mathcal{D}'}^{-1} \widetilde{\Sigma}_{\mathcal{D}' x_t}$$
(4)

such that $\widetilde{\mu}_{\mathcal{D}'}$ is a column vector with mean components $\widetilde{\mu}_x$ (i.e., defined similarly to (3)) for all $x \in \mathcal{D}'$, $\widetilde{\Sigma}_{x_t \mathcal{D}'}$ is a row vector with covariance components $\widetilde{\sigma}_{x_t x}$ (i.e., defined similarly to (3)) for all $x \in \mathcal{D}'$, $\widetilde{\Sigma}_{\mathcal{D}' x_t}$ is the transpose of $\widetilde{\Sigma}_{x_t \mathcal{D}'}$, and $\widetilde{\Sigma}_{\mathcal{D}' \mathcal{D}'}$ is a matrix with covariance components $\widetilde{\sigma}_{xx'}$ (i.e., defined similarly to (3)) for all $x, x' \in \mathcal{D}'$.

Theorem 2. *Computing* (4) *incurs* $\mathcal{O}(\tau |\mathcal{S}|^2)$ *time at time steps* $t = N\tau+1$ *for* $N \in \mathbb{Z}^+$ *and* $\mathcal{O}(|\mathcal{S}|^2)$ *time otherwise. It requires* $\mathcal{O}(|\mathcal{S}|^2)$ *memory at each time step.*

So, GOSGP [3] incurs constant time and memory (i.e., independent of t) per time step.

4   Experiments and Discussion

In contrast to existing localization algorithms that train the GP observation model of a Bayes filter offline, GOSGP [3] is used to learn it *online* for persistent robot localization and the resulting algorithm is called *GP-Localize* [3]. The *adaptive Monte Carlo localization* (AMCL) package in the *Robot Operating System* (ROS) is run on a Pioneer 3-DX mobile robot mounted with a SICK LMS200 laser rangefinder to determine its trajectory (Fig. 1a) and the 561 locations at which the relative light measurements are taken using a weather board (Fig. 1b); these locations are assumed to be ground truth. For empirical evaluation of GP-Localize with other real-world datasets, refer to [3].

The localization performance/error (i.e., distance between the robot's estimated and true locations) and scalability of GP-Localize are compared to that of two sparse GP localization algorithms [3]: (a) The *Subset of Data (SoD)-Truncate* method uses $|\mathcal{S}| = 10$

most recent observations (i.e., compared to $|\mathcal{D}'| < \tau = 10$ most recent observations considered by GOSGP [3] besides the assimilated summary) as training data at each time step while (b) the *SoD-Even* method uses $|\mathcal{S}| = 40$ observations (i.e., compared to the support set of $|\mathcal{S}| = 40$ possibly unobserved locations selected *prior* to localization and exploited by GOSGP [3]) evenly distributed over the time of localization. The scalability of GP-Localize is further compared to that of GP localization algorithms employing full GP and offline PITC. GP-Localize, SoD-Truncate, and SoD-Even achieve, respectively, localization errors of 2.1 m, 5.4 m, and 4.6 m averaged over all 561 time steps and 3 runs. Fig. 1c shows the time incurred by GP-Localize, SoD-Truncate, SoD-Even, full GP, and offline PITC at each time step. GP-Localize is clearly much more scalable (i.e., constant time) than full GP and offline PITC. Though it incurs slightly more time than SoD-Truncate and SoD-Even, it can localize significantly better.

Acknowledgments. This work was supported by Singapore-MIT Alliance for Research and Technology Subaward Agreement No. 41 R-252-000-527-592.

References

1. Csató, L., Opper, M.: Sparse online Gaussian processes. Neural Comput. 14, 641–669 (2002)
2. Quiñonero-Candela, J., Rasmussen, C.E.: A unifying view of sparse approximate Gaussian process regression. JMLR 6, 1939–1959 (2005)
3. Xu, N., Low, K.H., Chen, J., Lim, K.K., Özgül, E.B.: GP-Localize: Persistent mobile robot localization using online sparse Gaussian process observation model. In: Proc. AAAI (2014)

Distributional Clauses Particle Filter

Davide Nitti[1], Tinne De Laet[2], and Luc De Raedt[1]

[1] Department of Computer Science, KU Leuven, Belgium
[2] Tutorial services, Faculty of Engineering Science, KU Leuven, Belgium
{davide.nitti,luc.deraedt}@cs.kuleuven.be, {tinne.delaet}@kuleuven.be

Abstract. We review the Distributional Clauses Particle Filter (DCPF), a statistical relational framework for inference in hybrid domains over time such as vision and robotics. Applications in these domains are challenging for statistical relational learning as they require dealing with continuous distributions and dynamics in real-time. The framework addresses these issues, it supports the online learning of parameters and it was tested in several tracking scenarios with good results.

Keywords: statistical relational learning, probabilistic programming, particle filters, sequential monte carlo, tracking.

1 Introduction

Robotics and vision have made a lot of progress in state estimation, planning and learning, often employing probabilistic techniques [7]. However, the majority of the techniques used in these domains cannot easily represent relational information, i.e., objects, properties and the relations that hold between them. This calls for the use of probabilistic programming and statistical relational learning techniques (SRL) [1], which have integrated rich relational representations with uncertainty reasoning. Even though many such formalisms are described in the literature, only few of them have been applied to robotics or vision, especially in an online setting. The main challenges are dealing with the dynamics of the environment, continuous distributions and the real-time aspect. This paper reviews the Distributional Clauses Particle Filter (DCPF) framework [5,6] that addresses these issues and that has been applied in [6,3,4].

2 The Probabilistic Language: Distributional Clauses

The DCPF is based on a dynamic variation of Distributional Clauses [2], a language that extends logic programming formalism to define random variables. A *distributional clause* is of the form $h \sim \mathcal{D} \leftarrow b_1, \ldots, b_n$. Informally speaking, whenever the conditions in the body $b_1, \ldots, b_n$ hold, a random variable h is defined with distribution $\mathcal{D}$. A distributional clause is a powerful template to define conditional probabilities; indeed b_i, h, and $\mathcal{D}$ can contain logical variables that parametrize the clause. Consider the following examples:

T. Calders et al. (Eds.): ECML PKDD 2014, Part III, LNCS 8726, pp. 504–507, 2014.

$$n \sim \text{poisson}(6). \tag{1}$$

$$\text{pos}(P) \sim \text{uniform}(1, 10) \leftarrow \text{between}(1, \simeq(n), P). \tag{2}$$

$$\text{type}(A) \sim \text{uniform}([\text{magnet}, \text{ferromagnetic}, \text{nonmagnetic}]) \leftarrow \text{object}(A). \tag{3}$$

Clause (1) states that the number of people n is governed by a Poisson distribution with mean 6; clause (2) models the position $\text{pos}(P)$ as a continuous random variable uniformly distributed from 1 to 10, for each person P such that $\text{between}(1, \simeq(n), P)$ succeeds (i.e., $1 \leq P \leq n$ with P integer). Thus, if the outcome of n is 2, there will be 2 independent random variables $\text{pos}(1)$ and $\text{pos}(2)$. The term $\simeq(d)$ represents the value of the random variable d. Finally clause (3) describe a uniform distribution over 3 possible types for each object A.

Dynamic Distributional Clauses (DDC) extend Distributional Clauses towards temporal domains. They define a discrete-time stochastic process following the same idea of a Dynamic Bayesian Network. We need clauses that define: 1) the prior distribution: $h_0 \sim \mathcal{D} \leftarrow \text{body}_0$, 2) the state transition model: $h_{t+1} \sim \mathcal{D} \leftarrow \text{body}_t$, 3) the measurement probability: $h_{t+1} \sim \mathcal{D} \leftarrow \text{body}_{t+1}$, and finally, 4) clauses that define a random variable at time t from other variables at the same time: $h_t \sim \mathcal{D} \leftarrow \text{body}_t$. For example, to describe that the next position of every ball is equal to the current position plus gaussian noise we write:

$$\text{pos}(A)_{t+1} \sim \text{gaussian}(\simeq(\text{pos}(A)_t), cov) \leftarrow \text{ball}(A). \tag{4}$$

3 Inference and Parameter Learning in DCPF

Given a set of DDC clauses, the DCPF performs filtering, that is, it estimates the current (non-directly observable) world state through the observations obtained from sensors. Formally, filtering or state estimation computes the probability density function $p(x_t | z_{1:t}, u_{1:t})$, where x_t is the current state, $z_{1:t}$ is the set of observations, and $u_{1:t}$ the actions (inputs) performed from time step 1 to t.

Given a model defined as a set of DDC clauses, DCPF performs inference based on particle filtering [7], a Monte-Carlo technique to perform filtering in temporal models. Thus, DCPF is a relational particle filter where each particle $x_t^{(i)}$ is an interpretation, i.e., a set of ground facts for the predicates and values of random variables at time t. A key advantage of the DCPF is that it exploits the relational representation to optimize inference. Rather than working with full interpretations (that list the values for all state variables), DCPF propagates *partial* interpretations (Fig. 1), these are *partial* world descriptions in which the many state variables have been marginalized. This significantly improves the performance with respect to classical particle filters that keep the full state [5].

DCPF supports online parameter learning, that is, state estimation of static variables. Learning can be considered as a state estimation problem, adding the parameters to learn in the state. However, this solution produces poor results due to the degeneracy problem in particle filters. To solve the problem we focused on two simple techniques that have a limited computational cost: artificial dynamics and a variation of resample-move. Details can be found in [6].

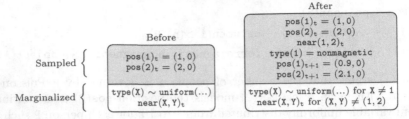

Fig. 1. Partial particle example in the magnetic scenario, before (left) and after (right) the propagation step. Initially the particle contains only the position of the two objects of interest, marginalizing over all other variables. The model states that distant objects do not interact, while close objects interact according to their types. Thus, to sample the next position, DCPF needs to check whether the objects are close. This is the case, so DCPF needs to sample the type of the objects to determine the possible interaction. Object 1 is nonmagnetic in this example, therefore there is no interaction and we can sample the next positions without sampling the type of the second object.

4 Applications

The DCPF framework has been applied to several tracking scenarios [1]. In all scenarios the objects are marked so that their position and orientation can be easily recognized.

Magnetism Scenario [5]: there is a table with objects that can be either permanent magnets, ferromagnetic, or non-magnetic objects. The goal is to track the objects and estimate their type from interactions of pair of objects. To reason about the types of the objects, a theory of magnetism is provided. At the high level it describes interactions, e.g., that two magnets attract or repulse each other. At the lower level, it describes how the positions of the objects evolve over time given the interactions between them.

Box Scenario [6]: the goal of this scenario is to track objects moved by a human during a packaging activity with boxes (Fig. 2a-d). The model provided implements principles such as: an object may fall inside the box if it is on the box in the previous step; if the box is rotated upside down the objects inside will fall down with a certain probability and so on. The framework is able to keep track of objects inside boxes, even objects inside a box inside another box.

String Scenario [6]: we have a table with several objects possibly connected by strings (Fig. 2e). The goal is to track the objects, estimate the current object directly moved by human and learn online the length of the strings between objects. To perform inference and learning we provide a model in DDC that describe the behavior of objects connected by a string.

Distributional Clauses (the static version) have also been used for modeling affordances in manipulation tasks [3] and occluded object search [4].

---

[1] Videos available at https://dtai.cs.kuleuven.be/ml/systems/dc/

(a) cube on the box (b) cube inside the box (c) rotated box on a beige box (d) cube and box inside the beige box (e) String scenario

Fig. 2. (a-d) packaging scenario. The bottom images represent moments of the experiment, while the top images show the corresponding estimated objects' positions, where each colored point represents an object in a particle. The cube is in blue, the small box in fuchsia and the big box in beige (e) string scenario. The top figure represents the estimated objects' positions (yellow and grey), and the estimated string length in red.

5   Experiments and Conclusions

We proposed a flexible representation for hybrid relational domains in temporal models and provided an efficient inference algorithm for filtering and online learning. This framework exploits the relational representation and the (in)dependence assumptions to reduce the particle size (through partial interpretations) and the inference cost. DCPF is particularly suited for (probabilistic) relational models that involve objects and relations between them. It was empirically evaluated and applied in several tracking scenarios with good results. The results show that DCPF outperforms the classical particle filter, and is promising for more complex robotics applications. The code, papers and videos (of all these scenario's) are available at https://dtai.cs.kuleuven.be/ml/systems/dc/.

References

1. De Raedt, L., Frasconi, P., Kersting, K., Muggleton, S. (eds.): Probabilistic Inductive Logic Programming, Theory and Applications. Springer, Heidelberg (2008)
2. Gutmann, B., Thon, I., Kimmig, A., Bruynooghe, M., De Raedt, L.: The magic of logical inference in probabilistic programming. Theory and Practice of Logic Programming (2011)
3. Moldovan, B., De Raedt, L.: Learning relational affordance models for two-arm robots. In: International Conference on Intelligent Robots and Systems (2014)
4. Moldovan, B., De Raedt, L.: Occluded object search by relational affordances. In: IEEE International Conference on Robotics and Automation, ICRA (2014)
5. Nitti, D., De Laet, T., De Raedt, L.: A particle filter for hybrid relational domains. In: International Conference on Intelligent Robots and Systems, IROS (2013)
6. Nitti, D., De Laet, T., De Raedt, L.: Relational object tracking and learning. In: International Conference on Robotics and Automation, ICRA (2014)
7. Thrun, S., Burgard, W., Fox, D.: Probabilistic Robotics (Intelligent Robotics and Autonomous Agents). The MIT Press (2005)

Network Reconstruction for the Identification of miRNA:mRNA Interaction Networks

Gianvito Pio[1], Michelangelo Ceci[1], Domenica D'Elia[2], and Donato Malerba[1]

[1] University of Bari "A. Moro" - Via Orabona, 4 - 70125 Bari, Italy
[2] ITB-CNR, Via Amendola 122/D, 70126, Bari, Italy
{name.surname}@uniba.it, domenica.delia@ba.itb.cnr.it

1 Introduction

Network reconstruction from data is a data mining task which is receiving a significant attention due to its applicability in several domains. For example, it can be applied in social network analysis, where the goal is to identify connections among users and, thus, sub-communities. Another example can be found in computational biology, where the goal is to identify previously unknown relationships among biological entities and, thus, relevant interaction networks. Such task is usually solved by adopting methods for link prediction and for the identification of relevant sub-networks. Focusing on the biological domain, in [4] and [3] we proposed two methods for learning to combine the output of several link prediction algorithms and for the identification of biological significant interaction networks involving two important types of RNA molecules, i.e. microRNAs (miRNAs) and messenger RNAs (mRNAs). The relevance of this application comes from the importance of identifying (previously unknown) regulatory and cooperation activities for the understanding of the biological roles of miRNAs and mRNAs. In this paper, we review the contribution given by the combination of the proposed methods for network reconstruction and the solutions we adopt in order to meet specific challenges coming from the specific domain we consider.

2 Learning to Combine Link Predictions

In the literature, several approaches for link prediction can be found, but they often fail in simultaneously considering all the possible criteria (e.g. network topology, nodes properties, autocorrelation among nodes). In [4] we presented a method for *learning to combine* the scores returned by several link prediction algorithms (which are based on one or few of the possible criteria) for the identification of interactions between miRNAs and mRNAs. In such case, some issues have to be taken into account: *i)* very few interactions are experimentally validated and can be considered as "stable" examples; *ii)* only positive examples are generally available; *iii)* prediction algorithms consider similar features and their combination can lead to collinearity problems.

In order to face *i)* and *ii)*, we propose a semi-supervised learning algorithm, which considers both positively labeled examples of interactions and the huge

T. Calders et al. (Eds.): ECML PKDD 2014, Part III, LNCS 8726, pp. 508–511, 2014.
© Springer-Verlag Berlin Heidelberg 2014

set of unlabeled (unknown) instances. As for *iii)*, the collinearity problem is alleviated by considering as features the scores (outputs) obtained by the prediction algorithms (instead of original features), resorting to a solution which is similar to meta-learning algorithms. The advantage of applying a machine learning method to the outputs of prediction algorithms consists in automatically adapting to unknown patterns of the outputs and performing more reliable predictions when these patterns occur. The proposed method consists in three main steps:

1. Each example of interaction is represented by a vector of scores, obtained by prediction algorithms, and is associated with a label representing the fact that it is labeled as positive (i.e. experimentally validated) or unlabeled.
2. A probabilistic classifier is learned to compute the likelihood that an example of interaction is labeled (known) / unlabeled.
3. A new probabilistic classifier which also exploits (*à la Bayes*) the likelihood computed in the step *2)* is learned. Such classifier associates a score to each interaction to decide whether this interaction is true.

In step *3)*, scores are computed by exploiting the assumption that all the labeled examples are taken randomly from all the positive examples. In other words, the probability of an existing interaction to belong to the set of labeled examples is independent of the specific interaction. Formal definitions can be found in [4].

It is noteworthy that steps *2)* and *3)* require to learn a classifier from a highly unbalanced dataset. Indeed, the set of labeled (in the first case) and positive (in the second case) examples is significantly smaller than the set of unknown examples and negative examples, respectively. Thus, we adopt an ensemble-based approach. In particular, K classifiers are learned by considering as training set the whole set of positive examples and a subset of negative examples, built through a random sampling with replacement. The score associated to each example is computed by averaging the output of all the classifiers that considered it during the learning phase. Further (formal) details can be found in [4].

3   Identification of Relevant Interaction Networks

In [3] we proposed the biclustering algorithm HOCCLUS2 for the identification of miRNA:mRNA interaction networks from the identified interaction scores. Although, in the literature, the application of biclustering techniques to biological data has already been proposed [1,2], some specific aspects are not considered. In particular, identified networks should be: *a)* possibly overlapping, since mRNAs and miRNAs can be involved in multiple interaction networks; *b)* hierarchically organized, allowing biologists to better interpret results and to distinguish between miRNAs involved only in specific pathways or in many biological processes; *c)* highly cohesive, i.e. miRNAs and mRNAs in the same network should be highly related and show only reliable interactions. HOCCLUS2 takes into account these aspects and allows the user to identify the most promising biclusters through a ranking based on a statistical test comparing intra- and inter-bicluster similarity in the Gene Ontology. In the following we describe the first two steps, whereas details about the ranking step can be found in [3].

The **first step** requires a threshold value β, i.e. the minimum score for a miRNA:mRNA interaction to be considered as reliable. The algorithm builds biclusters in the form of bicliques, by considering: avg_mirna - the average number of miRNAs which target each mRNA, with a score greater than β; abs_min_mrna and min_mrna - the *absolute* and the *outlier-proof* (respectively) minimum number of mRNAs which are targeted by each miRNA, with a score greater than β. min_mrna (*outlier-proof*) is computed by discarding the lowest 0.15% values (possibly outliers, according to the 3σ rule), by assuming a Normal distribution. The algorithm builds an initial set of bicliques, each consisting of a single miRNA and of the set of mRNAs associated with a score greater than β. The algorithm, then, iteratively aggregates two biclusters C' and C'' into a new bicluster C''' as follows: $C'''_r = C'_r \cap C''_r$; $C'''_c = C'_c \cup C''_c$, where C_r and C_c are mRNAs and miRNAs in C, respectively. Necessary conditions for aggregating are: $C'_r \cup C''_r \geq min\_mrna$; $C'_c \cap C''_c \leq avg\_mirna$. The basic idea is that a good biclique should contain approximately avg_mirna miRNAs, while keeping the highest possible number of mRNAs (at least min_mrna). Moreover, since we want to obtain a set of highly cohesive bicliques, among the possible aggregations of pairs of bicliques $\langle C', C'' \rangle$, we select the pair which maximizes $jaccard(C'_r, C''_r) * q(C''', A)$, where $jaccard(C'_r, C''_r) = \frac{|C'_r \cap C''_r|}{|C'_r \cup C''_r|}$, $q(C, A)$ is a cohesiveness measure defined as $q(C, A) = (|C_r| * |C_c|)^{-1} * \sum_{x \in C_r} \sum_{y \in C_c} A_{x,y}$ and A is the adjacency matrix containing the score associated to each interaction. The same iterative process is repeated starting from bicliques containing a single mRNA and the two sets of identified bicliques are merged into a single set.

The **second step** consists of an iterative process in which overlap identification and merging are performed. The assumption behind the overlap identification is that two non-overlapping biclusters should be separable in the space. Given two biclusters C' and C'' (belonging to the same hierarchical level), we identify two optimal separating hyperplanes between C' and C'' by learning an SVM model for each dimension (miRNAs and mRNAs). Objects in C' and C'' are used as both training and testing set. Misclassified objects are those which possibly belong to both the biclusters and are added to the bicluster which previously did not contain them. As regards the merging, we assume that miRNAs and mRNAs are normally distributed and consider the distance between pairs of biclusters. In particular, two biclusters C', C'' are candidates for merging if: $dist(C'_r, C''_r) - 2\sigma(C'_r) - 2\sigma(C''_r) \leq 0$ **or** $dist(C'_c, C''_c) - 2\sigma(C'_c) - 2\sigma(C''_c) \leq 0$, where $dist(w, z)$ is the Euclidean distance between the centroids of the clusters w and z, and $\sigma(w)$ is the standard deviation of the cluster w. Intuitively, two biclusters are candidates for merging if they are close according to at least one dimension. A pair of biclusters C', C'', candidate for merging, is merged if the quality constraint $q(C''', A) > \alpha$ is satisfied, where $C'''_r \leftarrow C'_r \cup C''_r$, $C'''_c \leftarrow C'_c \cup C''_c$ and α is a user-defined threshold. Low values of α facilitate merging, decreasing cohesiveness. Since a bicluster can be a candidate for multiple merging, we perform that resulting in the bicluster with maximum cohesiveness.

4 Discussion and Conclusions

The proposed methods have been applied for the identification of miRNA:mRNA interaction networks. In particular, our combination approach has been applied to validated data in miRTarBase 2.5 (4,270 interactions, available at: `mirtarbase.mbc.nctu.edu.tw`), and on the scores returned by 10 prediction algorithms in mirDIP (> 5,000,000 interactions, available at:`ophid.utoronto.ca/mirDIP`). We evaluated the accuracy in terms of the AUC measure on an independent testing set, i.e. TarBase (> 65,000 examples, available at `www.microrna.gr/tarbase`), comparing the results with those obtained by single prediction algorithms and by baseline combination approaches based on score averaging. Moreover, we applied HOCCLUS2 with different values of its parameters to the set of predictions obtained by our combination approach and by baseline combination strategies to evaluate the significance of the extracted networks. In this case, the evaluation was performed in terms of cohesiveness and of a statistical test that takes into account the intra- and inter- bicluster similarity with respect to the classification in Gene Ontology. The evaluation in terms of AUC showed that the proposed approach is able to identify a set of more reliable predictions with respect to the considered competitive approaches. This is also confirmed by the higher significance of the interaction networks extracted by HOCCLUS2, both in terms of the considered evaluation measures and in terms of a biological evaluation performed by a domain expert. These results prove that the proposed approach is able to better filter out false positives and let HOCCLUS2 focus on more reliable predictions so to obtain more significant interaction networks. Details about quantitative and biological analysis can be found in [3,4]. Download links: semi-supervised system: `www.di.uniba.it/~ceci/micFiles/systems/semisupervised_` `HOCCLUS2/`; HOCCLUS2: `www.di.uniba.it/~ceci/micFiles/systems/HOCCLUS/`;biological query system: `comirnet.di.uniba.it`.

Acknowledgements. We would like to acknowledge the support of the European Commission through the project MAESTRA (Grant number ICT-2013-612944).

References

1. Cheng, Y., Church, G.M.: Biclustering of Expression Data. In: Proc. of ISMB 2000, pp. 93–103 (2000)
2. Deodhar, M., Gupta, G., Ghosh, J., Cho, H., Dhillon, I.S.: A scalable framework for discovering coherent co-clusters in noisy data. In: Proc. of ICML 2009, p. 31 (2009)
3. Pio, G., Ceci, M., D'Elia, D., Loglisci, C., Malerba, D.: A Novel Biclustering Algorithm for the Discovery of Meaningful Biological Correlations between microRNAs and their Target Genes. BMC Bioinformatics 14(S-7), S8 (2013)
4. Pio, G., Malerba, D., D'Elia, D., Ceci, M.: Integrating microRNA target predictions for the discovery of gene regulatory networks: a semi-supervised ensemble learning approach. BMC Bioinformatics 15(S-1), S4 (2014)

Machine Learning Approaches for Metagenomics

Huzefa Rangwala, Anveshi Charuvaka and Zeehasham Rasheed

Department of Computer Science,
George Mason University,
Fairfax, Virginia, USA
rangwala@cs.gmu.edu, {acharuva,zrasheed}@gmu.edu

Abstract. Microbes exists everywhere. Current generation of genomic technologies have allowed researchers to determine the collective DNA sequence of all microorganisms co-existing together. In this paper, we present some of the challenges related to the analysis of data obtained from the community genomics experiment (commonly referred by metagenomics), advocate the need of machine learning techniques and highlight our contributions related to development of supervised and unsupervised techniques for solving this complex, real world problem.

1   Background

Advances in genome-sequencing have transformed the manner of characterizing large populations of microbial communities, that are ubiquitous across several environments. The process of "metagenomics" involves sequencing of the genetic material of all organisms co-existing within ecosystems ranging from ocean, soil and the human body. (can be referred to as community genomics). Orthogonally, proteomics and mass spectrometry allow the study of bio-transformations due to these microbial communities in the form of metaproteomes and metabolomes, respectively. Several researchers and clinicians have embarked on studying the pathogenic role played by the microbiome (i.e., the collection of microbial organisms within the human body) with respect to human health and disease conditions. In a similar effort, other groups of researchers are using the metagenomics technology to characterize different ecological environments across the planet (also referred by "Earth Microbiome").

Annotating microbial sequences (reads or quasi-assembled contigs) within a sample is a challenging task due to the unknown, diverse and complex nature of microbial communities within the different environments. There is a critical need to develop mining methods that can characterize metagenome data in terms of taxonomy, function and metabolic potential, and correlate the multi-modal, microbial data to clinical or environmental metadata.

We present our ongoing efforts that have lead to the development of novel supervised learning approaches and scalable clustering methods to solve these real world challenges.

T. Calders et al. (Eds.): ECML PKDD 2014, Part III, LNCS 8726, pp. 512–515, 2014.

2   Large Scale Metagenome Clustering

The sequencing technologies of today do not provide the complete genome for the micro-organisms, but produce short, contiguous subsequences (referred to as reads) that are fragmented from random positions of the entire genome. The problem of metagenome sequence assembly involves stitching together different reads (e.g., overlapping the prefixes and suffixes of smaller subsequences) to produce organism-specific contiguous genomes. Other challenges are introduced due to the varying abundance, diversity, complexity, genome lengths of previously uncultured (or never sequenced before) microbes within different communities. Genomic technologies also produce large number of sequence reads, and reads that may have varying error idiosyncrasies [5]. As such, the metagenome assembly and analysis problem is complex and challenging [2]. Targeted metagenomics or 16S rRNA gene sequencing provides a first step for the quick and accurate characterization of microbial communities. 16S sequences are marker genes, which exists in most microbial genomes and have a conserved portion for detection (primer development) and a variable portion that allows for categorization within different taxonomic groups [7]. Targeted metagenomics are also effective in detecting species with low abundances. However, they may not be good in discovering unique species (orphans) that have never been sequenced before.

Several algorithms have been developed to analyze targeted metagenomes (16S rRNA marker gene) and whole metagenome samples [5]. Clustering/binning approaches involve the unsupervised grouping of sequences that belong to the same species. Successful grouping of sequence reads has several advantages: (i) it improves the metagenome assembly, (ii) it allows computation of species diversity metrics and (iii) it serves as a pre-processing step by reducing computational complexity within several work-flows that analyze only cluster representatives, instead of individual sequences within a sample.

Contributions: We have developed a locality sensitive hashing (LSH) for binning 16S sequences [11,10] called MC-LSH. We further extended the approach using minwise hashing [1] (called MC-MinH[9]) to operate on unequal length sequences and evaluate the approach for both, 16S and whole metagenome sequences. We also extended the minwise hashing algorithm to develop a scalable Map-Reduce based algorithm for metagenome clustering. We refer to this approach as MrMC-MinH [8]. The key contributions of this work included the development of a distributed map-reduce based implementation of clustering algorithm and the ability to perform hierarchical agglomerative clustering instead of greedy clustering as in MC-LSH and MC-MinH.These developed methods provide key biodiversity estimation metrics that are used by biologists.

3   Multiple Hierarchical Classification

The relationship between the microbial communities and human health (or environment) is characterized by first identifying the content, abundance and variance of the microbes across different samples. For an understanding of the microbial-host interaction, it is also crucial to determine the "functionality" and "metabolic potential" induced by these microbes. As such, there is a need to develop methods that annotate the metagenome in terms of "taxa" content and further characterize the ORFs (predicted from metagenomes [12,4]) and transcript sequences in terms of functional and metabolic activity.

The past decade has also seen an explosion in the number of diverse databases that are curated and maintained by different researchers with varying expertise and interests. These databases have a unique characteristic i.e., the data is structured as a hierarchy. We seek to develop approaches that can benefit from jointly learning the prediction models for taxonomy, function and metabolic potential. Different hierarchical databases have implicit similarities between them. For example, the Gene Ontology database has 27 different mappings from other annotation databases, defined using manual and semi-automated procedures. The basis for these mapping include use of sequence, literature search, evolutionary information and structure information.

Contributions: Towards this end, we have developed regularized multi-task learning models [3] that leverage the existing hierarchical structure present in the annotation databases. The models also leverage the implicit relationships between the different databases available for the same annotation problem (e.g., KEGG and MetaCyc for metabolic potential).

Given, multiple hierarchical source databases; within our formulation the objective is to classify an instance accurately across all the multiple hierarchies.

For each of the different classes across multiple hierarchies, we define a binary classification task. These classification tasks predict whether an example belongs to the particular class or not. However, instead of training each of these tasks independently (single task learning), the training for all these tasks are combined using the MTL approach [3] The rationale for the proposed approach is that each of the binary tasks are related "within" the hierarchy due to the explicit structure in the databases. Across the hierarchical sources, it is expected that if the underlying relationships are modeled well, it will benefit the generalization performance for individual annotation problems. Further, the MTL approach is also suited for tasks (classes) that have scarce training examples. This MTL approach leveraged the underlying relationships between the multiple hierarchies and significantly outperformed traditional prediction models for classifying sequence data within multiple hierarchical annotation databases. We also extended this approach to classify text documents across large archives like WikiPedia and DMOZ Web directory [6].

4   Conclusion and Future Work

In summary, we presented a set of clustering and classification approaches to analyze metagenome associated data. Using the annotated sequences, we plan to extract an aggregated taxonomic, functional and metabolic activity profile for the microbial samples. These profiles will allow us to compare metagenome samples and correlate the information to clinical or environmental metadata; and train supervised phenotypic classifiers. All the developed tools are freely available and integrated within bioinformatics work-flows that are easy to use for the biology community.

Acknowledgment. Support by NSF Career to HR, IIS 1252318. All software are linked via project webpage at http://www.cs.gmu.edu/~hrangwal/microbiomeCareer

References

1. Broder, A.Z., Charikar, M., Frieze, A.M., Mitzenmacher, M.: Min-wise independent permutations. In: Proceedings of the Thirtieth Annual ACM Symposium on Theory of Computing, pp. 327–336. ACM (1998)
2. Charuvaka, A., Rangwala, H.: Evaluation of short read metagenomic assembly. BMC genomics12 (suppl. 2), S8 (2011)
3. Charuvaka, A., Rangwala, H.: Multi-task learning for classifying proteins with dual hierarchies. In: IEEE International Conference on Data Mining (ICDM), Brussels, Belgium, pp. 834–839. IEEE (December 2012)
4. Delcher, A.L., Bratke, K.A., Powers, E.C., Salzberg, S.L.: Identifying bacterial genes and endosymbiont dna with glimmer. Bioinformatics 23(6), 673–679 (2007)
5. Hugenholtz, P., Tyson, G.W.: Microbiology: metagenomics. Nature 455(7212), 481–483 (2008)
6. Naik, A., Charuvaka, A., Rangwala, H.: Classifying documents within multiple hierarchical datasets using multi-task learning. In: 2013 IEEE 25th International Conference on Tools with Artificial Intelligence (ICTAI), pp. 390–397. IEEE (2013)
7. Petrosino, J.F., Highlander, S., Luna, R.A., Gibbs, R.A., Versalovic, J.: Metagenomic pyrosequencing and microbial identification. Clinical Chemistry 55(5), 856–866 (2009)
8. Rasheed, Z., Rangwala, H.: A map-reduce framework for clustering metagenomes. In: Proceedings of the 12th IEEE International Workshop on High Performance Computational Biology (HiCOMB), Boston, MA. IEEE (May 2013)
9. Rasheed, Z., Rangwala, H.: Mc-minh: Metagenome clustering using minwise based hashing. In: SIAM International Conference in Data Mining (SDM), Austin, TX. SIAM (May 2013)
10. Rasheed, Z., Rangwala, H., Barbara, D.: Efficient clustering of metagenomic sequences using locality sensitive hashing. In: SIAM International Conference in Data Mining, Anaheim, CA, pp. 1023–1034. SIAM (April 2012)
11. Rasheed, Z., Rangwala, H., Barbara, D.: LSH-Div:species diversity estimation using locality sensitive hashing. In: IEEE International Conference on Bioinformatics and Biomedicine (BIBM), Philadelphia, USA. IEEE (October 2012)
12. Zhu, W., Lomsadze, A., Borodovsky, M.: Ab initio gene identification in metagenomic sequences. Nucleic Acids Research 38(12), e132 (2010)

Sampling-Based Data Mining Algorithms: Modern Techniques and Case Studies

Matteo Riondato

Brown University, Providence, RI 02912, USA
matteo@cs.brown.edu

Abstract. Sampling a dataset for faster analysis and looking at it as a sample from an unknown distribution are two faces of the same coin. We discuss the use of modern techniques involving the Vapnik-Chervonenkis (VC) dimension to study the trade-off between sample size and accuracy of data mining results that can be obtained from a sample. We report two case studies where we and collaborators employed these techniques to develop efficient sampling-based algorithms for the problems of betweenness centrality computation in large graphs and extracting statistically significant Frequent Itemsets from transactional datasets.

1 Sampling the Data and Data as Samples

There exist two possible uses of sampling in data mining. On the one hand, sampling means selecting a small random portion of the data, which will then be given as input to an algorithm. The output will be an approximation of the results that would have been obtained if all available data was analyzed but, thanks the the small size of the selected portion, the approximation could be obtained much more quickly. On the other hand, from a more statistically-inclined point of view, the entire dataset can be seen as a collection of samples from an unknown distribution. In this case the goal of analyzing the data is to gain a better understanding of the unknown distribution. Both scenarios share the same underlying question: how well does the sample resemble the entire dataset or the unknown distribution? There is a trade-off between the size of the sample and the quality of the approximation that can be obtained from it. Given the randomness involved in the sampling process, this trade-off must be studied in a probabilistic setting. In this nectar paper we discuss the use of techniques related to the Vapnik-Chervonenkis (VC) dimension of the problem at hand to analyze the trade-off between sample size and approximation quality and we report two case studies where we and collaborators successfully employed these techniques to develop efficient algorithms for the problems of betweenness centrality computation in large graphs [8] ("sampling the data" scenario) and extracting statistically significant frequent itemsets [10] ("data as samples" scenario).

2 The Sample-Size/Accuracy Trade-Off: Modern Techniques

There exist many probabilistic techniques to study the trade-off between accuracy and sample size: large deviation Chernoff/Hoeffding bounds, martingales,

T. Calders et al. (Eds.): ECML PKDD 2014, Part III, LNCS 8726, pp. 516–519, 2014.

tail bounds on polynomials of random variables, and many others [3, 5]. These classical results bound the probability that the measure of interest (e.g., the frequency) for a single object (e.g., an itemset) in the sample deviates from its expectation (its true value in the dataset or according to the unknown probability distribution) by more than some amount. An application of the union bound is then needed to get simultaneous guarantees on the deviations for all the objects. The so-obtained sample size or quality guarantee then depends on the logarithm of the number of objects. Due to the number of objects involved in many data mining problems (e.g., all possible itemsets or all nodes in a graph), the sample size may be excessively loose and the benefit of sampling could be lost or not enough information about the unknown distribution may be extracted. In a sequence of works [6–10] we investigated the use of techniques based on the Vapnik-Chervonenkis (VC) Dimension [11] to study the trade-off between accuracy and sample size. The VC-dimension of a data mining task is a measure of the complexity of that problem in terms of the richness of the set of measures that the task requires to compute. The advantage of techniques involving VC-dimension is that they allow to compute sample sizes that only depend on this combinatorial quantity (see below), which can be very small and independent from the number of objects and from the size of the dataset. The techniques related to VC-dimension are widely applicable as we show in our case studies.

Definitions and Sampling Theorem. A *range space* is a pair $(D, \mathcal{R})$ where D is a domain and $\mathcal{R}$ is a family of subsets of D. The members of D are called *points* and those of $\mathcal{R}$ are called *ranges*. The VC-dimension of $(D, \mathcal{R})$ is the size of the largest $A \subseteq D$ such that $\{R \cap A : R \in \mathcal{R}\} = 2^A$. If ν is any (unknown) probability distribution over D from which we can sample, then a finite upper bound to the VC-dimension of $(D, \mathcal{R})$ implies a bound to the number of random samples from ν required to approximate the probability $\nu(R) = \sum_{r \in R} \nu(r)$ of each range R simultaneously using the *empirical average* of $\nu(R)$ as estimator.

Theorem 1 ([4, 12]). *Let d be an upper bound to the VC-dimension of $(D, \mathcal{R})$. Given $\varepsilon, \delta \in (0, 1)$, let $\mathcal{S}$ be a collection of independent samples from ν of size*

$$|\mathcal{S}| \geq \frac{1}{\varepsilon^2}\left(d + \ln\frac{1}{\delta}\right) . \tag{1}$$

Then, with probability at least $1 - \delta$, we have

$$\left|\nu(R) - \frac{1}{|\mathcal{S}|}\sum_{a \in \mathcal{S}} \mathbb{1}_R(a)\right| \leq \varepsilon, \text{for all } R \in \mathcal{R} .$$

The bound on the deviations of the estimation holds *simultaneously* over all ranges. The sample size in (1) depends on the user-specified accuracy and confidence parameters ε and δ and on the bound to the VC-dimension of the range space. If the latter does not depend on the size of the D, then neither will the sample size. This is a crucial and very intriguing property that allows for the development of sampling-based data mining algorithms that use small samples and are therefore very efficient. The main obstacles in developing such algorithms are: *1.* formulate the data mining task in terms of range spaces and unknown distributions; *2.* compute (efficiently) an upper bound to the VC-dimension of the task at hand; *3.* have an efficient procedure to sample from the unknown distribution. It is possible but not immediate to overcome these obstacles as we did for different important data mining problems.

3   Case Studies

In line with the nature of this Nectar paper, we present two case studies where we and collaborators successfully used VC-dimension to develop efficient sampling-based algorithms for important data and graph mining problems.

Betweenness Centrality In [8] we developed a sampling-based algorithm to compute guaranteed high-quality approximations of the betweenness centrality indices of all vertices in a large graph. We defined a range space $(D, \mathcal{R})$ where D is the set of all shortest paths in the graph and $\mathcal{R}$ contains one range R_v for each node v in the graph, where R_v is the set of shortest paths that pass through v. A shortest path p between two nodes u and w is sampled with probability $\nu(p)$ proportional to the number of nodes in the graph and the number of shortest paths between u and w. With this definition of the sampling distribution, we have that $\nu(R_v)$ is exactly the betweenness centrality of the node v. We showed that the VC-dimension of this range space is at most the logarithm of the diameter of the graph. Thus, through Thm. 1, the number of $s-t$-shortest path computations to approximate all betweenness values depends on this quantity rather than on the logarithm of the number of nodes in the graph as previously thought.

True Frequent Itemsets. In [10] we introduced the problem of finding the *True* Frequent Itemsets from a transactional dataset. The dataset is seen as a sample from an unknown distribution ν defined on all possible transactions and the task is to identify the itemsets generated frequently by ν, without reporting false positives (i.e., non-frequently-generated itemsets). We formulated the problem in terms of range spaces and computed its VC-dimension in order to use (a variant of) Thm. 1. The domain D is the set of all possible transaction built on the set of items. For each itemset A we define the range R_A as the set of transactions in D that contain A. The frequency of A in the dataset is now the empirical average of the "true frequency of A", i.e., the probability that ν generates a transaction that contains A. We showed that the (empirical) VC-dimension of $(D, \mathcal{R})$ is tightly bounded from above by a characteristic quantity of the dataset, namely the maximum integer d such that the dataset contains at least d transactions of length at least d forming an antichain. A bound to this quantity can be computed with a single linear scan of the dataset. A more refined bound to the empirical VC-dimension can be computed by solving a variant of a knapsack problem. These bounds allow us to compute a value ε such that the itemsets with frequency in the dataset greater than $\theta + \varepsilon$ have, with high probability, true frequency at least θ. The use of VC-dimension allows us to achieve much higher statistical power (i.e., to identify more true frequent itemsets) than methods based on the classical bounds and the Bonferroni correction.

4   Future Directions and Challenges

Sampling will always be a viable option to speed up the analysis of very large datasets. The database research community, often an early adopter of modern storage technologies, is showing a renovated interest in sampling [1, 13]. There is thriving research to develop stronger simultaneous/uniform bounds to the deviations of sets of functions by leveraging modern probability results involving the Rademacher averages, the shatter coefficients, the covering numbers, and the many extensions of VC-dimension to real functions [2]. The major challenges in using these techniques for more and more complex data mining problems are *1.* understanding the best formulation of the problem in order to leverage the best available bounds to the sample size, and *2.* developing bounds to the VC-dimension (or other combinatorial quantities) to be able to use sampling theorems similar to Thm. 1. There is huge room for additional contributions from the data mining community, to show how powerful theoretical results can be used to develop efficient practical algorithms for important data mining problems.

References

[1] Agarwal, S., Mozafari, B., Panda, A., Milner, H., Madden, S., Stoica, I.: BlinkDB: Queries with bounded errors and bounded response times on very large data. In: EuroSys 2012 (2012)

[2] Boucheron, S., Bosquet, O., Lugosi, G.: Theory of classification: A survey of some recent advances. ESAIM: Probability and Statistics 9, 323–375 (2005)

[3] Dubhashi, D.P., Panconesi, A.: Concentration of Measure for the Analysis of Randomized Algorithms. Cambridge University Press (2009)

[4] Har-Peled, S., Sharir, M.: Relative (p, ε)-approximations in geometry. Discr. & Computat. Geom. 45(3), 462–496 (2011)

[5] Mitzenmacher, M., Upfal, E.: Probability and Computing: Randomized Algorithms and Probabilistic Analysis. Cambridge University Press (2005)

[6] Riondato, M., Akdere, M., Çetintemel, U., Zdonik, S.B., Upfal, E.: The VC-dimension of SQL queries and selectivity estimation through sampling. In: Gunopulos, D., Hofmann, T., Malerba, D., Vazirgiannis, M. (eds.) ECML PKDD 2011, Part II. LNCS, vol. 6912, pp. 661–676. Springer, Heidelberg (2011)

[7] Riondato, M., DeBrabant, J.A., Fonseca, R., Upfal, E.: PARMA: A parallel randomized algorithm for association rules mining in MapReduce. In: CIKM 2012 (2012)

[8] Riondato, M., Kornaropoulos, E.M.: Fast approximation of betweenness centrality through sampling. In: WSDM 2014 (2014)

[9] Riondato, M., Upfal, E.: Efficient discovery of association rules and frequent itemsets through sampling with tight performance guarantees. ACM Trans. Knowl. Disc. from Data (in press)

[10] Riondato, M., Vandin, F.: Finding the true frequent itemsets. In: SDM 2014 (2014)

[11] Vapnik, V.N.: The Nature of Statistical Learning Theory. Springer, New York (1999)

[12] Vapnik, V.N., Chervonenkis, A.J.: On the uniform convergence of relative frequencies of events to their probabilities. Theory of Prob. and its Appl. 16(2), 264–280 (1971)

[13] Wang, J., Krishnan, S., Franklin, M.J., Goldberg, K., Kraska, T., Milo, T.: A sample-and-clean framework for fast and accurate query processing on dirty data. In: SIGMOD 2014 (2014)

Heterogeneous Stream Processing and Crowdsourcing for Traffic Monitoring: Highlights

François Schnitzler[1], Alexander Artikis[2], Matthias Weidlich[3], Ioannis Boutsis[4], Thomas Liebig[5], Nico Piatkowski[5], Christian Bockermann[5], Katharina Morik[5], Vana Kalogeraki[4], Jakub Marecek[6], Avigdor Gal[1], Shie Mannor[1], Dermot Kinane[7], and Dimitrios Gunopulos[8]

[1] Technion - Israel Institute of Technology, Haifa, Israel
[2] Institute of Informatics & Telecommunications, NCSR Demokritos, Athens, Greece
[3] Imperial College London, United Kingdom
[4] Department Informatics, Athens University of Economics and Business, Greece
[5] TU Dortmund University, Germany
[6] IBM Research, Dublin, Ireland
[7] Dublin City Council, Ireland
[8] Department of Informatics and Telecommunications, University of Athens, Greece

Abstract. We give an overview of an intelligent urban traffic management system. Complex events related to congestions are detected from heterogeneous sources involving fixed sensors mounted on intersections and mobile sensors mounted on public transport vehicles. To deal with data veracity, sensor disagreements are resolved by crowdsourcing. To deal with data sparsity, a traffic model offers information in areas with low sensor coverage. We apply the system to a real-world use-case.

Keywords: smart cities, crowdsourcing, event pattern matching, traffic, stream processing, big data.

1 Introduction

New technologies related to mobile computing combined with sensing infrastructures distributed in a city or country are generating massive, heterogeneous data and creating opportunities for innovative applications. Levering such data to obtain a detailed and real-time picture of traffic, water or power networks, to name a few, is a key challenge to achieve better management and planning.

In this context, the goal of the INSIGHT project[1] is to support city or country managers in the detection of interesting events. The present work, originally presented in [3], gives a high-level overview of a traffic monitoring application in Dublin City, Ireland. Two particularly interesting features of this work for the machine learning and data mining communities are as follows.

- We present the general framework of an advanced smart city monitoring system leveraging large scale and heterogenous streams of sensor measurements, and the challenges that come up from a real application.

---

[1] www.insight-ict.eu/

T. Calders et al. (Eds.): ECML PKDD 2014, Part III, LNCS 8726, pp. 520–523, 2014.

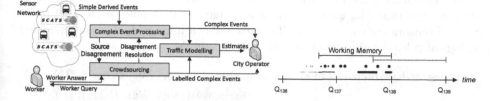

Fig. 1. Architecture overview   **Fig. 2.** RTEC event recognition

- We used real data streams coming from the buses and vehicle count SCATS
  sensors of Dublin city that we made publicly available[2]. The bus dataset
  includes 942 buses. Operating buses emit every 20-30 seconds. The SCATS
  dataset includes 966 sensors transmitting information every few minutes.
  They were collected during January 2013 and totalize 13GB of data.

The system architecture is schematized in Fig. 1. **Inputs** consist in the aforementioned sensors. Additional inputs can be requested from volunteering citi-
zens through a **crowdsourcing** component (Sec. 4). The system **outputs**, in
real time, a set of **complex events** (CEs) (Sec. 3), and **congestion estimates**
for every intersection (Sec. 5). The architecture is implemented as a streaming
system, using the **Streams framework** (Sec. 2).

2 Stream Processing

The *Streams* framework [4] is the backbone of our system. It provides a XML-
based language to describe data flow graphs that work on sequences of data items
(key-value pairs, i.e. attributes and their values). Nodes of the data flow graph
are processes that comprise a sequence of processors. Processes take a stream
or a queue as input and processors apply a function to the data items in a
stream. These concepts are implemented in Java. Adding customized processors
is realized by implementing the appropriate interfaces of the Streams API.

3 Complex Event Processing

For complex event processing, we use the Event Calculus for Run-Time reasoning
(RTEC) [1,2], a Prolog-based engine. Event Calculus is a logic programming
language to represent and reason about events and their effects.

In RTEC, event types are represented as n-ary predicates of the form
event(Attribute1,...,AttributeN). The occurrence of an event E at time T is
modeled by the predicate happensAt(E, T). The effects of events are expressed
by means of *fluents*, i.e. properties that may have different values at different
points in time, for example holdsAt$(F = V, T)$.

---

[2] www.dublinked.ie

In collaboration with domain experts, CEs have been defined over the input streams. For example, an intersection is congested if at least n ($n > 1$) of its SCATS sensors are congested, or if busses suffer a high delay. CEs are modeled as logical rules defining event instances, for example,

$$\text{happensAt}(delayIncrease(Bus), \ T) \leftarrow \text{happensAt}(move(Bus, Delay'), \ T'),$$
$$\text{happensAt}(move(Bus, Delay), \ T),$$
$$Delay - Delay' > d, \quad 0 < T - T' < t.$$

A $delayIncrease(Bus)$ CE is recognized when the delay value of a Bus increases by more than d seconds in less than t seconds.

At query times Q_i, RTEC recognizes CEs within a specified 'working memory' (WM) interval, based on data items received during the WM. Overlapping WMs allow to process, at Q_i, data items generated in $[Q_i - WM, Q_{i-1}]$ but arrived after Q_{i-1}. This is illustrated in Fig. 2. We performed both 'static' recognition, taking into consideration all sources, and 'self-adaptive recognition', where noisy sources are detected at run-time and temporarily discarded.

4   Crowdsourcing

We use crowdsourcing to ameliorate the veracity problem of the data. When the bus and SCATS sensors disagree about a congestion, the CE processing component requests additional inputs from the crowdsourcing component that queries human volunteers, or 'workers', close to the location of the disagreement.

Workers are presented with a set of possible labels (such as 'no congestion' or 'traffic jam') and select one. A key problem is to estimate the reliability of each worker, which we model by p_i, the probability that worker i provides a wrong label. Estimating $\Theta \equiv \{p_i\}_i$ is typically done in batch mode, for example using the Expectation-Maximization (EM) algorithm. In order to estimate Θ on streaming data, we use an online EM based on stochastic approximation.

We employ the MapReduce programming model to communicate queries to the workers without effort from the user to reach him and to achieve real-time and reliable communication [5]. MapReduce allows processing parallelizable tasks across distributed nodes by decomposing the computational task into two steps, namely map and reduce. In our system, the crowdsourcing query engine communicates the queries to the workers (map task), and aggregates the results (reduce task). The interface of the mobile application is illustrated in Fig. 3.

5   Traffic Modeling

Large parts of the city are not covered by the sensors available. A Gaussian Process regression provides operators with a picture on the entire city [6,7] .

To each vertex v_i in the traffic graph $\mathcal{G}$ corresponds a latent variable f_i, the true traffic flow at junction v_i. We assume that any finite set $\mathbf{f} = f_j$ has a multivariate Gaussian distribution $P(\mathbf{f}) = \mathcal{N}(0, \hat{K})$. $\hat{K} = \left[\beta(L + I/\alpha^2) \right]^{-1}$ is

Fig. 3. Interface of the mobile crowd-sourcing application

Fig. 4. Traffic Flow estimates. Green dots indicate low traffic, red dots congestions

the regularized Laplacian kernel function, with hyperparameters α and β. Zero mean is assumed without loss of generality. $L = D - A$ is the Laplacian, A the adjacency matrix of $\mathcal{G}$, and D a diagonal matrix with entries $d_{i,i} = \sum_j A_{i,j}$.

We also assume observations are affected by Gaussian noise: $y_i = f_i + \epsilon_i$, $\epsilon_i \sim \mathcal{N}(0, \sigma^2)$. A joint distribution over observed and unobserved traffic flows can be defined, and the distribution of the unobserved flows conditionally on the observed ones computed. Results visible to operators are illustrated in Fig. 4.

Acknowledgments. This work is funded by the following projects: EU FP7 INSIGHT (318225); ERC IDEAS NGHCS; the Deutsche Forschungsgemeinschaft within the CRC SFB 876 "Providing Information by Resource-Constrained Data Analysis", A1 and C1.

References

1. Artikis, A., Sergot, M., Paliouras, G.: Run-time composite event recognition. In: DEBS, pp. 69–80. ACM (2012)
2. Artikis, A., Weidlich, M., Gal, A., Kalogeraki, V., Gunopulos, D.: Self-adaptive event recognition for intelligent transport management. In: Big Data, pp. 319–325. IEEE (2013)
3. Artikis, A., Weidlich, M., Schnitzler, F., Boutsis, I., Liebig, T., Piatkowski, N., Bockermann, C., Morik, K., Kalogeraki, V., Marecek, J., Gal, A., Mannor, S., Gunopulos, D., Kinane, D.: Heterogeneous stream processing and crowdsourcing for urban traffic management. In: EDBT, pp. 712–723 (2014)
4. Bockermann, C., Blom, H.: The streams framework. Tech. Rep. 5, TU Dortmund University (December 2012)
5. Kakantousis, T., Boutsis, I., Kalogeraki, V., Gunopulos, D., Gasparis, G., Dou, A.: Misco: A system for data analysis applications on networks of smartphones using mapreduce. In: MDM 2012, pp. 356–359 (2012)
6. Liebig, T., Xu, Z., May, M., Wrobel, S.: Pedestrian quantity estimation with trajectory patterns. In: Flach, P.A., De Bie, T., Cristianini, N. (eds.) ECML PKDD 2012, Part II. LNCS, vol. 7524, pp. 629–643. Springer, Heidelberg (2012)
7. Schnitzler, F., Liebig, T., Mannor, S., Morik, K.: Combining a gauss-markov model and gaussian process for traffic prediction in dublin city center. In: EDBT/ICDT Workshops, pp. 373–374 (2014)

Agents Teaching Agents in Reinforcement Learning (Nectar Abstract)

Matthew E. Taylor[1] and Lisa Torrey[2]

[1] Washington State University, School of EECS, Pullman, WA, USA
mtaylor@eecs.wsu.edu
eecs.wsu.edu/~taylorm/
[2] St. Lawrence University, Math and Computer Sciences, Canton, NY, USA
ltorrey@stlawu.edu
myslu.stlawu.edu/~ltorrey/

1 Introduction

Using reinforcement learning [4] (RL), agents can autonomously learn a control policy to master sequential-decision tasks. Rather than always learning *tabula rasa*, our recent work [5,7,8] considers how an experienced RL agent, the *teacher*, can help another RL agent, the *student*, to learn. As a motivating example, consider a household robot that has learned to perform tasks in a household. When the consumer purchases a new robot, she would like the student robot to quickly learn to perform the same tasks as the teacher robot, even if the new robot has different state representation, learning method, or manufacturer. Our goals are to: 1) Allow the student to learn faster with the teacher than without it, 2) Allow the student and teacher to have different learning methods and knowledge representations, 3) Not limit the student's performance when the teacher is sub-optimal, 4) Not require a complex, shared language, and 5) Limit the amount of communication required between the agents.

Our approach was influenced by learning from demonstration [1] (LfD) and transfer learning [6] (TL). LfD methods typically do not achieve goals 3 and 5, limiting an agents' performance to that of the demonstrator, and requiring many trajectory demonstrations. The majority of TL methods assume that the trained agent knows the new agent's learning method or knowledge representation, failing to meet goal 2, and assumes direct access to the the "brain" of the student agent, failing goals 4 or 5.

We investigate how an RL agent can best teach another RL agent using a limited amount of advice, assuming that the teacher can observe the student's state and that the student can receive (and execute) action advice from the teacher. The teacher can give advice a fixed number of times, but cannot observe or change anything internal to the student. This paper presents three of our teaching algorithms and shows a selection of results in the Ms. Pac-Man domain, although our work has also evaluated our methods in the Mountain Car and StarCraft domains. A key insight is that the same amount of advice, given at different moments, can have different effects on student learning. Results show our teaching methods can achieve all five of the above goals.

2 Teaching on a Budget

In this setting, a teacher RL agent has learned a (potentially sub-optimal) policy π_T for a task. Using this fixed policy, it will assist a student agent. As the student learns using

T. Calders et al. (Eds.): ECML PKDD 2014, Part III, LNCS 8726, pp. 524–528, 2014.

RL, the teacher will observe each state s the student encounters and each action a the student takes. The teacher has a fixed budget — it may suggest an action at most b times for the student's current state using the teacher's policy: $\pi_t(s)$. Theoretically calculating how the teacher should best spend its advice is difficult except in the simplest of RL problems. We instead take an experimental approach to this question, proposing and testing several algorithms for deciding when to give advice.

Early Advising. Students should benefit more from advice early on, when they know very little. Early Advising serves as a baseline where the teacher always provides advice for the first b states the student encounters.

Importance Advising. When all states in a task are equally important, Early Advising should be an effective strategy. However, we hypothesized that some states are more important for learning than others, and saving advice for more important states would be a more effective strategy. In some situations, the wrong action could cause catastrophic failure, while in other situations, any action may be acceptable and none are disastrous. A teacher that is conscious of state importance could give advice only when it reaches some threshold th. We call this approach *Importance Advising* (Fig. 1, left).

When th is 0, this becomes equivalent to Early Advising, assuming importance values are non-negative. In this work, we consider teachers that learn Q-functions. If the teacher estimates the Q-values for all actions in s to be the same, it does not matter which action is selected.[1] However, if some actions have larger Q-values than others in s, the action selected matters and the state has some importance. We therefore define state importance as: $I(s) = \max_a Q(s, a) - \min_a Q(s, a)$.

Predictive Advising. The final algorithm builds upon Importance Advising by attempting to provide advice only when 1) the teacher evaluates the current state to be important and 2) the teacher believes the student will execute a sub-optimal action. Although the teacher cannot access the student's policy, it may be able to infer the policy through observation — by observing s, a pairs the student has executed, the teacher can train a classifier to predict student actions, π_s.

If a teacher's action predictor performs perfectly, the teacher will never "waste" advice on a state where the student would execute the action that would have been advised. Inaccurate predictions may waste advice, or miss opportunities for the teacher to give useful advice. Our implementation used the *SVM-Light* software package [2] where the SVM used the teacher's state representation of observed states as input and observed student actions as classification labels. The teacher trains a new SVM after each episode using training examples from the previous episode.[2] This classification task is inherently challenging: a student's behavior is non-stationary and includes exploration, and differences between teacher and student state representations mean that the classifier's hypothesis space may not even be able to represent the student's policy.

procedure IMPORTANCEADVISE(π_t, b, th)
  for each student state s **do**
    if $b > 0$ and $I(s) \geq th$ **then**
      $b \leftarrow b - 1$
      Advise $\pi_t(s)$

procedure PREDICTIVEADVISE(π_t, b, th)
  for each student state s **do**
    if $b > 0$ and $I(s) \geq th$ and
      $\pi_s(s) \neq \pi_t(s)$ **then**
      $b \leftarrow b - 1$
      Advise $\pi_t(s)$

Fig. 1. Both teaching algorithms use a teacher's policy, the budget, and a threshold

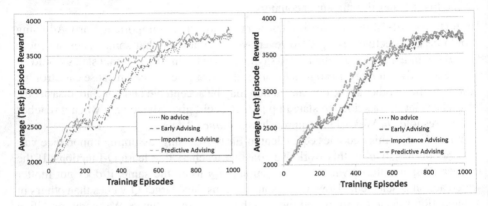

Fig. 2. Students with the high-asymptote feature learn from high-asymptote teachers (left) and low-asymptote teachers (right)

3 Experiments in Ms. Pac Man

Our experiments use an implementation based on the Ms. Pac-Man vs. Ghosts [3] competition. Pac-Man can move in the four cardinal directions but only some of these actions are available in many states of the maze. Points are awarded for eating the small food pellets, larger power pellets, or ghosts shortly after eating a power pellet. The episode ends if any ghost touches Pac-Man, Pac-Man eats all pellets, or after 2000 steps. Useful high-level features tend to describe distances, such as "the distance from Pac-Man to the nearest food pellet," detailed in the released code.[3]

We defined two feature sets. One representation uses 16 features; this "low-asymptote" feature set allows agents to average 2,250 points per episode after training. The other uses 7 heavily-engineered features; this "high-asymptote" feature set allows agents to average 3,380 points after training because they typically learn to eat at least one ghost. Teachers have an advice budget of 1000, roughly half the number of steps in a single well-played episode. The RL parameters that agents use for learning with the SARSA(λ) algorithm are $\epsilon = 0.05$, $\alpha = 0.001$, $\gamma = 0.999$, and $\lambda = 0.9$.

---

[1] $I(s)$ is computed with the teacher's learned Q-function, leveraging the teacher's knowledge without requiring any knowledge of the student's policy or internal representation.

[2] Training the SVM required approximately 1 second — in other domains, it may be possible to update the SVM during episodes or use incremental update methods to improve performance.

[3] http://www.eecs.wsu.edu/~taylorm/13ConnectionScience.html

In Fig. 2, students in both figures use a high-asymptote feature set. In the left figure, the teacher also uses the high-asymptote feature set. In this setting, Early Advising fails to improve over no advice, but Importance Advising significantly outperforms both, and Predictive Advising significantly outperforms all methods. In the right figure, the teacher uses the low-asymptote feature set. In this case, only Predictive Advising significantly outperforms learning without advise.[4] Even though the teacher has a different representation of state and an average performance of only 2,250 points, it is able to significantly improve student learning performance.

Teachers provide advice in only a small fraction of the training steps but can still significantly improve student learning. How quickly teachers provide advice is partly controlled by the importance threshold, th. Teachers in Fig. 2 (right) perform best by using their advice quickly before the students surpass them ($th = 50$). Teachers in Fig. 2 (left) perform better by using less frequent advice over longer periods ($th = 250$).

4   Conclusions and Future Work

As more problems become solvable by agent-based methods, it is important for agents to be able to work together, even if they have very different implementations. RL agents succeed at learning control policies for specific tasks, and allowing them to serve as teachers for these tasks can significantly improve the speed and applicability of RL for fielded agents. Our experimental results, a sample of which were presented in the previous section, lead us to the following conclusions about teaching with an advice budget. First, student learning can be improved with a small advice budget. Second, advice can have greater impact when it is spent on more important states. Third, when teachers can successfully predict student mistakes, they can use their advice budget more effectively. Fourth, students can benefit from advice even from teachers with less inherent ability, different representations, and different learning methods.

There are many exciting directions for future work. The concept of state importance could benefit from further investigation: there may exist better domain-specific ways to measure state importance or effective strategies for automatically selecting and adjusting importance thresholds. The teaching framework could be extended to include multiple teachers and/or students. Students currently always execute actions suggested by the teacher — in the future, the student could decide to ignore advice, or proactively ask the teacher for advice. We are interested in testing our method on other student learning methods, and modifying our methods to work other teacher learning methods. Finally, we are also excited about testing our teaching algorithms with human students.

Acknowledgements. This work was supported in part by NSF IIS-1149917.

References

1. Argall, B., Chernova, S., Veloso, M., Browning, B.: A survey of robot learning from demonstration. Robotics and Autonomous Systems 57(5), 469–483 (2009)
2. Joachims, T.: Making large-scale SVM learning practical. In: Scholkopf, B., Burges, C., Smola, A. (eds.) Advances in Kernel Methods - Support Vector Learning. MIT Press (1999)

---

[4] In both experiments, the SVM embedded in the Predictive Advising achieves an 86% accuracy.

3. Rohlfshagen, P., Lucas, S.M.: Ms Pac-Man versus Ghost Team CEC 2011 competition. In: Proc. of the IEEE Congress on Evolutionary Computation (2011)
4. Sutton, R.S., Barto, A.G.: Introduction to Reinforcement Learning. MIT Press (1998)
5. Taylor, M.E., Carboni, N., Fachantidis, A., Vlahavas, I., Torrey, L.: Reinforcement learning agents providing advice in complex video games. Connection Science 26(1), 45–63 (2014)
6. Taylor, M.E., Stone, P.: Transfer learning for reinforcement learning domains: A survey. Journal of Machine Learning Research 10(1), 1633–1685 (2009)
7. Torrey, L., Taylor, M.E.: Towards student/teacher learning in sequential decision tasks (poster). In: Proc. of the Int't. Conf. on Autonomous Agents and Multiagent Systems (2012)
8. Torrey, L., Taylor, M.E.: Teaching on a budget: Agents advising agents in reinforcement learning. In: Proc. of the Int't. Conf. on Autonomous Agents and Multiagent Systems (2013)

Author Index